INTERNATIONAL SERIES IN

HEATING, VENTILATION AND REFRIGERATION

GENERAL EDITORS: N. S. BILLINGTON AND E. OWER

VOLUME 11

NOISE, BUILDINGS AND PEOPLE

OTHER TITLES IN THE SERIES IN
HEATING, VENTILATION AND REFRIGERATION

Vol. 1. OSBORNE—Fans

Vol. 2. EDE—An Introduction to Heat Transfer Principles and Calculations

Vol. 3. KUT—Heating and Hot Water Services in Buildings

Vol. 4. ANGUS—The Control of Indoor Climate

Vol. 5. DOWN—Heating and Cooling Load Calculations

Vol. 6. DIAMANT—Total Energy

Vol. 7. KUT—Warm Air Heating

Vol. 8. BATURIN—Fundamentals of Industrial Ventilation

Vol. 9. DORMAN—Dust Control and Air Cleaning

Vol. 10. CROOME and ROBERTS—Airconditioning and Ventilation of Buildings

NOISE, BUILDINGS AND PEOPLE

BY

DEREK J. CROOME, B Sc, M Sc, C Eng, M CIBS, MASHRAE, M Inst P, M Inst F, FIOA

*Senior Lecturer with responsibility for
Environmental Engineering for Buildings,
Department of Civil Engineering
Loughborough University of Technology*

PERGAMON PRESS

OXFORD · NEW YORK · TORONTO · SYDNEY
PARIS · FRANKFURT

U.K.	Pergamon Press Ltd., Headington Hill Hall, Oxford OX3 OBW, England
U.S.A.	Pergamon Press Inc., Maxwell House, Fairview Park, Elmsford, New York 10523, U.S.A.
CANADA	Pergamon of Canada Ltd., 75 The East Mall, Toronto, Ontario, Canada
AUSTRALIA	Pergamon Press (Aust.) Pty. Ltd., 19a Boundary Street, Rushcutters Bay, N.S.W. 2011, Australia
FRANCE	Pergamon Press SARL, 24 rue des Ecoles, 75240 Paris, Cedex 05, France
WEST GERMANY	Pergamon Press GmbH, 6242 Kronberg/Taunus Pferdstrasse 1, Frankfurt-am-Main

First edition 1977

Library of Congress Cataloging in Publication Data

Croome, Derek J.
Noise, buildings, and people.

(International series in heating, ventilation, and refrigeration; v. 11)
Bibliography: p.
Includes indexes.
1. Soundproofing. 2. Noise control. 3. Environmental engineering (Buildings) I. Title.
TH1725.C76. 1976   693.8′34   75-40156
ISBN 0-08-019690-X
ISBN 0-08-019816-3 flexicover

Printed in Great Britain by A. Wheaton and Co., Exeter.

Purely untutored humanity interferes comparatively little with the arrangements of nature; the destructive agency of man becomes more and more energetic as he advances in civilization.

(G. P. MARSH, Man and Nature, 1864)

We have a crisis in mental health today, all over the world, because technology has thrown humanity out of the window.

(FOUDRAINE, 1974)

Aquí está encerrada el alma de Anne

MAN, NOISE AND MUSIC

This collage is based on an experience of the author. In a well-known modern theatre musicians and actors request the ventilation system to be turned off during performances, the only time intended for its use, because of the noise. Behind the music and vignette of Chopin, the adventures in sound environments of Stockhausen, which are music for some, noise for others, recede from the rating curves used as noise criteria in buildings.

CONTENTS

PREFACE   xiii

ACKNOWLEDGEMENTS   xix

PART I. INTRODUCTION

1. BUILDING ENVIRONMENTAL ENGINEERING   3

1.1.   Some Problems of Architecture Today   3
1.2.   Building Design Now and in the Future   7
1.3.   Patterns of the Human Mind   11
1.4.   Systems Design Thinking   19
1.5.   The Need for Appraisal of Buildings   23
1.6.   Education   24
1.7.   Environmental Engineers or Technologists   28
1.8.   The Role of Acoustics in Building Environmental Engineering   29

PART II. NOISE AND ITS CONTROL IN BUILDING DESIGN

2. MAN AND THE ACOUSTICAL ENVIRONMENT   33

2.1.   Comfort   33
     2.1.1.   Sensory Perception   36
     2.1.2.   Towards an Understanding of Comfort   44
     2.1.3.   Other Factors   51
2.2.   Physical and Subjective Qualities and Quantities in Acoustics   61
     2.2.1.   Sound Levels   62
     2.2.2.   The Response of the Human Ear   64
2.3.   Quality of Life: Understanding Stress   78
2.4.   Noise and People   83
     2.4.1.   Physiological Response of Man to Noise   83
     2.4.2.   Noise and Sleep   101
     2.4.3.   The Response of the Human Body to Low-frequency Vibrations   109
     2.4.4.   The Psychological Response of People to Noise (including some community noise criteria)   113
     2.4.5.   Noise Criteria in the Future   150
2.5.   The Effect of Noise on Human Performance   153
2.6.   Noise and Speech Communication   166
     2.6.1.   Characteristics of Speech   166
     2.6.2.   Noise Level Criteria for Speech Communication   179
     2.6.3.   Landscaped Offices   185
     2.6.4.   Open-plan Schools   206

3. NOISE SOURCES IN BUILDINGS   207

3.1.   General Aspects   207
3.2.   Sound Generation in Airflow Systems   208
     3.2.1.   Sound Generation Mechanisms   209

ix

3.3.   Sound Generation in Waterflow Systems   210
3.4.   Sound Generation by Airconditioning and Heating Equipment   214
       3.4.1.   Fans   214
       3.4.2.   Heat Transfer Coils   228
       3.4.3.   Pumps   229
       3.4.4.   Electric Motors   232
       3.4.5.   Compressors   236
       3.4.6.   Cooling Towers   244
       3.4.7.   Boilers   245
       3.4.8.   Combustion Equipment Noise   245
       3.4.9.   Chimneys   255
       3.4.10. Steam and Gas Turbines   255
3.5.   Sound Generation Along the Air-Distribution Network   256
       3.5.1.   Straight Ducts   256
       3.5.2.   Butterfly Dampers   257
       3.5.3.   Transition Pieces   257
       3.5.4.   General Conclusions   261
3.6.   Air Terminal Devices   265
       3.6.1.   Diffusers and Grilles   266
       3.6.2.   Sound Propagated from Vibrating Surfaces   270
       3.6.3.   Induction Units   270
       3.6.4.   High-pressure Terminal Equipment   271

4. CONTROL OF AIRBORNE AND STRUCTUREBORNE SOUND   272

4.1.   Basic Principles of Sound Control   272
4.2.   Sound Insulation   273
4.3.   Urban Planning and Noise Control   280
       4.3.1.   The Distance of Buildings from Noise Sources   280
       4.3.2.   Noise Reduction by Barriers   281
       4.3.3.   Landscaping   283
       4.3.4.   Town Planning   284
       4.3.5.   Tall Buildings   285
4.4.   The Rôle of the Building Structure in the Protection of Buildings from External
       Sources of Noise   286
4.5.   Noise Control within Buildings   300
       4.5.1.   Plant-rooms   301
       4.5.2.   Distribution of Spaces within buildings   307
       4.5.3.   Internal Sound Insulation and Isolation   309
4.6.   Noise Control Using Attenuation Equipment   333
       4.6.1.   Absorption of Sound   333
       4.6.2.   Applications of Sound Absorption in Practice   337
       4.6.3.   Attenuators for Airflow Systems   339
4.7.   Control of Structureborne Sound   346
4.8.   Some Examples of Noise and Vibration Control in Building Design   350
       4.8.1.   Case Study 1: Snape Concert Hall, Aldeburgh   350
       4.8.2.   Case Study 2: Sound-proof Booths   352
       4.8.3.   Case Study 3: Designing against Sonic Boom   354
       4.8.4.   A Picture Gallery of Sound Control in Action   357

5. SOME ACOUSTICAL DESIGN TECHNIQUES FOR BUILDINGS   363

5.1.   Acoustical Design Briefs and Specifications for Buildings   363
5.2.   Assessment of the External Noise Environment   364
5.3.   Control of Road Traffic Noise   373
       5.3.1.   Noise Screens   374
       5.3.2.   Selection of Building Structure to Satisfy Environmental Criteria   377
5.4.   Acoustic Design Problems of Internal Spaces   382
5.5.   Acoustical Design Procedure   386
       5.5.1.   Planning Stage   386
       5.5.2.   Assessment of Noise Control Equipment Required   387
       5.5.3.   Selection of Noise Control Equipment   387

5.6. Designing for Sound Privacy in Offices   387
      5.6.1. Small Offices   387
      5.6.2. Landscaped Offices   389
5.7. Sound Quality in Auditoria   391
5.8. Noise from Airconditioning Systems   393
      5.8.1. Case Study 1: Noise Conditions in University Lecture Rooms   393
      5.8.2. Case Study 2: Research on a Variable-speed Ventilation System Serving a
              Lecture Room   398
      5.8.3. Estimation of Background Sound Pressure Level in a Room Served by
              Airconditioning or Ventilation Systems   405

PART III. SOME FUNDAMENTALS OF ACOUSTICS

6. CAUSES AND PATHWAYS OF MECHANICAL VIBRATIONS   417

6.1. Sources and Pathways of Sound   417
6.2. Mass, Stiffness and Damping   419
      6.2.1. Mass   419
      6.2.2. Stiffness   420
      6.2.3. Damping   421
6.3. Impedance Model of Sound Sources and Pathways   427
6.4. Analogues   430

7. FUNDAMENTALS OF WAVE MOTION   436

7.1. Mathematical Representation of a Wave   436
7.2. Complex Signals   438
7.3. The Behaviour of Waves in a Bounded Medium   439
7.4. Wave Propagation in Pipes and Airconditioning Ducts   441

8. PROPAGATION OF SOUND   443

8.1. What is Sound?   443
8.2. Sound Propagation in Gases   445
8.3. The Wave Equation (including the effect of damping)   447
8.4. Energy in a Progressive Sound Wave   451
8.5. Energy in a Standing Wave   452
8.6. The Fundamental Nature of a Simple Pulsating Spherical Sound Source   453
8.7. Propagation of Sound through the Atmosphere   455
8.8. Examples   457

9. STRUCTUREBORNE SOUND   463

9.1. Undamped, Free Vibrations   463
9.2. Viscous Damped, Free Vibrations   466
9.3. Periodic Excitation in Linear Systems   468
9.4. Vibration Isolation Theory   472
      9.4.1. Forced, Undamped System   473
      9.4.2. Damped Active System   473
9.5. Limitations of Theory   475
      9.5.1. Non-linear Systems   475
      9.5.2. Multi-degree of Freedom Systems   480
9.6. Vibration Isolation Practice   485
      9.6.1. Vibration Isolators   485
      9.6.2. Viscoelastic Materials and Damping Applications   489
      9.6.3. Vibration Control by Reducing the Exciting Forces   494
9.7. Vibration Design Criteria for Buildings   494

10. THE BEHAVIOUR OF SOUND IN ROOMS   496

    10.1. The Nature of Sound Fields   496
    10.2. Growth and Decay of Sound Fields   505
          10.2.1. Sound Intensity Arriving at a Surface   505
          10.2.2. Sound Decay   506
          10.2.3. Growth of Sound   507
          10.2.4. Sabine and Eyring Formulae   507
          10.2.5. Other Aspects of Sound Growth and Decay Curves   509
    10.3. Reverberant Field Energy   514
    10.4. The Resultant Sound Pressure Level in a Room   515
    10.5. Multiple Noise Sources and Sound Distribution in Low-height Rooms.   516
    10.6. Impulse Response Analysis of Room Acoustics   519
    10.7. Designing Spaces for Music   520
    10.8. Case Study: Sydney Opera House   544
          10.8.1. General Approach   547
          10.8.2. Appraisal of Acoustical Conditions in the Completed Concert Hall and
                  Opera Theatre   549
          10.8.3. Drama Theatre   551
          10.8.4. Orchestral Rehearsal Hall and Recording Studio   551
          10.8.5. Sound Insulation   552
    10.9. Examples   554

11. THE TRANSMISSION OF SOUND THROUGH STRUCTURES   560

    11.1. The Direct Transmission of Airborne Sound Through a Massive Wall   560
    11.2. The Direct Transmission of Airborne Sound Through a Sprung Wall   562
    11.3. Resonance   563
    11.4. Sound Reduction Index   564
          11.4.1. Coincidence Effect   565
          11.4.2. Leakage of Sound Through Holes and Cracks   568
          11.4.3. The Effect of Room Absorption on the Sound Field   569
          11.4.4. Random Incident Sound Field   570
    11.5. Multi-layer Structures   571
    11.6. Examples   573

APPENDIX I.   List of British Standards and Codes of Practice Relating
                 to Sound and Vibration   575

APPENDIX II.   Health and Safety at Work   578

APPENDIX III.   Selected Reference Bibliography   580

APPENDIX IV.   Noise Advisory Council Publications published by
                 HMSO up to May 1976   581

REFERENCES   582

AUTHOR INDEX   597

SUBJECT INDEX   604

PREFACE

BUILDING design depends on the creative and analytical forces of architecture and engineering. A building is conceived from a visual inspiration within the creative and imaginative mind, and is brought to fruition by a series of steps leading from one to the next by a process of analysis, synthesis and evaluation. Buildings are designed by people for other people to work and live in. The act of worship to God inspired architects and builders in bygone ages to design and erect noble and splendid edifices which will always be awe-inspiring to look at. But they are more than slabs and columns of stone: they have presence, they create atmosphere for worship. The size, the shape, the surface textures, the space, the depth and quality of light in a church or cathedral all interact in some mysterious way to create an atmosphere of quiet and peace necessary for prayer or meditation.

Times have changed. A greater diversity of building types—factories, offices, schools, hospitals, universities, hotels, restaurants, houses—are built in ever-increasing numbers for increasing populations of people to work and live in. More people, more buildings, more transport bring more pollution and more stress.

The Dutch psychiatrist, Foudraine (1974), has evidence showing that one in six women and one in nine men can expect to be an in-patient in a mental hospital at least once in their lives, and admissions are rising. New York mental health authorities consider that the problem of mental stress for city dwellers is much worse than this. In a world where speed of travel, inflow and outflow of information, growth of economic anxieties are beginning to outstrip the biological clocks of Nature and an individual's consumption capacity, care of mind is paramount. Medical advancement over the centuries has been, and still is, largely concerned with the body rather than the mind. Noise is just one of many causes of neuroses that are infecting man's mind.

Today there is shortage of time and money, and this can stultify the imaginative and creative origins of architecture. The interface between architecture and engineering often suffers from a lack of communication so that the common purpose of architects and engineers—*buildings for people*— is lost. Building design requires a resplendent union of art and human engineering. Nowhere are these things more amply demonstrated than when considering the acoustics of buildings. Airconditioning systems in buildings may be inoperable because of the noise they generate; the amenity of a neighbourhood can be spoilt because of intrusion by traffic or aircraft noise; a concert hall may dullen music heard and felt so gloriously in buildings ages past. The size, the shape, the surface textures and transparencies of building spaces all control the response of the space to internal and external sound. The materials, the size, the shape of buildings have visual *and* functional values. It is vital for the acoustical environment of buildings, just as for the thermal, the visual, the structural, the spatial and the social aspects, that integrated decision making between clients, architects and engineers occurs from design inception.

xiii

This book briefly covers the physics of acoustics necessary only to understand the analytical aspects of acoustical design and noise control in buildings. The major part of the book is devoted to the problems of noise and people, whilst other chapters cover features of noise control in and around buildings. It is hoped that the contents will be adequate to serve the needs of undergraduates and postgraduates in the disciplines of building environmental engineering, architecture and regional urban studies besides being useful to those practising in these areas of building design. References are given to other treatises on various aspects of building acoustics. Acousticians must excuse the cursory treatment of the fundamental theories in acoustics and also the exclusion of parts not directly relevant to building environmental engineering; it is essentially a book that is applied and selective from the vast acoustics field. Until the last decade the problem of the noise environment has been neglected in the education of building engineers and architects, and yet it deserves more equal treatment with thermal and lighting aspects, not only because noise has become a more acute problem, but because as human beings we perceive a total environment, not one that is conveniently fragmented into heat, light and sound.

As this book goes to press the first acoustical engineering undergraduate course in the world is commencing in the Faculty of Engineering and Applied Science at the University of Southampton (see *Institute of Sound and Vibration Research Review*, Autumn, Number 11, 1974, page 5). The building environmental engineering undergraduate courses spoken of in Chapter 1 contain a thorough grounding in noise related to buildings and people, this book forms the backbone to one such course.

In October 1971 the Conférence Genérale des Poids et Mésures recommended the use of the Pascal as a unit of acoustic pressure instead of the N/m^2 unit. They are numerically equal. This recommendation has now been legally adopted throughout Europe. The reference sound pressure of 2×10^{-5} N/m² becomes 20 μPa. An attempt has been made to adopt the Pascal unit throughout this book, but some lapses in N/m^2 may remain.

Noise and vibration have spread like environmental cancers in this age of over population, high industrial machine power and traffic density. Town planners try to shield office workers and home dwellers from the irritation caused by road and air transport vehicles. This attempt to isolate people from the external noise and dirt climate has called for sealed buildings and has been an important factor in accelerating the airconditioning need. Industry, too, provides an ever-increasing number of noise problems. There is now a greater realisation for quieter living. Adaptation is not the answer because the human mind can be stressed without the individual realising it.

The Noise Abatement Act of 1960 although still active has been superseded in many aspects by other acts and regulations. *The Motor Vehicles (Construction and Use) Regulations of 1973* prohibit the causing of excessive noise by motor vehicles and trailers, and control the use of horns. *The Road Traffic Act of 1974* provides for a fine of up to £100 in respect of any vehicle making excessive noise. Responsibility for policy on the control of aircraft noise rests with the Department of Trade, the main source of statutory power is Section 29 of the *Civil Aviation Act 1971.* Under Section 58 of the *Control of Pollution Act of 1974* local authorities have powers to deal with general noise nuisances; Section 62 covers the use of loudspeakers and ice-cream chimes in public places; Sections 60 and 61 give local authorities the powers to impose requirements on the way in which construction, demolition and similar works are to be carried out to minimise noise; Section 63 is concerned with the power to designate noise abatement zones.

The 1974 Annual Report of HM Chief Inspector of Factories (published by HMSO Cmnd. 6322) states that the noise problem facing industry is a serious one, not only because noise reduction is often difficult, but because the number of workers exposed to potentially damaging noise is extremely large. A pilot survey of noise measurements in a random sample of 100 factories selected from manufacturing industry was carried out in 1971. From this it was estimated that out of about 6.4 million workers some 590,000 were exposed to noise of 90 dBA or more for more than 6 hours per day, and a further 570,000 for at least some of the time. The same factories were re-visited in 1973–74 and it was concluded that conditions were about the same; a further survey of a larger random sample was carried out during 1975 and 1976. The *Health and Safety at Work Act* of 1974 provides for the protection of workers against risk to their safety and health, including the risk of hearing damage. British Research Establishment Digest 184 of December 1975, covers the problems of assessing and controlling demolition and construction noise.

The Royal Commission on Environmental Pollution, in its Fourth Report entitled *Pollution Control: Progress and Problems* (published by HMSO as Cmnd 5780 in 1974), paints an incisive picture of the noise problem. At least 600,000 people are possibly suffering progressive loss of hearing due to the noisy conditions at their work places. The Department of Employment has issued a Code of Practice (published by HMSO in 1972) for reducing the exposure of employed people to noise. The Control of Pollution Act enables local authorities to rule that existing noise levels must be kept steady and possibly be reduced in the future. In 1970, 8 million people were subjected to traffic noise levels exceeding the absolute limit of acceptability for residential developments laid down by the Noise Advisory Council as $L_{10} = 70$ dBA from 6.00 a.m. to midnight. *If traffic growth continues unabated, then this number will increase to 29 million people by 1980.* Noise Insulation Regulations of 1975 (No. 1763) exist which state the conditions under which people are allowed a sound-insulation grant, but with the possibility of over half of the population being exposed to excessive noise this provides no real answer. Prevention is better than cure. The Royal Commission point out that whereas in this country freight is transported mainly by road (65·5%) with help from the railways (17·3%), inland waterways (0·15%), coastal shipping (14·8%) and pipeline (2·2%), in West Germany which has a similar population and land area to Great Britain, freight is loaded more evenly on to its transportation system—40·5% being conveyed by road, 30·6% by rail, 20·2% by inland waterways and 8·7% by pipeline. Heavy goods vehicles are very noisy.

Any change in transport strategy clearly depends upon many factors but less road freight would help the pollution problem. Aircraft noise troubles less people compared with road traffic, and most of the 2·5 million people affected live near Heathrow Airport. It is reported that unilateral action is being taken in a number of countries to encourage aircraft manufacturers and the airline operators to seek ways of making planes and landing and take-off operations quieter.

These comments are generally in agreement with those made by a working panel set up by the Noise Advisory Council to report on noise in the seventies. A report—*Noise in the Next Ten Years*—was published by HMSO in 1974. In it the following plan of action is proposed:

Road traffic

Statutory reduction in goods vehicle noise limits of at least 5 dBA before 1976.
Research into reduction of traffic noise levels in towns.
Advice to planners on the environmental aspects of traffic management.

Aircraft

Retrofit research programme for British aircraft including flight tests.
Progressive lowering of acceptable noise levels in the area around urban airports.
Review of environmental and operational aspects of civil flying policy with a view to seeking international agreeement on methods for achieving higher noise standards.

Neighbourhood

Support of legislation implementing the main recommendations of the Scott report, *Neighbourhood Noise*.
Establishment of a Quiet Town Experiment.†

Occupational

Technical manpower to support the Department of Employment's *Code of Practice*, with legislative backing if necessary.
Publication of the proposed *Code on New Machinery*.
Establishment of an Advisory and Information Service on occupational noise.

Much more incentive should be given to manufacturers who make quiet operation a prime feature of their designs. The message is clear. Planning and management can contain the present noise levels, but decisive governmental action is needed to *reduce* the levels of noise emitted by cars, lorries, aircraft and machinery. Much time and money is spent fining people for parking offences but only infrequently are noisy motorbikes or "hotted up" cars, for example, even cautioned about their offensive onslaught on peoples' nerves. All can help—the government, the manufacturers, the users, the planners and the officers of the law.

In a report, *Noise in Public Places*, the Noise Advisory Council point out that ordinary good manners can make the difference between a noise nuisance and an acceptable level of noise. They believe that many people do not willingly create disturbance for others, but too many do not think about the effects of their noisy

---

†Subsequently Darlington has been chosen for this experiment (see Acoustics Bulletin, October, 1976, page 9).

behaviour. It is only if they are similarly disturbed themselves that they learn the value of peace and quiet. Anything that can be done to encourage good noise manners would be eminently worthwhile.

The report, by the Secretariat of the United Nations at Palais des Nations, Geneva, on the United Nations Human Environment Conference held in Stockholm on 5–16 June 1972 contained the following statement:

Many speakers from both developing and developed countries agreed that the ruthless pursuit of gross national product, without consideration for other factors, produced conditions of life that were an affront to the dignity of man. The requirements of clean air, water, shelter and health were undeniable needs and rights of man.

To these requirements we may add the need and right of man for peace and quiet of mind and body.

DEREK CROOME

ACKNOWLEDGEMENTS

I SHOULD like to express my gratitude to the many people and organisations who have helped me in preparing this work. In particular my thanks are due to:

A. G. Aldersey-Williams (Design Consultants, London)
Professor G. R. C. Atherley (University of Aston in Birmingham)
A. Bain (Design Consultants, London)
Edwin Benroy (Benroy, Strickland and Partners)
Dr. Keith Attenborough (Open University)
Professor J. W. L. Beament (University of Cambridge)
Dr. N. T. Bowman (City of Leicester Polytechnic)
Alan Brook (Environmental Engineering Honours Graduate of Loughborough University of Technology and sponsored by Sound Attenuators Ltd) for his contribution towards sections 2.6.1 and 2.6.3 in Chapter 2.
Dr. M. E. Bryan (Salford University)
Professor L. Cremer (Technische Universität, Berlin)
Michael Carver (Environmental Engineering Honours Graduate of Loughborough, University of Technology for Fig. 5.1)
David Eason (Technician, Civil Engineering Laboratory, University of Technology, Loughborough)
General Manager of the Philharmonie Hall, Berlin
General Manager of the Sydney Opera House
Professor E. Grandjean (Eidgennössiche Technische Hochschule, Zürich)
Dr. P. Humphrey (Department of Psychology, Brunel University)
The late Dr. Hans Jenny (Institut für Schwingsforschung, Dornach, Switzerland)
Frau H. Jenny
Dr. V. L. Jordan (Acoustic consultant in Denmark)
A. D. Liquorish (Textile Industrial Components Ltd., TICO)
T. J. Lynn (James Walker & Co. Ltd.)
Roger Mills (Executive Producer, BBC Television, Documentary Programmes)
Dr. J. R. McKay (University of Sheffield)
Dr. H. McRobert (University of Salford)
Les Mustoe (Department of Mathematics, University of Technology, Loughborough)
E. Mynn (*The Times*, Photographic Library)
Mr. D. Rolfe (Noise advisory Council)
Dr. I. Oswald (Department of Psychiatry, Edinburgh University)
Michael Page (Environmental Engineering Honours Graduate of Loughborough University of Technology)

xix

D. F. Percy (Director, Industrial Acoustics Co. Ltd.)

Dr. S. A. Petrusewicz (Bath University)

B. Pomfret (Deputy Chief Engineer for Laurence, Scott & Electromotors Ltd)

C. G. Rice (Institute of Sound and Vibration Research, Southampton University)

B. M. Roberts (Drake & Scull Engineering Co. Ltd.)

M. E. Scott (Promotion and Marketing Manager, Sound Attenuators Ltd.)

Dr. T. J. B. Smith (Technical Director, Sound Research Laboratories Ltd.)

Dr. R. W. B. Stephens (formerly Reader in Acoustics at Imperial College of Science and Technology)

P. T. Stone and Julian.

D. Sugden (Arup Associates)

John Taylor & Co., The Bell Foundry, Loughborough

Jan Wenczka, artist in Woodley, Reading.

R. I. Woods (Managing Director, Sound Attenuators Ltd.)

K. Worthington (Metzeler Ltd., Peterborough)

Scott Young

Undergraduates on the Environmental Engineering Honours course at Loughborough University of Technology for their interest and diligence in studying with me this important area of building environmental engineering

The Institute of Sound and Vibration Research, University of Southampton for giving a non-acoustician the opportunity to develop an active interest in the problem of noise, buildings and people

Lastly, but not least, I would like to thank Mrs. Elinor Tivey, Mrs. Rowena Steele and Mrs. Gladys Maggs for their patience and goodwill in part preparation of the manuscript.

Finance for some of the research described in Chapter 5 was given by the Science Research Council.

PART I

Introduction

CHAPTER 1

BUILDING ENVIRONMENTAL
ENGINEERING

*My most important aim is to work toward
the humanization of modern Technology*

(ROBERT JUNGK, in *The Big Machine*, published
by Charles Scribner's Sons, New York)

THE title of this book is *Noise, Buildings and People.* Just as for the volume in this
series entitled *Airconditioning and Ventilation of Buildings* it is important in this
opening chapter to relate the specialism of this book to the broad discipline of
building environmental engineering. It will be seen in Chapter 5 how building
acoustics design interrelates with the thermal and lighting aspects of environmental
design. This is not surprising, because the variables available for design—materials,
building shape and form, site location and orientation, building services—are for a
given climate and specified building use, the common elements at our disposal to
design a building.

1.1. Some Problems of Architecture Today

It is my contention that today too much is expected of the architect. My reasons
for this I hope will become clear as this chapter evolves. There is no doubt that in
civilised countries all over the world there is a pride in their architectural heritage,
but will tomorrow's people consider the buildings of this century as part of these
heritages? What is it about the cities like Bath, York, Edinburgh or Paris; about the
village scene; about the rambles around the village churches accompanied perhaps
by some eloquent words from Sir John Betjamen; about the impact of Le Sacré
Coeur seen high up in Montmarte, the Kremlin in Moscow, the domes in Delhi, the
Western Front of Wells Cathedral, that makes people look in awe, that stirs up their
emotions and uplifts their spirits? There are, no doubt, many things. In these places
man has blended his own creations with those of nature. Walk around the Royal
Crescent in Bath overlooking the park to feel the meaning of this statement. There is
the setting. Then there are the buildings themselves—their scale, their proportions,
their detail. There has been insight in planning and great skill in the craft. The
perception of architecture is not just a visual one, it is an emotional one too.

Buildings are important in the present for the people using them and living near
them; they are important in the future, as in the past, as landmarks of progress in

3

The Royal Crescent in Bath.

civilisation. In the words of Le Corbusier:

Architecture is conditioned by the spirit of an epoch,
and the spirit of an epoch is made up of the depth of history,
the idea of the present and discernment of the future.

Today there is a confused idea of the present and little time for discernment of the future.

So much comment today criticises many modern buildings for their appearance, their lack of humaneness, their inability to function satisfactorily (see *Crisis in Architecture* by MacEwen, 1974). In an age where knowledge and information are in abundance, many buildings do not create that sense of well-being for people outside or inside their environs. In a forward-looking age where ultra-sophisticated systems operate super-technology in space, in nuclear physics, in missile defence systems, we look back to feel the poise and functional elegance of good architecture.

Architecture is a social art; it is an industrial art; it is there for all eyes to see and for all to use.

Architecture is the scientific, accurate and magnificent play of masses brought together in light. Our eyes are made to see forms in light; light and shade reveal these forms; cubes, cones, spheres, cylinders and pyramids are the great primary forms which light reveals to advantage.

(LE CORBUSIER, *Vers une Architecture*, Paris, 1923)

The innate nature of architecture demands a blend of practical and imaginative skills with a feeling for human needs and the balance required between man and his environment, together with a clear vision of the present and intuition about the future. Architecture demands all these things if it is to be successful.

In early times an architect (the word is derived from the Greek words *archi* (chief) and *tekton* (builder)) was employed to design buildings, to work out the science and engineering of them, besides being responsible for their social and visual values.

Today archaeologists and anthropologists mark the cultures and achievements of civilisations long past by tracing the ruins of their towns and dwellings. The insignia of the rich cultural heritage left to us from the Romanesque, Gothic, Baroque, Rococo and Neoclassic eras remain with us, but modern architecture has rucked the eyebrows and will not nestle comfortably in our consciences. Some voices will declare that change always arouses controversy. But architecture is something different from a mere progressive movement like that in the visual arts with its Hepworths, Hockneys or Moores, or the *avant-gardes* of music like Stockhausen and Boulez, because we all *have* to work and live in, and with, buildings; we can *choose* to live with modern music, painting, poetry or sculpture. By the very nature of architecture, buildings will not satisfy all the expectancies that we have of them.

Today there are many different types of industrial, commercial, social and residential buildings. There is shortage of time to think about what we are creating; there is shortage of money. It is these stark realities upon which John Gloag (1950) comments:

> *No wonder the streets are unsmiling. Architecture conceived between balance sheets is born dead or feeble-minded; while under bureaucracy, that contraceptive for all creative impulses, architecture is not conceived at all.*

and more recent words:

> *May I offer just three constraints that conspire against architectural inspiration and delight. First, building legislation is now so complex that even the size and number of windows is covered by prescribed formulae rather than aesthetic considerations. There is hardly a historic masterpiece existing that would conform to today's Building Regulations. Second, planning procedures are so extensive and stultifying that it is rarely the case that the architect's original design escapes untouched from either the dead hand of officialdom or the conservatism of lay committees. Imagine a Turner painted under the direction of a committee of butchers, bakers and candlestick makers. Third, and most important of all, are the imposed cost constraints. It is a truism that you get what you pay for, and if proof is needed just consider a town the size of mediaeval Salisbury and then wonder at the proportion of its wealth that it must have devoted to its cathedral. The fine buildings of the past were dependent on cheap labour and educated patronage. Today a labourer can earn more than his site manager; the major patrons are speculative developers, whose only interest is in highest return for lowest investment, and the public sector, controlled by bureaucrats whose main concern is to balance books.*

> (GEORGE OLDHAM, Vice-President of the Royal
> Institute of British Architects in a letter to the
> *Sunday Times*, 18 May 1975.)

Unfortunately, today too much is demanded of the architect. Creative building design, economical and functional building performance, design and construction project co-ordination—are all things we expect to stem from an architect. If we do not like the look of a building; if the building fails to satisfy the users functionally; if the building is too hot, too cold, too draughty, too stuffy, too noisy, too quiet; whatever the faults they are always traced back ultimately to the architect. And yet all these things may be due to lack of creative thought, lack of analytical thought, bad building, poor communication, poor management, some of which may be due to the architect but not all will be, and sometimes not any.

In an age in which money dictates false values, time spent on work is too rushed, the information rate from the environment is exceeding man's ability to cope with it, building designers need to mix and blend their skills if the built environment is to fulfil the functional, social, economic and aesthetic values that society demands of it.

The feelings of many are echoed in the following remarks:

> *If the architecture of a culture may be said, like any other art, to illuminate that culture, then the outlook for our future is dismal indeed as it appears under the light which is shed by current examples of city building. We seem to be lost in a welter of*

conflicting ambitions and desires, goals, aims, needs, and aspirations. The city now appears more often as a great battlefield, than a place of co-operation and exchange. Neighbourhoods and communities, classes, races, sexes and ages are set one against the other, each striving to obtain power, which means power over others, to impose its own set of rules as the law of the city. The idea of the city as a place which can and should accommodate variety is in the process of disappearing.

(SHADRACH WOODS *The Man in the Street*, Penguin.)

... protecting Jerusalem from the contemporary world-wide architectural scourge, which is destroying the aesthetic and historic aspects of all our beautiful cities.

(YEHUDI MENUHIN, in a letter to *The Times*, 5 December 1974.)

An attempt has been made to identify some of the basic causes which give rise to this kind of comment from people who work and live among, and in, buildings. A greater concentration is given in the following sections to some of the fundamental aspects that can improve the situation.

1.2. Building Design Now and in the Future

The built environment is a crystallisation of the technical, social and moral qualities of the society that does the building, of the society's way of life, showing how it spends its substance, what it loves and admires and cherishes, as well as that to which it pays no heed.

(PERCIVAL GOODMAN, Life in the year 2000, *RIBA Journal*, July, 1973.)

Why do we need buildings? Mainly for people to work and live in. But how many buildings reflect the necessary interdependence of building structure and the systems that make them function? How many suit human needs? The answer to both of these questions is—not many. Several field studies have shown that many buildings waste energy through over-heating, or neglecting those valuable sources of heat in buildings—lights and people. Only too often does one encounter the familiar story, be it a theatre with a ventilation system which is switched off during the performances because of noise; a cafeteria with a ventilation plant-room having no noise control but situated on the roof only yards from houses; a concert hall failing on completion to meet its principal requirement, i.e., to give a pleasing acoustical environment for performers and their audiences; strikes by workers and students trying to work in places which are more like bakehouses due to over-heating than factories or lecture rooms which they are supposed to be; a building costing as much as fifteen times its predicted cost. Often money has to be spent to correct malpractice, or the system is switched off and the capital cost allowed to rot. And yet, occasionally a building does fulfil its requirements and does so economically.

Much can be learnt from existing buildings, but unfortunately buildings are rarely systematically appraised in use.

Man's Culture reflected in his Buildings

Old houses in the town of Dubai where merchant families still live. Shaikh Rashid has ordered that some must be preserved as part of the country's history. The rooftop "boxes" are an early form of airconditioning designed to catch a breeze coming from any direction (with grateful acknowledgement to *The Times* in which this picture appeared in a Special Report on 23 May 1974).

The traditional process of building design has been under question for some time. A client requiring a building usually consults an architect. The process of design begins, inspirations occur and plans to bring them into practice are harnessed by the architect from structural engineers, services engineers and the like. A linear chain of communication has begun, and ingrained therein is the first problem. Is it not better for a client to meet a building design team—architect, building environmental engineer, building economist and structural engineer—together from the outset? This forms the basis of collective decision making, ensuring that a complete brief and specification with sensible budget allocations for structure and services is prepared. Various solutions may then be offered to the client on a cost–benefit basis. The communication process is now an iterative one of analysis, synthesis, evaluation—from design inception to completion, and continued through into building in use. In the case of speculative building, the client is undefined, the building design team is incomplete, and success can only be by luck or by giving the future client great flexibility.

Even when communication *has* been improved there is a further basic question: Does a *complete* building design team exist? Many will argue that it does. Style in architecture results when inspirations in building forms, materials and shapes become a reality by a fusion of intellectual, imaginative, spiritual, technical and economical attitudes. Le Corbusier speaks of "the engineer's aesthetic and architecture are two things that march together and follow one from the other". There is, of course, a common interface between architecture and building engineering which in practice often remains unexplored. How can human needs be assessed? Have building psychologists a part to play in the building design team?

A building structure creates a modified climate for people; its materials control the interchange of light, heat, solar radiation, moisture, ions and noise between the external and internal environments. Building shape and form not only control the passage of wind, smoke and noise around buildings but they also control the spatial–temporal patterns of sound, light and air movement inside the building; the influence of shape on sound distribution is discussed in Chapter 10, whereas its effect on air movement is detailed in Chapter 11 of *Airconditioning and Ventilation of Buildings* (Volume 10 in this Series). Surface textures and colours affect the aural and visual fields. The reverberant sound field has an analogy with the reflectance field in lighting. Hard surfaces reflect sound; white surfaces reflect light. Of course, structural strength and aesthetics must not be forgotten, but building materials, forms and shapes are selected from a matrix that includes not only these parameters but all the environmental ones too. Buildings rarely fail because of structural problems. It is the environmental, the functional and the social problems which are the scourge of architecture.

There is also the very practical and relevant issue of saving energy. With the advent of the energy crisis the government departments responsible for building regulations have at last realised the folly of continuing with poor insulation standards in this country although this was pointed out to us by the Egerton Committee as long ago as 1944. But it is not only a question of using less energy—it is also a matter of not wasting energy.

Comfortable environments for man to work and live in need lighting, moisture, air purity, sound, heating and cooling levels compatible with the job the person is doing. In Chapter 2 it will be seen that the physical environment has an effect on arousal level, on peoples' emotional attitudes and that these factors have a significant role to play in work efficiency. Comfort has been interpreted as the pleasant, stimulating quality of an environment rather than in the traditional sense of it being one of relaxed, "soft cushion", neutral contentment.

Environmental services help to keep the building comfortable for people to work and live in throughout Nature's short- and long-term cycles. There must be no mismatch between the building structure and form and these services. In practice there often is because the common architectural–engineering interface has been ignored. It can be argued, too, that cost–benefit analysis techniques, which would enable the total cost of building engineering solutions to be weighed against human benefit and hence work productivity, have been slow materialising in building design because architects and engineers do not jointly subjectively and objectively appraise their buildings in action, and thus we do not have enough data which allows a true cost–benefit analysis to be made.

Every building tells a story. Classification of user experience should be standardised and the experience be available for all rather than a few. The building design team needs a feedback man. Experience shows that this could result in cost savings far in excess of the expense in employing him. A recent study on government office buildings carried out by the Property Services Agency (part of the Department of the

Environment) showed that the use of optimum start control systems, in which the natural exponential temperature decay and growth curves for a given building structure are used to set pre-heat times, resulted in annual fuel savings ranging from 15% to 40%. The need for building appraisals is demonstrated clearly by the work of Markus *et al.* (1972).

Quality design means that buildings must work reliably over a long period of time. They must be easy and economical to use and maintain; they must offer flexibility because a building will be occupied by generations of different users. The client should be offered building design on a *total cost* basis. Designers still need to convince most clients that acceptance of building designs in capital cost terms only is unwise and more costly in the long term. A total cost basis means that operating costs, building and plant depreciation costs, even inflation estimates, are part of the cost assessment. Already cost–benefit analysis has been advocated in which this total cost is related to human needs, but usable data on this remains scarce.

The division of the capital cost budget is often based on traditional cost analysis of $x\%$ for heating or $y\%$ for lighting irrespective of the fact that the existing building, forming the basis of cost comparison for the future building may have (a) been unsatisfactory in use, and (b) the priority needs of the users in the two buildings may be very different. Noise control is often neglected on the pretext of being too expensive, whereas it really is because there is a fear that it *might* be expensive. Only in recent years has it become the practice to adopt acoustics design as a consideration in *all* buildings, and hence there is little cost experience. In truth there are two important issues. Firstly, noise control *always* needs consideration even if it is later ranked as being of secondary importance, and, secondly, it is not expensive in buildings where the noise level lies between about 30 dBA and 70 dBA. Beyond these limits it becomes progressively more expensive, but in terms of human and functional needs it also becomes critical. The adage *prevention is better than cure* has sense in building design because prevention is better in human *and* in cost terms.

Architects and engineers have remained apart in education too, but changes are taking place. Degrees may now be taken in building environmental engineering; some architectural schools now give more attention to building science. On April 23rd 1976 the Chartered Institution of Building Services was formed, the professional home of the building environmental engineer, in the first instance replacing the former Institution of Heating and Ventilating Engineers. If no attempt is made to co-ordinate building design at the learning stage for all the members of the building design team, how can these minds embark upon common projects in real life practice with little or no understanding of the rôles and aims of each other?

Summing up we have briefly stated that:

Many buildings

* do not fulfil human or social, functional and environmental needs;
* waste energy;
* waste money;

because of

* insufficient time for thinking;
* poor communication processes;

* neglecting the fact that materials, shape and form have visual (mainly felt by the architect) *and* functional aspects (engineering and architectural involvements);
* education and professional imbalance between architecture and building engineering;
* assessment of human needs and priorities rarely carried out in existing buildings.

Solutions

* act to dispel above causes—but how?

It is the aim of this chapter to take a very fundamental look, albeit very briefly, at the building design process and suggest some avenues which may be worth pursuing. The present role and organisation of the professional bodies will not be examined; MacEwen (1974) in his book *Crisis in Architecture* has written a critique on the architect and the Royal Institute of British Architects, but it does not attempt to look at the other professional roles and their interrelationship at education and professional levels. A common theme underlies this work. Any solutions to present problems may only be achieved by the *integrated building design* defined here as a proper consideration of the building materials, shape and form related to the functional, environmental, structural, economical and social needs of a building and its users by an integrated building team which will include the client, the architect, the environmental engineer, the structural engineer, the building economist and the builder, on a continuous basis with other specialists (e.g. interior designer, socio-psychologist, existing users) making contributions as necessary throughout the project; after construction a building will be commissioned for use and appraised when in use. Needless to say this is true for city, urban and regional development, but our concern here is for individual buildings within such developments.

Nervi has stated that *the gulf between architects and engineers is catastrophic*, but this need not remain so. By collective decision making; by assessing the priority order of environmental, functional and social factors; by crediting energy and total cost requirements; by complete appraisals of buildings in use; by recognising the role of building environmental engineering in education, as well as professionally, better buildings with less money and energy waste and expenditure will come to pass, with a more complete attention being given to human and functional needs.

1.3. Patterns of the Human Mind

An individual can benefit in life by gaining some insight into how he or she approaches a problem, whether it be in technical work, in learning, or in communication. Individuals who have to work together can make this undertaking easier and increase the likelihood of success by trying to appreciate each other's way of thinking and viewing a problem.

Few minds can encompass the breadth and style of thought and vision needed to design a building inside out, from the creative forces that fuse building materials, shapes and forms into a visual entity; to the logic required to recognise and analyse the three-dimensional flow of people and information throughout a building; to the

numerative assessment, imbued with a commonsense economy, of energy generation and distribution through a building and across its structure; to the understanding
of the human needs, to management and construction of the design. This required
breadth of thinking vision and skills in building design is acknowledged in practice by
the use of design and construction teams.

If we can accept some of the current findings in brain research, all this division of
thought, so often talked about and experienced in practice, at professional and
educational levels is not surprising. Recent research on the brain by Ornstein (1974)
concludes that the brain can be considered in two parts. The left-hand hemisphere
deals with the analytical processes, e.g., those used in understanding mathematics
and verbal activities; the right-hand hemisphere takes responsibility for the creative
functions which play such a vital role in the conceiving and perceiving of sculpture,
poetry, painting, music, space† and environment. Some people are mainly creative,
others mainly analytical in their outlook, as pictured by the jobs and hobbies they do;
few people are all-rounders, and work equally well at either extremes of these
domains.

An engineer requires spatial ability to a certain extent in laying out systems, but it is
usually considered as paramount in an architect's training. One quickly deduces that
since in designing a building creative and analytical skills are necessary, then a mixture
of types of minds are needed in a design project to combine these skills. An architect is
expected to be creative, whilst building engineers should analyse the
environmental performance of building forms created by the architect (similarly, a
structural engineer analyses the structural performance). But both need to study and
to practise the interface of architecture and engineering in building design, i.e. the
common purpose of buildings for people, the common sense of economy, good
management and construction and good communication.

The opening words of the presentation of the Dom-Ino project in the *Oevre
Complete 1910–29* by Le Corbusier are: *L'intuition agit par éclairs inattendus*
(intuition works in sudden flashes).

What happens in the mind of an architect when someone has asked him to design a
building? What is the process of thinking which brings a bare request to reality? At
one moment nothing exists except some hazy notions of a client; sometime later
shapes and forms emerge on to paper. A building is born. Musical composition is
another marvel of the mind to fathom.

Exactly how the creative mind works is not known, but psychologists have
advanced our understanding by observing the performance and personality characteristics of many individuals in different professions (see Vernon, 1970; Lee, 1972).

A dictionary definition of the verb *to create* is, *to bring into being or form out of
nothing*; the noun *imagination* is stated to be the artist's creative power, or the
faculty to form images in the mind. In Chapter 13 of the book *Creativity*, edited by P.
E. Vernon (published by Penguin in 1970), C. R. Rogers writes: "The creative
process is the emergence in action of a novel relational product growing out of the

†Stringer (see Honikman, 1971) defines spatial ability as that required to call up and manipulate two- or
three-dimensional configurations in one's mind. It comprises (i) *spatial orientation*: ability to comprehend
the position and the configuration of objects in space, primarily with subject's body as spatial reference
point; (ii) *visualisation*: ability to rotate, turn, fold or twist the images of objects or parts of objects
according to an explicit set of movements, and to compare the consequent manipulated image with
drawings.

uniqueness of the individual on the one hand, and the materials, events, people or circumstances of his life on the other."

We still lack an exact understanding about creativity, but it is certain that some individuals are distinctly more creative than others. Even though it is a characteristic which cannot be defined or measured precisely, it is known from experience that some people have very original and free-ranging ideas, whereas others do not. In recent years it has been found that above an intelligence quotient (IQ) level of about 120 there is no correlation between creativity and intelligence. Table 1.1 attempts to convey the broad meanings of personality and intelligence.

TABLE 1.1. BROAD MEANINGS OF PERSONALITY AND INTELLIGENCE

	Facet	Definition
Personality	Character	Qualities of an individual defined by society
	Temperament	Qualities relating to level, mood and style of life
	Interests	Activities sought and those avoided
	Attitudes	Reactions to people, phenomena and concepts of society
Intelligence	Ability	What an individual can do
	Aptitude	What an individual can learn
	Achievement	What an individual has learnt
	Analytical ability	Verbal and numerical reasoning
	Creative ability	Spatial and perceptual abilities

It is thought that *divergent thinking* accounts for some but not all of the intellectual components of creativity. In Table 1.2 the principal characteristics of the convergent–divergent spectrum are illustrated by listing words and phrases that have been found to be associated with either type of thinking.

It must be emphasised that Table 1.2 represents the bi-polar ends of a spectrum of convergent–divergent attitudes, whereas in real life people often show a mixture of these attitudes. Besides, there are various degrees of originality, logicality and so on. Some people expose different facets of their personality in different situations. Nevertheless, this spectrum of convergent–divergent thinking with its associated behaviour patterns does help in understanding how people approach problems, what conditions are more likely to motivate them, and it has fundamental implications throughout education and in the organisation of work for people.

The final line in Table 1.2 equates convergency with an analytical thinking style and divergency with a creative one. This picture of the story is incomplete. Convergency–divergency accounts for some but not all of the intellectual components of the analytical–creative styles; on the other hand, this pattern does also seem to be concordant with that of the analysis–creative brain model proposed by Ornstein (1974) mentioned near the beginning of this section. Either view gives credence to the other.

Studies carried out to ascertain why some people reach eminence in their chosen field of work show that above a certain intelligence level success in work is more dependent on differences in personality rather than differences in intelligence. A

TABLE 1.2. WORDS ASSOCIATED WITH CONVERGENT–DIVERGENT SPECTRUM

Convergent thinking	Divergent thinking
Prefers specific goal problems requiring one solution	Attempts open-ended problems producing a number of solutions
Predictable	Original
Authoritarian	Liberal
Unemotional	Emotional
Sober	Humorous
Prefers a rigid logic	Free-ranging, flexible, adaptable
Prefers well-known methods	Will chance unusual, different methods
Conventional	Novel
Dislikes controversy	Likes controversy
Certainty	Uncertainty
More limited imagination	Elaborative imagination (develops original ideas to the full)
More limited range of ideas	Fluency and fertility of ideas
Dependent	Independent
Analytical (or vertical) thinking	Creative (or lateral) thinking

capacity for hard work, the will to persist, high motivation, self-sufficiency and independence of mind are some other important characteristics that are commonly found in successful people. Highly imaginative people probably have an alert subconscious which is almost continually thinking about problems. It is a common experience for an idea or a solution to a problem to dawn at a most unlikely moment, a time when no conscious thought is being given to the problem. The interplay of genetic and environmental influences on our learning and personality development has been debated for a long time. Even if creativity is inborn, many people still have to learn how to use it.

Mackinnon (1961, 1967, 1970) in America, and Edberg (1974) in Sweden have carried out research which has discovered some aspects about creativity in architects. The 124 American architects in Mackinnon's study underwent many types of personality tests. In very general terms these architects were found to be strongly original, self-assertive, independent, imaginative and sensitive when compared with groups of other subjects from different professions. At the time of writing the following words, Mackinnon pointed out that more research data had to be analysed:

Our image of the creative architect is fast becoming several images of different types of creative architect. But if I were to summarize what is most generally characteristic of the creative architect as we have seen him, it is his high level of effective intelligence, his openness to experience, his freedom from petty restraints and impoverishing inhibitions, his aesthetic sensitivity, his cognitive flexibility, his independence in thought and action, his high level of energy, his unquestioning commitment to creative endeavor and his unceasing striving for creative solutions to the ever more difficult architectural problems which he constantly sets for himself.

Edberg (1974) concluded from his research that architectural students judged to be creative did not have the highest grades when assessed in a conventional manner, marks being given for their examinations and coursework. It has already been noted from the work of others that creative people often do not perform as well at closed-type intelligence tests but do better when attempting open-ended type of tests.

How do people get ideas? In architecture Leonardo da Vinci copied Alberti, Lewerentz studied and copied manmade objects, Le Corbusier used Nature as his model, Bach copied out the music of Vivaldi. The work and ideas of others, objects or Nature itself, can fire the imagination. Edberg concluded that in some of his architectural students their thought processes were oriented towards objects, while in others their minds were turned towards the human being. The *Modulor* concept (Fig. 1.1) of Le Corbusier is an outstanding example of an architect thinking in terms of the human being.

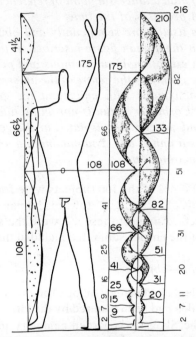

FIG. 1.1. The Modulor Concept (see *Le Corbusier* by Cresti, 1970). Le Corbusier defines the Modulor as a *range of harmonious measurements to suit the human scale, universally applicable to architecture and to mechanical things.* The Modulor is a system of measurement based on the stature of man with his arm upraised (226 cm), which gives an unlimited series of related proportions. For example, 226 cm = 113 + 70 + 43 cm; 113, 70 and 43 are measurements based on the Golden Sections and fit many characteristic positions of the human body in space. A seated man corresponds to 43 cm, a man leaning at a table to 70 cm, or against a parapet, 113 cm: or else 113 + 70 = 183 cm = man's average height. For a more thorough study refer to the two volumes by Le Corbusier, *Le Modulor* and *Modulor 2*.

Cresti (1970) writes:

> *On this occasion too* [in the summer of 1945] *Le Corbusier lived up to his vocation, perceiving and registering the tensions related to the stresses of the moment; sensitive interpreter of the situation, he agreed that* "in order to work out the answers to the formidable problems presented by our time and with regard to its 'equipment', there is only one acceptable criterion which will reduce every

question to a common denominator: man". *It is significant that at the time he was collecting together in "The Modulor" the results of many years' study, a proportional scale based on human stature, to be used as a guide-line at the planning stage. Although inspired by the principles of the golden section, immutably defined as the* "fundamental expression of a unitary universe ... a cardinal point of reference, echoing and re-echoed in all things from the smallest to the largest, and harmonising each separate part with the whole". *Le Corbusier's modulor concept cannot be considered as a passive, scholarly revival of classicism. Nor is it an approach imitating the speculation of Renaissance theory. The latter imposed, dogmatically, through determinations made possible by the study of perspective, numbers and ratios for the anthropometric definition of architecture in order to make the proportions of man a universal point of reference and answer in figurative terms the philosophical premises of the time.*

In fact nothing lends itself more significantly and obviously to the purpose of universal harmony than the human figure inscribed in the perfect limits of circle and square. Man as an exceptional, autonomous and perfect individual therefore represented, in his static geometry the centre of the universe, and his symmetry and proportion become synonymous with unity and the guarantee of harmonious reference. However, Le Corbusier's man is not an archaeological imitation of the homo ad quadratum *and* ad circulum *isolated in the geometric microcosm of a simple, flat figure. He is a united and dynamic man, a unit consciously related to the whole and to the environment.*

In Edberg's experiments 74% of the 100 subjects were found to have their thought processes oriented towards objects, 12·5% towards human aspects, and 4% towards objects *and* human aspects. Amongst these 4% were the students with the highest creativity grades. Some of the entries submitted included the following comments:

Orientation of thought—objects

* one of the students had made a hillock for children's play in the middle of the site—this idea had been inspired by the large mound topped by graves and trees in the Skogskyrkogärden cemetery in Stockholm;
* another student had a cymbal at home suspended on a string and this had given him the idea of a rope for the children to hang on and swing from, that in its turn gave him the idea of a suspension bridge: this called for mounds and hollows on the site and these were therefore also included in the sketch;
* still another student had been inspired by the barnyard at home;
* and another again had had a French farm in mind;
* an entry by a student from Sigtuna transformed the gasworks building into a ruin, using the ruins in Sigtuna as a model.

Orientation of thought—human

* I have tried to provide an interior, an intermediate and an exterior home for the children with courtyards where they can experience a sense of security;
* one entry was primarily concerned with supervision of the children.

Edberg confesses surprise at these results in view of the fact that architecture is generally considered as being human oriented. This pinpoints one of the conflicts in architecture—it is a visual art but it is one that has to meet the needs of people. The most successful buildings are those which can blend both of these attributes; today the further needs for economy in terms of capital and operating costs and a minimum energy usage and wastage restrain the abandon of creative architects even more. There is also often too little time to think.

The ideas of Ornstein (1974) and Vernon (1970) can be used to further our understanding of the rôles of the various disciplines in building design in practice as well as at an educational level. In Fig. 1.2 an attempt is made to identify some of the key rôles and activities in building design with a higher or lower bias towards creative–analytical thinking. The client hopes that the building he needs will be imaginatively thought out, will be suited to his needs, will be completed on time, will be a fair cost, will be economical to run. The first step at design inception must be highly creative as needs and constraints are translated into visual forms; divergent thinking is very active at this stage. These forms require detailed analysis until a design is complete and costs are estimated. The process of estimating is an example of convergent thinking.

The activities in building design and construction require a blend of creative and analytical thought. These activities are carried out by professional people spanning the range shown in Fig. 1.2; hence any building design and construction organisation needs minds that cover this spectrum. One example of the important common interfaces which occur throughout building design and construction is shown, in this case for the architect and the environmental engineer. There are few "all-rounders", i.e. people who can perform equally well at either end of the divergent–convergent thinking continuum, but their value at a high decision-making level in a company would be inestimable. The education system for the building industry has to cater for the full range of thinking rôles illustrated in Fig. 1.2. Integrated design teams help to co-ordinate and unify the different thought processes of individuals and make communication easier.

Let some words from a letter written to *The Times* (21 February, 1973) by Sir Ove Arup close this section:

I have been a contractor, a structural engineer, even a client. I have also, together with my partners, been responsible for the total design of buildings. This controls all the physical properties of a building and therefore also its architectural and functional qualities.

Such a total design has to satisfy many, partly conflicting, demands. Its success depends even more on a happy integration of its constituent parts than on the excellence of each part taken by itself. The aim is to create an artistic whole which is the most favourable compromise between conflicting desiderata.

This synthesis used to be the sole prerogative of the architect. Now he needs many different kinds of specialist to help him to realize the total design. The importance of the contribution of the structural engineer is now widely understood, but the growing importance of the rôle of the services engineer is only just being appreciated. The services of the building, heating and ventilation, airconditioning, lighting, domestic machinery of all kinds are now in some cases outstripping all the other parts of a building in importance. In this situation of rapid growth and

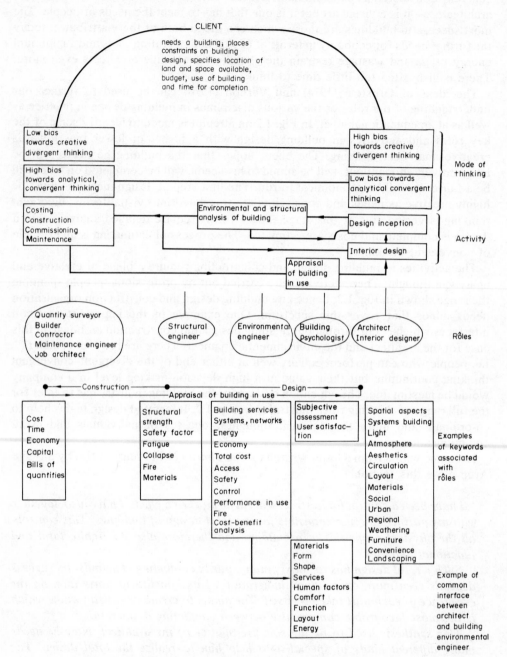

FIG. 1.2. Convergent and divergent thinking rôles in the building design process.

complexity we need designers of a wider outlook than that provided by industry-based practical specialists, although these are needed too. We need "synthesizers" of building services in general who can make a useful contribution to the basic decisions.

In our terms the "synthesizers" of building services are building environmental engineers.

1.4. Systems Design Thinking

. . . what I have been looking for, and have found very little activity in fact, was devising new visions and devising new concepts.

I think that you have to combine the two things—analysis plus imagination. Out of analysis plus imagination you might create a new modern world which would be not anti-technological but where the technology would be directed into another direction, would be handled very differently . . . the technology we have nowadays is mirroring our society and our state of mind. Technology has been going through historical phases and this is just one phase of technology. It is a phase of very inhuman and very muddled technology . . . . I feel that the philosophers have failed, and the humanists have failed—in not telling the engineer what he might do, how he might actually employ what he knows to better goals than what he has done so far.

(ROBERT JUNGK, in Open University Technology
Foundation Course Units 32–34, 1972.)

Conditions in society must not continue to stifle the creative aspects of architecture. Equally the follow up to this—analysis of building performance at the design stage—must measure up to the needs of economy and functional requirements in buildings.

The data used for the analysis and the consequent synthesis of building design solutions is based on experimental evidence such as the empirical data of Beranek and van Os for noise criterion and noise rating curves, or on a rationalist approach. Management control relies on personal experience, and by reasoning with this and related experiences of others coming to a rational judgement; there is often an absence of measured, observed empirical data. The process of reasoning relies on *deductive logic* in which the truth of the premisses guarantees the truth of the conclusions to an argument, and *inductive logic* in which the truth of the premisses does not necessarily guarantee the truth of the conclusions. Logic is mainly concerned with mapping out a valid pathway from the premisses of an argument to the conclusions. In deduction there is a clear-cut conclusion, true or false, whereas with induction we may only speak of probable outcomes. In the environmental design of buildings the outcome of a decision is blurred by subjectivity and the complex interrelationships between all the factors. Increase the sound level in a space—some people will notice it, some will not, depending on their sensitivity and also on their state of concentration at that moment. To bias the likely outcome of a decision in the direction required, a logical process is needed. Systems thinking provides such a logical process ensuring that interacting elements of a design are identified and that there is a logical sequence of events in designing the building. The

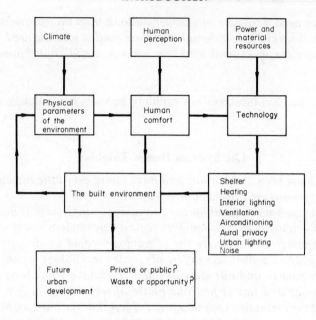

FIG. 1.3. Conceptual diagram of the built environment as used in the Open University Social Science
Course DT 201, Urban Development, Units 15–18 (1973).

word technology is derived from the Greek *tecknologia* meaning systematic treatment, and *tekne* meaning arts or craft. Figure 1.3 shows a conceptual diagram which
clearly identifies the various inputs and outputs which comprise this interpretation of
the term—built environment.

Another conceptual model is shown in Fig. 1.4. The Building Performance
Research Unit (see Markus *et al.*, 1972) state that this model is the basis of a complex
system comprising a number of subsystems—the objectives system, the activity

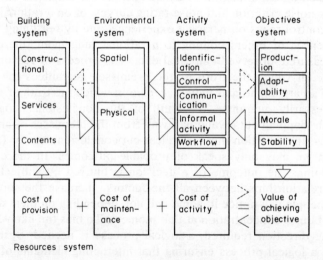

FIG. 1.4. Conceptual model of the system of building and people (Markus *et al.*, 1972).

system, the environmental system, the building system and the resources system— and the total system is open to the influence of politics, economics, culture, climate, the urban setting, the social and the commercial context.

The building universe exists within this total system but must relate to these influences. People within a building do not wish to be totally isolated from the world outside; the organisation owning the building has to blend its objectives not only with those of the people working for it but also with the physical and social environment outside.

The system in Fig. 1.4 identifies some fundamental problems.

(a) Objective System

An organisation exists to provide goods and services for people and to make a profit. Individuals work in the organisation to earn money to buy goods and services and yet surely this is a shallow, materialistic view? People need to busy their brains in an interesting way fulfilling their potential. In practice the objectives of an organisation and the individuals working within it are not always the same. Production is best met when people have a high morale and high motivation. The built environment can contribute towards this. Buildings also need to be adaptable to meet the changing needs of an organisation.

(b) Activity System

Workflow must be continuous and operate smoothly. Communication routes and systems must be clearly identified. Informal activities are important in helping to make up the difference between the aspirations of the organisation and those of the individuals working within it.

(c) Environmental System

The physical and spatial aspects of the environment have to be designed to suit the nature of the activity system. For example, noise can hinder communication, but can help to give privacy and increase the arousal level of an individual. In boring tasks this increase in arousal level can be an advantage, but for work in which quiet conditions are necessary for concentration, it will be a disadvantage. Distribution of sound and air movement depends on the volume and the shape of a space. The visual and psychological qualities of a space are also important.

(d) Building Systems

A building is an environmental filter controlling the amplitude and phase differences between heat, light, moisture and sound incident upon the internal and external surfaces of a building. Materials, form, the shape of a building all contribute towards the environmental and structural performance of a building. The environmental and utility services, often referred to as the mechanical and electrical services, or building services, make a building function by providing heat, cooling, sanitation, lighting, sound control, electricity and communication and other systems. The type of system must be related to the type of building. Type of building refers to the use and the architectural style of the building. Finally, a working and living space needs furnishings, fittings and finishes.

(e) Resources System

The resources of an organisation are the land and space available, the manpower to operate the organisation and the finance available to provide, maintain and operate the building and the people. The value of achieving the objectives must exceed or at least be equal to the total cost. The fallacy of relying on capital cost (i.e. cost of provision) for acceptance of a building design is clearly evident. Cheap first cost may mean a high maintenance cost, and also difficulty in controlling the activity cost if, for example, staff turnover becomes high due to a low-quality/built environment.

Decision making itself can be defined by the system diagram shown in Fig. 1.5 which summarises many of the things we have already declared to be important for a venture to succeed. There must be time to *think* creatively and analytically, out of which some decisions will be crystallised enabling *action* to be taken; further decisions after *communication* with others, further evaluation and decisions lead back to more thought. Decisions punctuate the links between the activities of thinking, action and communication. Thus a professional organisation, or an educational course, must ensure that all of these fundamental activities reach a high standard of achievement.

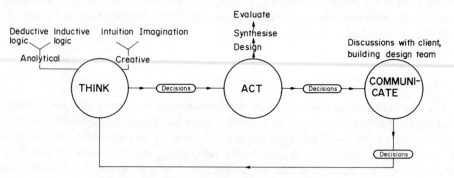

FIG. 1.5. System diagram of decision making.

Crickmay (1972) lists the obstacles to innovation within an organisation as:

(a) *rigidity of thought and behaviour in persons responsible for change;*
(b) *delays that prevent innovation proceeding at the pace of thinking, and consequently inhibit it altogether;*
(c) *lack of time for uninterrupted thought and discussion;*
(d) *inability to stimulate spontaneous thinking on problems to which there is no conventional solution.*

1.5. The Need for Appraisal of Buildings

Terotechnology is described as a combination of management, financial en-gineering and other practices applied to physical assets in pursuit of economic life cycle costs; it is concerned with the specification and design for reliability and maintainability of plant, machinery, equipment, buildings and structures, with their installation, commissioning, maintenance, modification and replacement, and with feedback information on design performance and costs. It aims to promote improved understanding and co-operation between the user, designer, contractor, operation and maintenance staff, and other professional advisers concerned with buildings, structures, sites and services in order to achieve the most satisfactory use of resources.

(IAN K. CARPENTER, *Building Services Engineer*, **42**, December, 1974, A18.)

Ideally a building should satisfy all the functional requirements of the users (including aesthetics) at a minimum total cost besides consuming a minimum amount of energy and wasting no energy. In practice it is difficult to arrive at a solution which is a true optimisation of all these factors because not only is there limited data about user requirements and cost effectiveness but workable processes of optimisation are only just being realised. If a building design fails, four possibilities exist:

(a) the designers have used the wrong criteria and these have been achieved in practice;
(b) the designers have used no criteria;
(c) the designers have used the correct criteria but these have not been achieved in practice;
(d) the building structure and environmental services are incompatible so that no acceptable design criteria exist.

The causes of these possibilities are various—an inadequate brief and specification; a genuine lack of data on certain aspects; poor communication between the client and the design team, and also between the members of the design team themselves; inadequate analysis or synthesis of solutions; unpredictable quirks of human behaviour resulting in poor installation or maintenance may each, or all, contribute towards a complete or part failure in building performance.

The Building Performance Research Unit at Strathclyde University (see Markus *et al.*, 1972) found that many schools had very low daylight factors; lights were left on when there was sufficient daylight for classroom activities in order to reduce contrast between the ceiling and the sky; many classrooms were over-reverberant and noisy; specified sound attenuation performance of partitions were often defeated in practice by flanking sound transmission pathways; many ergonomic factors were ignored in classroom design (e.g. recommended desk heights ignored; equipment used which was not suited to its purpose).

These facts and many others would not have been discovered unless a systematic feedback of objective and subjective data had been carried out. Appraisal of buildings is essential if building design and construction is going to improve because

(a) the behaviour of human beings is variable and often unpredictable, and (b) the performance of systems is also variable; calculations made to predict performance are defeated by non-quantifiable factors such as poor materials, bad construction, incorrect design.

In an energy-conscious age the virtues of monitoring source, distribution and terminal energy expenditure in heating, airconditioning, water and electricity systems is obvious. But the true meaning of *terotechnology* goes beyond this. It reaches out to establish the pattern of user behaviour in buildings. It attempts to answer such questions as: What do people find distracting, pleasurable, boring and annoying in their everyday work and living? What do people value most and least? What are people's needs?

Experimental psychology has an important rôle to play which has been demonstrated by Mehrabian and Russell (1974) in America, by Markus *et al.* (1972), Canter (1974) and Croome (1975) in England, and by Acking, Kuller, and Hesselgren in Sweden (see Honikman, 1971). A terotechnology case study is described by Lovelock (1974); the subjective measurements in this study have been oversimplified but the article still remains a landmark in building design because it forms a comprehensive picture of a building in action from which all can learn.

> *It is interesting to report that, in the author's experience, space designers rarely, if ever, start from a base of objective data on user needs. Requirement surveys, and the like, are of course routine, as in their project, but objective attempts to predict and evaluate are never conducted by design teams. At first glance, the usefulness, utilisation, and applicability of these data is argumentative: without prior experience how can these data be translated into hardware? What constitutes a perception of "utility" or "sociability"? Are these factors functions of the appearance of the office environment, or of work procedures, or manifestations of management policy, or what? Clearly much work must be done in answering these questions, but it is possible that repeated studies of this nature can contribute to a valid experimental design philosophy.*

> (BROOKES, 1972.)

1.6. Education

> *Most of us spend most of our waking life in buildings, and the quality of the environment in which we live and work is, to a great extent, the result of the skill of the heating, ventilating and lighting engineers whose work has gone into the design and construction of the building.*

> *For the last twelve years, as Professor of Architecture and later as Head of the School of Environmental Design at University College London, I have been concerned to develop education in the necessary skills, and research into the serious and fascinating problems of environmental control in buildings. What we have done in London has been matched by similar developments in many other universities in Great Britain. There is now a substantial body of highly specialised skill and serious research on human comfort in buildings, and on the means by which it can be provided efficiently and economically.*

> (LORD LLEWELYN-DAVIES, in a letter to *The Times*,
> 20 February 1973.)

The demand for comfort in building and the rôle of the building services engineer have expanded greatly over the last two decades. He has assumed a major responsibility for the whole range of environmental provision within the integrated building design process. He now must have a wide knowledge of physiology, illumination, acoustics and fire protection as well as his traditional skills in heating and ventilating systems. His work can account for 30–50 per cent of the total cost of a modern office or factory.

> (ALEX GORDON (Past President of Royal Institute
> of British Architects),
> KENNETH SEVERN (Past President of Institution
> of Structural Engineers),
> BASIL SAPCOTE (Past President of Institute of
> Building),
> in a letter to *The Times*, 2 February 1973.)

We educate one another and we cannot do this if half of us consider the other half not good enough to talk to.

> (GEORGE BERNARD SHAW.)

Interdisciplinary working at an educational and professional level is the remedy for today's confused state of thinking and it is in such an interdisciplinary team that I can envisage the environmental engineer's contribution being made. I have always found that there has been some degree of overlapping with all other professions based mainly on our duty to the public and because we are all striving for the same purpose in satisfying the basic needs of people. The educational system, our petty jealousies and the lack of time we often experience have sometimes separated us and, therefore, communication is breaking down.

> (EDWIN BENROY, The meaning of environmental
> engineering: viewpoint of an architect, *Architect
> and Surveyor*, March–April 1974.)

We have had artificial division all over the place and what we will have to do is to bring things together again and stop these artificial separations. It is the same in the universities, you have all sorts of faculties of law and biology and technology, natural sciences and medical sciences. In fact we find more and more that all these things interrelate very much, and if we don't dare to put them all together it is because we feel that our brains are not big enough to do this, that one man cannot see it all without exploding because it is too much for him. Now maybe there is another reason for that; I think we actually can have human beings who can see the whole rather than the parts alone.

> (ROBERT JUNGK, in Open University Technology
> Foundation Course T100 Units 32–34.)

Earlier we have stated that it is important to follow the creative impulse of the architect with a penetrating analysis of building performance. The rôle of the building environmental engineer can be seen in Fig. 1.2 to be essentially to provide this analysis of the building's environmental performance.

In 1897 the Institution of Heating and Ventilating Engineers was formed some 69 years after the first professional engineering institution of the many now existing in Britain. The job of a heating and ventilating engineer was to design and install the heating and airconditioning plant and distribution network in a building (Fig. 1.6a). Over the years this rôle has expanded to include the building services in general, simply because heating, cooling, electrical, lighting, sound, sanitation, circulation and communication services compete for space in a building, all contribute towards the working environment for people and all consume energy (Fig. 1.6b).† In other words there is a strong interaction between all these services. During the last decade there has been further development. The choice of services system type depends on

Design and installation of:

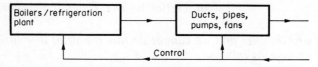

(a) Rôle of heating and ventilating engineer.

Design and installation of:

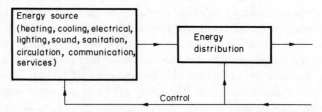

(b) Rôle of building services engineer.

Joint involvement with client and architect on design and installation of:

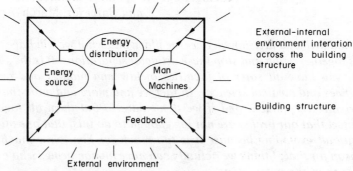

(c) Rôle of the building environmental engineer
(or environmental building technologist)

FIG. 1.6. The evolution of the building environmental engineer.

---

†On 23 April, 1976 The Institution of Heating and Ventilating Engineers became the Chartered Institution of Building Services.

the building type (i.e. activity, materials, form and shape); hence the architect–engineer bond needs to be firmer and more closely linked than in former times. In a similar vein section 1.3 closed with a statement made by Sir Ove Arup in which he calls attention to the need for "synthesisers" of building services. It is the building environmental engineer that has evolved to take this role. Such an engineer, or technologist, helps the architect to match the services to the building by considering a building–energy–man environment system like that shown diagrammatically in Fig. 1.6c. It will be quickly appreciated that the roles shown in Fig. 1.6a and b are incomplete systems because they do not consider man or the building-services interaction in sufficient depth. The meaning of building environmental engineering is illustrated in Fig. 1.7.

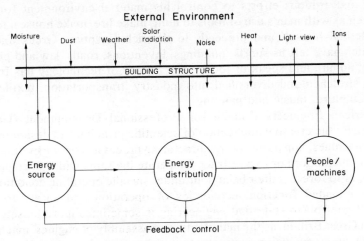

FIG. 1.7. The meaning of building environmental engineering.

One of the principal difficulties in building design has been that the education in the various disciplines has been uneven. Whatever criticisms are made of university architectural education, it is comparable to medicine in its extensiveness and thoroughness. Structural engineers follow three- or four-year degree courses or their equivalent. But who has been educated with such thoroughness on environmental matters? No one.

In an attempt to rectify this, some architectural schools are biasing their studies strongly towards building science. The arguments against this pathway of action are that creative architects are still required, and, in any case, are most architectural students motivated to study analytical aspects of the building environment? Further, the building environment needs to be studied extensively and for three or four years, not as only part of a course, and to commence at an early age just as the disciplines of architecture and medicine do.

Building environmental engineering degree courses do exist now in British universities and polytechnics. Care is needed in balancing the low number of such graduates required compared to traditional engineering disciplines with the number of universities that can operate such courses with viable numbers.

1.7. Environmental Engineers or Technologists

The differences between engineers and technologists is often debated. The word engineer appears to come from an old French and mediaeval English word *engyneour*. In early times the architect fulfilled the architectural and engineering rôles in building. The first time that the rôle of an engineer is distinguished as such is in the late seventeenth century by the military engineer. The Industrial Revolution dawned in the eighteenth century, bringing with it much specialisation and diversification of skills. The Institution of Civil Engineers was the first professional body of engineers to be formed in this country in the year 1828.

It has been seen already that the word technologist has Greek origins in words relating arts, crafts and systematic treatment. The history of technology is a story of man's long and arduous efforts to control his material environment for his own benefit. It begins with man's use of tools to hunt, make fire, make houses, to carve in caves, to defend himself and proceeds to the rich Egyptian, Greek and Roman cultures which have left us sports, buildings, inventions, roads, law and philosophy that we use, admire and speak of today. The history of technology has from these early times stretched and diversified into industry, transportation, textiles, space, medicine, sculpture, music and painting.

The American Engineers Council for Professional Development (US) define engineering as: "the creative application of scientific principles to design or develop structures, machines, apparatus, or manufacturing processes, or works utilising them singly or in combination; or to construct or operate the same with full cognizance of their design; or to forecast their behaviour under specific operating conditions; all as respects an intended function, economics of operation and safety to life and property." It goes on to state that engineering is sometimes more loosely defined, especially in Great Britain, as the manufacture or assembly of engines, machine tools and machine parts, including instruments and associated measuring and control devices.

The *Encyclopaedia Britannica* (1969) defines a term *architectural engineering* as "the art and the technique of building" and distinguishes between *structural engineering* (i.e. structural analysis, structural materials, building structures) and *environmental control* (i.e. thermal, visual acoustic and urban environments, sanitation).

Other dictionary definitions are:

Engineering—the application of science and mathematics by which the properties of matter and the sources of energy in nature are made useful to man in structures, machines, products, systems and processes; the sciences of applying knowledge of the properties of matter and the natural sources of energy to the practical problems of industry.

Technology—the totality of the means employed to provide objects necessary for human sustenance and comfort; the practice of any or all of the applied sciences that have practical value or industrial use; technical methods in a particular field of industry or art.

The creation of building is part of our technological history, whereas the specific task of designing an airconditioning system is an engineering one. One would expect that the chief decision makers in building and construction firms would be technologists with architectural or building engineering, forming a major part of their skill in providing suitable working and living environments for people in buildings.

1.8. The Rôle of Acoustics in Building Environmental Engineering

This is the subject of this book.

It is hoped that from it the reader will begin to feel the important rôle that the sound environment for people in buildings has to play in the wider context of building environmental engineering described in this chapter before embarking on a detailed study of sound and noise problems in buildings. The ideas that need to be discussed with particular reference to buildings are:

(a) the characteristics of sound sources in buildings (Chapter 3);
(b) the propagation of vibrations through fluids and solids (e.g. through ductwork systems and building structures) (Chapters 6, 7, 8, 9 and 11);
(c) the subjective response to noise (Chapters 2 and 10);
(d) designing for optimum sound conditions (e.g. noise control, isolation of structureborne sound, behaviour of sound in spaces) (Chapters 4, 5, 9 and 10).

These ideas form the contents of this book. The book has been divided into three parts reflecting the philosophy and thought discussed in this chapter which forms Part I, an *Introduction* followed by Part II (*Noise and its Control in Building Design*) and Part III (*Some Fundamentals of Acoustics*).

PART II

Noise and its Control in Building Design

NOISE-MEASURING TRUCK of the City of New York, 1929. It was employed in observing and recording "noise levels" at one hundred and thirty-eight positions in the city.

Anthrop (1973) notes that the first really comprehensive city-wide noise survey ever made appears to be the one undertaken in New York between November of 1929 and May 1930 by the New York Noise Abatement Commission. This Commission was formed in the Autumn of 1929 as a result of mounting citizens' complaints over noise, and published the results of its findings in a report, "City Noise" in 1930. Results from a questionnaire indicated that traffic noise was responsible for 36% of the complaints, public transportation for 16%, radios (in homes, stores and streets) for 12%, collections and deliveries for 9%, whistles and bells for 8%, construction for 7.5%, and miscellaneous noise sources for 11.5%.

CHAPTER 2

MAN AND THE ACOUSTICAL ENVIRONMENT

We can state two things about noise: the first is that, on the whole, noise, however intense it is, is not very likely to disturb and people who make it, or even the people whose exposure to it is regular and long continued; and the second is that probably the disturbing effects of noise are at their maximum for people who have to do mental work, but are for some reason bored, tired, forced into a job which is a bit too difficult for them, or not quite difficult enough, or in which they are only moderately interested. If any community contains a considerable number of people of whom these things are true, complaints of noise will be common. As a distraction, however, while its effects do not justify the sensational statements that are often made, it is certainly harmful enough to provide a justification for all the efforts that can be made towards its reduction.

(SIR FREDERICK BARTLETT, *The Problem of Noise*, Cambridge University Press, 1934.)

Doctors are definitely convinced that noise wears down the human nervous system, so that both the natural resistance to disease and the natural recovery from disease are lowered. In this way noise puts health in jeopardy, the most intelligent folk can understand this from its effect upon themselves.

(LORD HORDER, *Quiet* 1, (July 1937) 5–9.)

2.1. Comfort

A comfortable environment is one in which there is freedom from annoyance and distraction, so that working or pleasure tasks can be carried out unhindered physically or mentally. Everyday experience tells us that there are a host of factors which are relevant to this concept. Not only do sound, lighting, colour, temperatures, humidity, air movement and air purity play a part, but psychosociological factors also have an important role. The attitudes of people around us, the social organisation in which man works, the organisation of space, colour schemes and many other factors all can have an influence on our mood and work output. Since there is an interaction between all these factors, the problem is complicated further.

A deficiency in one of the physical factors can spoil the balance of the environment; equally so, surroundings which contain disturbing social or psychological aspects can be uncomfortable.

Although human beings are very adaptable, adaptation in itself may be more of a philosophical acceptance of conditions rather than a purely physiological returning to them. The conditions may still be harmful to the human system, although the person may be unaware of this. A well-known example is the workers in the hosiery and textile industries working in high levels of noise, ranging from 90 to 110 dBA, such that the deterioration in their hearing at the age of 30 is more like that of someone aged over 70 who has worked in normal acoustic climates of, say, 35–60 dBA. Even harder to measure than physiological and pathological damage is the stress imposed on the nervous system due to the quality and pace of the environment with the consequent effects on moods arising from increased irritability and anxiety. For too long, action has been considered necessary only when people complain, but only certain types of people will actually express their feelings to authority. McKennell and Hunt (1970) comments that in their experience only a very small fraction of those people annoyed actually complain; most people, equally annoyed as the complainants, remain silent.

Noise can also be helpful by masking other sounds when desirable, providing auditory cues in alarm systems for example, and even increasing the performance of some work tasks by raising the arousal level of the individual to an optimum value. An optimum, not a minimum, amount of sound is required in a space for aural comfort. Deciding on suitable noise criteria for buildings is difficult because not only the level, but the pattern and quality of the sound are also important; the level, the spatial–temporal pattern, and the quality of sound all depend on subjective assessment in a particular situation.

Few studies have been carried out to ascertain the effect of stress imposed on people by their environment. The question of noise as an environmental stress has been reviewed by Shepherd (1975); Freeman (1975) also points out that it can be assumed that there are many cases of less psychiatric disorder that are partly due to excessive noise. The World Health Organisation define health in a total way as:

> *Health is a state of complete physical, mental and social well-being and not merely an absence of disease and infirmity.*

Disease and infirmity are self-evident. Physical illness can be diagnosed fairly easily but the assessment of mental and social well-being is difficult. Strong evidence now exists which demonstrates that noise adversely affects people mentally and socially. Some social surveys are discussed in Chapters 2, 3 and 4.

In a report entitled "The Noise Around Us" issued by the United States Department of Commerce (COM71-00147) in 1970, reference is made to evidence in Europe, the USSR and the USA which suggests that noisy environments adversely affect the number of miscarriages, the death rate and the incidence of heart attacks. Ando and Hattori (1970, 1973) have suggested that noise affects the human embryo and that there is a higher incidence of low weight births in noisy areas. Abey Wickrama *et al.*, (1969) have shown that there is a significantly higher rate of admission to a psychiatric hospital from areas near Heathrow than from outside this noise area. A few years later replication of this study by Gatonni and Tarnopolsky (1973) demonstrated that this trend still existed.

Ferguson (1973) in a study carried out on telegraphists and mail sorters working in Australian cities, found that the only occupational characteristics that could be identified strongly with neurosis were the negative attitudes to the job and supervi-

TABLE 2.1. COMMONEST STATED CON-
TRIBUTORY INFLUENCES IN 171
NEUROTIC SUBJECTS (FERGUSON, 1973)

Stated influence	%
Job beyond ability[a]	44
Noise[b]	37
Monotony	31
Family or own health	29
Other domestic difficulties	29
Sense of inadequacy	27
Housing finance[b]	24
Responsibility[b]	18
Job unsatisfying	16
Supervisory inadequacy	14
Poor ventilation	14
Dislike of job	12

[a]A greater problem for men over 40 years of age.
[b]A greater problem for men under 40 years of age.

sion, and the noise in the work place. Neurotic subjects were invited to state influences which they thought may have contributed to their symptoms (see Table 2.1). Multiple influences were commonly mentioned. Only 4% could not account for their symptoms by some perceived external or internal stress.

Herridge (1972, 1974) proposes a *4 A's Hypothesis* to explain the effect of noise on mental health. Briefly, people are well when their actions produce effects which improve their life and they display the *a*ggressive, the *a*rticulate and the *a*ctive parts of their nature to a satisfactory level, but when their sense of well-being is undermined, by noise for example, people become *a*pathetic and their aggression is turned inwards and there is an increased likelihood of producing psychiatric breakdown. Adaptation does not overcome noise apathy.

Winder (1976) and Tarzi (1976) also take the view based on considerable experience that noise is a potent factor in the production and exacerbation of stress. Noise does not just contribute towards occupational stress. It can affect the quality of sleep, our attitudes to neighbours and to the district we live in; even the noise of one's own children can be stressful. And yet what soothing and calming qualities sounds can possess when they play like music on the ear and then the spirit. But music, like noise, is a word which is subjective—one man's music, is another man's noise.

The work of the building designer is complex because he has to conceive the environment as a totality and refer to the environmental scientist, ergonomist, sociologist and psychologist for guidance. This book is confined to a study of acoustical environmental systems which control the physical properties and distribution of sound in working spaces. Firstly some general ideas about sensory perception will be described before dealing specifically with the acoustical environment. Sound is only one attribute of an environment. It is important to consider comfort, or the absence of stress, in a general way before specifically considering noise. Comfort is important because the individual is more likely to be happy and also because much research has shown that the mental capacities are fully and most efficiently used in that state.

2.1.1. SENSORY PERCEPTION

Knowledge does not reside in the
impressions but in our reflections upon them. (PLATO.)

Knowledge is gained by man through the senses but perception goes beyond this. James (1892) stated that while part of what we perceive comes through our senses from the environment, another part always comes from within our own mind. There is an interplay between our immediate intake of information and that information which has been stored in the memory from previous experiences. Consider how language is intelligible to us in the course of learning; a memory store containing many sound patterns has been built up. If, by comparing the incoming sound patterns with those in the memory store, there is a correspondence, then we understand the language; if not, then the opportunity exists to acquire new data for the memory but comprehension is lacking.

Perceptions are constructed by complex brain processes, from fleeting
fragmentary scraps of data signalled by the senses and drawn from the brain's
memory banks—themselves constructions from snippets from the past. On this
view, normal everyday perceptions are not part of—or so directly related to—the
world of external objects as we believe by common sense. On this view all
perceptions are essentially fictions: fictions based on past experience selected by
present sensory data.
(GREGORY, in *Concepts and Mechanisms of Perception,*
published by Duckworth, 1974.)

Immersed in a physical environment, the sensory system in man has the opportunity to select many of the stimuli presented to it. A host of factors determines the probability that any particular stimulus will be transmitted through a sensory channel—the nature of the stimulus, the existing events taking place in the sensory system, the arousal level of the individual, the previous experience of the individual with respect to the stimulus. The meaning of "arousal level" is indicated by the continuum:

sleep   drowsiness   relaxation   attention   excitement   anxiety
LOW _____ MIDDLING _____ HIGH
AROUSAL   AROUSAL   AROUSAL

Berlyne (1960) defines arousal level as "a measure of how awake the organism is and how ready it is to react".

Similarly, Corcoran (1963a) describes the term as meaning "the inverse probability of the subject falling asleep".

Arousal level is determined by the environment (i.e. changes, information rate, task being performed) and by internal human states (i.e. level of anxiety, drowsiness, alcohol, drugs). Mehrabian and Russell (1974) factor-analysed the responses to various questions given to 530 subjects. A principal component solution was obtained that gave the intercorrelations shown in Table 2.2 for the various factors of arousal.

TABLE 2.2. INTERCORRELATIONS AMONG THE VARIOUS FAC-
TORS OF AROUSAL-SEEKING TENDENCY (MEHRABIAN AND RUS-
SELL, 1974)

Factors	2	3	4	5
1. Arousal from change	0·60	0·65	0·44	0·53
2. Arousal from unusual stimuli		0·53	0·57	0·45
3. Arousal from risk			0·41	0·40
4. Arousal from sensuality				0·27
5. Arousal from new environment				

Correlations in excess of 0·15 are significant at the 0·01 level.

It seems likely that arousal level increases when an increase in activation of the reticular system occurs.

Electroencephalographic recordings (EEG), variation in blood pressure and pupil dilatation are examples of some of the physiological measures which are related to changes in arousal. More recent work reported by Mehrabian and Russell (1974) concludes that this unitary description of arousal is too simple, and defines *electrocortical arousal* as that dependent on the reticular system, *behavioural arousal* as dependent on the hypothalamus and *autonomic arousal* as dependent on other parts of the nervous system. In general all forms occur simultaneously, but are independent of one another.

The concept of arousal level does help to explain some of the personal differences in reaction to various environments. There is some evidence to suggest that females generally have a higher arousal level than males, and can hence tolerate sensory deprivation situations better. People who are extrovert, or those who tend to be neurotic and anxious, are often more arousal-seeking than introvert personalities. Introverts, however, are often very sensitive to variations in the arousal quality of environments.

Sensory pathways take the form shown in Fig. 2.1. The environment provides a stimulus input ϕ_i, which is received by a receptor (e.g. ear, eye, thermoreceptors in the skin) where a transducer action takes place and the incoming signal is transformed from a sound wave, light quanta and temperature, or whatever, into a series of neural impulses which then travel through the central nervous system.

Processing of the incoming data takes place in the cortex and subcortical regions of the brain after which a response occurs. We may make an active response to the environment (e.g. be distracted by a noise, or dazzled by a light or shiver in a cool room) or a passive response in which a *status quo* is preserved. The description so far has been a neurophysiological one, but the response has physiological and psychological components, as our opening quotation from the work of James suggested; thus

$$\phi_o = f(\phi_i, \psi),$$

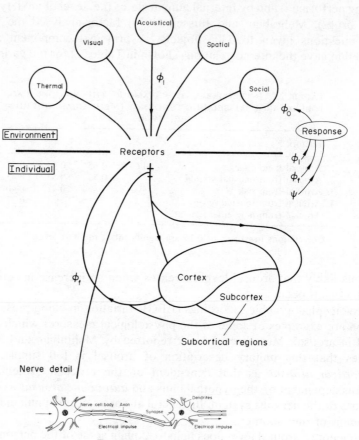

FIG. 2.1. Sensory pathways.

where ψ is a *psyche function*. *Feedback*, represented by the multivariable ϕ_f in Fig. 2.1, occurs between the central nervous system and the sense receptors and is brought about by *adaptation*, a decrease in amplitude of the neutron discharge level for a given input (Adrian, 1928), or *fatigue*, in which tiredness or excessive stimulus levels can produce short-term lapses, and age long-term lapses, in the efficiency of the physiological system. The sensory system may also undergo long-term adaptation in alien environments, called *acclimatisation*. Thus ϕ_i can be expressed as

$$\phi_o = f(\phi_i, \phi_f, \psi).$$

This function is a complex one, and current research is aimed at obtaining a more complete knowledge about it.

In the sleeping state, $\phi_i \to 0$ and

$$\phi_o = f(\phi_f, \psi).$$

The output response is a psycho-physiological one, the brain electrical activity being measured to show sleep disturbance (K-wave bursts), deep sleep (δ-waves) or waking state (α-waves). Dreams occur at regular intervals throughout sleep when rapid eye movements can be observed; their use is possibly to tidy up the long-term memory which has been recording information during the conscious state. Another possibility that may sometimes happen is the excitation of dream patterns by random electrical noise signals resonating in the sensory system. Sleep is discussed in section 2.4.2.

Classical psychophysics attempted to relate ϕ_o and ϕ_i in a simple way, the just-detectable change in any environmental response being related to a given fraction k of the input stimulus change

$$\delta\phi_o = k \frac{\delta\phi_i}{\phi_i}.$$

If k is assumed constant, then on integration the logarithmic law due to Weber and Fechner is established:

$$\phi_{o,1} - \phi_{o,2} = k \ln \left(\frac{\phi_{i,1}}{\phi_{i,2}} \right).$$

The application of this in acoustics has been shown by Riescz (1928) (Fig. 2.2). The differential sensitivity is greatest at around 2500 Hz which happens to be the same frequency range where the ear has its maximum absolute sensitivity to sound. Weber's law states that $\Delta\phi_i/\phi_i$ is constant; it can be seen that this is true for a limited range of frequencies and sensation levels. This logarithmic law is, nevertheless, the fundamental reason for scaling acoustical qualities logarithmically.

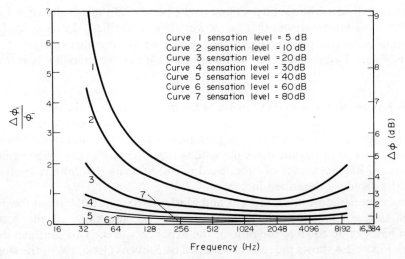

FIG. 2.2. Differential intensity discrimination (Riescz, 1928).

It has been long recognised that the various sensory modalities exhibit marked similarities (refer to Geldard, 1972). Lord Rayleigh wrote in his famous treatise of 1877, *The Theory of Sound*: "It appears that the streams of energy required to influence the eye and the ear are of the same order of magnitude, a conclusion already drawn by Toepler and Boltzmann." Stevens (1953) used subjective scaling procedures to compare loudness and brightness and found that both sensations increased as the cube root of the stimulus intensity (or in the case of sound, the

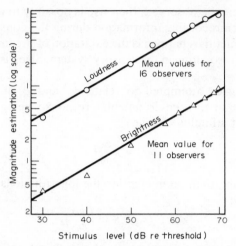

FIG. 2.3. Magnitude estimations of the loudness of a 1000 Hz tone and the brightness of a luminous, achromatic 5 degree disc. In both these early experiments (Stevens, 1953) the subject was presented with the strongest stimulus (70 dB) and was told to call it 100. He was then given the various stimulus levels in random order and asked to assign numbers proportional to the apparent magnitude. The lines through the data have a slope equal to 1/3 in the log–log coordinates, i.e. they are cube-root functions. The reference levels for the decibel scales are 20 μPa and 10^{-10} lambert. Later studies with the method of magnitude estimation showed that it is usually better to omit the naming of a standard (Stevens, 1972a).

loudness increased as the sound pressure to the power 2/3); this is shown in Fig. 2.3. Thus the possibility arises that a power law describes the psychophysical relationship between the perceived response ϕ_o and the stimulus strength ϕ_i better than the Weber–Fechner law. If the threshold value of the stimulus is ϕ, then

$$\phi_o = k(\phi_i - \phi)^\beta.$$

When ϕ_i represents the sound intensity, or the brightness, then the power index β is 1/3. Stevens (1972a) has shown that the power law does not apply to the subjective loudness evaluation of all sounds, in particular white noise and other broadband signals. In order to overcome these deviations from the power law when estimating the perceived loudness level of broadband noise, Stevens (1972b) has evolved the Mark VII procedure described in section 2.2.2.

Further interesting work by Stevens and Marks (1965) has established the general principle that governs cross-modality matching, i.e. the slope of the matching function has a value equal to the ratio of the exponents of the two continua that are matched. Figure 2.4 shows the results obtained by Stevens (1966, 1969); the slopes of the lines give a direct indication of the wide range of exponents that govern the various sense modalities. The value of this work lies in its application to subjective variables that need appraisal.

What is the likelihood of a stimulus being accepted by the sensory system? Poulton (1971) describes the filter model of the sensory system first proposed by Broadbent (1958a). Figure 2.5 shows a series of inputs to the working memory taken in from the world about us; in any particular sensory model there may be several simultaneous inputs, besides the inputs occurring simultaneously in other modes. The brain is featured in this model as having a short-term working memory, which

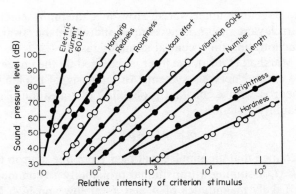

FIG. 2.4. Equal sensation functions obtained by matches between loudness and various criterion stimuli; the relative positions of the functions are arbitrary, but the slopes are those determined by the matching data from Stevens (1966), Stevens (1972a).

holds a representation of what the person is looking at, listening to or thinking about. Any visual impression remaining in it after about half a second is retained by selecting items from the sensory impression, whereas sounds can take several seconds before fading away; an input selector feeds messages one at a time to the box marked "computer" which can then "rehearse" the signal, thus retaining it in the short-term memory, and if learning occurs establishing the information in the long-term memory and also drawing on any possible past experience of the signal from the long-term memory before selecting the appropriate output response. The computer works relatively slowly and accounts for more than half of the reaction time needed for the response of an individual to act. Low signal-to-noise ratios in the sensory system, an overburdened sensory system, fatigue, complex tasks all make the computer take longer to process the incoming information. Highly practised skills need less computer time, the input selector merely selecting the appropriate data from the long-term memory and passing it through the computer to the output.

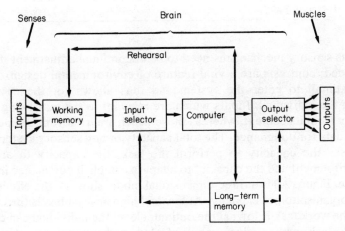

FIG. 2.5. A simple theoretical model which shows what the brain does (modified from Broadbent, 1958a, by Poulton, 1971).

But several questions remain unanswered. How does the brain select relevant inputs whilst rejecting irrelevant ones? How can an individual be aware of peripheral stimuli whilst concentrating on a central stimulus task?

In the work of Marshall (1967a), the filter is replaced by the concept of *stochastic dominance* in the central nervous system. An input signal is pictured as stepping from neuron to neuron with a constant probability of transition p along a sensory pathway. A probability function

$$p_k = p(1-p)^{k-1}$$

is defined as the chance that an impulse will pass from one neuron to another in an interval of time $k\Delta t$, assuming a transitional probability p common to all neuron interconnections. Likely magnitudes of p and k for low and high levels of p_k are given in Table 2.3.

TABLE 2.3

Characteristic	Likely values of	
	p_k level	p and k
Load on neural network; e.g., if system heavily burdened	low ($p_k \to 0$)	$(1-p)^{k-1} \to 0$ for $k \sim 11$ $\left.\right\}$, $k \sim 2$ $\left.\right\}$ $p = 0{\cdot}1$, $p = 0{\cdot}9$
Arousal state of individual: (a) drowsy (i.e., too low)	low	k high ($\sim 2 \to 11$)
(b) optimal	high ($p_k \to 1$)	$\left[1 - \dfrac{\ln p}{\ln (1-p)}\right] \to k$ for $k > 0$, $p > 0{\cdot}5$ $p = 0{\cdot}6$ $\left.\right\}$, $p = 0{\cdot}9$ $\left.\right\}$ $k = 0{\cdot}44$, $k = 0{\cdot}95$
(c) anxious (i.e., too high)	low	k high ($\sim 2 \to 11$)

The various sensory mechanisms need to be in continual adjustment if monotony is to be avoided; contrasts are a vital feature of environmental design. This model permits all stimuli to enter the system and thus allows for the environmental awareness of the individual. Inputs which are relevant to the task being performed by the subject form its context, whereas those which are irrelevant compete with the task sensory data for dominance. The total capacity of any sensory system at a given time comprises the capacity to perform the task, the capacity to adapt to the changing environment and the capacity to adapt to establish dominance in a selected sensory mode. Figure 2.6 is a three-dimensional model showing the effect of arousal state on the organisation of capacity. At low and high arousal levels the capacity for performing the work task is low; at the optimum level the individual can concentrate on the work task while being aware of, but undistracted by, the peripheral environmental stimuli into the sensory system (e.g. noise may alert a sleep-deprived

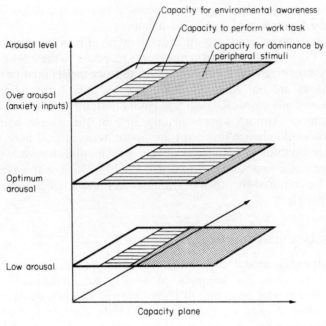

Capacity for environmental awareness
Capacity to perform work task
Capacity for dominance by peripheral stimuli
Arousal level
Over arousal (anxiety inputs)
Optimum arousal
Low arousal
Capacity plane

FIG. 2.6.

subject), but such stimuli will be distracting if they overburden and compete successfully with the central stimuli. Incidentally, at the optimum arousal level it is likely that concentration will be at a maximum. The ease of attaining this ideal and remaining at the optimum arousal level for lengthy work periods depends on the personality type; introverts can reach and remain at the optimum more easily than extroverts. Morgenstein *et al.* (1974) concludes that introverts work less efficiently in the presence of distractions whilst the converse is true for extroverts.

A topological model of the sensory input–output system has been proposed by Zeeman (1965) and shows similar features to the stochastic–dominance model just described. The essential details of a neural pathway are shown on one of the insets in Fig. 2.1. Zeeman postulates that the pattern of activity of the brain determines thought, and the strength of the synapses determines the direction of thought flow; hence the synapses determine memory which catalogues our experiences. He imagines the cortex to be a 10^{10} dimensional phase space C (the brain contains 10^{10} neurons), each dimension corresponding to the rate of firing between each neuron. The strength of all the synapses are represented by a vector field $m(C)$ (i.e. memory), clothing the neuron space in the cortex. Similarly, $m(T)$ represents the strength of all the synapses in the subcortical region. The cortex field $m(C)$ is difficult to change, whereas $m(T)$ is easily altered because the subcortical region includes the emotional centres and the reticular formation (i.e. arousal and attention centres).

Interconnections between the cortex and subcortex mean that the fields $m(C)$ and $m(T)$ are strongly intercoupled. Sensory inputs cause $m(T)$ and $m(C)$ to respond, and if resonance occurs between them, then it is suggested that this corresponds to short-term memory; the coupling simultaneously causes the decay of short-term memory and the formation of long-term memory.

These models contribute towards an understanding of how the personality and the arousal state of the individual figure in a person's response to the environment. Some people are very sensitive to their environment, sensory inputs modify the subcortex field easily; others are not so easily influenced by their surroundings. To cite a well-known noise study by McKennell and Hunt (1961, 1963) made in the vicinity of London's Heathrow Airport, approximately 30% of the people when asked the question—if you could change just one thing about living around here, which would you choose?—could think of nothing to change in their environment even at the very highest levels of noise exposure (i.e. over 103 PNdB). This apparent insensitivity to the noise may be coloured by other environmental factors which were important to that group of people.

2.1.2. TOWARDS AN UNDERSTANDING OF COMFORT

A physical dynamic model of comfort will now be described which helps to explain in a basic way the meaning of terms like freshness, environmental awareness, environmental contrast, all ingredients of the *gesamt* term—comfort.

Imagine the mind, defined in this context to be the container of all the thought processes, to be made of a flexible membrane which encloses the part, or whole, of the mind concentrating on a particular work task termed the *central stimulus*; this is an *attention sphere*, somewhat similar to the thought sphere of Zeeman but more localised. This sphere is continually being bombarded with stimuli from the world around us, called the *peripheral stimuli*; they are the patterns of light, temperature, air movement, space, sound and social behaviour which surround us. The peripheral stimuli form a random pattern in time and space. The general idea is shown in Figs. 2.7a and b for subjects having optimum and suboptimum arousal levels. At low arousal levels the attention sphere is stiff and the environmental awareness is low, whereas at high levels of arousal concentration on the central stimulus is practically impossible due to the easily distorted attention sphere; if at any instant the amplitude or some other defined characteristic of one or more such stimuli punctures the membrane, instead of indenting it then bouncing off, distraction and/or annoyance occurs, the attention of the subject on the central stimulus breaks down, and the environment is no longer satisfactory (Fig. 2.7c). One's attention can, of course, be diverted by pleasant features in the environment, but in this case a voluntary change of the central stimulus occurs. This model, although artificial, treats the environment as a whole and relates to the work or leisure involved. Lying on a beach is comfortable with a lot of sunshine producing levels of solar radiation, which would be uncomfortable if presented to someone working inside a building.

Clearly, each person has his own range of comfort values in different activity situations. The task of the environmental engineer and the architect is to create environments which are acceptable to most of the people living inside buildings; this is difficult when large groups of people are involved. Current research is attempting to establish a valid methodology for correlating subjective responses to physical measures so that degrees of temperature, decibels of sound and lux of light have real meaning (Canter, 1968, 1969; Woods and Canter, 1970; Hesselgren, 1971; Croome,

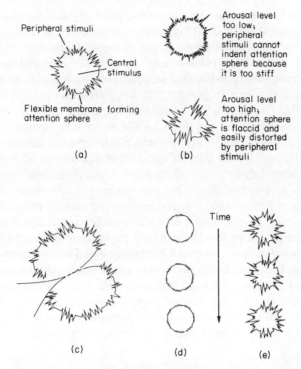

Peripheral stimuli

Central stimulus

Flexible membrane forming attention sphere

(a)

Arousal level too low; peripheral stimuli cannot indent attention sphere because it is too stiff

Arousal level too high; attention sphere is flaccid and easily distorted by peripheral stimuli

(b)

Time

(c)   (d)   (e)

FIG. 2.7. (a) The environmental situation for a person at an optimum arousal level; (b) as (a) but for a person at a sub-optimum arousal level; (c) peripheral stimulus punctures attention sphere, distraction and/or annoyance occurs; (d) the stimulus pattern in a boring environment; (e) the stimulus pattern in a stimulating environment.

1971). Environmental design today often begins by looking at other similar situations which give clues to interaction effects and patterns of social behaviour within organisations. With our model a number of ideas can be explained. It can be supposed that the membrane will have properties which represent facets of the personality and the psychological make-up of an individual; if the membrane material is very flexible, peripheral stimuli can indent the attention sphere easily and the environmental awareness of the individual is high, so that distraction will be less likely (e.g. pleasant variations in air movement). For any individual the size of the attention sphere varies according to his or her arousal level. Figure 2.7b shows that when the environmental awareness is low, the peripheral stimuli either cannot enter or, alternatively, can only confuse the attention sphere. In a boring situation (Fig. 2.7d) the stimulus pattern is the same over a period of time, whereas in a stimulating environment large variations occur (Fig. 2.7e). In the latter situation the contrast pattern recognition is said to be high. One could argue that in the case of a monotonous task the central stimulus does not take a well-defined form, hence there is no attention sphere for the peripheral stimuli to impinge upon.

In itself, comfort may evoke no comment from people in the environment, for it possesses a neutral, stable, acceptable quality. People expect their surroundings to allow them to pursue their work activity unhindered. On the other hand, an environment which is uniform in space and time is monotonous—the sensory information chain between the outside world, the body and the brain becomes idle due to sensory deprivation; the saying "being bored to death" has some truth, a fact verified by scientists investigating the death of monkeys on prolonged space flights. Rathbone (1967) stated her fundamental hypothesis of building environmental engineering aesthetic theory to be, *that satisfaction which is obtained when pattern is perceived*. Contrast plays an important rôle in pattern recognition because changes are required within certain limits to assure the brain that the body is working properly. These contrasts are provided by the peripheral stimuli.

Figure 2.8 is a proposed comfort model which attempts to summarise the ideas that have just been discussed. Any state of comfort or discomfort is determined by four cardinal dimensions: *the physical sensation level*; *the emotional sensation level*; *the arousal level*; *the distraction level*. One can interpret these dimensions as follows. Arousal level defines how attentive a person *feels*; the scale of physical and emotional sensations are what a person *is*; distraction level indicates the probability of *concentration* being high or low. In truth all these dimensions are concerned with individual personality defined by genetic endowment and environmental conditioning.

The levels of all these cardinal dimensions affect work performance because they influence the mind state (mental health) or body (physical health). If any one or more of these levels falls outside the comfort zone a person may experience discomfort, but there are exceptions. For instance at the optimum arousal level the concentration of an individual is likely to be high; it can be so high that it is difficult to cause

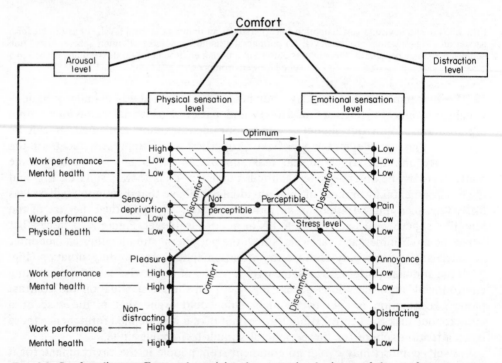

FIG. 2.8. Comfort diagram. Factors determining the proportional mixture of the comfort parameters: work task; personality susceptibility to environment; physical sensitivity; emotional sensitivity; intelligence; concentration ability; past experience of event; expectancy; quality of stimulus; information; content of stimulus; interaction with other factors.

distraction or, initially, even physical stress. The exact limits of comfort zone and the proportional distribution of the four levels depends on the work task; the personality of the individual; the susceptibility to the environment; the degree of physical and emotional sensitivities; the past experience of the event and the consequent expectancy of it; the concentration of the individual (i.e. the ease to reach the remain at the optimum arousal level); the quality, not just the level, of the stimulus; and, lastly, interactions with other factors. It may be argued that an *aesthetic sensation* level scale should also be included, but it has been omitted here for simplicity. Most of these factors will be discussed in various parts of this chapter, but to begin susceptibility and expectancy are considered.

It has already been seen in the noise study made by McKennel and Hunt (1963, 1966) that 30% of the sample questioned appeared to be insensitive to the very high noise levels in their neighbourhood. Beranek (1971) has drawn attention to the 1961 noise survey carried out in London. A susceptibility rating was derived from the answers to six questions on a forty-item questionnaire that evoked statements from the interviewees about their sensitivity to noise as follows:

(a) Does noise ever bother, annoy, or disturb you in any way?
(b) When you hear the noise most annoying to you in your home do you feel very annoyed, moderately annoyed, a little annoyed, or not at all annoyed?
(c) Would you say you were more sensitive or less sensitive than other people to noise?
(d) Is there too much or too little fuss made about noise nowadays?
(e) How far would you agree that noise is one of the biggest nuisances of modern times?
(f) Could you sum up your opinion by saying whether you find noise in general: very disturbing, disturbing, a little disturbing, or not at all disturbing?

The results shown in Fig. 2.9a suggest that 10% of the population were extra-sensitive to noise, whereas about 25% of the people were comparatively insensitive. The most troublesome noise sources were road and air traffic, although for people rated as having average susceptibility the difference in the degree of nuisance between these sources and noise from neighbours' dwellings was small (Fig. 2.9b). It was concluded that annoyance is very strongly related to the susceptibility rating, but the validity of applying the questionnaire approach must itself be questioned (Phillips, 1971). Perhaps some of the apparently insusceptible group would say they had become used to noise. Now although this may suppress their emotional feelings about the problem, the stress on the nervous system may remain deteriorative. Likewise, the acceptability of the environment in which we live as judged by complaints is invalid because many people concerned about events around them do not complain (McKennel and Hunt, 1970). Beranek (1971), Schultz (1970) and Webster (1969) have drawn attention to the differences in noise tolerances in colder and warmer countries; people in warm climates leaving windows open for ventilation seem more tolerant of noise by 5–10 dBA. This, however, from the point of view of the nervous system, may be a false tolerance.

By environmental conditioning we become to expect a certain outcome from an event. If this expectancy is not fulfilled a new conditioning may occur and the initial

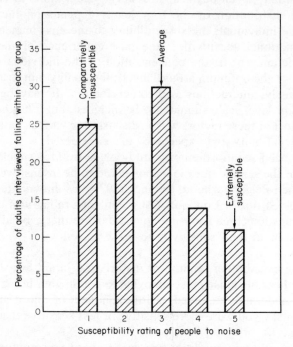

(a) This bar graph shows the percentage of 1377 adult residents interviewed in depth in a 1961 Central London community survey for each of five categories of noise susceptibility rating.

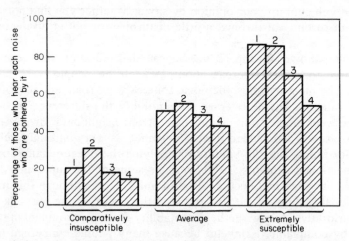

(b) This bar graph shows the percentage of the 1377 adults interviewed in Central London who said they are bothered by particular outdoor noises that they hear when they are in their homes as related to their susceptibility rating to noise. Outdoor noises heard indoors: 1, road traffic noise; 2, aircraft noise; 3, noises from neighbours' dwellings (children's and adults' voices, radio/TV, bells, footsteps, banging, etc.); 4, noise of pets.

FIG. 2.9. Susceptibility of people to their environment (Beranek, 1971).

uncertainty is resolved. On the other hand, the original imprint may never be forgotten as Goethe relates in beautiful style in one of his Roman Elegies written during the years 1786–1800. The barriers of language are formidable in the case of poetry, so the original German is included here; the prose translation is by David Luke and may be found in *Selected Verse by Goethe* published by Penguin in 1964.

Manche Töne sind mir Verdruß, doch bleibet am meisten
  Hundegebell mir verhaßt; kläffend zerreißt es mein Ohr.
Einen Hund nur hör ich sehr oft mit frohem Behagen
  Bellend kläffen: den sich der Nachbar erzog.
Denn er bellte mir einst mein Mädchen an, da sie sich heimlich
  Zu mir stahl, und verriet unser Geheimnis beinah.
Jetzo, hör ich ihn bellen, so denk ich mir immer: sie kommt
  wohl!
Oder ich denke der Zeit, da die Erwartete kam.

Many noises annoy me, but barking of dogs I detest above all;
  my ears are split by their yelping.
There is one dog only whose high-pitched bark often fills me with gladness,
  and that is the dog my neighbour has reared.
For he once barked at my sweetheart when she was stealing to me on the sly,
  and nearly betrayed our secret.
And now, when I hear him bark, I think always:
  she must be coming!
Or I remember a time when I waited for her, and she came.

It is worth while comparing the comfort model proposed in Fig. 2.8 developed independently by the author with that by Mehrabian and Russell (1974). They propose a theory, in their own words:

  that physical or social stimuli in the environment directly affect the emotional state of a person, thereby influencing his behaviors in it. Three emotional response variables (pleasure, arousal, *and* dominance) *summarize the emotion-eliciting qualities of environments and also serve as mediating variables in determining a variety of approach-avoidance behavior such as physical approach, work performance, exploration, and social interaction.*

These three variables have their bases in the fundamental nature of human behaviour. Arousal describes the level of activation or *activity*; dominance and its antonym submissiveness define the level of aggression or *potency* of the human interaction with the environment; the pleasant–unpleasant dimension can be thought of as the *evaluation* of an environment. These factors can be described by a series of semantic differentials described by Mehrabian and Russell (1974), e.g. for pleasure, happy–unhappy, pleased–annoyed; for arousal, excited–calm; for dominance, autonomous–guided. To picture these variables combining in situations, consider the environments described in Table 2.4.
At an enjoyable party one is in high spirits and all the variables are high; but at an

TABLE 2.4

Environment judged to be	Level of variables		
	Pleasure	Arousal	Dominance
Comfortable	High	Optimum	High
Boring or fatiguing	Low	Low	Low
Exciting	High	High	High
Stressful	Low	High	Low

interview the personality is bristling to interact with that of the interviewer in a favourable way; the activity level or arousal is high in a situation where the interviewer has to submit to the questions and probing of the interviewer; it may turn out to be pleasurable but at the beginning it certainly is not.

The model in Fig. 2.8 includes dimensions of arousal and of pleasure. In my view dominance works itself out in the judgments made on these dimensions. Clearly individuals differ physiologically as well as psychologically, so a physical sensitivity dimension must be included in an overall view of comfort. Distraction is an important aspect of work performance; it can be thought of as being related to arousal level in a manner described in the text related to Fig. 2.6. There are some differences in these views; nevertheless, a similar attitude has been taken towards the interaction of man with his environment. Mehrabian and Russell (1974) also recognise that it is not only the intensity level and frequency of the temperature, the noise or whatever which are important, but the spatial–temporal pattern of the stimuli too. They use the term *information rate* to define this pattern which depends on such factors as the novelty, the expectancy, the interaction, the intermittency and the contrast of the stimulus field. In order to relate the important environmental psychology research of Mehrabian and Russell (1974) to the built environment, their own summary of Chapter 4 in their book *An Approach to Environmental Psychology* is now given:

> The object of this chapter was to review the available evidence relating important aspects of physical stimuli to feelings. The large number of studies that related colors to emotions led to the conclusions that brightness and saturation are directly correlated with pleasure. For hues, the following colors are ranked in descending order of pleasantness: blue, green, purple, red, and yellow.
>
> Color "warmth" is correlated with arousal and ranges from red, orange, yellow, violet, blue, to green, with green being the coolest. Color warmth is also directly correlated with saturation and inversely correlated with brightness.
>
> The results for thermal stimulation are complicated because several variables influence thermal comfort. The temperature range of 60° to 68° Fahrenheit is optimal for comfort (pleasure), and temperature gradients from floor to head level reduce comfort. Arousal increases with large deviations in temperature (away from body temperatures), small mean radiation temperature, alterations in air speed, and decreasing humidity.
>
> For light intensity, the findings show that increased intensity of lighting is pleasant, but that discontinuities in lighting (glare) are unpleasant. Light intensity is a correlate of arousal.

For sound stimulation, most of the findings are based on noise or pure tones and show that pleasantness increases with decreasing loudness and/or duration, intermediate frequencies, simplicity of the sound spectrum, less variability, and more expected qualities of noise and pure tones. In other words, in the case of noise, pleasantness increases with lower information rates. On the other hand, arousal increases directly with the loudness of noise or music.

The remainder of this chapter consisted of the detailed analysis of one comprehensive set of adjective pairs for describing environments. The factor analysis of these physical descriptors yield nine intercorrelated factors. These were the pleasant, bright and colorful, organized, ventilated, elegant, impressive, large, modern, and functional qualities of environments. The expressions of each of these factors in terms of basic emotional reactions showed that all but one of them were highly correlated with pleasure. These concepts, taken from the vocabulary of architects and environmental designers, not only primarily reflected evaluative attitudes but were also restricted mostly to the visual sense. Specific equations did, however, show some interesting differences in the other two emotions, arousal and dominance, elicited by each of the nine qualities of physical stimulation.

Presumably arousal and pleasure increase with variable air movement. Fresh environments are those in which the arousal and the pleasure dimensions will be rated highly. Comfort is defined by Mehrabian and Russell on the pleasure dimension, whereas the author views comfort in a more overall way as summarised on Fig. 2.8.

2.1.3. OTHER FACTORS

Job satisfaction is essential if we are to work efficiently. How can this be assured? Herzberg (1959) in his book *The Motivation to Work*, classifies the main contributory causes from his research as achievement, recognition, work interest, responsibility, possibility for growth in status and salary, relationships with people at a higher, similar and lower status level, policy and administration of the organisation, working conditions, personal life and job security. Although these are listed in an approximate order of importance, the priority depends upon the particular situation and the interactions that occur between these factors.

The total environment in buildings is the resultant of the interactions between the relevant factors—warmth, light, noise, vibration, air movement, humidity and air purity—as moderated by psychological and sociological factors, all of which have an important influence on man's comfort, well-being and efficiency at work. They cannot, therefore, be treated independently in building design, which should be aimed at achieving an optimal balance between them.

The deeper significance of the social and psychological factors which exist and their interplay with the sensory communication systems are now being realised. An example of this is the *Bürolandschaft* concept—open space, landscaped office planning with desks and office equipment arranged in functional and non-regular patterns. Everyone, except the chief executive, shares the same large open space, has the same grade of furniture and carpeting, enjoys the plant décor and other accessories. What influence will this type of luxurious work space have on productivity? Will people become more discriminating about their physical environment? Will people communicate with one another more easily? Can privacy be preserved? Sound levels in some offices have been found to be too low (Croome, 1969a).

The arguments advocated in favour of landscaped offices are:

(a) advantages of organisation, increased flexibility, faster communication and easier exchange of information (Kyburz, 1968);

(b) increase of effective floor area due to elimination of corridors and united rooms; Schmidt (1967) suggests that the effective floor area is 83% of the total for a landscaped office compared with 57% for a conventional office layout; Zeitlin (1969) quotes a 40–50% reduction in space requirements;

(c) Kyburz (1968) states that the accessibility of superiors is a prerequisite for a co-operative form of leadership that results in discipline and mutual consideration becoming normal modes of behaviour; Gubler (1967) points out the favourable effects of mutual supervision as well as improved team work due to the elimination of possible rivalry;

(d) with bureaucracy and group selfishness thus done away with, Lappat (1969) is of the opinion that employees take a greater interest in the firm and show improved manners and co-operation;

(e) measurements by Einbrodt and Beckmann (1969) suggest that the work output in landscaped offices is improved due to improved climatic conditions; Zeitlin (1969) mentions a 10–20% increase in productivity;

(f) Zeitlin (1969) records claims of 20% decrease in maintenance costs, and 95% reduction in set-up and renovation times;

(g) improved staff morale and decreased absenteeism.

As a counterweight to these arguments are the following:

(a) some people have reported a lack of aural and visual privacy, and difficulty in concentration due to noise and visual distractions (Nemecek and Grandjean, 1971, 1973);

(b) Heusser (1968) comments that many people object to the unification, as well as the minimisation, of personal participation in their environment; he maintains that it is easier to integrate into a smaller community than into landscaped offices where the feeling of security and support develops more slowly.

Riland (1970) relates an American experience. Communication and social contact was better, and employees liked the improved psychological climate—such things as colour, attractiveness and atmosphere, but temperature control, specific equipment, storage facilities were criticised and the landscape was too public for many employees.

Hundent and Greenfield (1969) summarise the general state of research into the landscaped office environment as follows:

Clearly, the space must influence behaviour—at the most elementary level, by interactions of occupants with columns, barriers, desks, etc. But to what extent physical inconveniences do interact with work organisation, work throughput, and group cohesiveness, is not shown in these field studies. Are there thresholds of inconvenient or unaesthetic environments for example, beyond which degradation of performance occurs? Or to what extent can the office environment be sensed by its occupants as reflecting the policy of management, and for that reason impinge on behaviour and performance? Or are style of surroundings and performance independent of one another?

To what extent these contentions are valid may be considered by briefly reviewing a survey carried out by Nemecek and Grandjean (1971, 1973a) of the climatic conditions in fifteen landscaped offices throughout Switzerland and who at the same time questioned 519 employees about their experience of working in them. Table 2.5 shows the range of environmental conditions.

TABLE 2.5. ENVIRONMENTAL CONDITIONS IN
FIFTEEN LANDSCAPED OFFICES (NEMECEK
AND GRANDJEAN, 1971, 1973a)

Air temperature	21–23.5°C
Relative humidity	35–56%
Air movement	0·05–0·18 m/s
Background noise (L_{50})	
(a) General	47–52 dBA
(b) Technical research department	38 dBA
(c) Punched card department	57 dBA
Frequent peaks (L_{10})	51–65 dBA
Illumination intensity	390–2000 lux

Noise disturbance, particularly the content rather than the sound level of conversations, was mentioned by 35% of the respondents; managers and research staff were more disturbed than the administrative staff. No correlation between disturbance and any of the 50%, 90%, 10% levels could be established; however, recent work by Hay (1973) has shown a correlation between dissatisfaction and the (L_{10}–L_{90}) levels (see section 2.6.3). Table 2.6 shows the frequency response to the question which attempted to identify the most distracting noise sources.

Most people (74%) preferred an airconditioned office; one-fifth complained about the large inside-to-outside temperature differential, draughts and unsatisfactory air quality; and 15% found it annoying not to be able to open the windows. The principal

TABLE 2.6. DISTRACTING NOISE
SOURCES (NEMECEK AND GRANDJEAN,
1971, 1973a)

Noise source	Percentage of people disturbed
Conversations	46
Office machines	25
Telephones	19
People walking about	7
Others	3

conclusions that Nemecek and Grandjean drew from their field study were:

(a) better communication and improved personal contact are predominant factors; this agrees broadly with some of the views of Kyburz (1968);
(b) people experience a loss of concentration and a loss of privacy due mainly to noise. Involuntary attention switching occurs between neighbouring groups of workers having conversations (i.e. it is the meaningful information content that is important rather than the level of the noise);
(c) of the people questioned, 63% considered working in a landscaped office easier and more practical than in a conventional one;
(d) overall, 59% of the workers preferred a landscaped office and 37% a conventional one.

A summary of the principal characteristics of the landscaped offices investigated by Nemecek and Grandjean (1973a) and as assessed by the users in given in Table 2.7.

Another survey of 2575 people working in thirty-eight landscaped offices at thirty-three private companies or centres of public administration in Sweden, by Wolgers and Wiedling (1971), found major differences—primarily in the physical environment—between the various premises. Considering the qualities of sound, light, air and layout (note these correspond to the 4L qualities stressed by Swedish designers, *ljud, ljus, luft* and *layout*) no office design was found to have been successful in all these qualities, most achieving only two or three of them to the satisfaction of the users. Lighting and layout obtained the highest number of satisfactory responses, acoustics and ventilation the least. Smaller offices tended to show a fairly good acoustic climate but poor air movement, while for the larger offices the converse of this was generally true. The following points are some of those received in reply to the survey questionnaire:

Aspects Requiring Improvement

(a) the need for more screens; start with too many rather than too few;
(b) noise disturbance from conversations (telephone and face-to-face) and office machines; disturbance sensitivity found to increase as the education level of the respondent became higher, similarly as for the Nemecek–Grandjean study;
(c) air too hot and dry in summer and of uneven quality in winter; women were found to be more concerned about this than men;
(d) work areas should have generous dimensions and should not be reduced to introduce new staff.

Creditable Aspects

(a) office equipment;
(b) clean and spacious layout;
(c) better collaboration and communication either way between the hierarchial levels inside and outside a department;
(d) efficiency of routine tasks appeared to be improved, whereas specialised work requiring deep concentration was impaired by acoustic and visual distractions.

TABLE 2.7. THE ADVANTAGES AND DISADVANTAGES OF LANDSCAPED OFFICES (NEMECEK AND GRANDJEAN, 1973a)

"What are the advantages of a large office?"

	Total N = 519	Men N = 412	Women N = 107	Managers N = 76	Graduates N = 98	Employees N = 289	Unqualified employees N = 56
Number of responses (≡ 100%)	631	532	99	110	139	334	48
Advantages	Distribution of responses (%)						
Better communication	40	42	29	43	42	40	25
Personal contacts	28	27	35	19	24	30	46
Work flow, supervision, discipline	15	16	9	17	17	15	10
Better relations between colleagues	6	6	9	6	6	6	6
Equipment, comfort	4	4	5	5	6	3	4
Space and room	4	3	9	3	4	4	6
Miscellaneous	3	2	4	7	1	2	3

"What are the disadvantages of a large office?"

	Total N = 519	Men N = 412	Women N = 107	Managers N = 76	Graduates N = 98	Employees N = 289	Unqualified employees N = 56
Number of responses (≡ 100%)	518	418	100	105	110	257	46
Disadvantages	Distribution of responses (%)						
Disturbances in concentration	69	71	62	55	70	74	70
Confidential conversations impossible	11	12	7	24	15	5	7
No privacy	6	5	10	6	2	8	7
Poor climatic conditions	5	4	8	3	5	5	4
Too many people together	3	2	6	3	2	4	4
Various other data	6	6	7	9	6	4	8

For the total statements χ^2-tests of the differences between men and women are significant at $p < 0.05$ and between the employee categories at $p < 0.001$.

The latter two aspects correspond with the observations of Nemecek and Grandjean. Of the total, 71% claimed they felt more tired in their new premises compared with their previous ones, and 30–40% felt more tense and irritable than before, while 35–45% felt they were "more supervised and checked upon than before".

Brookes (1972) describes a research field study carried out in America in a pilot landscape office consisting of 120 people occupying 2800 m² of floor space. Subjects completed a questionnaire besides being interviewed in depth; this procedure was carried out in the office occupied before the move, and afterwards in the landscaped office. A comparison of the responses received in these surveys is given in Table 2.8; the factors involving acoustics are indicated by asterisks. Clearly aural privacy and distraction become a problem in a landscaped office, the higher noise level in the modern office perhaps not being such a problem as the noise information content. Notice that in the conventional office there are ten items of dislikes with a frequency response of over 10%; in the new office there are eight such items. On the items that are liked, seven items are ranked over 10% in the old office and nine items in the new office. There are a large number of items liked and disliked in the new office.

TABLE 2.8. THE FREQUENCY OF RESPONSES TO AN OPEN-ENDED QUESTION ASKING SUBJECTS TO LIST THE THREE MAJOR THINGS THEY LIKE AND DISLIKE. ITEMS RANKED ACCORDING TO FREQUENCY OF MENTION (BROOKES, 1972)

Former conventional office		Nine months after moving to the new landscaped office the subjects reported the following major "likes" and "dislikes".	
Dislikes (15 Items)	Likes (16 Items)	Dislikes (22 Items)	Likes (19 Items)
Crowding of desks, 47	Good lighting, 37	*Noise of conversation, 47	Colourful design, 34
*Noise of conversations, 43	Good adjacencies and layout, 34	*Lack of privacy, 42	Comfortable chairs/furniture/workplace, 34
Poor adjacencies and layout, 37	Good furniture and equipment, 29	Crowding of desks, 43	Attractive décor, 26
*Poor HVAC, 31	*Adequate privacy, 27	Flat-top desk too small, 26	Indoor plants, 18
Poor lighting and glare, 24	Sufficient room, 16	Lack of drawer space for personal storage, 24	Good lighting, 18
Drab décor, 21	Location in building, 11	Adjacencies and layout, 24	Cheerful atmosphere, 13
Inadequate workspace storage/shelving, 20	Pleasant co-workers, 11	*Poor HVAC, 17	*Carpet on floor, 13
Inadequate filing spaces, 20	Cleanliness of workspace, 7	Lack of shelf space, 16	*Lack of noise, 12
*Lack of privacy, 18	Décor, 7	Poor lighting, 8	Adequate adjacencies and layout, 11
Lack of provisions for visitors, 13	Close to canteen, 7	Too many plants and insects, 7	Modern appearance, 10
No windows, 8	*Adequate HVAC, 6	Cheapness of furniture construction, 7	Adaptability of furniture, 7
Lack of machines and business equipment, 8	Efficient atmosphere, 4	Lack of filing space, 7	*Adequate privacy, 6
Need for coat racks, 5	Other people's offices, 2	Drab décor, 6	Small conference areas/provisions for visitors, 5
Lack of flexibility in arrangement of space, 3	*Quiet, 2	In traffic flow, 5	Ease of keeping workplace clean and tidy, 4
Lack of room for expansion, 1	Impression made on visitors, 2	Uncomfortable chairs, 4	Better filing space, 4
	Good pin-up space, 1	Lack of electrical outlets, 3	More personal contact, 4
		Getting lost, 2	Good impression on visitors, 3
		No windows, 2	Out of traffic flow, 1
		No provision for visitors, 2	Easy to supervise secretary, 1
		Poor telephone location at desk, 2	
		Poor expression of status, 1	
		No buzzer on telephone, 1	

*Factors involving acoustics.
HVAC denotes heating, ventilation and airconditioning services.

By listing similar items side by side under the dislikes–likes headings as in Table 2.9 the improvement or otherwise of a particular item can be seen at a glance, as can also any new aspects or redundant ones which appear or disappear in the new and old situations. Notice that crowding of desks is a problem in conventional and landscaped offices. It may also be another factor related to privacy. Privacy depends on a person's sense of enclosure, besides the aural and visual contacts possible with the surroundings, and the need or sense of privacy varies from person to person and with job to job.

Brookes (1972) made the following conclusions:

> *In summary, it* [landscape office] *looks better but it works worse.*
>
> *The Bürolandschaft concepts of privacy, noise control, visual control of space, are to be challenged by these data.*
>
> *It does not seem to be enough to know you cannot be overheard. The security of privacy seems to demand a solid visual barrier. And the removal of fixed barriers does not necessarily seem to change office functioning—why should it, unless original conditions are extreme?*
>
> *This study, as with the other reported, was unable to distinguish the positive effects of office landscape from those that might just as easily have been ascribed to a modern conventional office—good atmosphere, decent aesthetics; but the latter style would have the plus of noise control and privacy. Clearly, in the extreme case of office cubicles—a prison—sociability must decline; but in the office many people leave open their doors in any case, while having the option to seclude themselves if necessary.*

Table 2.10 is a psarchigraph using a semantic scale scored from 0, implying least association between the scale and the perception of that attribute, to 4, indicating the most association of word with perception. It shows how well the design team was able to meet the subjects needs. In fact the new office did *not* meet any of the functional needs, was poor in respect of privacy; the aesthetic and geometric factors were moderately successful whilst sociability was only fairly satisfactory.

A more optimistic view of landscaped offices is reached by Boyce (1974). He describes three questionnaire surveys taken in a group of different types of conventional offices, and one landscaped office with integrated environmental services design; all built in England. The principal conclusions drawn are:

* the design and installation of airconditioning is not as straightforward as often appears, and great care is necessary in design if a good thermal environment is to be experienced in practice;
* do not attempt to put in more staff than the space was designed for; overcrowding has already been mentioned in reference to the work of Brookes (1972);
* the requirements of different groups of people in one office for environmental and social conditions may be very different, and this needs taking into account throughout the building design process;
* communications between departments improve in landscaped offices.

TABLE 2.9

Dislikes (percentage frequency responses)

Item	Office Old	New
Crowding	47	43
*Noise conversations	43	47
Poor adjacencies and layout and	37	24
*Poor heating, ventilating and airconditioning	31	17
Poor lighting and glare	24	8
Drab décor	21	6
Inadequate workspace/storage/shelving	20	16–24
Inadequate filing space	20	7
*Lack of privacy	18	42
Lack of provision for visitors	13	2
No windows	8	2
Lack of machines and business equipment	8	—
Need for coat-racks	5	—
Lack of flexibility in arrangement of space	3	—
Lack of room for expansion	1	—
Too many plants and insects	—	7
In traffic flow	—	5
Flat-top desks too small	—	26
Cheapness of furniture construction	—	7
Uncomfortable chairs	—	4
Lack of electrical outlets	—	3
Getting lost	—	2
Poor telephone location at desk	—	2
Poor expression of status	—	1
No buzzer on telephone	—	1

Likes (percentage frequency responses)

Item	Office Old	New
Good adjacencies and layout	34	11
*Adequate heating, ventilating and airconditioning	6	18
Good lighting	37	26
Attractive décor } Colourful design	7	34
Better filing space	—	4
*Adequate privacy	27	6
Provision for visitors	—	5
Indoor plants	—	18
*Quiet	2	12
Impression on visitors	2	3
Out of traffic flow	—	2
Good furniture and equipment } Adaptability of furniture	27	34
Sufficient room	16	7
Location in building	11	—
Close to canteen	7	—
Pleasant co-workers	11	—
More personal contact	7	4
Cleanliness	7	—
Easy to supervise secretary	—	1
Efficient atmosphere	4	—
Cheerful atmosphere	—	13
*Carpet on floor	—	13
Good pin-up space	1	—
Modern appearance	—	10

TABLE 2.10. HOW GOOD WAS THE DESIGN TEAM AT SATISFACTORY PERCEPTUAL NEEDS? A COMPARISON OF THE NEW LANDSCAPE (SOLID LINE) WITH THE PREDICTED NEEDS (BROKEN LINE). MEAN SCORES FOR ALL SUBJECTS. (BROOKES, 1972).

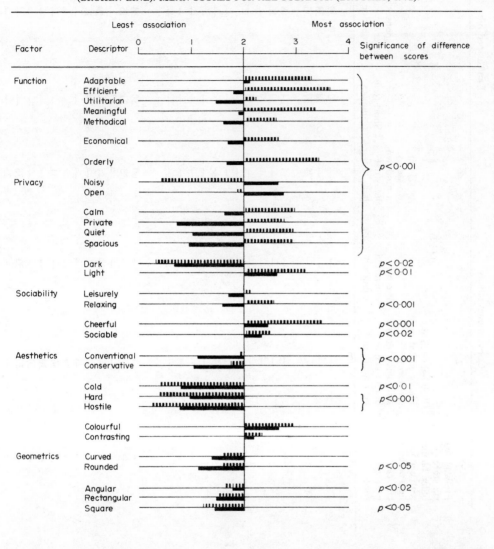

A list of the principal good and bad features is given in Table 2.12; this is compared with the data of Nemecek and Grandjean (1973a) and Brookes (1972) in Table 2.13. A direct question on draughts in the Swiss study produced a 30% level of complaint although draughts did not appear on the list of bad features. Notice that in general Boyce (1974) finds more good and less bad features in the landscaped office than in the conventional offices, but as people settle down in the landscaped office fewer good features are mentioned, whereas people seem to have become more aware of bad features. Particularly noticeable is a decrease in enthusiasm for the aircondition-

TABLE 2.11. PERCENTAGE FREQUENCY RESPONSES TO THE QUESTION, NAME (a) THREE GOOD THINGS AND (b) THREE BAD THINGS ABOUT THE BUILDING; RESPONSES UNDER 15% LEVEL NOT INCLUDED (BOYCE, 1974)

Factors	Conventional offices				Landscaped office	
	1. Pre-war heavy six-storey office block with central light well; in city centre near main road	2. 1950 lightweight highly glazed six-storey slab office block, in city centre near main road	3. Pre-war converted suburban house	4. As (2) but two-storey in rundown industrial area	Short-term responses	Responses made after a year of occupation
Good features						
*Airconditioning	—	—	—	—	42	29
Furnishing and decoration	—	—	—	—	38	38
Lighting	—	—	—	25	37	33
Communications with other departments	—	—	—	—	—	—
Cleanliness	—	—	—	15	24	18
Space available	—	—	—	—	19	16
*Comfort and working conditions	—	—	—	—	15	—
Convenience for shopping	56	43	33	—	—	—
Convenience for journey to work	39	41	45	—	—	—
Location	45	24	—	—	—	—
Trolley service	—	—	21	—	—	—
*Lack of noise	22	—	15	—	—	—
View	—	—	15	—	—	—
Style of accommodation	—	—	—	—	—	—
Bad features						
*Noise	64	43	—	60	30	34
Location	—	—	—	33	21	15
*Distractions	—	—	—	—	17	19
Lack of privacy	—	—	—	—	19	15
Draughts	—	—	21	—	15	—
Car-parking facilities	30	—	—	—	—	—
Cleanliness	24	39	—	—	—	—
*Ventilation conditions	17	30	—	31	—	—
Heating system	—	30	—	—	—	—
Lighting	—	20	—	—	—	—
Space available	—	20	33	38	—	—
Smells from outside	—	—	—	23	—	—

*Factors involving acoustics.

TABLE 2.12. ENVIRONMENTAL CRITERIA FOR LANDSCAPED OFFICE (BOYCE, 1974)

Parameter	Average value	Range
Air temperature[a]	21·1°C	19·7–21·8°C
Relative humidity	45%	
Air velocity	0·12 m/s	0·05–0·3 m/s
Illuminance	800 lux	
Glare index	18	
Noise level	L_{90} = 54 dBA	L_{10} = 62 dBA
Window area (glazed area/wall area, %)	13	
Space	9·4 m^2 per person	

[a]The temperature in offices should not now (since January 1975) exceed 20°C.

ing; draughts and a lack of privacy have become other points of criticism. Noise has a bad feature in all the buildings except the suburban offices.

The environmental criteria (Table 2.12) adopted for the landscaped office produced generally satisfactory conditions except that draughts were often experienced in cold weather.

Clearly the building environmental engineer has an important rôle to play if offices are to be successful; these studies show that air movement or ventilation aspects of airconditioning designs besides the acoustics are factors needing particular attention. The acoustical aspects of landscaped offices will be discussed in detail in section 2.6.3. These studies have also shown how useful feedback information is when assessed from buildings in use, and that environmental solutions will only be successful if they are related to man's behaviour patterns at work.

2.2. Physical and Subjective Qualities and Quantities in Acoustics

Music and many, but not all, sounds in Nature form a euphonious acoustical text; noise (probably derived from the Latin *nausea* meaning disgust, or *noxia* meaning hurt) is sound unwanted by the recipient. Noise can be distracting, it can be annoying, and at high levels can produce physiological damage of a temporary or permanent nature. What is pleasing and agreeable to one person may not be to another, or even to the same person at a different time. To classify subjective qualities and to correlate them with physical factors is a major difficulty in establishing economic and comfortable environments in buildings.

What are the common elements which make up a sound whether judged to be noise or music or something in between these extremes? To discover these we must look at the spatial–temporal patterns of sound. A tuning fork emits a sinusoidal waveform which mirrors the movement of the air particles caused by the acoustic pressure originating from the acoustic source. Its spatial characteristics are wavelength and intensity amplitude; the temporal characteristics are frequency and duration (Fig. 2.10a). Musical instruments display a more complex pattern (Fig. 2.10b); not only is the fundamental frequency present but the harmonics of this frequency too.

The harmonic content of a sound gives it *timbre* (or *quality*) and enable distinctions to be made between one type of sound and another; a horn has a

TABLE 2.13. A COMPARISON OF VARIOUS SURVEYS ON LANDSCAPED OFFICES

	Boyce (1974) (see Table 2.12)	(%)	Nemecek and Grandjean (1973a) (See Table 2.7)	(%)	Brookes (1972) (See Table 2.8)	(%)
Good features	Airconditioning Furnishing and decoration	29	Communications	40	Colourful design Comfortable	34
		38	Personal contacts	28	furnishings	34
	Lighting	33	Workflow, supervision and discipline	15	Attractive décor	26
	Communications	18			Indoor plants	18
	Cleanliness	16			Lighting	18
Bad features	*Noise	34	*Disturbances in concentration	69	*Conversation noise	47
	*Lack of privacy	19			*Lack of privacy	42
	Location	15			Overcrowding	43
	Draughts	15			Desk too small	26
					Lack of personal storage space	24
					Layout	24
					Poor airconditioning	17
					Lack of shelf space	16

*Factors involving acoustics.

spectrogram which will probably include the first four harmonics, whereas a violin or viola may display even the sixteenth or eighteenth harmonic (Figs. 2.10e and f). Tone quality of musical instruments may be judged from their spectral content (Fig. 2.10d). A decrease in the harmonic amplitude does not necessarily follow with an increase in the harmonic order number. A quite different picture results when the sound from an air jet is recorded (Fig. 2.10c). Here a random distribution of all frequencies is evident and is referred to as *white noise* or *Gaussian noise*. Most industrial noise sources have discrete components superimposed on white noise characteristics.

2.2.1. SOUND LEVELS

Already we have seen that Weber–Fechner proposed a logarithmic relationship between stimulus strength and perceived response which represents the auditory sensory modality fairly well in non-extreme conditions of sound frequency, intensity and duration. Another good reason for using a logarithmic scale to express acoustical quantities is that the ear is a very sensitive receptor and responds to acoustic power as low as 10^{-12} W but as high as 10^{2} W. Thus we define *levels* of sound power, sound intensity and sound pressure in decibels as:

$$\text{sound power level,} \qquad L_w = 10 \lg \left(\frac{W}{W_o} \right) \text{dB,} \qquad (2.1)$$

$$\text{sound intensity level,} \quad L_I = 10 \lg \left(\frac{I}{I_o}\right) \text{dB,} \qquad (2.2)$$

$$\text{sound pressure level,} \quad L_p = 20 \lg \left(\frac{p}{p_o}\right) \text{dB,} \qquad (2.3)$$

where W_o, I_o and p_o are reference quantities ($W_o = 10^{-12}$ W, $I_o = 10^{-12}$ W/m², $p_o = 2 \times 10^{-5}$ Pa or N/m²)† corresponding to the average auditory thresholds of sound energy and pressure. L_p can be derived from L_w by substituting $(p^2/\rho c)$ and $(p_o^2/\rho c)$ for W and W_o in the expression for sound power level, ρc being the acoustic impedance.

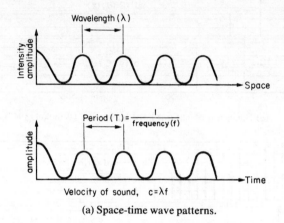

(a) Space-time wave patterns.

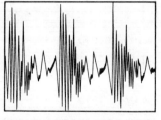

(b) Waveform of a musical note.

(c) Waveform of Gaussian or white noise.

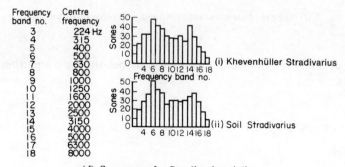

(d) Sonograms for Stradivarius violins.

FIG. 2.10. Sound waveforms.

†The units of acoustic pressure are Pascals or Newtons/m²; since 1971 Europe has adopted the Pascal.

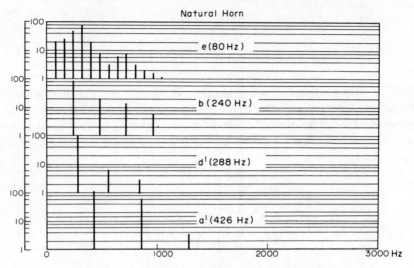

(e) Spectrum of natural horn harmonics (Winckel, 1967).

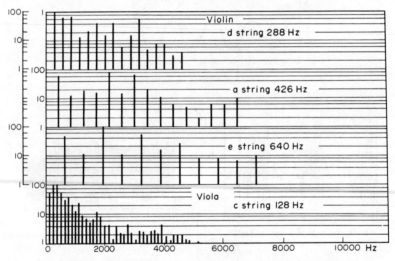

(f) Spectra of harmonics of violin and viola tones (Winckel, 1967).

FIG. 2.10. (*Continued*).

Note that decibels are not units or dimensions but they are logarithmic ratio scale values.

2.2.2. THE RESPONSE OF THE HUMAN EAR

(i) Loudness

Evaluating the loudness of a complex noise is difficult because the human attention may be shared between the various frequencies comprising the noise. One type of experiment, called magnitude estimation, relies on subjects making comparison judgements. A pure 1000 Hz tone is set to a given loudness level; another tone is compared with this until they are judged to be at the same loudness level. The results amassed from a large number of experiments are shown in the equal loudness level curves (Fig. 2.11a due to Fletcher and Munson, 1933; Churcher and King, 1937; Robinson and Dadson, 1956). These curves show that:

* the ear is a non-linear device having a maximum sensitivity in the region of 3–4 KHz;
* the ear can tolerate higher loudness levels at lower frequencies;
* the loudness level in phons is equal to the sound pressure level in decibels (re 20 μPa) at 1000 Hz (note: the experimental conditions were binaural listening to pure tones);
* as the loudness level increases the degree of non-linearity decreases.

Measuring instruments are designed to respond in a similar way to the human ear by having electronic weighting networks, known as A, B, C and D weightings (i.e. measurements are recorded in dBA, dBC, dBD which simulate the non-linearity; the A, B and C weightings are derived from the equal loudness level curves as indicated on Fig. 2.11a and b), whereas the D weighting is equivalent to the perceived noise level in PNdB (see Fig. 2.52). An approximate equivalent is loudness level in phons $\equiv$ dBA + 12.

Stevens (1972a) has suggested an E weighting network (Fig. 2.12) measuring sound levels in dBE equivalent within a decibel or two to the *perceived level* (PLdB) calculated by his Mark VII procedure described later in this section. The principal advantage of using dBE is that the reference tone is 3150 Hz, a much more sensitive frequency region of the ear than the 1000 Hz reference tone used in the work on loudness just described. German standards use DIN-phon curves: curve 1 applies to the 60–130 dB range, curve 2 to the 30–60 dB range and curve 3 below 30 dB; these dB values are at 1000 Hz on Fig. 2.11a.

Another type of experiment, magnitude production, requires the subject to estimate ratios of loudness sensations. In this case the loudness curve (Fig. 2.13) is derived which enables the loudness of pure tones to be directly compared (Garner, 1954; Robinson, 1957; Stevens, 1959; Hellman and Zwislocki, 1961, 1963, 1964, 1968; Kryter, 1966). This curve shows that:

* a loudness of 1 sone is equivalent to a loudness level of 40 phons;
* above a loudness level of 40 phons, doubling the loudness requires an increase of about 9 to 10 dB in the loudness level, thus:

$$\text{loudness level (phons)} = 40 + 10 \lg_2 (L), \qquad (2.4)$$

where L is the subjective loudness in sones;
* in contrast to loudness level, loudness values are additive.

Subjective loudness decreases with time due to auditory adaptation, and we can conclude that it is a complex function which can be expressed as

$$L = \phi_L (L_p, f, \phi_f, \psi), \qquad (2.5)$$

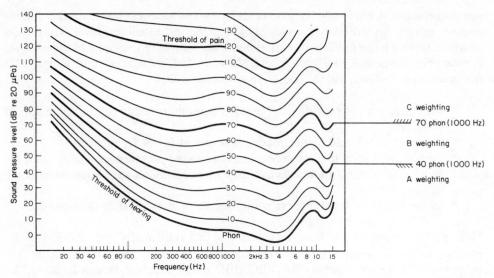

(a) Equal loudness level contours for pure tones and normal threshold of hearing for persons aged 18–25 years, using free-field listening (from ISO Recommendation R226).

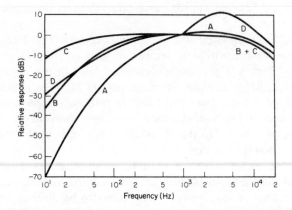

(b) Frequency response of weighting networks in a sound level meter.

FIG. 2.11. Equal loudness level and weighting network contours.

where L_p is the sound pressure level, f is the frequency, ϕ_f is the feedback function, and ψ is the psyche function discussed in section 2.1.1. The nature of ψ, i.e. the aspects of personality which affect loudness judgements, has been studied by several researchers (Hood, 1968; Reason, 1968; Barbenza *et al.*, 1970; Stephens, 1970; Stevens, 1972a, b). Work by McRobert (1973) suggests that aspects such as introversion, neuroticism and anxiety do not correlate with loudness evaluation; a possible correlate is excitability. It is too early to make positive conclusions on this matter which so far has relied on the personality inventories of Hathaway *et al.* (1951) and Eysenck *et al.* (1968). It is much more certain that noise annoyance depends on personality, and this is discussed later in section 2.4.4.

In practice we wish to calculate the loudness of a complex noise which may

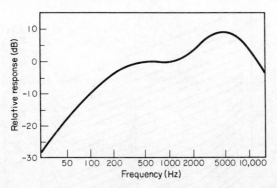

FIG. 2.12. Calculated response of a network for a sound level meter designed to give a suitable approximation to the "ear weighting"; such a meter could be calibrated to read dBE. For many common noises dBE would approximate the perceived level in PLdB (Stevens, 1972a, b).

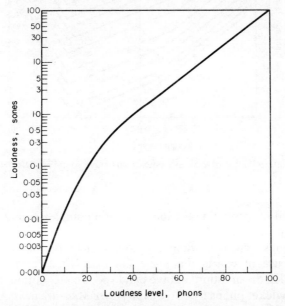

FIG. 2.13. Plot of the transfer function (loudness in sones) as a function of loudness in phons (Fletcher and Munson, 1937).

include several pure tones. The Stevens (1957) method is normally employed. The total loudness in sones is

$$S_t = S_{max} + F\left(\sum S - S_{max}\right),\tag{2.6}$$

where S_{max} is the highest value of the loudness index, S is the sum of the loudness indices of all the octave ($F = 0.3$), one-third octave ($F = 0.15$) or one-half octave bands ($F = 0.20$). The loudness index for each value of the band sound pressure level is read from the contours shown in Fig. 2.14. This method is also described by the International Standards Organisation in their publication ISO/R-532 (1967).

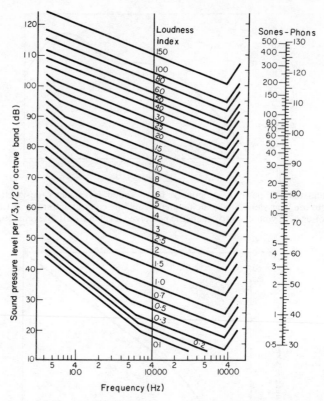

FIG. 2.14. Chart for determination of loudness index from octave sound pressure level values (ISO/R-532, 1967).

Another method of estimating the loudness of a complex sound is due to Zwicker (1960). It has been suggested that his method is preferable when the sound spectrum contains pure tones and also when the sound yield is non-diffuse. The method necessitates the use of charts that have been designed for free or diffuse field conditions on which the one-third octave band spectra are plotted; the loudness level is expressed in Zwicker phons (see *Acoustic Noise Measurements*, 2nd edition, Brüel and Kjaer, 1971, pp. 174–184, for a comparison of the Stevens and Zwicker methods of loudness calculations). Only the Zwicker method is standardised in Germany (DIN Norm 45631).

Reference must be made here to the latest method for estimating loudness level advocated by Stevens (1972a, b)—the so-called *Mark VII procedure*; at this point in time (May 1974) it has not yet been accepted by the International Standards Organisation. By assembling some twenty-five sets of frequency weighting data from eleven different laboratories in England and America measured over the past quarter of a century, Stevens (1972a) found a contour of equal perceived magnitude (Fig. 2.15a), (the term magnitude can refer to loudness annoyance or acceptability, no systematic difference being found to occur in the contour), which forms the outline of the Mark VII sone curves shown in Fig. 2.15b. Notice that the equal loudness contours have been extended to low frequencies where they appear to converge

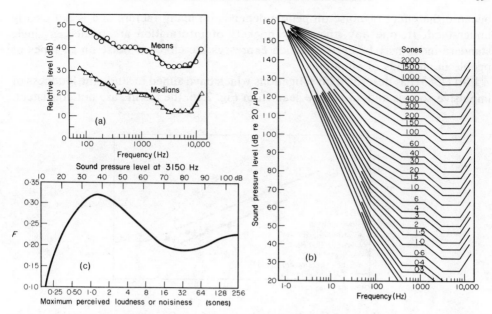

(a) The results of averaging 25 separate contours of equal perceived magnitude; the 5-section weighting function of Mark VII has been fitted to both kinds of averages (Stevens, 1972a, b).

(b) Mark VII countours of equal perceived magnitude in sones; this family of contours can be used for the calculation of loudness and noisiness. The sound is presumed to be measured in octave or one-third bands in decibels re 20 μPa. Values for these contours are tabled in Stevens, 1972b (Stevens, 1972a).

(c) Showing how the fraction F depends upon level. The value F stands for the fractional loudness contributed by each one-third octave band. The value of F used in the calculation of perceived level is determined by the value in sones of the loudness or noisiness of the one-third octave band that produces the maximum perceived magnitude. Rule for octave bands: subtract 4·9 dB from the loudest one-third octave band; determine from (b) the sone value with which to read the F-value from the graph; then double the F-value. Values of F are tabled in Stevens, 1972b (Stevens, 1972a).

FIG. 2.15. Stevens' Mark VII procedure for estimation of perceived loudness level.

towards a common point at about 1 Hz and a sound pressure level of 160 dB. Further research is needed to establish the truth concerning this convergence point. Thus the sone can be defined as the *perceived level* of a one-third octave band noise centred at 3150 Hz and having a sound pressure level of 32 dB (re 20 μPa). Using the data in Fig. 2.15a–c, the perceived level in PLdB can be calculated in the same way as described for equation (2.6).

In the plethora of perceived loudness descriptions developing in this field, confusion understandably arises. In approximate terms the following equivalences apply (note PNdB are defined in section 2.4.4 and Fig. 2.52):

$$PLdB \equiv L_{(Zwicker)} - 13 \qquad\qquad PNdB \equiv L_{(Zwicker)} - 4$$

$$PLdB \equiv PNdB - 9 \qquad\qquad PNdB \equiv PLdB + 9$$

$$PLdB \equiv L_{(Stevens, 1957)} - 8 \quad \text{or} \quad PNdB \equiv L_{(Stevens)} + 1$$

$$PLdB \equiv dBA + 4 \qquad\qquad PNdB \equiv dBA + 13$$

The reason for the apparent propagation of variously designated noise levels is because the effect of noise on people depends on many factors and is not clearly understood. If, one day, the many pockets of information are unified, a single standard index may be derived which expresses the effect of noise on all types of people in all kinds of situations.

Port (1963) has described investigations which have aimed to study the loudness of impulsive sounds and, as can be seen from Fig. 2.16, the results are not consistent.

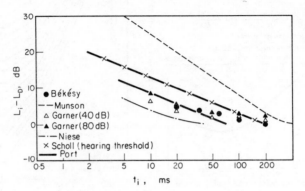

FIG. 2.16. Comparison of measured results on the loudness of impulsive sounds as reported by various investigators (Port, 1963).

The ordinate represents the sound pressure level difference between the impulse level L_i and the level of steady sound L_D judged to be equally as loud. If the impulse has a duration greater than 50 ms, then $(L_i - L_D)$ tends to zero, but as the duration time decreases the level of the pulse has to be increased to give the same loudness sensation as the steady sound. The auditory sensory system takes a finite time to receive and process acoustical stimuli from the outside world and transmit them from the ear to the brain producing a subjective judgement of, in this case, loudness; the neuron transmission rate in the auditory nerve takes place at a velocity in the order of 10–100 m/s which suggests time delays of 10–100 ms for processing of acoustical data in the sensory system. These times are comparable with the duration of impulsive sounds. If the ear responds to the sound energy averaged over a given response time, then energy is proportional to the square of the particle pressure, and hence is a function of the sound pressure level; the loudness sensation ϕ_o may be expressed in a functional form as

$$\phi_o = \frac{1}{\Delta t} \int_{t_1}^{t_2} \phi(L_p)dt. \tag{2.7}$$

It can be seen that if the duration time of the impulse $(t_2 - t_1) < \Delta t$, then there will only be a response comparable with that for a continuous sound if the energy function $\phi(L_p)$ has a high enough amplitude.

The perception of a complex sound containing more than one pure tone is dependent not only on loudness and pitch but also on the harmonic content of the tones, the transient behaviour of the harmonics and any phase differences which may exist between the tones. If the frequencies of both tones lie within a *critical*

TABLE 2.14

Centre frequency (Hz)	Critical bandwidth (Hz)	Centre frequency (Hz)	Critical bandwidth (Hz)
50	100	1850	280
150	100	2150	320
250	100	2500	380
350	100	2900	450
450	110	3400	550
570	120	4000	700
700	140	4800	900
840	150	5800	1100
1000	160	7000	1300
1170	190	8500	1800
1370	210	10500	2500
1600	240	13500	3500

bandwidth, then the resultant loudness is not the addition of that for each tone. The critical bandwidths are listed in Table 2.14.

The ear acts as a frequency analyser of complex tones, the limit of its resolving power being the critical bandwidth. Classical texts refer to *Ohm's acoustical law*, which simply states that the ear can distinguish many of the harmonic constituents in a complex tone as long as the critical bandwidths of the fundamental frequencies do not overlap. Thus when the sounds lie in different critical bandwidths they are heard simultaneously but sensed as separate tones, and the resultant loudness is the addition of that for each frequency. Little needs to be said about the well-known beat phenomenon experience when two tones with a very small frequency difference are sounded together. An explanation of this is offered again from the work of Stevens. Howes (1950) plotted the loudness, or excitation pattern, of each tone against the mel scale of pitch (the mel is defined in the following section on pitch) as shown in Fig. 2.17. All equally loud tones show similar patterns because the mel

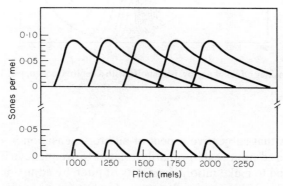

FIG. 2.17. Patterns of excitation for equally loud tones spaced 250 mels apart. When plotted in sensation coordinates, sones per mel against mels, the patterns all have the same form. The area under a curve is proportional to the perceived magnitude. At the lower perceived level, 2·5 sones or 44 PLdB, there is no overlap of the pattern, and the loudnesses add. For the upper curves, for 25 sones or 74 PLdB, the overlap is extensive and the total perceived loudness is less than the sum of the separate components (Howes, 1950; Stevens, 1972a).

scale provides a linear map of the cochlea in the ear. When the patterns do not overlap, the loudness of the components add: when the patterns overlap the tones exert a mutual inhibition on one another and the total loudness becomes less than the sum of their separate loudnesses.

When the tones lie close together, however, another rule obtains. If the tones fall within a critical band, which is about 100 mels wide (a hectomel), then the energies of the tones summate. For this case, Fletcher and Munson (1953) have stated: "one must first add together all the intensities of the components in a critical bandwidth and treat the combined intensity as a single component". For tones separated by more than a hectomel, the simple formula developed by Howes (1950) gives a good prediction of the loudness of the multicomponent tones.

Masking occurs when one sound drowns another, and is particularly prevalent when the frequencies lie in a bandwidth equal to 0·4 of the critical bandwidths listed in Table 2.14; a temporary threshold shift occurs around the frequency of the masking tone as illustrated in Fig. 2.18 for a narrow band noise having a centre frequency of 1200 Hz.

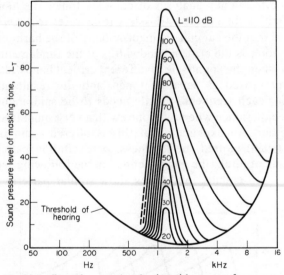

FIG. 2.18. Masking effect of narrow band noise with a centre frequency of 1200 Hz.

(ii) Pitch

Pitch is defined as that subjective aspect of auditory sensation in terms of which sounds may be ordered on a scale primarily related to frequency. Experimentally, subjects are requested to make ratio judgements of pitch by adjusting the frequency of one audio oscillator until it is twice the pitch of another one. In this way a scale of pitch is developed (Fig. 2.19). A reference value of 1000 mels has been chosen as the pitch of a 1000 Hz tone with a sound pressure level of 60 dB (re 20 μPa).

Fletcher (1934), Stevens (1935) and Snow (1936) have shown that pitch is not completely dependent on frequency. Research has established that the effect of

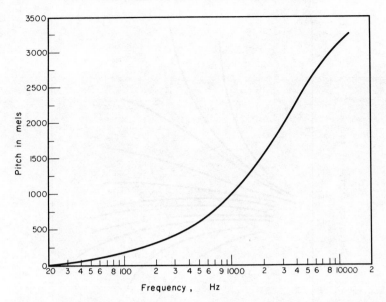

FIG. 2.19. Relation between subjective pitch expressed in mels and frequency. Note that subjective pitch increases more and more rapidly as frequency is increased logarithmically. The musical scale, by comparison, is a logarithmic scale; that is to say, an octave is a doubling of frequency. (*After Stevens and Volkman, Am. J. Psychol.* **53**, 329–353 (1940).)

intensity on pitch is negligible for some people, but pure tones at varying sound levels in the range 40–120 dB may cause pitch changes of as much as 35% for others. Stevens (1935) concludes that a low-pitched note has its pitch lowered when its intensity is raised, whereas a high-pitched note will have its pitch raised; in fact the auditory system displays a phenomenon reminiscent of the Bezold–Brücke effect in vision. Just as increasing or decreasing the intensity of most spectral lights produces shifts in hue, so low tones are lowered and high tones raised in pitch when the sound pressure level of the tone is increased. Stevens (1937) derived the isopleths of pitch shown in Fig. 2.20. There is not as yet any clear explanation for this effect. Fortunately this pitch variation is only noticeable for pure tones and is not detected in music, the richness of the overtones in some way clouding any variation that might exist. Some very effective demonstrations of this phenomenon were given by Professor James Beament in the BBC Radio 3 series called The Deceptive Ear broadcast at 7 p.m. on 31 December 1974, 7, 14 and 21 January 1975.

A possible explanation is offered by Munson (1957). Sound excites the hair cells distributed along the basilar membrane. High frequencies produce neuron pulses at the basal end of the basilar membrane, whereas low frequencies excite the far end (apex). As the sound level of a pure tone increases, more hair cells are affected; this spreading effect has been found to be more predominant at low frequencies. Perceived loudness can be considered as an integration of all the neuron impulses. Thus

$$L = \int \frac{dL}{dn} \, dn,$$

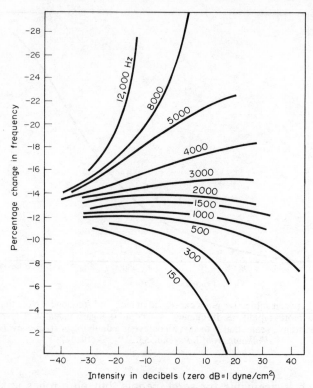

FIG. 2.20. Contours of equal pitch for variations in intensity. The ordinate scale was arbitrarily chosen so that a contour with a positive slope shows that pitch increases with intensity (Stevens, 1937).

where L is the loudness, dL/dn is the loudness contribution from a small element of the frequency and dn is the bandwidth of the frequency element. Pitch in mels is roughly proportional to the number of nerve endings stimulated along the basilar membrane. Since dn is also proportional to the number of nerve endings, pitch may be related to n. Pitch is also related to loudness because increased sound levels cause more nerve endings to be stimulated extending the bandwidth dn and possibly causing a shift in the frequency maximum. The effects of overtones and attack on a note (impulsive or gentle) is not clearly understood.

Winckel (1967) also refers to this interdependence between pitch and loudness in his treatise *Music, Sound and Sensation*, and Helmholtz (1863), in *On the Sensations of Tone*, (both works are published by Dover). Békésy (1960) measured the phase difference between the stapes and a point along the basilar membrane 30 mm from the stapes as the amplitude of a pure tone was varied. The phase difference was found to decrease as the amplitude increased, thus causing a displacement towards the apex end of the basilar membrane, thus corresponding to a lowering of pitch. For a 200 Hz tone with amplitudes of $1 - 3 \cdot 5 \times 10^{-5}$ m the frequency shift varied up to 10%. Békésy concludes that most of the known changes of pitch with changes in loudness have their origin in neural rather than mechanical processes.

It can be concluded that although pitch is primarily a function of frequency, it is also dependent on the sound pressure level and the spectral content of the sound, besides the feedback and psyche functions; in fact pitch can be expressed in a similar way to subjective loudness (see equation (2.5)) thus:

$$P = \phi_P(f, L_P, \phi_f, \psi).$$

In general, a variation in frequency, intensity or waveform can affect any one of the subjective sensations of loudness, pitch or timbre.

Harris (1952) established the differential frequency discrimination of the ear by comparing two successive tone bursts. The difference limen (Δf) for pitch was shown to be dependent on the absolute tonal frequency and sound intensity. The results are shown in Fig. 2.21.

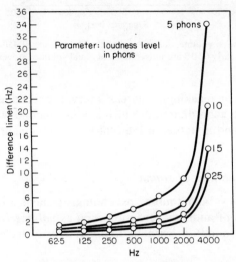

FIG. 2.21. The difference limen for pitch (Δf) as determined by frequency and intensity variations (Harris, 1952).

Just as for loudness, the duration of a sound features in the subjective impression it leaves on the listener. Tones lasting for less than about 200 ms lose their rich, tonal character, and if they last for only 100 ms or less they seem more like a brittle, atonic click.

The loudness of several tones sounded simultaneously has already been discussed. There also exist *difference* and *summation* tones, the former sometimes being referred to as *Tartini's tones* in memory of the great violinist who observed the presence of a third tone in double-stopping. The pitch number for any combination tone is $N = mh + nl$, where h is the higher frequency, l is the lower frequency and m and n being integers.

The *consonance* and *dissonance* effects of several tones played simultaneously seem to arise from harmonic interference. The beating between the harmonics of dissonant intervals is felt and produces a "jarring" or "roughness" impression, whereas the harmonics of consonant intervals coincide and reinforce each other. Recent work suggests that most dissonant intervals are those corresponding to frequency differences of approximately one-quarter of the critical bandwidth (see

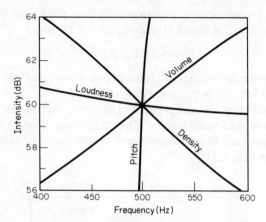

FIG. 2.22. Isopleths of loudness, volume, density and pitch. Thus a tone of 450 Hz and 58 dB and one of 550 Hz and nearly 62 dB are judged to be of equal volume (Stevens, 1934).

the previous paragraph on masking), whereas intervals exceeding this bandwidth are judged to be consonant; according to this idea a minor second would be judged to be dissonant but a ninth and a sixteenth interval to be consonant.

(iii) Other subjective qualities of sound

Besides pitch, loudness and timbre, psychologists have attempted to quantify several other attributes of auditory sensation such as *volume* (related to the spatial quality or extensity of sound), *brightness, tonality, density* (related to the compact-

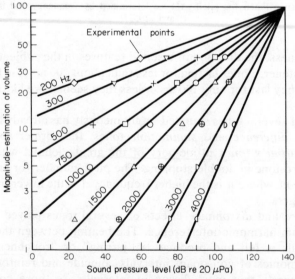

FIG. 2.23. Power functions, for nine different frequencies, relating tonal volume to intensity (Terrace and Stevens, 1962).

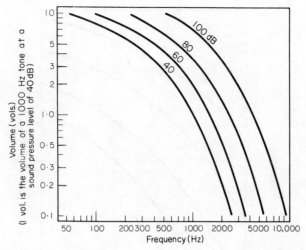

FIG. 2.24. Tonal volume as a function of frequency (Terrace and Stevens, 1962).

ness of sound) and *vocality* (vowel similarity). These features are not clearly understood although they were mentioned by Boring in 1942, and he discussed tonal brightness as early as 1936. Experiments on vocality were carried out by Köhler in 1910.

Stevens (1934) has attempted to show isopleths of volume and density (Fig. 2.22 which includes loudness and pitch contours also) although the real meaning of these subjective qualities remains debated. Terrace and Stevens (1962) have concluded that tonal volume is a power function of sound pressure and is frequency dependent (Fig. 2.23); by plotting a log–log graph of volume and frequency a curvilinear function results (Fig. 2.24), disputing the nearly linear one implied in Fig. 2.23.

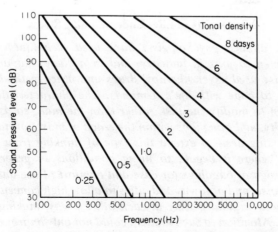

(a) Contours of equal density; unit density (1 dasy) has been selected as that of a 1000 Hz tone of 40 dB SPL (Guirao and Stevens, 1964).

FIG. 2.25.

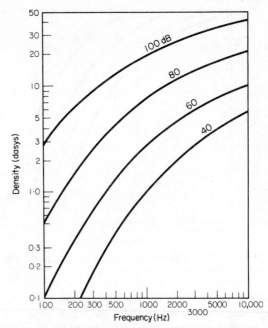

(b) Density as a function of frequency at four intensity levels; narrow bands of noise were used in this experiment, though tones give much the same result (Guirao and Stevens, 1964).

FIG. 2.25. (*Continued*).

Similar work carried out by Guirao and Stevens (1964) on sound density is summarised in Fig. 2.25a and b. Reference should be made to Geldard (1972) for a more extensive account about these lesser understood attributes of sound.

2.3. Quality of Life: Understanding Stress

Traditionally, applied psychologists have used techniques of selection and training to fit men to jobs in industry, and in a somewhat analogous manner psychiatrists have used psychotherapy, drugs and other methods of treatment to enable patients to cope with life's demands. The characteristic ergonomic approach has been to modify the task rather than the man, so as to remove focal points of difficulty, and bring jobs within the reach of most normal people. There is currently a pressing need to extend this type of thinking from industry and the armed services where it began, to living conditions in general including, for example, the design of buildings for ease and comfort of use and for privacy and freedom from noise, and of towns and cities to allow high concentrations of people to move and work without getting in each other's way, making contact only by mutual consent. Attention to such matters could not only improve the quality of life for normal people, but could make a crucial difference to those abnormally intolerant of stress who at present fall into psychiatric care.

(WELFORD, 1973.)

In the world today materialistic philosophies prevail. Profit motive is all important, and yet at the United Nations Human Environment Conference held in Sweden in 1972 the Indian Prime Minister, Mrs. Gandhi, declared this to be one of the worst pollutants of our civilisation; Schumacher (1973), in his book *Small is Beautiful*, also questions the growth of materialism. Money can bring pleasure but not necessarily happiness or a sense of well-being. Surely we should be seeking to reduce the conflict between people, between people and their environment, between man and himself? These matters are concerned with stress.

McGrath (1970) has emphasised that stress is the result of an imbalance between an organism's demand in a particular activity determined by the environment and social conditions, and an organism's capacity basically dependent upon genetic endowment as refined by training mental and physical health.

The comfort diagram featured in Fig. 2.8 shows diagrammatically the stress–strain relationship between man and his environment. It is seen to depend on the *past experience of the event* and therefore the individual's expectancy, if any, of the event and its information content; the *arousal level* of the individual and therefore the personality, the degree of attention and the concentration ability of the person (i.e. distraction level); the suitability of the work task for the individual; the *physical and emotional sensitivities* of the individual; the intelligence of the individual. Any stress model must include these factors.

Arousal level has been referred to already. When the arousal level is too low the response of an organism to the environment is lethargic, insensitive and inert, but if too high then the behaviour becomes tense and disorganised. Monotony, predictability and concordance tend to lower the arousal level, whereas sensory stimulation, novelty and conflict tend to raise it. For optimum arousal and best level of performance and least stress, moderate levels of these factors are required. The oft-quoted tenet *moderation in all things* does sensibly, it seems, apply to life quality. Motivation can tip this fine balance towards achieving and maintaining optimum arousal, but factors such as lack of short-term or long-term aims, an effort required beyond a person's capacity or willingness, can easily dim any motivation.

Welford (1973) finds more meaning in a model of stress–arousal based on signal detection theory rather than one based on the inverted-U hypothesis described later in section 2.5. In this model the brain is conceived as being spontaneously active in such a manner that incoming signals have to be distinguished from a background of random neural noise. Assumptions are made that as arousal increases, the cells of the brain are more likely to fire and that the neural noise and the signal plus neural noise distributions are Poissonian. Referring to Fig. 2.26, it can be deduced that at low arousal all cases of no signal will be reported correctly but cases of signal will be missed. As the distributions move to the right of cut-off, more cases of signal will be reported correctly, but eventually at high arousal, the number of false positives will overtake the correct responses, thus limiting the number of correct responses again. Below optimum arousal people are underactive, and any errors which occur are mainly ones of omission due to drowsy and inert performance: above optimum arousal, people are overactive, and any errors which occur are mainly ones of commission due to rash and ill-considered actions lacking caution and restraint brought about by anxiety and tension. Evidence supporting these contentions will be discussed in section 2.5.

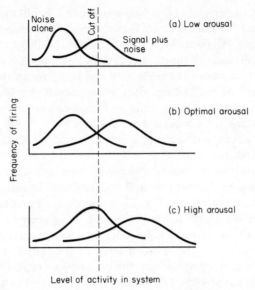

FIG. 2.26. Poisson distributions of noise and signal plus noise (Welford, 1973); above cut-off subject treats everything as signal, and below it treats everything as either noise alone or no signal.

In physiological terms arousal is thought to be concerned with the efficiency of information transfer between the cortex and the subcortical regions which house the arousal centres. Wilkinson and Haines (1970) have attempted to relate the patterns of evoked cortical responses to arousal aspects of attention (Fig. 2.27) and McCallum (1969) states that there is much evidence supporting the view that the contingent negative variation in these patterns is related to attention, and changes in amplitude and form of the CNV are closely related to moment-to-moment changes in conscious attention. The activation in the brain which accompanies arousal appears to be correlated with autonomic activity (Davies and Krkovic, 1965).

Individuals differ widely in their reaction to stress. Their capacities and the demands made upon them by circumstances and their own aims vary. Welford (1973) discusses some of the implications of this in everyday life. For instance introverts tend to seek peace and quiet, favour solitude or restricted companionship, are often sensitive, restrained and self-driving, whereas extroverts often enjoy noisier environments, more social contacts, and tend to be less restrained and less self-driving. A point already mentioned is the fact that introverts are more likely to reach and maintain their optimum arousal level, whereas extroverts find it more difficult to reach it and are more easily distracted from it. Many retardates seem to show extrovert characteristics, and many psychiatric cases introvert ones. There are no definite, rigid rules that classify people into groups and communities because there are always exceptions, but one conclusion that can be drawn from the evidence behind these general tendencies is that work organisation, educational processes and building environments should be more adaptable and flexible to achieve stated objectives in different ways suited to the particular group of people.

A quantitative picture of stress can be developed by extending the idea of the mind capacity attention sphere described in section 2.1.2 (see also Fig. 2.7). Stress is

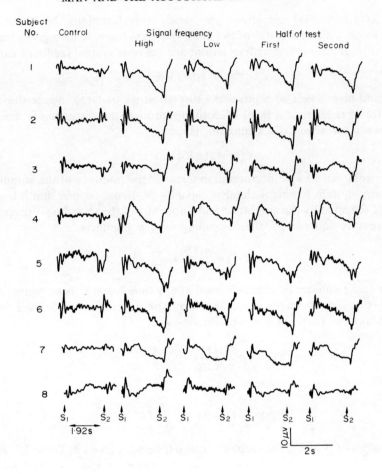

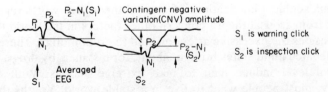

FIG. 2.27. Patterns of average evoked cortical responses for 8 people (Wilkinson and Haines, 1970).

imposed on the individuals by the world around us, and this produces a degree of strain in the individual. The stress–strain pattern will depend on the psychological and the physiological make-up of the individual and will show short-term and long-term variations. An example of such stress–strain patterns has been described by McPherson (1973) and Wyon *et al.* (1968). Using the psyche function described in section 2.1.1, we may say

$$X(t) = f(\psi) Y(t),$$

where $X(t)$ and $Y(t)$ are stress and strain time functions. Now strain can be interpreted as a distortion dV, of the mind sphere volume V, for a given individual at a given arousal level a and with an autonomic nervous system feedback function ϕ_f.

$$Y(t) = (dV/V)_{\psi, a, \phi_f}.$$

At low and high levels of arousal the distortion dV is large, hence the strain and hence stress, is high; at a level of optimum arousal the distortion is defined to be small, hence the strain is a minimum. Hence

$$X(t) = f(\psi)(dV/V)_{\psi, a, \phi_f}. \tag{2.7}$$

But the stress can also be interpreted in terms of the intensity of the stimuli entering the sensory system having a channel capacity N. If we assume that intensity I (or energy as reflected by the amplitude and frequency of the stimulus) is proportional to stress pressure squared p^2, then summing over n channels

$$X(t) = \frac{n(t)}{N} \left(C \sum_{n=1}^{n=N} I \right)^{1/2},$$

where n is the number of channels used at any time t and C is a channel resistance factor. Using the stochastic dominance probability function p_k defined on page 39, the likelihood of stress in a given situation is

$$X(t) = p_k \frac{n(t)}{N} \left(C \sum_{n=1}^{n=N} I \right)^{1/2}$$

or

$$X(t) = p(1-p)^{k-1} \frac{n(t)}{N} \left(C \sum_{n=1}^{n=N} I \right)^{1/2}. \tag{2.8}$$

Likely values of p and k for various arousal levels are given in Table 2.3. Equations (2.7) and (2.8) suggest the interdependence between the stimulus level and quality, the sensitivity and capacity of the autonomic nervous system and the personality of the individual.

The philosophical implications of arousal theory are worth repeating. Lessons to be learnt from arousal theory concern other factors in the world that are pernicious to man besides the excessive attention paid to profit motive. These are boredom and falsely elevated mind states brought about, for instance, by drugs. The extreme ends of arousal level induce man to take extreme actions. With extremism comes instability; most people want to live in a stable world. Youths that are bored may turn to varying degrees of violence for stimulation; perhaps men with boring jobs seek variety in living by striking; married couples bored with life hope for stimulation by odd, but not uncommon, social diversions. To many wives, housework is boring and the level of neuroticism among them has been found to be high; their husbands, too, may have very boring jobs on assembly lines. Environmental design in the broadest sense should attempt to avoid setting the seeds of boredom in society and avoid nourishing man's genetic endowment to take pathways which bypass work. If working hours are to continue to decrease, what will people be able to do during the increased leisure time? A much greater variety of interesting things to do, which are easily accessible to people, must be made available to communities in the future.

Jobs, too, must be made interesting and give each man and woman some sense of achievement.

> *Seeking a balance in things perceiving that balance is a sign of health, seems to come very naturally to us. We seek it in banking, walking, eating, argument—even, in times of political sickness, in coalitions. Excess, on the other hand, even of good things like strawberries and cream or aspirin, is so evidently harmful that talking of excessive health is a contradiction in terms.*
>
> *And yet, at this point in our development I believe we live in a gross state of imbalance—it's a difficult one to define, but for the sake of convenience I like to think it stems from neglect of....*

(JACKY GILLOT, in the 6 October *Sunday Times*, 1974.)

Jacky Gillot goes on in her article to suggest that we need to nurture more gentleness and patience and to evolve a dynamic but more caring society.

2.4. Noise and People

2.4.1. PHYSIOLOGICAL RESPONSE OF MAN TO NOISE

The human auditory system is shown in Fig. 2.28 and is a particular example of the sensory pathways described in section 2.1.1. Detailed physiological explorations and

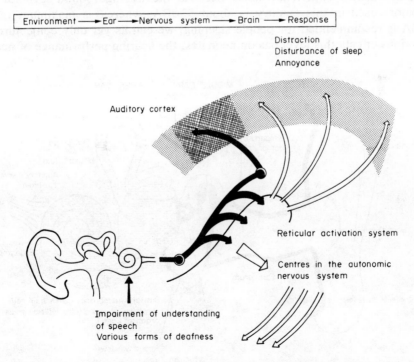

FIG. 2.28. Diagram of the auditory tract and its connections with different sections of the brain, where the most important effects of noise are localised (Grandjean, 1973a).

reasoning about the ear's response to sound waves with varying intensities and frequencies have often appeared in textbooks and journals during the last 50 years. In particular, I would like to pay tribute to the distinguished work of Georg von Békésy (1960, 1964). This chapter is more concerned with the annoyance, disturbance and distractive effects of noise on people, so apart from the brief account of the ear's response characteristics in section 2.2.2, the theories of hearing which attempt to explain the ear's response will not be discussed.

The mechanism of the ear is shown in Fig. 2.29. Sound waves from the air around are collected by the *pinna*, travel down the *meatus*, and are conducted to the *cochlea* via the three auditory ossicles (i.e. the malleus, the incus and the stapes which act as an impedance device, matching the sound wave impedance in the air to that in the basilar fluid) and the oval window. The ossicular chain produces a pressure amplification of about 1.3. The vibrations conducted in the basilar fluid cause groups of hair cells along the basilar membrane to move; this motion induces piezoelectric action and the mechanical energy is converted to an electrical pulse which travels along the auditory nerve to the brain.

Clearly, if this mechanism is damaged in any way either by excessive sound levels or by diseases which affect the brain, the auditory nerve or the auditory ossicles, then hearing is impaired. Intense sound levels are experienced in many industries and can cause temporary or progressively permanent deafness. Measures are now being taken to combat excessive noise levels in factories, (*Code of Practice for Reducing the Exposure of Employed Persons to Noise*, HMSO, 1972). Even premature babies experience noise because the average sound level of many incubators used in this country is about 57 dBA, whereas a maximum value of 35 dBA is recommended for people sleeping; we can as yet only conjecture what deleterious effects this level has on neonates; the hearing performance of neonates

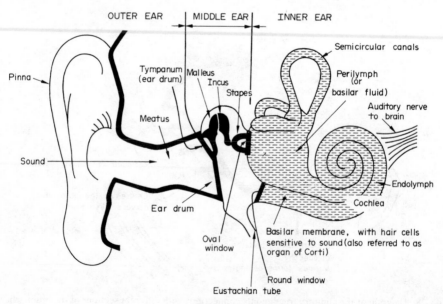

FIG. 2.29. Mechanism of the ear.

needs careful observation over the first few years of their lives; besides audiometer tests, electroencephalograms and plethysmographic measurements may be worth while.

The reader should refer to the treatises by Burns (*Noise and Man*, 1968) and Kryter (*The Effects of Noise on Man*, 1970) for detailed accounts of deafness and physiological hearing damage. Most people live and work in environments with noise levels less than 85 dBA—the starting point where damage risk is at present thought to be imminent. With regard to this level of 85 dBA, brief mention should be made of work by Schwetz *et al.* (1970) which shows that hearing at a level of 75 dBA may take 2 hours or more to return to normal; the need for lowering this 85 dBA starting point should be considered. For most people the danger of exposure to high noise levels arises from traffic, and for the teenagers from discothèques as well.

Normally people experience some hearing loss particularly at high frequencies which increases with age but Rosen (1962) states that Mabaan tribesmen in Sudan show hardly no hearing loss with age which suggests that presbycusis is partly caused by occupational noise.

At lower levels the psychological effects of noise become more prominent, and it is these aspects which are brought into focus in this chapter. As will be seen, much about them remains unknown.

In considering environmental perception in section 2.1.1, it was suggested that any model of man's response to the environment could use the function

$$\phi_o = f(\phi_i, \phi_f, \psi)$$

as a starting point, where this represents the physiological and the psychological behavioural responses to input stimuli that succeed in entering the body. This credits belief in the dualistic philosophical view of man's nature described as the physical self (i.e. the interaction of the ϕ_i and the ϕ_f functions) and the psyche self (the interaction of the ϕ_f and ψ functions). Some philosophers argue that one cannot distinguish between the physical and psyche selfs; this is a *monism* viewpoint. The alternatives are set out by Hanfling (1973) in Table 2.15.

It is useful to recount the linguistic interpretation of the words mind, soul and personality.

Mind (from the Latin *mens*—the mind) is defined as that which knows, thinks, feels and wills; consciousness; involves memory, judgement, opinion, intellect, soul, personality.

Soul (from the German *Seele* meaning mind, spirit *or* soul; there is an old English form *sawol*) refers to the innermost being or the moral and emotional nature with which one identifies oneself.

Personality (the Latin *persona* means a player's mask) may be defined as the integrated organisation of all the psychological, intellectual, emotional and physical characteristics of an individual.

The $\phi_o(\phi_i, \phi_f, \psi)$ function can be used to signify many things. ϕ_i and ϕ_f can be identified in the reception areas of the brain, along the nervous system and at the sensory receptors whereas ψ is identified by some (monists) as the brain itself, by others (dualists) as something apart from the brain. Pleasure centres have been

TABLE 2.15. PHILOSOPHICAL VIEWPOINTS OF THE HUMAN MENTAL AND PHYSICAL STATES (HANFLING, 1973)

Dualism	Monism
Interactionism	*Materialism (or Physicalism)*
Two-way causal interaction between the physical and the mental: physical events cause mental events, and vice versa.	(a) Mental processes are really brain processes and therefore physical. (b) Talk about mental processes is really talk about people's behaviour.
Parallelism	*Idealism*
Mental and physical events run in concurrent series, but events of either kind are not caused by events of the other.	All that exists is really mental. The notion that there are material things outside our consciousness is an illusion.
Epiphenomenalism	*Double Aspect Theory*
Mental events are a kind of byproduct of a physical system, the human body, which from a causal-scientific point of view is self-sufficient. Human beings can be regarded as machines which would function in just the same way without the puzzling "extra something" that we call "mind" or "consciousness"; but this is *not* to deny that the latter exists.	The mental and the physical are two aspects under which we experience the same underlying reality.

Two contrasting philosophies and currently upheld about the concept of mind. *Dualism* states that reality is mental and physical, whereas *monism* defines reality as basically of one kind.

located in the brain and emotional centres in the reticular formation. It is possible that the psyche function ψ has some physical reality as well as representing the non-identifiable aspects of personality.

A Christian believing in life after death would represent death by the function

$$0 = \phi_o(0, 0, \psi)$$

in which ψ denotes the soul, whereas an agnostic would write

$$0 = \phi_o(0, 0, 0)$$

the soul survival after death having no meaning.

A pathological condition will mean that the reception of the input signal may be disturbed so that $\phi_i \to \tilde{\phi}_i$ hence

$$\phi_o = \phi_o(\tilde{\phi}_i, \phi_f, \psi)$$

(note: $\sim$ has been used to denote a disturbance).

If a person is deaf, then most acoustical stimuli, although some lower frequency vibrations may enter the system, will be ineffective, i.e. $\phi_i = 0$, hence in the auditory mode

$$\phi_o = \phi_o(0, \phi_f, \psi).$$

Beethoven was totally deaf by the time he was writing his greatest music. There is, of course, a tragedy and also much to marvel at in this particular example. His career as a pianist had to end, and how could he rehearse an orchestra when he had to be turned around to see the ecstatic reception at the *première* of his Choral Symphony which he conducted? But the composition of music is a different matter from performing; composition will need some past experience of sounds (could someone born deaf compose music?), but that is all because it is the mind which is composing. In our functional notation a music performer responds to sound in a way which may be defined by a function

$$\phi_o = \phi_o(\phi_i, \phi_f, \psi),$$

whereas a composer perceives his inner world of sound by a relationship

$$\phi_o = \phi_o(0, \phi_f, \psi).$$

Nervous diseases will disturb the feedback function, thus $\phi_f \to \tilde{\phi}_f$ and

$$\phi_o = \phi_o(\phi_i, \tilde{\phi}_f, \psi).$$

Psychiatric states will change ψ thus

$$\phi_o = \phi_o(\phi_i, \phi_f, \tilde{\psi}).$$

Clearly physiological and mental illnesses can interact with one another.

During sleep the subconscious may be interpreted by the function

$$0 = \phi_o(0, \phi_f, \psi).$$

This suggests that defining the mind as a state of consciousness is incorrect because the mind component of the psyche function is also active during sleep.

It has also been shown that the response of an individual has diurnal variations which depend on the arousal level pattern for an individual and longer term variations depending on the environmental experiences of the individual. In other words behavioural response originates from the genetic being and it is developed and modified throughout life by environmental conditioning.

The factors controlling the stress–strain relationship between man and his environment have been discussed in section 2.3. Various measures have been used in attempts to measure some of these factors; these are listed in Table 2.16.

There is clearly a wide variety of possible measures but as yet there is no clearly defined basic measure that can be used to predict the effect of noise on people. The work of Conrad (1973) suggests that photoplethysmographic measures may correlate with annoyance, whilst that of Atherley *et al.* (1970) suggests that noise of high subjective importance causes increased adrenal medullary activity (i.e. increased adrenaline levels) and at the same time diminished adrenal cortical response (i.e. reduced urinary 17-ketosteroid levels); Gibbons (1970) feels that the skin resistance increase curve, after sound has been experienced, can be correlated with subjectivity. Each stress experiment is superimposing the stress under investigation on stresses already existing within the individual; often, too, we are subject to not one stress but several in combination. Would some measure of the activity in the autonomic nervous system be a better and more reliable indicator of stress? Perhaps there is a large rôle for the neurophysiologists to play in this field. For instance, stress can be interpreted as an increase in arousal level, and this may be correlated with features

TABLE 2.16. MEASURE OF STRESS–STRAIN RELATIONSHIP BETWEEN MAN AND ENVIRONMENT

Stress measures	Measures of strain	Factors governing stress–strain relationship	
		Factor	Possible measures
Thermal, visual, acoustical, olfactory and pollutant aspects of the environment can be measured using instrumentation, but social and spatial characteristics are more difficult to quantify	*Physiological* (Davis *et al.*, 1955; Kryter and Grandjean, 1961; Jansen, 1967; Gibbons, 1970)	Arousal level	Evoked cortical response (Wilkinson and Haines, 1970)
	Cardiovascular response (Lehmann *et al.*, 1956, 1958, 1965; Meyer-Delius, 1957; Etholm and Egenberg, 1964; Jansen, 1967; Raab, 1968; Keefe and Johnson, 1970)	Effort	Sinus arrythmia (Kalsbeeck *et al.*, 1965)
	Blood cell count (Selye, 1950; Atherley *et al.*, 1970)	Personality	Personality tests (Hathaway *et al.*, 1951; Eysenck, 1968)
	Adrenocorticotrophic hormone (ACTH) secretion from pituitary gland (Selye, 1950)	Intelligence	Intelligence quotient
	17-Ketosteroids (Sakamoto, 1959; Hoagland, 1957; Atherley *et al.*, 1970)	Experience of stress	Question, interview, indirect observation
	Digestive action (Laird, 1929, 1930, 1932; Stern, 1964)		
	Metabolism (Laird, 1929; Harmon, 1933; Uglow *et al.*, 1937; Stevens, 1941; Bugard, 1951; Karrasch, 1952; Givoni and Goldman, 1971)		
	Respiration (Corbeille and Baldes, 1929; Poole *et al.*, 1966; Semczuk, 1968)		
	Muscle tension (Morgan, 1916; Davis, 1932; Kennedy 1936; Stevens, 1941; Bugard, 1951; Steinmann and Jaggi, 1955; Davies *et al.*, 1955)		
	Plethysmography (Oppliger and Grandjean, 1959; Jansen, 1967; Conrad, 1973)		
	Skin resistance (Davis, 1932; Davis *et al.*, 1955)		
	Electroencephalography (Darrow, 1947; Kluge and Friedel, 1953; Richter, 1966a; Jansen, 1967; Wilkinson and Haines, 1970)		
	Skin temperature (Silink *et al.*, 1941)		
	Eye-pupil dilation (Jansen, 1967)		
	Subjective (references given throughout Chapter 2)		
	Questionnaire		
	Interview		
	Indirect observation		
	Performance (for references see Table 2.41 and throughout Chapter 2)		
	Productivity		
	Errors		
	Speed of working		
	Absenteeism		
	Labour turnover		
	Concentration		

General references covering many aspects of the work referred to in the above table: Åstrand and Rodahl (1970), Borg and Møller (1973), Broadbent (1971), Dean and McGlothen (1962), Levi (1972), Mackworth (1950), Morgan (1965), Selye (1950), Suggs and Splinter (1961), Teichner *et al.* (1963), Viteles and Smith

of the evoked cortical response. The arousal and emotional centres lie in the reticular activating system of the brain, and this system not only affects the central nervous system but the sympathetic nervous system also. Overstimulation of the reticular activating system results in constriction of the cutaneous blood vessels, some eye-pupil dilation, increase in heart rate due to the direct sympathetic nervous stimulation of the organs concerned, together with increased adrenaline flow. These effects, the so-called N-complex somatic responses, have been measured for many years but many findings have been paradoxical; more basic measurements on the nervous system may be needed. The most useful measures will be those from which psychological as well as physiological conclusions can be made.

Table 2.17 summarises some of the research that has been carried out on the physiological effects of noise on human beings but excludes work on temporary or permanent deafness.

Vasoconstriction seems to depend on the sound *level* irrespective of whether the sound is noise or music; it also seems to be independent of age or the past noise conditioning of the individual. This suggests that it is an unlikely indicator of subjective response, and yet the recent work of Conrad (1973) must not be dismissed.

The paradoxical findings of Atherley *et al.* (1970) are worth discussing. Electrodermal activity (e.g. changes in skin resistance) is thought to be associated with activity in the sympathetic division of the autonomic nervous system (Hume, 1966). This activity in turn excites the adrenal medulla resulting in the release of adrenaline/noradrenaline. This sympatho-adrenomedullary interaction is thought to play an important rôle in the general response to stress (Levi, 1967). Increased adrenal cortical activity due to stress is indicated by an increased adrenocorticotrophic hormone (ACTH) (Selye, 1950) level; urinary 17-ketosteroid levels are now considered to be a reliable indicator of increased or decreased adrenal cortical activity level (Atherley *et al.*, 1970). A substantial body of opinion considers that prolonged sympathetic activity reflects in some way the arousal level of the individual, and hence also the levels of attention and anxiety.

Now Atherley *et al.* (1970) found no evidence of increased ACTH production; in fact their results suggested the reverse. They suggested that noise of high subjective importance caused increased adrenal medullary activity and at the same time diminished adrenal cortical response. Subjects exposed to the aircraft and typewriter noises did speak of feeling tired and irritable. It might be possible, the authors concluded, that response to noise having high subjective importance results in a mild type of *anxiety–depression* syndrome in which the anxiety is associated with increased sympathetic activity and depression with reduced adrenal cortical activity. The work of Hoagland (1957) (Table 2.17) lends some support to this speculation.

TABLE 2.17. PHYSIOLOGICAL RESPONSE OF MAN TO NOISE

Research topic and source	Observation
Cardiovascular response Lehmann and Tamm (1956)	Figure 2.30a shows the results of tests carried out over a period of 6 months on factory workers that had experienced the factory noise environment of 90 DIN-phon (200–400 Hz) over 2–40 years; vasoconstriction occurs, the blood pressure increases, but the heart beat and systolic blood volume decrease; note the transient effects when the noise is switched on–off.
Lehmann and Meyer-Delius (1958)	Figure 2.30b shows that vasoconstriction occurred when 43 min of white noise at a level of 90 DIN-phon was received by 10–34 subjects of different ages; the maximum resistance to blood flow occurred at about 6 min after the noise was switched off. Figure 2.30c shows vasoconstriction again occurring, as depicted by a decrease in the finger-pulse amplitude, when noise (90 DIN-phon, 3200–6400 Hz) is applied with durations of 10–40 s to 14 subjects in each of 20 tests; maximum decrease occurs within 10 s of noise being switched off; original blood circulation level restored at a time of about 8 times the noise duration period after the noise has been switched off; reactions of children and adults were similar, but subjective effects troubled adults more; these latter effects could not be correlated with the blood circulatory changes.
*Jansen (1967)	Figure 2.31a–c shows that vasoconstriction occurs when noise is perceived and is more for broadband than for pure tone sounds; the effect is almost instantaneous, but the recovery period may be as long as twice or three times the noise duration; the effect depends on the sound level and becomes marked above about 75 dB. Figure 2.31d suggests that vasodilation may occur during work periods; blood circulation during work periods is higher than in rest periods; noise reduces blood circulation, but the change is often less marked than the change brought about by work; some adaptation to the noise occurs but the noise durations were too short to make firm conclusions. Clearly the thermoregulatory system interacts with any circulatory effects brought about by lighting, noise or work; Fig. 2.32a shows that at a room temperature of 40°C the noise vasoconstrictive effect almost balances the thermal vasodilation in the peripheral blood circulation, the initial transient remains. Figure 2.32b suggests that young children exhibit the same circulatory responses as adults but to a lesser extent. Figure 2.32c and d shows that the blood circulation in men from the Sudan, who have presumably been used to different noise and thermal climates than a European, responds to noise in a very similar way to that in men from Germany, and again age seems to have little effect. If 30 s bursts of noise are presented more than three times with intermittencies of 30 s to 3 min, then adaptation to a lower blood circulation level occurs.
Jansen (1964)	In this work the effects on the cardiovascular response of music (by J. S. Bach at a level of 98 DIN-phon) were compared with those of broadband noise at the same level, each lasting 7 min; the blood pressure and pulse frequency showed no significant changes, although blood volume flow rate did significantly increase for some subjects or decrease for others; no distinction could be demonstrated between the effects of noise and music.
*Reilly (1959), *Jansen (1961), *Andriukin (1961), *Shatalov et al. (1962), *Strakhov (1966)	Surveys carried out in various industries in Germany and Russia by these researchers clearly show that people working in very noisy environments (e.g. ball-bearing plants, lathe shops, foundries) experience more blood circulation, heart, digestive and neurotic trouble than those in less noisy ones; this view is also supported by earlier work in America by Davis (1958).
Rothlin et al. (1956)	Arterial pressure in 20 rats found to increase from about 125 to 158 torr when exposed to intermittent noise (90 phons, 700–1000 Hz).
Digestion Laird (1929, 1930,	The quantity of saliva, and the frequency and amplitude of peristaltic stomach action decreased in noise climates of 40–80 dB. Changes from low to moderate stimulation

TABLE 2.17(*Continued*).

Research topic and source	Observation
1931, 1932), Cannon (1929), Davis and Berry (1964), Stern (1964)	levels cause increased gastrointestinal mobility; decrease in stimulus from high to lower levels causes a decrease in mobility.
Metabolism Laird (1929), Harmon (1933), Uglow *et al.* (1937), Stevens (1941), Bugard (1951)	Metabolic rate found to increase accompanied by increases in pulse rate and respiratory rate.
Respiration Corbeille and Baldes (1929), Kennedy (1936), Bugard (1951), Steinmann and Jaggi (1955), Davis *et al.* (1955)	Increased respiratory rate; pulse frequency and arterial pressure sometimes increased, sometimes decreased.
Work output Karrasch (1952)	Three subjects pedalling an ergometer in an environment of 105 phons (1200–2400 Hz) showed a decrease of 5–25% in work output.
Skin resistance (galvanic skin response) Davis (1932), Davis *et al.* (1955), Helper (1957)	Skin resistance decreases with 110 dB of white noise, but an arithmetic task caused a greater decrease; noise and task caused greatest decrease.

After noise exposure skin resistance increases exponentially to original level; decay time of this curve shown to be a valid and reliable measure of subjectivity importance. His results showed:

*Gibbons (1970)

Sound	dBA	Noise pollution level (L_{Np} dB)	Objective ranking using L_{Np}	Subjective ranking using physiological measure
Baby	89·5	144·5	1	3
Aircraft	94·6	108·7	2	1
Bell	93·0	95·6	3	2
White noise	92·5	95·1	4	4

Research topic and source	Observation
Muscular tension (electromyographic (EMG) studies) Morgan (1916), Davis (1932), Stevens (1941), Davis *et al.* (1955), Helper (1957)	Muscular tension increases. An arithmetic task caused a greater increase in tension than 110 dB of white noise alone; muscle tension had similar value when the task was carried out in noise as when performed in much quieter conditions.
Eye-pupil dilation Jansen (1967)	Infrared photographic techniques used to measure differences of eye-pupil diameter under different noise conditions; subjects sat in anechoic chamber. Figures 2.33a–c show some of the principal findings. Above 55 dB there is a significant increase in pupil size. Attempts have been made to correlate pupil size with concentration but no conclusive results have yet been established.

TABLE 2.17 (*Continued*).

Research topic and source	Observation
Hand volume Oppliger and Grandjean (1959), Rossi *et al.* (1959)	Forty subjects were exposed to 6 s bursts of noise (motor-horn 240–4000 Hz) at levels rising in 5 dB steps from 60 to 105 dB. Using plethysmographic techniques the blood flow through the hand as indicated by the hand volume, was measured. Five subjects showed no vasomotory response. Figure 2.34a indicates that 30% of the subjects showed a significant reduction (2·5–5%) in hand volume when exposed to a 65 dB sound level; the percentage reaction rose to 56% at 105 dB. Figure 2.34b shows that below 80 dB only 18–28% of the tests on subjects 19 years old or under registered significant responses, whereas this frequency of response rose to 27–53% for subjects older than 19; the 20–30-year-old age groups gave the highest response, except at 70 dB. Hand skin cooling by 0·08 to 0·13°C also took place as the noise level increased (Fig. 2.34c); 68–92% of the tests showed significant increases. Later experiments by Rossi *et al.* (1959) were conducted on people in a noise climate of 70 dB (500 Hz); some of the subjects were exposed to a further noise of 80–100 dB (1 min duration, 2000 Hz) having a variable intermittency of about 1 min. The vasomotory responses were the same for both groups of subjects suggesting that adaptation to a background level has little influence on the physiological mechanism.
*Conrad (1973)	Sixteen university students worked in periodic and aperiodic 93 dBA broadband noise. Studies indicated negligible decremental effect of noise on short-term memory task performance. Physiological variables recorded were finger photoplethysmographic blood volume pulse amplitude, pulse rate and forearm electromyogram; subjective response measured by noise annoyance questionnaire. Finger blood volume pulse response increased (i.e. increased intensity of effort) in all the noise conditions, and was higher for subjects that were more highly annoyed than others less annoyed.
Electroencephalographic studies (EEG) Darrow (1947), Canac and Bladier (1953), Kluge and Friedel (1953), Perl *et al.* (1953), Hess (1954), Rowland (1957), Steinicke (1957), Jouvet (1963), Williams *et al.* (1964), Richter (1966a), Schieber (1967), Jansen (1967), Wilkinson and Haines (1970)	Noise can decrease the frequency of the 10 Hz α-brain waves typical of the conscious state to 5–7 Hz waves of greater amplitude; authors speak of psychophysiological adaptation between the cortex and subcortical regions of the brain. Effects of noise on sleep are discussed later in the text of this chapter. Possible correlation between evoked cortical response and arousal level.
Chemical changes in the blood and urine from glandular stimulation Hale (1952)	A group of workers and a group of volunteer test subjects both showed a decrease in their eosinophil counts after exposure to noise.
*Hoagland (1957)	Seventeen ketosteroid secretions found to be low for schizophrenic patients who were quiet and withdrawn; for aggressive patients the converse was true.
Sakamoto (1957, 1959)	Thirty-two female light factory workers in an environment of 90–95 dB, 28°C and 70% relative humidity were found to have a decrease in the 24 urinary 17-ketosteroid secretion; using two groups of workers the effects of the noise and thermal climates were assessed separately, and the 17-ketosteroid urinary levels of the workers decreased for those in the noisy factory but were not affected in the workers subjected to the thermal factors alone.

TABLE 2.17 (*Continued*).

Research topic and source	Observation
	Other experiments on 17 female light factory workers showed that if they received an adrenaline or adrenocorticotrophic hormone (ACTH) injection before working in the noise the eosinophils count decreased by 50% after the work period (or by 19% if the injection was given after the work period).
*Levi (1967)	Urinary excretion of adrenaline and noradrenaline increased by industrial (sorting steel balls) and office work, and by pleasant (two-hour comedy film) and unpleasant (two-hour horror film) stimuli; noise, light or task has less influence on excretion levels than does the subject's attitude; in general emotionally vulnerable people do not excrete more catecholamines than do normal people under experimental stress.
*Raab (1968)	The cardiovascular adrenergic and adrenocortical responses of 40 male subjects were measured during a 20 min period in which the subjects heard a rhythmically interrupted ringing telephone bell and saw a flickering bright light; in the second 10 min of that period they had two mental arithmetical problems to solve. Significant increases occurred in the plasma cortisol levels and were highest in the emotionally irritable subjects.
*Atherley *et al.* (1970) (also Gibbons, 1970)	Observations made were:

Stimulus (exposure time 7 h per stimulus)	Increased level	Decreased level	No change
Aircraft noise 95 dBA Typewriter noise 70 dBA (both of high subjective importance)	Lymphocytes Neutrophils White cell count	17-Ketosteroids Eosinophils Skin resistance (suggested decrease in ACTH)	
White noise 95 dBA (low subjective importance)			17-Ketosteroids Eosinophils Neutrophils White cell count
Expected change based on work of Selye (1950) on conventional stressors	17-ketosteroids Neutrophils White cell count ACTH	Eosinophils Lymphocytes	

Conclusions made were that exposure to noise of high subjective importance may lead to a syndrome that is characterised by people feeling tired and irritable with a decrease in the 24 h urinary 17-ketosteroid levels and with some changes in the composition of the blood. Noise of low subjective importance does not show these effects.

*Work attempts to correlate physiological measures with subjective response.

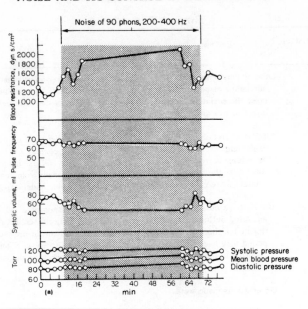

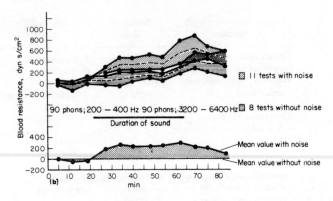

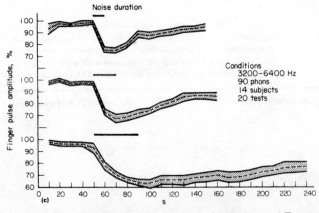

FIG. 2.30. Cardiovascular response of man to noise: (a) tests of Lehmann and Tamm (1956); (b) and (c) tests of Lehmann and Meyer-Delius (1958).

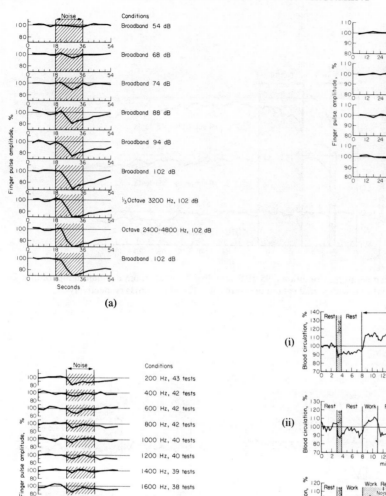

(a)

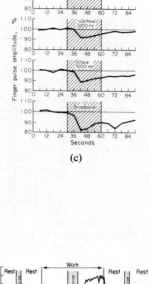

(c)

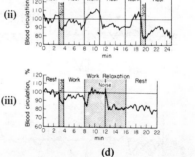

(d)

(a) Finger-pulse amplitude measurements for broadband sound at varying levels and of 18 s duration.
(b) Finger-pulse amplitude measurements carried out on eight people subjected to pure tones at a level of 90 DIN-phon and 30 s duration.
(c) Effect of noise frequency content on finger-pulse amplitude; the sound level was 95 DIN-phon and of 30 s duration; 17 subjects were used and 122 test runs were made.
(d) Effects of work, rest and noise on blood circulation (Jansen, 1964).
  (i) Influence of a 2 min noise effect during a 12 min work period. Noise: wide-band 95 DIN-phon; work: 5 MCP/s bicycle ergometer; 53 tests, 7 subjects.
 (ii) Influence of noise after termination of noiseless manual work. Noise: wide-band 95 Din-phon work: 5 MCP/s bicycle ergometer; 45 tests, 7 subjects.
(iii) Behavior of finger-pulse amplitude during noise in a work phase and in a following rest phase. Noise: wide-band 95 DIN-phon; work: 5 MCP/s bicycle ergometer; 36 tests, 8 subjects.

FIG. 2.31. Finger-pulse amplitude measurements made by Jansen (1967); a decrease in amplitude means that the peripheral blood circulation decreases, i.e. vasoconstriction occurs.

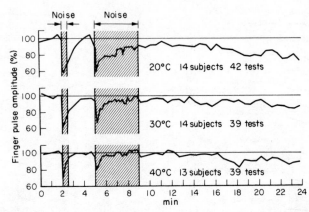

(a) Effect of room temperature. Conditions: 95 dBB broadband sound; room dry-bulb temperatures 20°, 30° and 40°C with partial vapour pressures of 8, 12 and 13 mmHg respectively.

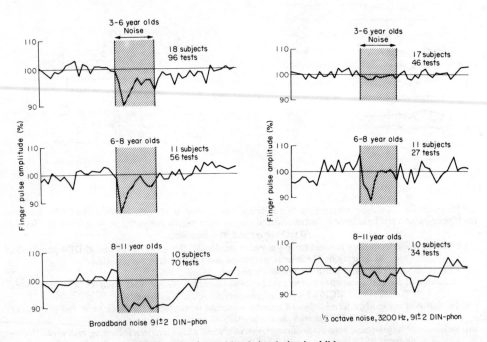

(b) Effect of noise on blood circulation in children.

FIG. 2.32. Influence of temperature, age and race on changes in blood circulation due to noise (Jansen, 1967).

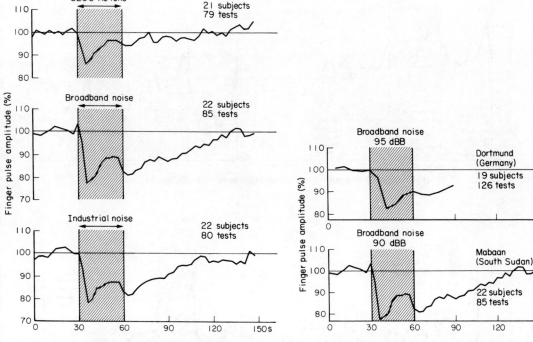

(c) Effect of 90 dBB noise on blood circulation of 20–35-year-old men from Mabaan (in south Sudan).

(d) Comparison of circulatory effects in men from Europe and Africa; age range 60–85 years.

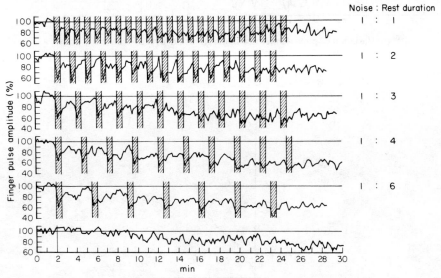

(e) Adaptation of peripheral blood circulation system to noise of 30 s duration repeated after rest periods of 30 s, 1, 1½, 2 and 3 min duration.

FIG. 2.32. (*Continued*).

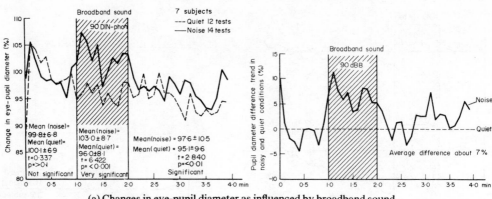

(a) Changes in eye-pupil diameter as influenced by broadband sound.

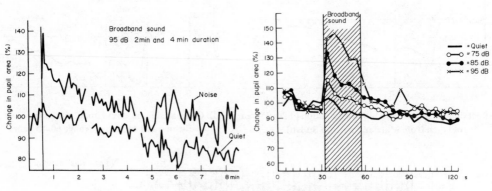

(b) Variation of pupil diameter with duration and level of broadband noise.

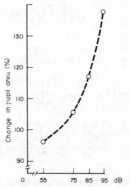

(c) Influence of various levels of broadband sound on eye-pupil size.

FIG. 2.33. Influence of noise on eye-pupil size (Jansen, 1967).

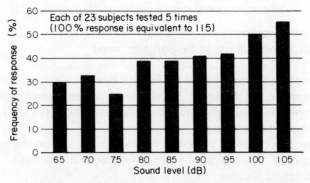

(a) Reduction in hand volume when subjects exposed to noise.

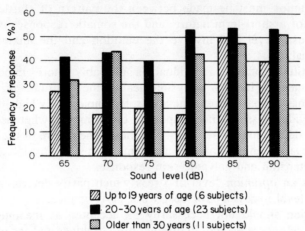

☑ Up to 19 years of age (6 subjects)
■ 20–30 years of age (23 subjects)
▨ Older than 30 years (11 subjects)

(b) Age analysis of vasomotory response to noise.

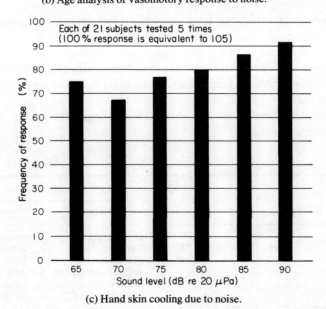

(c) Hand skin cooling due to noise.

FIG. 2.34. Plethysmographic studies carried out by Oppliger and Grandjean (1959).

99

When coming to interpret the valuable work that has been carried out in many countries several points have to be borne in mind:

* A variety of experimental methods have been used; the subjective studies use small numbers of subjects. Standardised international experimental procedures need to be adopted if valid comparisons between various pieces of research are to be made.
* The somatic responses reviewed have concentrated on the response to noise, but there is a pattern of somatic responses to light, heat and other environmental factors.
* The various somatic responses are interrelated.
* A distinction must be made between the pattern of startle–fear responses usually of a short-term nature, and the somatic responses which depend on the arousal level and involve the emotion and the anxiety levels of the individual.
* The attitude of the individual is probably more important than the environmental factors. Physiologically the attitude of the individual determines the base adrenaline levels. Psychologically high motivation makes the individual less dependent upon the environment, whereas boredom makes the individual more dependent upon them. The nature of the task also has an equal or larger effect than the environment on such physiological measures as muscle tension and galvanic skin response.
* There is an *optimum* level of stress which partly determines the optimum arousal level necessary for best work performance.
* Adaptation should not necessarily be accepted as meaning that optimum stress levels are achieved. People get used to things and accept things that are not necessarily good for their body and mind.
* Physiological measures have tended to emphasize the physical aspect of man's make-up, whereas the mental health reflected by the psyche state of the individual is just as important. The physical and psyche states have common starting points in the brain and the autonomic nervous system. For this reason the need for more research relating stress to the brain and the nervous system is iterated.
* The sound levels given in Table 2.18 are suggested as having the relevant significances noted.

TABLE 2.18

PNdB (Kryter, 1970)	dB (Jansen, 1967)	dBA (Author)	Somatic response	
			Physiological	Psychological
<30	<60	<20	Not detectable	Important in partly
30–80	60–75	20–55	Moderate	determining arousal
80–135	75–95	55–115	Distinctive; progressively more harmful effects on hearing system.	level. Stress effects on nervous system unknown in many situations.
>135	>95	>115	Pain	Physiological effects more important.

2.4.2. NOISE AND SLEEP

One dictionary defines sleep as a *natural state of unconsciousness and immobility recurring in man and animals at least once a day*. From time immemorial scientists, philosophers and poets have mused upon the need for, the meaning of, the security and the peace found in that mysterious limbo—slumberland—a land of fantasy, abandonment which sometimes is a fairy-tale and at others a nightmare. Richard Strauss composed a beautiful musical setting to Hesse's poem *Going to Sleep* in one of his *Four Last Songs*. The words convey so much within a few lines in a way that only poets or musicians know:

> *Now the day has tired me,*
> *I yearn for the starry night.*
> *May she receive me kindly,*
> *Like a tired child!*
>
> *Hands, leave your doing;*
> *Brain, leave your thinking—*
> *all my senses*
> *would now sink into slumber.*
>
> *And the unwatched soul*
> *wants to soar up freely*
> *to live a thousand times more intensely*
> *in the magic circle of night.*

(HERMANN HESSE, 1877–1962.)

The rest of mind and body, the release from the ties and the constraints of conscious living, the freedom from responding to the world about us; all these things are echoed in these words. Tender words by Friedrich Hebbel, that have been set to music by Alban Berg in his Opus 2 Lieder, reflect the desire for peace; for no intrusion by dreams or anything that might awaken us, acknowledging that even our sleep hours may be disturbed.

> *To sleep, to sleep, only to sleep!*
> *No awakening, no dream!*
> *Let the pains I had to bear*
> *be hardly remembered—*
> *so that, when the fullness of life*
> *sounds into my sleep*
> *I draw my sheet closer around me*
> *and hold my eyes more tightly shut!*

(FRIEDRICH HEBBEL, 1813–63)
from *Let Anguish have its Due*.)

Using the human response function

$$\phi_o = \phi(\phi_i, \phi_f, \psi),$$

then for sleep $\phi_i \to 0$ and $\phi_o \to 0$, hence

$$0 = \phi(\phi_f, \psi).$$

This suggests that during sleep there is still activity within the autonomic nervous system and that there is a strong interplay between this and the psyche. To ensure that the effect of any input stimuli is reduced to zero (Hesse's words—*Brain, leave your thinking—all my senses would now sink into slumber*) the eyes are closed, the body lies horizontally on a mattress reducing gravity forces, the normal auditory threshold when awake is raised by 20–80 dB depending on the state of sleep (Kryter, 1970; Williams *et al.*, 1964). The body has its own way of reducing the possibility of disturbance during sleep; man designs bedrooms to be quiet and lights are switched off, of course. With the rapid growth of air and road traffic the problem of what noise will disturb sleep is continually being raised. It is not even a simple matter of what level of sound will awaken people, because noise can disturb the quality of sleep without awakening the person; it can make it difficult for a person to fall asleep; it can awaken people too early in the morning.

When people are deprived of sleep they become irritable and may show irrational behaviour (West, 1967). Many scientists have emphasised the need for good sleep for health of mind and body (Grandjean, 1962; Jansen and Schulze, 1964; Lehmann in unpublished work; Richter, 1966a and b and in unpublished work).

Electroencephalographic measurements have been useful in revealing some of the attributes of the sleeping state. Figure 2.35 shows the EEG patterns recorded during sleep and the effect of introducing a noise. Deep sleep is characterised by the slow delta brain rhythm (about 1–3 Hz, 80–100 μV amplitude), and intermediate sleep levels by theta waves (5–6 Hz) combined with spindle muscle activity (14 Hz), in contrast to the 10 Hz, 50 μV α-rhythm that distinguishes the conscious or unconscious waking state. It is during this unconscious α-activity that sleep learning can occur (Simon and Emmons, 1954). A disturbance is seen to cause a series of transients, the so-called K-complex waves, after which α-waves appear and if the

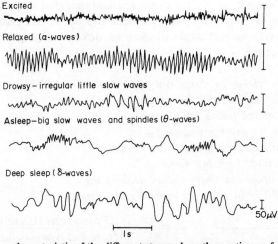

(a) Brain waves characteristic of the different stages along the continuum from excitement to deep sleep.

FIG. 2.35. Electroencephalograms (EEG) for sleeping people (Richter, 1966a, b, 1971; also unpublished reference given).

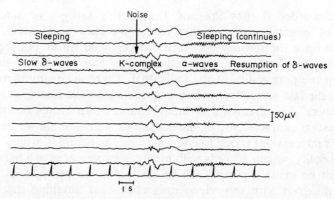

(b) Transitory wave bursts (K-complex) in sleep following acoustic stimulation (arrow: knocking twice on the window). Note: waking (alpha) rhythm (10 Hz) arises for 4 s whilst the person examined continues to sleep.

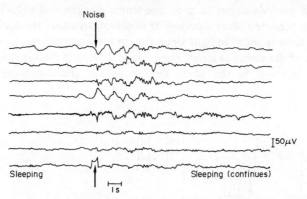

(c) K-complex following noise (arrow) in sleep as elicited by traffic noise (motor-cars, tramways) without waking up the sleeper. Note: stimulation was of short duration and low intensity.

FIG. 2.35. (*Continued*).

subject continues to sleep θ- or δ-waves reappear; if the subject consciously woke up the α-rhythm would continue.

There are different stages and depths of sleep. The deepest sleep is reached about 60–90 min after falling asleep. Between the succeeding levels of sleep that continue throughout the night we *unconsciously* wake up, and in between this and leaving one level and returning to another level of sleep, *dreaming* occurs. During dreaming rapid eye movements (REM) occur as well as other muscle responses. Dreams may or may not be remembered on awakening. The need for dreaming has been speculated upon for years. Jungian philosophy reasons that the psyche self is depicted by the nature of our dreams and from them we may learn more about ourselves and the reasons for some of our behaviour patterns. A more pragmatic view is that during our conscious state we have innumerable stimuli from the environment impacting onto our minds, and it is only during dream periods that our impressions are filed away if treated as important by the mind for recall when required or discarded if they are not. During this process numerous unrelated impressions of events come together and sometimes form vivid, inexplicable situations, but once the sensory impressions have been sorted out then everything seems to be in order once again. The efficiency of this process probably depends very much on the personality of the individual. There are many more attempts at explanations; the last one reflects a personal feeling on the matter at present; some sleep researchers view dreaming as unimportant. Sleep closes down the sensory perception system characteristic of the conscious state and allows time for sorting out each day's impressions trying to find a pattern, the nature of which may well be a basis of personality, which accords with previous experiences in a person's life. If a pattern cannot be made a person will be restless, even neurotic. However much others may disagree with this view, it is clear that anything that disturbs the sleep–dreaming pattern is detrimental to health.

Figure 2.36 shows a sleep–dream pattern recorded by Williams *et al.* (1964) for one subject. In parts R_1 and R_2 of this diagram the effects of sleep deprivation are illustrated. REM periods occur four or five times a night and each period lasts for about 15–20 min. Sleeping after a period of deprivation alters the time spent in each level; hardly no time is spent in light sleep (i.e. in the unconscious awake and stage 1 sleep), the stage of dreaming is lengthened, stage 2 sleep is decreased, stages 3 and 4 are lengthened but not to the same degree as for the dreaming period. Sleep

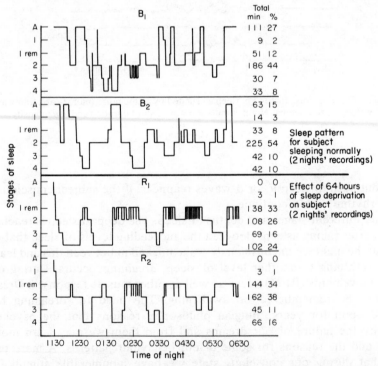

FIG. 2.36. Effect of sleep loss on the distribution of EEG stages of sleep for one subject. Sleep loss produced a decrease in the time taken to go to sleep as well as in the amounts of stages 1 and 2, but the amounts of stages 3, 4, and 1 rem increased (Williams *et al.*, 1964).

deprivation lowers the response to stimuli to an even lower level than that for normal sleeping subjects (Williams *et al.*, 1964).

Williams *et al.* (1964) and Jouvet (1963) have established that during the REM stage, man, in common with other animals, is usually insensitive to auditory or other stimulation. Thus during REM sleep our earlier expression is probably true, i.e.

$$0 = \phi(\phi_f, \psi).$$

but during other stages of sleep,

$$\phi_o = \phi(\phi_i, \phi_f, \psi).$$

The human response function is the same as for our conscious state except that the probability of ϕ_i entering the system is low, and consequently ϕ_o is low. This probability can be altered by other factors such as the familiarity or strangeness of the surroundings in which the person is sleeping; by suggestion, bordering on the territories of hypnosis; by stimulating the sleeping subject with auditory signals having significant meaning to them (e.g. speaking their name).

Work by Rowland (1957) and Williams *et al.* (1964) suggests that discrimination between certain sounds can be made during sleep as well as whilst awake. There may be some control exercised over the length of the stage of sleep. If more time is spent in the light stages of sleep (stages 1 and 2 on Fig. 2.36) than in deep sleep (stages 3 and 4 on Fig. 2.36), then the probability of hearing sounds will be more because Kryter (1970) suggests that the auditory threshold is raised by 30 dB in stage 2, by 50 dB in stage 3 and by 80 dB in stage 4 for sleep-deprived subjects; the magnitude of these threshold increases will be less for people sleeping normally but their trend will be similar.

Schieber (1967) suggests that the temporal variations in sound level may be more important than the actual level.

The research of Williams *et al.* (1964) shows that as the sound intensity is increased the brainwave activity and behavioural awake responses increase; the increase is less for sleep-deprived subjects; the REM period is least affected. The vasoconstriction of the peripheral blood vessels behaves somewhat differently in that it is not influenced very much by sleep deprivation, the response is very similar in all stages of sleep and exhibits an even higher response in REM than in stages of light sleep. These findings have been confirmed by Jansen and Schulze (1964).

Some results of Jansen's work are shown in Fig. 2.37. Clearly the duration and level of sound is important as the K-complex waves show differing forms. The effects of vasoconstriction are very marked even for short duration sounds, and generally the blood circulation seems to be more restricted by noise when people are asleep than when they are awake (refer to Table 2.17). To what degree the sleep–dream pattern can be allowed to change remains unknown.

Lukas and Kryter (1969) have found that older persons are much more likely to awake than younger persons. Figure 2.38 shows the effect of simulated sonic booms and recorded subsonic aircraft noise on old, middle-aged and young people. The proportion of people woken up by these noises seems to rise rapidly after middle age, and old people can even be awoken from REM very easily. People over 65 form roughly one-seventh of the population in Great Britain and this fraction is increasing;

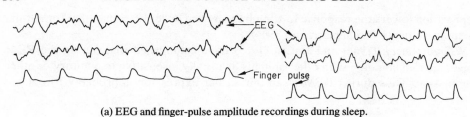

(a) EEG and finger-pulse amplitude recordings during sleep.

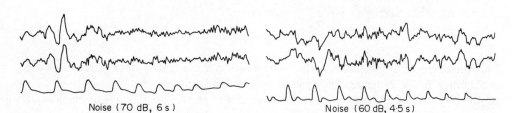

Noise (70 dB, 6 s)   Noise (60 dB, 4·5 s)

(b) Effect of broadband noise on brain activity and blood circulation during sleep.

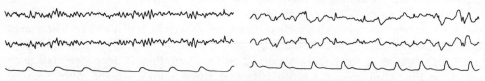

(c) Brain and finger-pulse activity during sleep after noise is switched off.

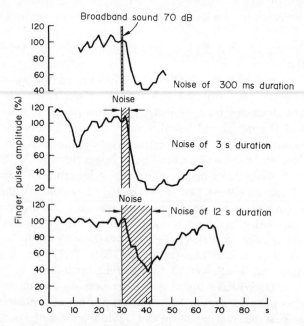

(d) Changes in finger-pulse amplitude during sleep with 70 dB broadband sound of different durations.

FIG. 2.37. Electroencephalogram (EEG) and finger-pulse amplitude recordings during sleep (Jansen, 1967).

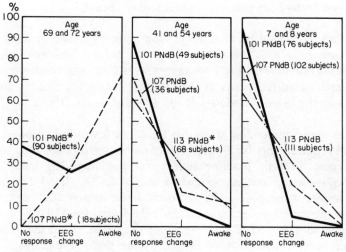

*Estimated flyover intensities as if measured outdoors

(a) Effect of subsonic aircraft noise on sleep.

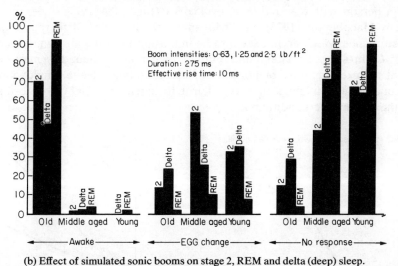

(b) Effect of simulated sonic booms on stage 2, REM and delta (deep) sleep.

FIG. 2.38. Results of exposure of three different age groups while asleep (Lukas and Kryter, 1969).

any standards should reflect this increased susceptibility of the old to noise during sleep.

Little work has been carried out on noise-induced sleep although Olsen and Nelson (1961) claim that a tone of 320–350 Hz calms crying babies. Likewise audioanalgesia has been effectively used in America with some dental patients (Gardener and Licklider, 1959). Patients relaxed by listening to music and switched to filtered random noise when they felt pain. Confidence between the patient and the dentist, and also expert use of the procedure were found to be two essentials for successful audioanalgesic application. Kryter (1970) concludes that when audioanalgesia is effective several explanations may apply:

* suggestion enhanced by distraction from pain by music or noise;
* music and noise may relax the patient and reduce false anxiety;
* the neural impulses in the auditory system take precedence over the pain impulses transmitted between the reticular formation and the higher nerve centres in the brain (Gardener *et al.*, 1960; Melzack, 1961).

When it comes to recommending standards for sound levels in bedrooms, one of the most stringent requirements for houses, flats and hotels, it is not easy to make firm recommendations. Present practice favours a maximum L_{10} level of 35 dBA. Steinicke (1957) carried out an extensive study on the level of sound which consciously awakens people; it has already been stated that this should not necessarily be the criterion for setting standards. Even if a 35 DIN-phon loudness level is comparable with a 35 dBA one (Grandjean, 1973a, states that it is) then 22% of the subjects were consciously awakened by the noise (Fig. 2.39a).

Jensen (1973) reports some Canadian research which has shown that there is a 10% probability of awakening and a 20% probability of causing a significant shift in sleep level of a person with a sound level of 45 dBA (Fig. 2.39b).

Work by Macpherson (1973) (Fig. 2.40) shows the influence of thermal factors on sleep disturbance which do not seem to be important until the air temperature rises above 25°C. In some climates conditions of high relative humidity may be more important than temperature or noise (people will willingly leave windows open in non-airconditioned buildings in noisy Beirut to catch any slight breeze that may relieve the oppressive conditions).

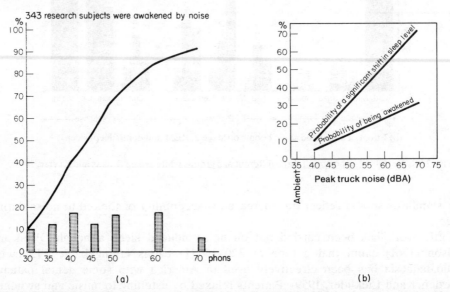

(a)   (b)

FIG. 2.39. (a) Threshold values of an experimental noise (60–5000 Hz) which awakened research subjects. Each sound was presented for 3 min. The graph is a summation curve, and the vertical columns indicate the percentage of people who were awakened by each level of noise (Steinicke, 1957). (b) Sleep interference (Jensen, 1973).

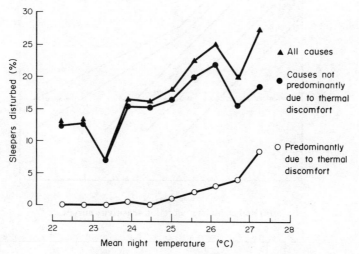

FIG. 2.40. The frequency distribution with respect to mean night temperature of the votes recording disturbed sleep expressed as a percentage of all the votes cast at each temperature (Macpherson, 1973). Causes of sleep disturbance are due to fretful babies, sick children, barking dogs, traffic noise, parties, besides thermal discomfort. Data was collected in Australia.

In designing and selecting internal partitions for hotel bedrooms, houses or flats, it should be remembered that people show extreme variations in bedtime hours. Meaningful noise such as television programmes or conversations from neighbours defy physical assessment because the ear and mind pick up too easily noises which physical measurements may show to be acceptable when compared to current standards. New standards need to be based on psychological as well as physiological needs of people. The study by Bitter and Weeren (1955) on the problem of sound nuisance and sound insulation in flats is a fine example of a more complete approach.

2.4.3. THE RESPONSE OF THE HUMAN BODY TO LOW-FREQUENCY VIBRATIONS

Table 8.1 lists the sensors tactile receptors in the skin, proprioreceptors in the muscles and the vestibular system that respond to low-frequency mechanical vibrations in the range 0–100 Hz which may arise from vibrations transmitted to the body from surrounding surfaces set into motion by airconditioning plants, distribution systems linked to the building structure, wind sway of tall buildings, footsteps or earthborne vibrations from traffic, seismic and other external sources. Their effect on the body is twofold. Firstly, mechanical stress may cause some cells in the body tissue to be destroyed directly or, by repeated strain, setting up metabolic fatigue; secondly, strain imposed on nerve receptors may cause disturbances in the autonomic nervous system. Some of the principal effects of vibration are given in Table 2.19 and are discussed in detail by Coermann (1970).

Human sensitivity to vibration has been defined by Dieckmann (1958) by an index

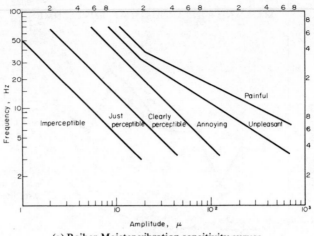

(a) Reiher-Meister vibration sensitivity curves.

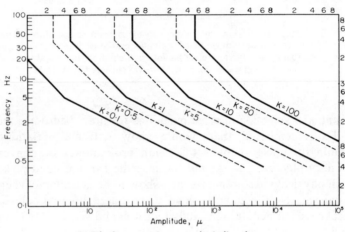

(b) Dieckmann values: vertical vibration.

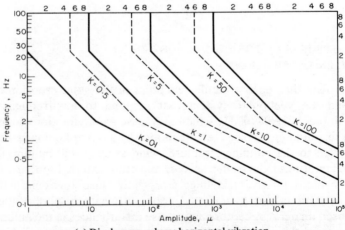

(c) Dieckmann values: horizontal vibration.

FIG. 2.41. Vibration criteria.

110

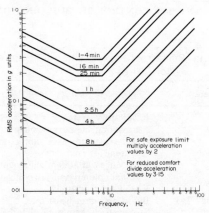

(a) ISO recommendation for vertical vibration exposure limits as a function for fatigue-decreased proficiency boundary.

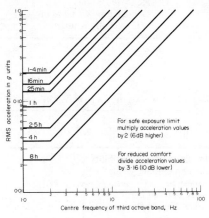

(b) ISO recommendation for horizontal vibration exposure limits as a function for fatigue-decreased proficiency boundary.

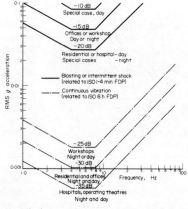

(c) Suggested limits to prevent excessive building vibrations.

FIG. 2.42. ISO recommendations.

111

TABLE 2.19. EFFECTS OF LOW FREQUENCY VIBRA-
TION ON HUMAN BODY

Frequency	Effect
1–3	Respiratory system; travelling sickness when whole body responds as a simple mass system
4–5	Resonances in abdominal system, spinal column and jawbones
5–10	Vision deterioration
6–10 (at 1·2g)	Blood pressure, heart rate, oxygen consumption disturbed
7	Severe chest pains
8–12	Lumbosacral pains
13–15	Resonance of pharynx and pelvis
12–20	Head sensations
40–100	Decrease in visual acuity if eyeballs resonate

TABLE 2.20. CLASSIFICATION OF K-VALUES

K-value	Degree of human sensitivity	Effect on work
0·1	Threshold of perceptibility	Nil
0·1–0·3	Just perceptible; easily tolerable	Nil
0·3–1	Easily perceptible; tolerable; unpleasant for durations > 1 h	Nil
1–3	Very noticeable; very unpleasant if duration > 1 h	Can be distracting; $K = 0$–1 allowable in industry for any period of time
3–10	Unpleasant	Considerable interference with work task concentration; $K = 10$ only permissible for a short period of time
10–30	Very unpleasant if duration > 10 min	Work barely possible
30–100	Tolerable for about 1 min	Work impossible
> 100	Intolerable	$K = 100$ represents the upper limit of strain for the average man

K which represents the degree of strain resulting from various intensity levels of vibration. A classification of the K-values is given in Table 2.20.

Calculations of K-values may be made according to the relationships in Table 2.21.

A comprehensive account of the effects of vibration on man has been written by Guignard and Guignard (1970).

Figure 2.41a–c shows the Reiher–Meister sensitivity scale and the Dieckmann values.

The International Standards Organisation has recently recommended vibration tolerance limits and these are described by Coermann (1970) and Ashley (1971); a summary of the ISO recommendations on vibration exposure limits is illustrated in

TABLE 2.21. CALCULATIONS OF K-VALUES

Vertical vibration	Horizontal vibrations
<5 Hz, $K = 0.001Af^2$	<2 Hz, $K = 0.002Af^2$
5–40 Hz, $K = 0.005Af$	2–25 Hz, $K = 0.004Af$
>40 Hz, $K = 0.2A$	>25 Hz, $K = 0.1A$

A is the amplitude of the vibration in microns ($1\,\mu = 10^{-6}$ m).
f is the frequency of vibration (Hz).

Fig. 2.42a–c. It can be seen that the body is most sensitive to vertical vibration (expressed in g units) in the frequency range 4–8 Hz, whereas the peak sensitivity occurs at 1–2 Hz for the horizontal component. These criteria have been constructed on the fatigue decreased proficiency (FDP) concept.

2.4.4. THE PSYCHOLOGICAL RESPONSE OF PEOPLE TO NOISE
(INCLUDING SOME COMMUNITY NOISE CRITERIA)

> ... noise (in dwellings) had catastrophic effects not only on nervous fatigue but also on social relations between families and within one family.
>
> (P. CHOMBART DE LAUWE, Proceedings of the First International Council for Building Research (CIB) Congress, Rotterdam, 1959, Elsevier, 1961.)

In section 2.1.2 a model of comfort has been proposed which uses four principal dimensions—arousal level, physical sensitivity level, emotional sensitivity level and distraction level (see Fig. 2.8). Each of these dimensions varies from individual to individual and each is dependent on such factors as: the work task and the organisation associated with it; susceptibility to the environment; past experience of the event and hence expectancy; stimulus level, quality and meaningful information content. These factors interact with one another. A person with a high motivation for the task he is doing is perhaps less easily distracted and less dependent on the physical environment around him. Arousal level theory is discussed in sections 2.3

and 2.5, the physical sensitivity of people to noise has been reviewed in section 2.4.1 and the general problem of stress in section 2.3.

What remains to be discussed now are the emotional and distractive effects of noise. What do we mean by *annoyance*? Dictionaries give the following synonyms and antonyms for *annoy*.

Synonyms	Antonyms
tease	soothe
vex	conciliate
irritate	appease
disturb	regard
affront	quiet
molest	accommodate
pain	study
disquiet	tend
incommode	foster
tantalise	cherish
bother	smooth
weary	gratify
inconvenience	
plague	
discommode	
harass	
chafe	
trouble	

In a similar fashion, *disturb* can have the meaning of breaking a person's quiet, peace or tranquillity, and possible synonyms and antonyms may be:

Synonyms	Antonyms
annoy	soothe
disquiet	quiet
derange	pacify
discompose	order
disorder	collocate
discommode	arrange
plague	compose
confuse	leave
rouse	
agitate	
trouble	
interrupt	
incommode	
worry	
vex	
molest	

When we interchange the words disturbance and annoyance this is arguably permissible, although annoyance possibly has a broader spectrum of shades of meaning. Finding antonyms is more difficult. In Fig. 2.8 annoyance/pleasure has been adopted on the emotional sensitivity bipolar scale. Pleasure is not a direct opposite of annoyance but it does mean a state of satisfaction or acceptability; amongst a list of suggested antonyms for pleasure is pain, which is also a synonym for annoyance. Pain has been chosen for the physical sensitivity scale; in the environmental sense the opposite end of this scale is a state of no stimulus, i.e. sensory deprivation.

Kerrick *et al.* (1969) have demonstrated the importance of defining the terms used in subjective experiments. Figure 2.43 shows that judgements between artificial sounds, traffic and music were indistinguishable on the loudness–noisiness scale, but differentiation is quite clearly made in judgements on acceptability–noisiness scales. The type of sound or stimulus quality, and the meaningful information content are important factors in determining what is acceptable (or not annoying); stimulus intensity alone is insufficient evidence for attributing annoyance values to environmental factors.

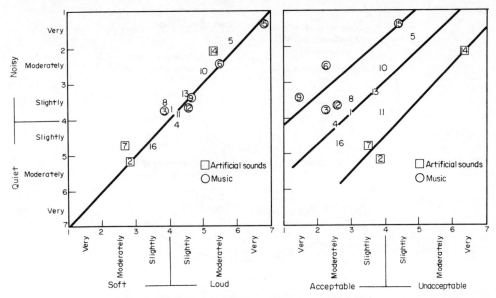

FIG. 2.43. Judgements of soft–loud and acceptable–unacceptable against noisiness: sound stimuli: (1) DC8 flyover, (2) octave-band centred at 1000 Hz, (3) Bernstein (jazz), (4) motorcycle passby, (5) helicopter flyover (plus 20 dB), (6) popular music, (7) shaped synthetic broadband noise, (8) auto passby, (9) folk music, (10) 720B flyover, (11) helicopter flyover, (12) Vivaldi (classical), (13) truck passby, (14) tone complex, (15) popular music (+20 dB), (16) rain (Kerrick *et al.*, 1969).

Distraction is taken to mean diverting attention or preventing concentration and should not be used as a synonym for annoyance or disturbance. The attention sphere model of comfort demonstrates distraction very clearly (see Fig. 2.7) but it is much more difficult to show the occurrence of annoyance. Again it is much easier to pinpoint a source of distraction but less easy to pinpoint the origin of annoyance. The origin of annoyance may be the person themselves, their state of being, or a combination of several causes. Noise criteria should be used which will ensure that annoyance, not only distraction, will not have its origin in the sound environment.

Noise has already been defined as sound, including non-audible low- and high-frequency mechanical vibrations, unwanted by a person, or sound which becomes a nuisance. Clearly music can be a nuisance as well as a benefit. The

difficulty of evaluating noise nuisance was discussed as early as 1934 by Bartlett and
more recently, for example, by Bottom and Croome (1969).

Early work, by Laird and Coye (1929) for instance, has demonstrated annoyance is
felt more for sounds of high frequency than ones of low frequency; annoyance is
more likely to be experienced when a person cannot locate the direction of a sound,
when the sound is intermittent and so on. A broader base for discussion is the idea of
annoyance shown in Fig. 2.44. Some of these aspects have been referred to already.

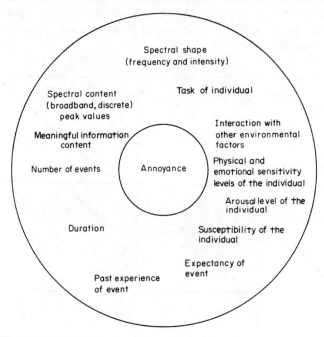

FIG. 2.44. Annoyance and dependent factors.

Nuisance value depends on the background against which noise is experienced. If
a man hears an unusual sound in his car he is bothered, whereas the slam of a door
signalling the exit of a bore may be gratifying. Wives of high-salaried air pilots may
welcome living near an airport; convenience of travel or schooling may make the
noise of an area a secondary consideration. The true harm of noise to mental health
can be disguised very easily by other things and is another good reason for
establishing a basic measure of noise response that is related to the nervous system.

Work by Bryan (1973), and Moreira and Bryan (1972) has attempted to unravel the
factors that feature in the subjective emotive response we call annoyance. Why is it
that some people seem to be insensitive and others extra sensitive to the
environment in which they live? Earlier in this chapter (section 2.1.2) comfort was
discussed and it was concluded that the personality has an important part to play in
the subjective response resulting from stimuli incident upon the human sensory
system. This is also one of the conclusions made by Moreira and Bryan (1972). In
their words they suggest that: "In order to make reliable prediction of annoyance of
an individual by noise it is necessary to take into account not only the noise level to

which he is exposed but also his personality." They exposed thirty-four subjects, reading material of their own choice for 45 min, to a random selection of street noise, aircraft noise and industrial noise at 55, 65, 75, 85 and 95 dBA sound levels. The data obtained from all the subjects was averaged and graphs were drawn of the mean annoyance score for the five different sound levels (Fig. 2.45a); the data was linearised by plotting the square root of the noise rating against noise level as shown in Fig. 2.45b. Notice that the judgements for the three kinds of noise source were not

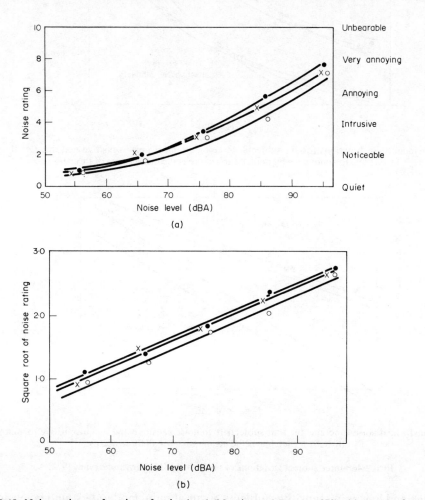

FIG. 2.45. Noise rating as function of noise level (Moreira and Bryan, 1972): (a) mean values for 34 subjects, for street (●———●), aircraft (○———○) and industrial noise (×———×), (b) linearised mean values for 34 subjects for street (●———●), aircraft (○———○) and industrial noise (×———×).

significantly different but the rating between subjects of the same noise varied considerably. The least noise-sensitive subjects could miss rating some of the quieter noises if absorbed in reading, whereas the most noise-sensitive found that even the quietest sounds intruded and distracted them. As an example of the intersubject variation look at Figs. 2.46a and b. It can be seen that the largest difference between

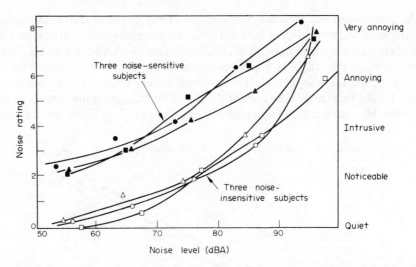

(a) Individual noise functions for six subjects (three of the most noise sensitive and three of the most insensitive to annoyance by noise); each curve is the mean rating of three noises.

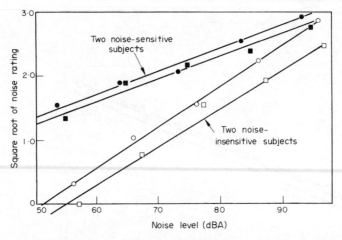

(b) Transformed noise functions for four subjects (two noise sensitive and two insensitive to annoyance by noise); each line is the mean rating for all three noises.

FIG. 2.46. Inter-subject variations of noise rating (Moreira and Bryan, 1972).

the high and low noise-sensitive subjects occur at quite moderate sound levels in the range 50–70 dBA.

In attempting to establish a crude character profile of noise-sensitive people from evidence collected in field surveys, Moreira and Bryan suggest that individuals with high noise susceptibility may be personality types that show a great interest in and have sympathy with others, have a great awareness of their environment and are likely to be intelligent and creative; their studies show that neuroticism is not consistently related to noise annoyance. Long ago the philosopher Schopenhauer (1890) wrote: "noise is torture to people of great intellect."

There was no correlation between factors like age, sex, educational level, job responsibility, home background sound level and noise susceptibility, although these experiments and surveys were confined to small selected groups of people. The noise rating scale used by Moreira and Bryan (1972) is not a pure annoyance scale because it mixes the words *annoying, intrusive* and *noticeable*, the last two words referring more to the distracting qualities of the noise; a person can be distracted but not necessarily annoyed.

What are the sources of noise that bother people in everyday life? Several surveys have been carried out over the years. Chapman (1948) questioned 2000 people about disturbance from the noise of neighbours. Some 80% of the people were aware of the noise produced in their own homes but only a quarter were disturbed by this noise; the commonest sources of noise were doors shutting and water systems. More people (41–57%) were about equally disturbed by the noise from neighbours and traffic, whereas in detached houses traffic noise was more troublesome.

The second British inquiry was conducted in 1952–3 by Gray *et al.* (1958); 1491 housewives in three groups of flats with different floor insulation values were questioned about the sounds they heard and any nuisance they experienced on account of them. As regards the most troublesome sounds this work agrees with that of Chapman (1948). Other findings were as follows.

(a) When the insulation value of the floors has reached a certain value, the location of sound in other quarters than in the adjoining houses are of almost equal importance. This agrees with the work of Bitter and van Weeren (1955), namely, that noise from upstairs neighbours is heard to a great extent than noise from downstairs neighbours and is heard least from side neighbours, although the latter also depends upon the house-planning scheme. Occupants of top flats complain less than occupants from bottom flats, whilst the latter complain less than those in half-way flats.

(b) Impact sounds were found to be at least as important as airborne sounds.

(c) The number of cases of annoyance caused by noise increases with the number of children in surrounding flats, but is mainly governed by the number of children of top-floor neighbours. The likelihood of finding a carpet in a dwelling decreases as the number of children increases.

(d) Quite a large number of persons were found willing to pay more rent in consideration of improved sound insulation.

Table 2.22 shows a comparison between the percentage of judgements in the surveys of Bitter and van Weeren (1955) and Gray (1958) rating the most disturbing sounds; in general the annoyance order was found to be similar in both studies.

Bitter and van Weeren (1955) suggested the following noise nuisance classifications shown in Table 2.23.

Clearly peoples' habits (i.e. the frequency and time of use) besides the quality of the sound will influence a person's annoyance assessment. The impulsive character of impact noise makes it nearly always troublesome; conversations have a high information content; radios can be a nuisance to people trying to sleep.

Bitter and van Weeren (1955) made a comparison between *sounds heard* and *annoying sounds* (Table 2.24). It is assumed, of course, that people know when

TABLE 2.22. PERCENTAGE OF PERSONS WHO
FIND A CERTAIN SOUND THE MOST ANNOYING

Sounds	Gray et al. (1958)	Bitter and van Weeren (1955)
Doors-slamming	21	18
Radio, television	13	11
Footsteps	10	5
Children	12	4
Plumbing noises	2	1
Adults' voices	5	5
Poking fire	2	0 (0·2)
Others	10	30[a]
Not disturbed (no most annoying sound)	25	26

[a]Vacuum cleaner, stamping, piano, dog, hammering, traffic sounds.

TABLE 2.23. CLASSIFICATION OF A NUMBER OF SOUNDS ACCORD-
ING TO THE NUISANCE VALUATION (BITTER AND VAN WEEREN,
1955)

Very annoying	Rather annoying	Generally not annoying
Doors-slamming	Children playing	Whistling kettle
Stamping on the floor	Radio	Plumbing noises
Sounds of things dropping	Talking	Vacuum cleaner
Walking	Walking	

TABLE 2.24. COMPARISON OF "PERCENTAGES HEARD" AND "PERCENTAGE ANNOYING" OF
UPSTAIRS AND DOWNSTAIRS NEIGHBOURS FOR SOME SOUNDS AS SHOWN BY THE ENGLISH
SURVEY (E) BY GRAY AND CARTWRIGHT (1958) AND THE DUTCH SURVEY (D) BY THE
RESEARCH INSTITUTE FOR PUBLIC HEALTH ENGINEERING TNO (BITTER AND WEEREN,
1955)

Sounds	Percentage of persons who hear a particular sound from upstairs neighbours		Percentage of persons who find a particular sound from upstairs neighbours annoying		Percentage of persons who hear a particular sound from downstairs neighbours		Percentage of persons who find a particular sound from downstairs neighbours annoying	
	E	D	E	D	E	D	E	D
Wireless or television	81	65	24	20	71	60	18	16
Grown-up's voice	78	58	12	12	70	52	9	11
Footsteps	89	79	27	25	—	—	—	—
Banging or hammering	81	64	42	38	—	—	—	—
Vacuum cleaning	36	74	9	12	18	40	2	5
Baby crying, child's voice	48	37	9	13	44	19	6	5
Total number of persons	1126	808	1126	808	1120	806	1120	806

sounds become annoying; the nervous system may be stressed *before* we can subjectively assess that we are. Notice that the differences between the results of the surveys are smaller for sounds judged to be annoying than those sounds heard.

It must be remembered that 16–25% of the respondents in these surveys were not disturbed by any noise. Table 2.25 summarises the replies indicating where people thought the sources of disturbing noise were located.

TABLE 2.25. PERCENTAGE OF PERSONS WHO LOCATE THE MOST ANNOYING SOUND AT CERTAIN PLACE OF ORIGIN, FOR THE ENGLISH SURVEY (E) (GRAY, 1958) AS WELL AS FOR THE DUTCH SURVEY (D) (BITTER AND VAN WEEREN, 1955)

Place of origin	E	D
Flat above	30	20
Flat below	11	8
Flat next door	4	6
Stairs	7	19
Somewhere else[a]	16	19
Not disturbed (no most annoying sound)	25	16
Don't know	7	2
	1491	1226

[a]Mainly more than one source, one of which is the flat above other neighbours + street and garden (14%), more than one place of origin (5%).

In 1961 and 1962 noise was measured at 540 points in London and 1400 people were asked for an opinion about it. The results are summarised in Table 2.26. Ten years ago noise was emerging as a significant factor in city people's lives and was more troublesome at home than when outdoors or at work; people's expectancy of the environment at home, and work or outside are very different.

Over the years the BBC programme Desert Island Discs has endeared itself to many listeners. In selecting music for a desert island, personalities are asked what would they be most glad to get away from, and countless times the replies have been the noise, the bustle and the scurry of life. If an Englishman's home is to remain his castle, he wants it to remain peaceful. Families need protection from the noise brought about by high density living and increased rates of flow in road and air traffic. People need a contrast between the rat-race occupational side of life and their out-of-work hours.

Since these measurements were made in 1961–2, things are much worse. The philosophy reiterated several times in this chapter is that even if adaptation to noise takes place, this ignores any stress in the nervous system which remains, stress that we as individuals may be unaware of because it can be masked by and interact with other stresses. It is worth noting the increase in disturbance due to external noise over the years 1948–61 shown in Table 2.27.

TABLE 2.26. MCKENNELL AND HUNT (1963) AND WILSON
(1963).

(a) Relation of noise to other factors

Chief cause of dissatisfaction	Percentage of those questioned
Public transport and services	14
Noise	11
The kind of people round about	11
The amount of traffic	11
Slums, dirt, smoke	10
Facilities for shopping and entertainment	7
Different reasons, or no reply	6
Not dissatisfied	30

(b) Noises which disturb people at home, outdoors and at work

Source of noise	Percentage of persons questioned who were disturbed		
	When home	When outdoors	When at place of work
Street traffic	36	20	7
Aircraft	9	4	1
Railway	5	1	4
Industry and building	4	—	4
Internal noises at home	4	—	4
Noise from neighbours	6	—	—
Children	9	3	—
Adults (talking)	10	2	2
Radio and television	7	1	1
Bells and sirens	3	1	1
Pets	3	—	—

TABLE 2.27. INCREASE IN DISTURBANCE DUE TO
NOISE (CHAPMAN, 1948; WILSON, 1963)

	Percentage response			
	External noise		Internal noise	
Response	1948	1961	1948	1961
Disturbed	23	50	19	14
Noise noticed but not disturbing	19	41	21	14

A survey on internal noise in 2593 flats was carried out in Sweden by Boalt (1965). More than 50% of those questioned named noise sources outside their flat disturbing, and yet neighbours' noise was found to be even more irritating than traffic noise; children shouting, doors slamming and water-flushing caused the most complaints.

An Open University study (Attenborough, 1973) gave qualitative evidence of a general relationship between noise annoyance and local environment. The results of the study based on 2000 responses from students aged 21–55 years living in twelve regions throughout England, Ireland, Scotland and Wales suggested the following conclusions.

(a) Children rate highly as an annoyance source to 26–50-year-old students when studying, but road traffic is the principal environmental source of noise especially for the age groups 21–25 and 55 who are less likely to have young children around them.

(b) Apart from children (22–39%), the ranking order and approximate ranges of percentage response for other most disturbing environmental noise sources was found to be road traffic (24–38%), aircraft (9–15%), lawn mowers (7–14%), railways (1–3%), industry (0·5–1·5%). Other miscellaneous sources accounted for a 12–18% range of response. The survey was carried out throughout the spring and the summer and may account for the prominence of lawn mowers in the responses.

(c) The percentage of people annoyed by traffic noise was larger for urban than for rural and suburban areas; aircraft noise was troublesome to a smaller number of people in urban and suburban regions than in rural ones. This suggests high-level background noise can reduce subjectively judged annoyance to other intermittent sources, although unconsciously high background noise levels may leave the nervous system in an undesirable state of tension.

(d) Students living within a 5 min walk of a motorway or trunk road feel much more strongly that noise is one of the biggest nuisances of our time than students living within a 10 min (or more) walk of a motorway.

(e) Cognitive abilities may be influenced by environmental noise.

The social problems of high rise living have received much attention in recent years. In a report called *The Social Effects of Living off the Ground* (HDD Occasional Paper 1/75) issued by the Department of the Environment, it is noted that in speaking of the advantages of living in high rise flats people often mention the lack of noise, especially the nuisance of children's noise, but aural privacy is linked with loneliness, and mothers strain to keep the children quiet more than in houses or flats at low level. Hird (1967) cites children's noise as a reason for some people wishing to move.

This study claims that 44% of tenants could hear noise from the flat above and 17% from the flat below. Sound insulation was often inadequate. Other sources of noise besides children at play were communal walk-ways, lift motor rooms, rubbish chutes, water and drainage systems, wind funnelling up between the blocks. Stewart (1970) found that several tenants in very high flats were convinced that sound outside travelled upwards and was more noticeable at higher levels.

On a Roehampton high rise development it was found that people were over-sensitive to noise because of isolation and a lack of visible stimuli (see *Municipal Journal*, 1969, 77, (38), September 19, 2367–2369). In a written report by the Housing Sectional Committee of the National Council of Women of Great Britain in 1969 and entitled 'Guidelines for happier living in high blocks' it is concluded that "tensions, anxiety and repression are suffered by flat dwellers because of their noisy surroundings".

Now the critical question arises: How can feelings of people be related to numerical objective measurements so that design criteria can be evolved? This is difficult because, as we have seen, the terms annoyance or disturbance, which we use to indicate when a stress becomes undesirable, are not pure. Annoyance is not confined to noise; it does not depend just on the event of the moment; it depends on personality so that it is difficult to find a reliable indicator of community response; attitudes change. The annoyance sphere shown in Fig. 2.44 attempts to assemble the principal parameters upon which annoyance depends: stimulus quality and information content; duration, past experience, expectancy and number of stimulus events; physical, emotional and arousal levels and other personality attributes of the individual; the activity of the individual whether sleeping, working or relaxing; interactions with other factors—all form a multi-dimensional array. This concept is a continuance of the comfort diagram shown in Fig. 2.8.

Unfortunately, no neat, crisp and tidy relationships are yet known between all these parameters, although the significance of the various ideas has already been discussed throughout this chapter. Designers cannot wait, so various criteria have evolved which attempt to correlate subjective and objective measurements; these will now be briefly discussed, and the section will end by looking at current research which questions their validity. Note should be made of a praiseworthy review of community noise ratings by Schultz (1972).

The simplest and most convenient indicator of subjective response is the dBA measure obtained using a sound level meter (Ford *et al.*, 1970). Various work is illustrated in Fig. 2.47 that attempts to achieve this but notice in particular three things:

* the subjective scales mix terms like acceptability, intrusiveness, distraction, annoyance and degree of noisiness;
* in most cases no indication is given of the percentage subjective response;
* no distinction is made between the mean and peak noise levels.

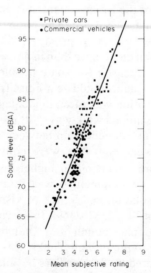

(a) Correlation of vehicle noise with subjective rating (Wilson, 1963).

FIG. 2.47. The evolution of dBA as a prediction of subjective response.

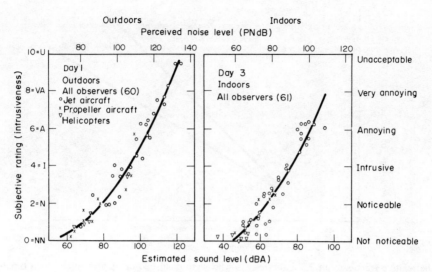

(b) Outdoor (left) and indoor (right) judgements of the category scale of intrusiveness plotted against sound level, dBA, and perceived noise level, PNdB (Robinson *et al.*, 1963a).

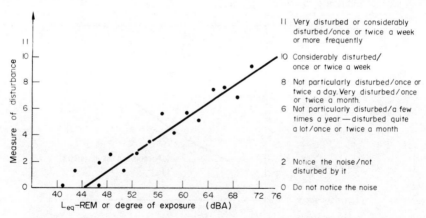

(c) The mean value of the measurement of disturbance as a function of degree of exposure i.e. the mean energy level corrected for distance and barriers (Fog *et al.*, 1968).

FIG. 2.47. (*Continued*).

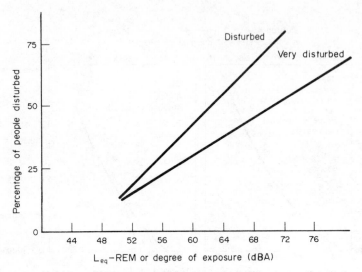

(d) The relative frequency of 722 interviewees who were disturbed or very disturbed as a function of degree of exposure i.e. the mean energy value for a 24 h period corrected for distance and barriers (Fog *et al.*, 1968).

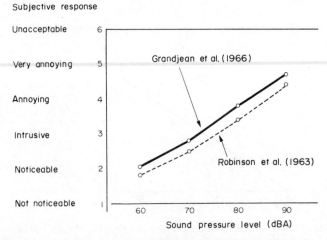

(e) Subjective response to aircraft noise for two situations—students listening to a lecture (Grandjean *et al.*, 1966) and workers watching a film during a meeting on aircraft (Robinson *et al.*, 1963).

FIG. 2.47. (*Continued*).

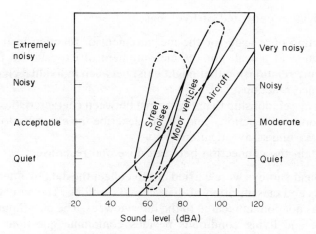

(f) Judgements of road, air traffic and street noise obtained in social survey in London (Wilson, 1963).

FIG. 2.47. (*Continued*).

In respect of Fig. 2.47e, Grandjean *et al.* (1966) inform us that 22% of their students listening to a lecture felt at least *distracted* by 60 dBA; 62% were similarly effected by 70 dBA and 89% by 80 dBA. The corresponding *very disturbing* figures for road traffic (Fig. 2.47d, as ascertained by Fog *et al.* (1968) are 30% at 60 dBA, 60% at 70 dBA and 18 dB (see also Fig. 2.48b).

An interesting point on Fig. 2.47b is the fact that subjects found the aircraft noise more intrusive when they were making judgements indoors than outdoors by some 18 dB (see also Fig. 2.49b).

The work of Fog *et al.* (1968) in Sweden is an example of a more comprehensive and more rigorous study carried out by a multidisciplinary team (an architect, a sociologist and an acoustician). Their investigation used the following model:

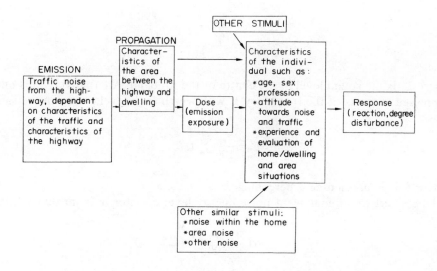

The investigation consisted of five phases:

1. construction of a scale for the measurement of reaction to traffic noise;
2. construction of a scale for the measurement of exposure to traffic noise;
3. test for any relationship that might exist between individual characteristics and disturbance reactions;
4. test for any relationship that might exist between characteristics of the situation and disturbance reactions, and also to describe the exposure to noise (existing) in the area under investigation;
5. to ascertain the connection between dose and response.

Extensive field surveys were carried out to collect the data by interviews, acoustic measurements and classification of district and dwelling. The actual purpose of the interviews was not communicated to the interviewees. The questionnaire inquired of general home and living conditions besides containing questions which allowed traffic noise to be mentioned spontaneously as a cause of lowered comfort.

Acoustic measurements were recorded in dBA, but from these the *mean energy value* was obtained. This idea of relating sound energy to noise load commenced in Germany and has led to the concept of *equivalent level of sustained noise* L_{eq}, which was designated as a level of moderate disturbance. In this way the disturbance from intermittent sources can be compared with that from continuous sources. The equivalent level of sustained noise L_{eq} is given by

$$L_{eq} = \frac{10}{\alpha} \lg \left(\sum_i \frac{t_i}{T} 10^{(\alpha L_i/10)} \right) \qquad (2.9)$$

where α is the equivalence constant, T is the total time of measurement, L_i is the noise level at the ith time interval, t_i is the duration of ith time interval, and $f_i = t_i/T$ is the frequency of occurrence of L_i.

The expression (2.9) can be derived by considering a sample of sound recorded over a time T. Divide this sample into equal time bands. For i bands each of t_i seconds the average sound level will be

$$L = \frac{\sum_i (L_i t_i)}{T},$$

where L_i is the sound level in the centre of the ith band. Alternatively, this may be expressed in terms of the frequency f_i, with which each level occurs in the sample, thus

$$L = \frac{1}{100} \sum L_i f_i,$$

where f_i is expressed as a percentage.

To convert to an equivalent noise energy level a factor α is introduced, thus

$$\alpha L = \frac{1}{100} \sum \alpha L_i f_i.$$

Averaging can be done by taking 10 lg (antilg peak L).*

The noise dose, or emission exposure, in the work of Fog *et al.* (1968) is given by the mean energy level for a 24 h period referred to as L_{eq} – REM, and is calculated by putting $\alpha = 1$ in equation (2.9) as shown in Fig. 2.47c and d. This was found to relate very well (correlation coefficient = 0·91) to a disturbance scale.

The scale used for the measurement of disturbance was based on the stated occurrence of subjective disturbance by traffic noise and the intensity and frequency of these disturbances. Fourteen different combinations based on these variables were made and the disturbing effect that they had was judged by eighty subjects who made paired comparisons of the different combinations. Afterwards the intervals between these combinations were estimated by forty-seven other persons.

This scale of response was then tested by relating it to the other expressions of disturbance which were given. It turned out to have a connection with physical symptoms given (e.g. headache), the occurrence of disturbance of activity (e.g. not able to listen to the radio), how well the area of residence was liked, spontaneous complaints about traffic noise and steps taken or planned to reduce the effect of this noise. The scale of disturbance which was constructed is therefore a summing up of several different expressions of inconvenience caused by traffic noise.

Sex, age and civil status did not seem to vary in connection with the disposition for being disturbed, a fact already brought to our attention by Moreira and Bryan (1972), but there was found to be a connection between the attitude towards traffic noise and noise disturbance. The more disturbed subjects had more negative attitudes towards traffic noise than those less disturbed; the more disturbed subjects also considered themselves more sensitive to noise. Figure 2.47d shows that for exposure on the other side of the windows with a mean energy value for a 24 h period of more than 55 dBA (traffic noise alone) the number of those very disturbed exceeds 20%. This corresponds to an L_{eq} – REM of 45–50 dBA in the room with the window open, or 30–35 dBA if the window was closed.

Figure 2.48 summarises the results of further studies on subjective responses to road and air traffic (Fig. 2.48a and b), and the background noise of the ventilation system, the lighting system and the corridor circulation in the case of Fig. 2.48c. The work of Andrews and Finch (1957) shown on Fig. 2.48a was carried out in a laboratory with recordings of traffic noise, and it can be argued that the results are not realistic. Throughout the studies again notice that the subjective scales use mixed semantics. The studies were carried out in different ways on people living in Europe and America. Figure 2.48b shows again in common with Fig. 2.47b that people indoors require a level of noise that is lower than that which is found acceptable outdoors.

Figure 2.48c shows the importance of interpreting noisiness scales in terms of satisfaction or acceptability if sensible building design criteria are to be established. Scales of warmth, noise or brightness are no good, if for a given task designers do not know what people mean by word descriptions such as quiet, noisy and so on.

Noise fluctuates over time. It has become common practice to distinguish between

---

*To take the mean sound level $\bar{L}$ for N events with levels at L remember that if $L = 10 \lg (W/W_0)$, then $(W/W_0) = 10^{L/10}$ and hence on average is

$$\frac{1}{N} \sum \frac{W}{W_0} = \frac{1}{N} \sum 10^{L/10} \quad \text{or} \quad \bar{L} = 10 \lg \left(\frac{1}{N} \sum 10^{L/10} \right) = 10 \lg \left(\frac{1}{N} \sum \frac{W}{W_0} \right).$$

background noise, defined in this country by the L_{90} level, as the noise level exceeded for 90% of the time, and the peak level, defined by the L_{10} level, as the noise level exceeded for 10% of the time. If the probability distribution of noise levels is approximately Gaussian, the L_{10} and L_{90} levels may be read from a plot of the cumulative distribution of levels (Fig. 2.49). Some countries define the peak level by the L_1 level, but clearly this is more difficult to measure, and any standards formulated using this level may be more expensive to achieve in practice. If, however, stress is found to critically depend on the peak level, it will be necessary to adopt this measure. Work by Keighley (1966) has shown that the fluctuating nature of noise (i.e. the peakiness of the spectrum) can be correlated with annoyance using the *transient peak index* (TPI). Keighley (1965) and Langdon (1966) describe the empirical evolution of the TPI as a means of scaling noise annoyance in offices. From the survey carried out in twelve offices with some 1200 respondents it was concluded that the subjective response to noise was dependent on the average noise

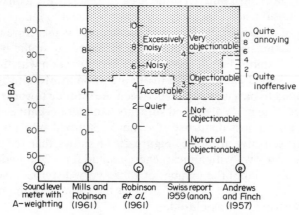

(a) Comparison of subjective and sound-level scales for motor-vehicle noise (Robinson *et al.*, 1961; Mills and Robinson, 1961).

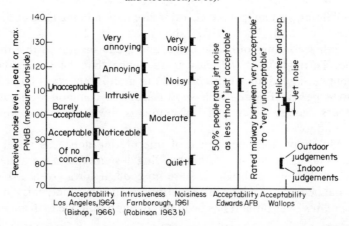

(b) Comparison between perceived noise level of aircraft flyovers and category scales of acceptability, intrusiveness and noisiness (Bishop, 1966; Robinson *et al.*, 1963; Kryter *et al.*, 1970).

FIG. 2.48. Subjective and objective noise scales.

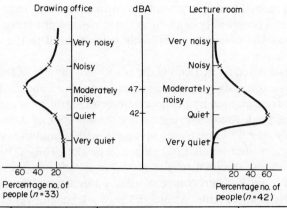

	Very unsatisfactory	Unsatisfactory	Satisfactory
Drawing office	1 person	12 people	20 people
Lecture room	0	4	38

(c) Correlation of noise level with subjective scales of noisiness and satisfaction (Francis, 1969; Croome, 1975).

FIG. 2.48. (*Continued*).

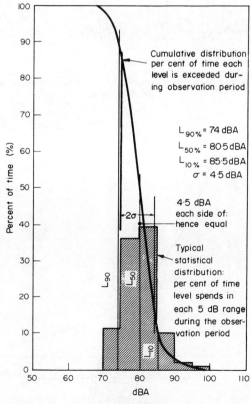

FIG. 2.49. Typical statistical distribution in form of histogram and cumulative distribution of A-weighted noise levels.

level, the degree of momentary fluctuation in level, and individual differences in noise tolerance. Acceptibility of the noise climate was governed by all three of these factors, whereas the noisiness rating was more related to the sound level and the individual's tolerance level.

The TPI is estimated by inspecting the peaks in a sample of the office noise. In Fig. 2.50a the peaks exceeding the average level by 5, 10, 15 and 20 dB are counted; thus the high peaks are weighted by being counted cumulatively at each 5 dB step, and the final TPI is the peak totals averaged over the number of noise samples recorded. Understandably, the TPI correlates inversely with sound pressure level because at high background sound levels the peaks are masked (Fig. 2.50b).

Figure 2.51 shows the final results of this work. Sound pressure level is plotted against the TPI and the percentage of office populations in each room finding the noise conditions unacceptable. Although this work has not yet been used extensively it remains important because it shows that the *quality* of the noise and not just the

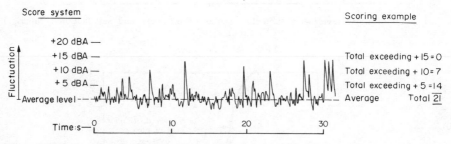

(a) Illustration of procedure for obtaining transient peak index from level recorder trace of office noise (Langdon, 1966).

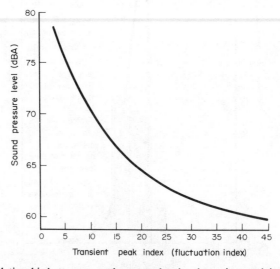

(b) Plot showing relationship between sound pressure level and transient peak index (Langdon, 1966).

FIG. 2.50. Transient noise index concept.

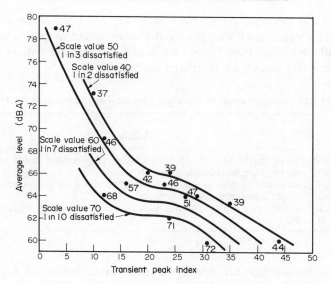

FIG. 2.51. Contours of equal acceptability fitted to data for 12 offices (Langdon, 1966).

level of it is important; *individual differences* have to be accounted for in rating people's acceptability of the environment, and, finally, the experimental method used for correlating subjective with physical measurements sets an example for future workers.

Hay (1973) has related the noise disturbance in landscaped offices to the difference between the L_{10} and L_{90} levels (see section 2.6.3). Griffiths and Langdon (1968) attempted to relate physical measurements of traffic noise to overall dissatisfaction and found that noise conditions correlate with a weighted combination of mean sound levels (dBA) exceeded for 90% (L_{90}) and 10% (L_{10}) of the sampling period and established the *traffic noise index* as (TNI)

$$TNI = 4(L_{10} - L_{90}) + L_{90} - 30. \tag{2.9a}$$

It is unlikely that this index will be used as a standard but it is interesting to take note of the psychosociological studies made to evolve it and to see how the (L_{10}–L_{90}) differential figures in relation to the L_{90} level as predictors of community dissatisfaction with the noise environment.

The scale of *equivalent sound level*, L_{eq}, is recommended by the International Organisation for standardisation (ISO Recommendation R 1996, 1971) and is alternatively referred to as *equivalent energy level* or *means energy level*. L_{eq} can be regarded as a notional sound level which would cause the same A level weighted sound energy to be received as that due to the actual sound over a period of time.

$$L_{eq} = 10 \lg \left\{ \frac{1}{100} \sum f_i \, 10^{L_i/10} \right\}.$$

For traffic noise Grandjean (1973b) gives the following relationship between L_{eq}, L_{50} and L_1 levels

$$L_{eq} = L_{50} + 0.43(L_1 - L_{50}). \tag{2.9b}$$

In Great Britain the average 10% level L_{10}, expressed in dBA for each hour between

6.00 a.m. and 12.00 midnight on a normal weekday has been recommended by the Noise Advisory Council and adopted by the government as giving a satisfactory correlation with dissatisfaction. Some values of L_{10} (18 h) and typical conditions in which they are experienced are tabulated in Table 2.28.

TABLE 2.28. NOISE ADVISORY COUNCIL LEAFLET—*a guide to noise units*, 1973

L_{10} (18 hr) dBA	Situation
80	At 60 ft (18 m) from the edge of a busy motorway carrying many heavy vehicles, average traffic speed 60 mph; intervening ground is grass
70	At 60 ft (18 m) from the edge of a busy main road through a residential area, average traffic speed 30 mph; intervening ground is paved
60	On a residential road parallel to a busy main road and screened by houses from the main road traffic

Criticisms of this criterion are, firstly, that in urban situations the L_{10} value can be very high due to traffic accelerating, decelerating and braking, whereas most data refers to freely moving traffic (e.g. the author has measured peak noise levels of 92 dBA at 20 m from lorries braking at traffic lights), and, secondly, people may have the quality of their sleep disturbed, but not necessarily be woken up, by night traffic. In Switzerland the emphasis for traffic noise criteria is placed on the desire to minimise sleep disturbance, and the 1% level L_1 is being advocated as a standard, which is even more difficult and expensive to measure than L_{10}, but it may be wiser to use such a standard (see Alexandre *et al.*, 1975 for a complete discussion on Road Traffic Noise).[†]

What do these various levels mean in terms of the number of people disturbed? Grandjean (1973b) gives the data shown in Table 2.29.

TABLE 2.29. PERCENTAGE OF PEOPLE CLEARLY DISTURBED BY ROAD TRAFFIC NOISE AT THE LEVELS STATED (GRANDJEAN, 1973b)

Country	Sound level	Percentage of people disturbed
Austria	*L_{eq} = 45–50 dBA[(a)]	30
Sweden	L_{eq} = 50–55 dBA	15–25
France	L_{50} = 60–65 dBA	Sharp increase
England	L_{50} = 50–68 dBA	40

[(a)]$L_{eq} = L_{50} + 0\cdot43(L_1 - L_{50})$.

The first step taken to relate the physical attributes of aircraft noise to subjective reaction was made by Kryter (1959). Using a similar basis to the Stevens method of computing loudness (see section 2.2.1) a scale of perceived noise level (PNL), measured in *perceived noise decibels (PNdB)*, was derived which gives more weight to the higher frequencies characteristic of aircraft noise. Kryter constructed a scale of noisiness, similar to the sone scale of loudness, using a unit called the *noy*, defined

[†]Also see Delany *et al.*, 1976, *J. Sound Vibrn.*, **48**, (3), 305–25.

as the noisiness in the 500–1000 Hz octave band at a sound pressure level of 40 dB (re 20 μPa). The PNdB level is derived as follows. From Fig. 2.52 read off the noy value for each octave band sound pressure level and calculate the total perceived noisiness N from the expression

$$N = n_{max} + 0.3\left(\sum n - n_{max}\right), \tag{2.10}$$

where n_{max} is the highest noy value and Σn is the sum of the noy values in all the octave bands.

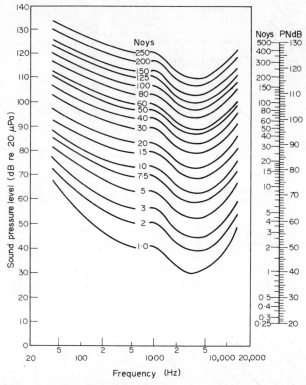

FIG. 2.52. Equal "noisiness" contours.

Convert N into perceived noise level using the expression

$$\text{PNdB} = 40 + 10 \lg_2 N. \tag{2.11}$$

The precise numerical difference between PNdB and dBA varies depending on the sound spectrum, but a rough guide is PNdB $\simeq$ dBA + 13. Stevens' (1972a, b) Mark VII loudness level estimation and the consequent perceived loudness concept were discussed in section 2.2; approximate equivalences between PNdB and PLdB were also given.

In 1968 the International Standards Organisation adjusted the PNdB levels to *effective perceived noise levels in EPNdB* so as to include the subjective influence of discrete tones and the duration of each higher noise level (also refer to their

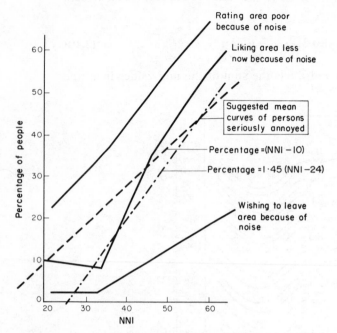

(a) Percentage of people annoyed by a given NNI as elicited from various questions asked (Richards, 1969).

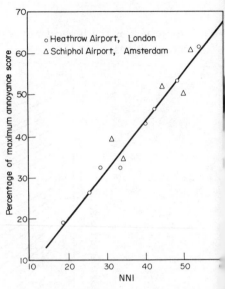

(b) Relationship between average annoyance response and NNI (Ollerhead, 1973).

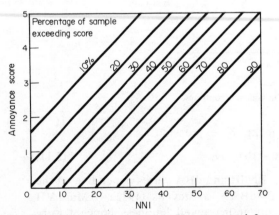

(c) Distribution of annoyance scores: 1, not annoyed; 2, a little annoyed; 3, moderately annoyed; 4, very annoyed; 5, severely annoyed; 6, difficult to tolerate (Ollerhead, 1973).

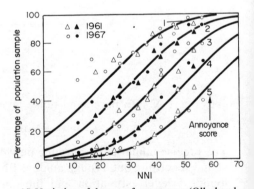

(d) Variation of degree of annoyance (Ollerhead, 1973).

FIG. 2.53. Noise and number index.

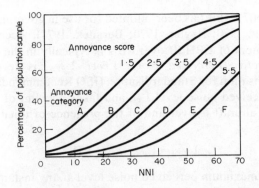

Annoyance category	Feelings about aircraft noise
A	Not annoyed; practically unaware of aircraft noise
B	A little annoyed; occasionally disturbed
C	Moderately annoyed; disturbed by vibration; interference with conversation and TV/radio sound; may be awakened at night
D	Very annoyed; considers area poor because of aircraft noise; is sometimes startled and awakened at night
E	Severely annoyed; finds rest and relaxation disturbed and is prevented from going to sleep, considers aircraft noise to be the major disadvantage to the area
F	Finds noise difficult to tolerate; suffers severe disturbance; feels like moving away because of aircraft noise and is likely to complain

(e) Classification of feelings about aircraft noise as a function of noise exposure (Ollerhead, 1973).

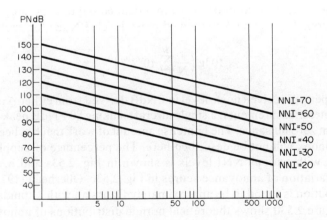

(f) PNdB and NNI equivalents: $NNI = L + 15 \lg N - 80$, where L is the average peak noise level in PNdB and N is the number of aircraft landings and take-offs over a specified time period (Grandjean et al., 1969).

FIG. 2.53. (Continued).

publication ISO/R-507). This has been adopted for use in government certification of aircraft (see Schultz, 1970; Kryter, 1970; Beranek, 1971). According to Ollerhead (1973) the improvement of EPNdB over dBA is typically in the order of only 1 dB. The scale of *Effective Perceived Noise Level*, L_{EPN}, is recommended by the International Organisation for Standardisation (ISO Recommendation 507, 1970) and is based on the perceived noise level L_{PN} and takes account of the duration of the noisiest part of the aircraft flyover and of the presence of tonal components in the noise spectrum.

$$L_{EPN} = L_{PN\,max} + C + D$$

where $L_{PN\,max}$ is the maximum perceived noise level at any instant of time during the flyover; C is a correction for discrete frequency components and D is a duration allowance, generally depending on the time interval during which the instantaneous value of the perceived noise level is within a specified value (not less than 10 dB) of the maximum value. C is calculated by reference to a normalising constant chosen to be 10 seconds. L_{EPN} has been adopted by the International Civil Aviation Organisation as the standard for noise evaluation measures used in aircraft noise certification.

The index *Weighted Equivalent Continuous Perceived Noise Level*, has the status of International Civil Aviation Organisation 'Recommended Practice' for international use in assessing the total noise environment from a succession of aircraft. The index is based on a scale of equivalent continuous perceived noise level, which is defined (as its name implies) similarly to L_{eq} but based on L_{PN} rather than on L_A. Differential weightings are applied for noise events which occur during the day, evening and night and also for events which occur during hotter months of the year when it can be assumed that people will be out of doors or will leave windows open.

The Wilson Committee (1963) introduced the *noise and number index (NNI)*.

$$NNI = 10\,lg\left(\frac{1}{N}\sum_1^N 10^{(L/10)}\right) + 15\,lg\,N - 80, \tag{2.12}$$

where N is the number of aircraft heard in a defined daytime period (usually 06.00 until 18.00 hours), L is the peak perceived noise level (PNdB) for each aircraft and

$$10\,lg\left(\frac{1}{N}\sum_1^N 10^{(L/10)}\right)$$

is the average peak perceived noise level in PNdB derived from either A or D weighted sound level meter measurements; a single aircraft making 80 PNdB peak noise level is taken to present zero nuisance. The immense amount of work that has been carried out to derive this index will not be described here. The percentage of people expressing dissatisfaction with various NNI levels is shown in Fig. 2.53a and b, but note the considerable variation of annoyance scores in Fig. 2.53c. Ollerhead (1973) comments that the distribution is found to be approximately Gaussian with a standard deviation of 20 NNI. Figure 2.53d shows theoretical normal distributions of annoyance scores corresponding to Fig. 2.53c.

A score of 3·5 was found to be a critical value; above this, aircraft noise was seriously detrimental to living. The curves in Fig. 2.53e have been derived by Ollerhead (1973) by shifting those curves in (d) by half a point based on a number of surveys carried out in Europe and America. NNI and their PNdB level equivalents are shown in Fig. 2.53f.

Ollerhead (1973) states the following general conclusions:

(a) for every 10 NNI increase in noise exposure the number of people expressing any particular degree of annoyance is increased by around 20% of the total number exposed;

(b) accepting a score of 3·5 as critical, the number of people who are seriously annoyed at any exposure level is approximately 2(NNI − 20) per cent of the total;

(c) a change in exposure equivalent of about 10 NNI is required to cause a "category change" in annoyance; for example from "a little annoyed" to "moderately annoyed" or to be annoyed by a further kind of disturbance; roughly 20% of people fall into each category at any exposure level; it is interesting that a change of 10 NNI (10 dB) corresponds to a doubling of loudness which implies that noise reductions of this order are required to produce a noticeable improvement in the noise climate and so to achieve a significant improvement, reductions of about 20 NNI are required.

The NNI concept has now been adopted by the government, and hence planning authorities, as an index of disturbance from aircraft noise at commercial airports. For planning purposes long-term average values of daytime NNI during the peak summer period (mid-June to mid-September) are used. Some values of NNI and an indication of the conditions associated with them are listed in Table 2.30. It must be remembered that even at NNI 35 at least 20% of the population are likely to be very annoyed (Large, 1971, 1972, 1973; Ollerhead, 1973).

Ollerhead (1973) describes some more interesting facts that are important for planners to bear in mind. There is considerably more annoyance in areas classified below NNI 35 except for the very highest annoyance category. This is because although the fraction of people affected decreases at lower noise levels, the total number of people exposed increases. The usual neglect by planners of areas below NNI 35 gives a very misleading impression of the problem. Ollerhead (1973) refers to work which shows that for each 10 NNI reduction the number of people seriously affected is reduced by about 50%.

Variants of the NNI concept have emerged in other countries. In Austria, Bruckmayer and Lang (1967) derived the $\bar{Q}$ function as an index of disturbance due to traffic noise. This is calculated by putting $\alpha = 0.75$ in equation (2.9). Thus

$$\bar{Q} = \frac{40}{3} \lg \sum_i \frac{1}{T} 10^{(3L_i/40)} t_i$$

or

$$\bar{Q} = \frac{40}{3} \lg \frac{1}{100} \sum 10^{(3L_i/40)} f_i.$$

Data were collected by making acoustic measurements indoors and outdoors and by questioning 400 people living or working in homes, offices and schools in Vienna. The difference between sound levels measured outside and those indoors was found to be 5–7 dB when windows were open and 10–20 dB when normal windows were

TABLE 2.30. NNI LIMITING VALUES FOR REGIONAL PLANNING
NEAR AIRPORTS
(Noise Advisory Council Leaflet, *A Guide to Noise Units*, 1973; and
Grandjean *et al.*, 1969)

NNI	Typical conditions
60	Close to airports; many over-flights at low altitude
	Noise levels interfere with sleep and conversation even in some sound-insulated dwellings
45	Mainly occurs near busy routes from airports
	Many aircraft are heard at noise levels which can interfere with conversation in ordinary houses
35	Overflying is irregular
	Noise levels are noticeable and will be intrusive within ordinary houses with regular overflights
10	Occasional aircraft flying overhead producing noise levels which cause disturbance out of doors but not within houses
Above 50	Industrial building with special attention to noise insulation and airconditioning in offices and administration buildings; warehouses; areas for military or agricultural use
41–50	Business premises with special sound insulation; industrial building
36–40	Mixed zone (industry, factories and some dwellings)
25–35	Residential areas with a reasonable amount of aircraft noise; schools; hospitals if they have special sound insulation
Below 25	Quiet areas for convalescence; hospitals without special sound insulation

closed. These differences depend on the amount of glazing in the building structure, on the room absorption and on the type of glazing. People were more disturbed at night than during the day and became more sensitive to sound when the windows were closed. Some of the results are summarised in Table 2.31.

In France data on road traffic noise has been collected notably by Lamure and Auzou (1964, 1966) and Lamure and Bacelon (1967): 420 people were questioned, all of whom lived within 10–150 m of a motorway. There was a good correlation between annoyance and the L_{50} level. A principal conclusion was that the noise levels in front of the buildings in the range 60–65 dBA for 50% of the time corresponded to critical levels; a rapid increase in the number of people disturbed occurred above these levels. As regards aircraft noise, La Commission du Bruit du Ministère des Affaires Sociales have recommended the use of an index R defined as

$$R = PNdB \text{ (average peak level)} + 10 \lg \frac{N}{2500}$$
$$= PNdB + 10 \lg N - 34,$$

where N is the number of aircraft landings and take-offs for 24 h.

Lundberg (1963) has described an *equivalent daytime disturbance number* (EDD)

TABLE 2.31. EQUIVALENT LEVEL OF SUSTAINED NOISE $\bar{Q}$ AND FREQUENCY OF REPORTS OF "VERY DISTURBING" AND "UNBEARABLY DISTURBING" (BRUCKMAYER AND LANG, 1967)

| $\bar{Q}$ in dBA (based on indoor sound levels) | Percentage of persons who were "very disturbed" and "unbearably disturbed" | | | |
| | Windows open | | Windows shut | |
	Day	Night	Day	Night
20–25	—	—	—	0
25–30	—	—	0	24
30–35	—	0	9	42
35–40	0	36	9	52
40–45	14	56	15	53
45–50	30	70	40	—
50–55	47	76	—	—
55–60	60	76	—	—
60–65	68	—	—	—
65–70	70	—	—	—

EQUIVALENT LEVEL OF SUSTAINED NOISE IN VARIOUS TYPES OF ROAD, AND THE FREQUENCY WITH WHICH THEY WERE JUDGED "VERY DISTURBING" TO "UNBEARABLY DISTURBING" (BRUCKMAYER AND LANG, 1967)

| Type of road | Windows open | | | | Windows shut | | | |
| | Day | | Evening and night | | Day | | Evening and night | |
	$\bar{Q}$ dBA	Percent disturbed	$\bar{Q}$ dBA	Percent disturbed	$\bar{Q}$ dBA	Percent disturbed	$\bar{Q}$ dBA	Percent disturbed
Residential	41–44	14	33–38	36	26–32	9	24–28	24
Side road	48–51	30	41–45	56	39–29	9	23–31	24
Main road	59–63	68	53–56	76	40–49	40	37–40	52
Side road in town centre	54–59	60	—	—	36–41	9	—	—
Main road in town centre	58–69	70	—	—	41–47	40	—	—

estimated from the peak levels and their frequency of occurrence. Swedish research has suggested the following maximum values:

EDD per year	Critical dBA level
500	95
1500	90
5000	85
15 000	80
50 000	75

For an EDD of 50 000 it is reckoned that 20% of people will feel very disturbed.

Murray and Piesse (in Wilson, 1963) have discussed the use in Australia of an *annoyance index* (AI) defined as:

$$AI = 10 \lg \sum 10^{L/10},$$

where L is a peak level in PNdB.

In America the *composite noise rating* (CNR) has been advocated for use in dealing with aircraft noise problems. The use of the CNR is described in great detail by Kryter (1970). This index can be defined in the form

$$CNR = PNdB + 10 \lg N - 12.$$

Notice the similarity in the expressions for the NNI as used in Great Britain, R as used in France and the American CNR basis. Table 2.32 is taken from Grandjean *et al.* (1969) and gives comparison figures for the three methods of assessment.

The *Noise Exposure Forecast (NEF)* also has its origins in America (Gallaway and Bishop, 1970; Kryter, 1970; Beranek, 1971). For a specific class of aircraft i on a flight path j, the NEF_{ij} is expressed as

$$NEF_{ij} = EPN_{ij} + 10 \lg \left(\frac{N_{D,ij}}{K_D} + \frac{N_{N,ij}}{K_N} \right) - C,$$

where EPN_{ij} = effective perceived noise level (see p. 135) produced at a point in the neighbourhood by an aircraft class i following a flight path j;

$N_{D,ij}$, $N_{N,ij}$ = numbers of landings and take-offs (7.00–22.00) and night-time (22.00–7.00);

$K_D = 20$, $K_N = 1{\cdot}2$, $C = 75$ — the choice of these constants indicates that one night flight contributes as much to the NEF as seventeen day flights; choice of C depends on establishing a range of NEF numbers which will not be confused with other composite noise ratings.

The total NEF at a given position is found by summing all the individual NEF_{ij} values on an energy basis thus:

$$NEF = 10 \lg \sum_i \sum_j \text{antilg} \left(\frac{NEF_{ij}}{10} \right).$$

The reaction of people to railway noise has been studied by Gilbert (1973) in France. He made a daytime survey of 350 people living in twenty locations within

TABLE 2.32. COMPARISON FIGURES FOR AIRCRAFT NOISE CALCULATION METHODS (GRANDJEAN et al., 1969)

N	80 PNdB			90 PNdB			100 PNdB			110 PNdB			120 PNdB		
	NNI	R	CNR	NNI	R	CNR	NNI	R	CNR	NNI	R	CNR	NNI	R	CNR
1	0	46	68	10	56	78	20	66	88	30	76	98	40	86	108
10	15	56	78	25	66	88	35	76	98	45	86	108	55	96	118
50	26	63	85	36	73	95	46	83	105	56	93	115	66	103	125
100	30	66	88	40	76	98	50	86	108	60	96	118	70	106	128
200	35	69	91	45	79	101	55	89	111	65	99	121	75	109	131
500	41	73	95	51	83	105	61	93	115	71	103	125	81	113	135

N = number of aircraft landings and take-offs, PNdB = peak noise level.

143

25 km of Paris. The percentage of people annoyed by train noise A was found to be given by

$$A = 2 \cdot 3(L_{eq} + 4N + 2B + 4T - 4Q) - 120,$$

where N is the ratio of number of rooms exposed to noise to the total number of rooms in the dwelling, B is a noise animosity factor, T is a railways animosity factor, Q is a neighbourhood satisfaction factor; B, T and Q may each take the values of 0, 1 or 2. If $N = 1/3$, then:

$$A \simeq 2 \cdot 3 L_{eq} - 107 \quad \text{if} \quad B = T = Q = 2,$$
$$A \simeq 2 \cdot 3 L_{eq} - 112 \quad \text{if} \quad B = T = Q = 1,$$
$$A \simeq 2 \cdot 3 L_{eq} - 117 \quad \text{if} \quad B = T = Q = 0.$$

Notice that depending on the type of person (i.e. in this case measured by the B, T, Q profile) annoyance increases rapidly above L_{eq} values of $(107/2 \cdot 3)$ dBA, $(112/2 \cdot 3)$ dBA, or $(117/2 \cdot 3)$ dBA; each increase in L_{eq} of 5 dBA will mean a further $11 \cdot 5\%$ of people are annoyed.

The level of noise emitted from industrial premises is currently assessed using the *corrected noise level* (CNL in dBA) described in British Standard BS 4142: 1967. The CNL consists of a basic level to which a correction has been added to allow for the tonal character (whine, hum, etc.), the impulsive character (bangs, clanks), the intermittency and the duration of the noise. Some values of the CNL and the situations in which they were found are given in Table 2.33.

TABLE 2.33. NOISE ADVISORY COUNCIL LEAFLET—*A Guide to Noise Units, 1973*

CNL (dBA)	Situation
85	At 30 ft (9 m) from a building housing an air-stream drop forge hammer
75	At 30 ft (9 m) from an air compressor housed in a building with louvred doors
60	At 50 ft (14 m) from a can-making factory. Continuous noise from stamping machinery and from handling of thin sheet metal

Gilgen (1970)[†] has proposed that international standards should be agreed for various types of noise—road traffic, aircraft, railways, industrial noise and so on. The disadvantage of this is that in practice various sources of noise operate simultaneously. Table 2.34 summarises the provisional standard limiting values of noise levels that have been worked out in Switzerland and other European countries. Notice there are differences. Austria makes more stringent recommendations about its noise climate than other countries. The present proposed standard in this country for residential areas is about equivalent to the Swiss recommendations for mixed zones.

Robinson (1969, 1971) has derived a noise pollution level (L_{NP}) which combines the traffic and aircraft noise components

$$L_{NP} = L_{eq} + 2 \cdot 56\delta,$$

[†] Also see *Environmental Factors in Urban Planning* by Grandjean and Gilgen, 1976 (Taylor & Francis).

TABLE 2.34. PROVISIONAL SWISS LIMITING VALUES FOR EXTERNAL NOISE-EMISSION MEASURED AT AN OPEN WINDOW; THE DESIRABLE LEVEL IS 10 dB BELOW THIS, WITH A MINIMUM OF 30 dBA (GRANDJEAN, 1973a)

Classification of noise zone	Background noise (L_{50}) (dBA)		Frequent peaks (L_{10}) (dBA)		Rare peaks (L_1) (dBA)	
	Night	Day	Night	Day	Night	Day
Quiet area, convalescence	35	45	45	50	55	55
Quiet residential area	45	55	55	65	65	70
Mixed zone	45	60	55	70	65	75
Commercial area	50	60	60	70	65	75
Industrial area	55	65	60	75	70	80
Main traffic artery	60	70	70	80	80	90

RECOMMENDED EXTERNAL SOUND LEVELS IN FRONT OF WINDOWS OF DWELLINGS IN VARIOUS COUNTRIES (GRANDJEAN, 1973b)

Country	Zone classification	Sound level $L_{eq}[= L_{50} + 0 \cdot 43(L_1 - L_{50})]$ Day (dBA)	Night (dBA)
Sweden	All residential areas	55	45
West Germany	All residential areas	55	40
	Mixed zones	60	45
Austria	Residential areas	46–51	36–43
	Mixed zones	56	50
Switzerland	Quiet residential areas	60	50
	Mixed zones	60	48

where L_{eq} is the energy mean noise level $= 10 \lg \frac{1}{100} \Sigma\, 10^{L_i/10} f_i$, L_i is the median sound level of the ith 5 dBA interval, f_i is the percentage time that a sound level is on in the ith interval, and δ is the standard deviation of the instantaneous sound level considered as a statistical time series over the same specified period.

To date this is the only noise rating method which takes into account the time and amplitude fluctuations of the noise emission from road traffic, aircraft and buildings. As yet no criteria have been advocated in terms of L_{NP} mainly because there are few studies that have correlated it with subjective reaction. In an investigation by Bottom (1970) he concluded, from empirical evidence collected around Heathrow Airport, that: "L_{NP} is a good predictor of median, general dissatisfaction for existing communities and existing noises. The results might be very different if either new noises or new populations or both were to be considered."

In order to find L_{eq} and δ, noise measurements are necessary which will enable a histogram and a cumulative distribution of the A-weighted levels to be plotted as shown in Fig. 2.49. If the probability distribution of the noise levels is approximately Gaussian, L_{10}, L_{50}, L_{90} and δ may be read from a plot of the cumulative distribution and the noise pollution level calculated from

$$L_{NP} = L_{eq} + (L_{10} - L_{90}) \quad \text{in dB(NP)} \tag{2.13}$$

or

$$L_{NP} = L_{50} + (L_{10} - L_{90}) + \frac{(L_{10} - L_{90})^2}{60}. \qquad (2.14)$$

If the A-level readings of the noise do not give a Gaussian distribution, the true standard deviation δ must be determined.

In America the Department of Housing and Urban Development has adopted a set of guideline criteria for the external noise climate near houses. It should be emphasised that these criteria are based on social surveys carried out in the United States, and in the absence of further information should not be used in this country; they are included here out of interest and to show a possible use for the noise pollution concept. The criteria are shown in Fig. 2.54 and show that the borderline criteria for external noise levels according to American experience are:

Clearly acceptable	$L_{NP} < 62$ dBNP
Normally acceptable	$L_{NP} = 62\text{--}74$ dBNP
Normally unacceptable	$L_{NP} = 74\text{--}88$ dBNP
Clearly unacceptable	$L_{NP} > 88$ dBNP

The limit of 74 dBNP between normally acceptable and normally unacceptable is the upper permissible limit for people sleeping with open windows.

The hope is that L_{NP} will be able to provide a unified system of noise assessment in the future (see Robinson, 1971).

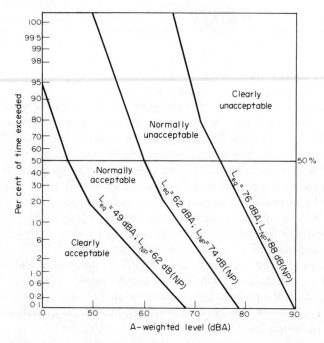

FIG. 2.54. External noise level criteria adopted by US Department of Housing and Urban Development for non-aircraft noise measured outdoors in residential areas (Beranek, 1971).

Clearly standards are being evolved in this country, so it is unwise to state rigid recommendations, but note should be made of any information that is forthcoming from the government. The Joint Circular 10/73 and 16/73, *Planning and Noise*, from the Department of the Environment and the Welsh Office, and bearing the date of 19 January 1973, published by HMSO, was issued to all local authorities and joint planning boards in England and Wales. In it *maximum* values of noise are given as showed in Table 2.35. In most cases new dwellings, schools or hospitals should be sited at least 300 m away from main roadways.

TABLE   2.35. MAXIMUM   LEVELS   OF
NOISE   EXTERNAL   AND   INTERNAL
LEVELS

	L_{10} (18 h)	Corrected noise level	
	dBA	dBA	
		Day	Night
Site standard	70[a]	75	65
Maximum noise level within dwellings with windows closed	68[b] 50	55	45
"Good standard" of noise within dwellings with windows closed	40	45	35

[a]See Department of Environments Design Bulletin, *New Housing and Road Traffic Noise*, HMSO 8 December 1972.
[b]Recommendation (a) superseded by *Calculation of Traffic Noise*, HMSO, 1975.

Environmental engineers and architects need to be aware of many features in the external climate. Noise is one of the many burdens of modern man's new wilderness. Air travel is doubling every 5 years and air freight is growing even faster. Indications are that traffic noise, the most prevelant external noise source in our community, will have increased by over 6 dB at the end of the century when the volume of traffic on the road will be more than threefold the present value; industrial noise output is increasing at a similar rate (see Table 2.37).

If again we compare the British recommendations with the ones being made for other countries (Table 2.36), one can conclude that perhaps the attempts to formulate noise standards for building planning and regional development in Great Britain should take a stricter line. Grandjean (1973b) writes about progress in Sweden, Norway, Austria, France and Germany, besides his own country Switzerland, and it is clear from this that our decisions at government level to date are very pale and meagre; this cannot be gratifying for the many people carrying out excellent research in this field.

The trends shown in Table 2.37 ignore hopeful developments in industrial design and the advent of more legislation to combat noise problems. Aircraft can be routed to

TABLE 2.36. MAXIMUM SOUND LEVEL RECOMMENDATIONS FOR QUIET RESIDENTIAL AREAS
(Values refer to measurements outside the building facade with windows closed; daytime means
06.00 to 22.00 hours, and night-time is 22.00 until 06.00 hours)

Noise source	L_{eq}[a] (Gilbert, 1973) Day (dBA)	L_{50}[a] (Grandjean, 1973b) Day (dBA)	L_{50}[a] (Grandjean, 1973b) Night (dBA)	L_{10} (UK) Day (dBA)	CNL (UK) Day (dBA)	CNL (UK) Night (dBA)
Motorway	65	50 $(L_1 - L_{50}) \leqslant 8$	40	68	75	65
Trains	70	—	—	—	—	—
Aircraft	65	—	—	—	—	—

[a] $L_{eq} = L_{50} + 5$, $L_{eq} = L_{50} + 0\cdot43\ (L_1 - L_{50})$.

TABLE 2.37. GROWTH OF NOISE

Cause	Increase in the noise environment (dB) 1965	1970	1975	1980	1985	1990
Subsonic flights near airports unlimited by passenger saturation (Richards, 1969)	0	$3 + 4$[a]	$6 + 4$	$9 + 4$	$12 + 4$	$15 + 4$
Road traffic	0	1	2	3	4	5
Industry	0	1	2	3	4	5

[a] The 4 dB factor allows for the changeover to jet aircraft.

fly over the sea and over lowly populated areas; car and machine manufacturers are becoming more aware of noise and vibration as important design features; the age of the fuel-cell-powered car may arrive; vertical take-off aircraft are being developed.

In order to assess the degree of sound protection needed, some assessment is needed of the external noise climate. Road and air traffic have been highlighted as major noise sources in the climate outside buildings, but significant contributions to the environment are also made from railways, road works, construction plants, factories, milk delivery and refuse disposal services. Sometimes, simple measures can be adopted; for example, plastic milk crates and dustbins are now often used as a simple means of reducing the noise caused by the last two when handling.

The *London Noise Survey* by Parkin *et al.* (1968) summarised the external noise levels exceeded for 10% of the time in various types of area; these are shown in Table 2.38.

These values can be reduced into more generalized terms by taking a 10% external noise level range of:

Range of 10% levels (dBA)	
Daytime	65–70
Evening	60–65
Night	55–60

TABLE 2.38

Case No.	Type of area	10% noise levels (dBA)		
		Daytime 7 a.m. to 7 p.m.	Evening 7 p.m. to midnight	Night midnight to 7 a.m.
1	Residential districts	65·2	60·5	52·7
2	Industrial	66·1	58·7	53·5
3	Shopping	70·2	66·1	58·2
4	Railways	68·2	63·9	56·5
5	Office	68·9	64·6	58·0
6	Open spaces	64·7	59·9	54·0
7	Commercial	65·6	58·1	53·6

Then the noise reduction required between the source and inside the building can be derived (Table 2.39).

TABLE 2.39

Building requiring internal noise rating	Required noise reduction (dBA)		
	Day	Evening	Night
NR 20	45–50	40–45	35–40
25	40–45	35–40	30–35
30	35–40	30–35	25–30
35	30–35	25–30	20–25
40	25–30	20–25	15–20
45	20–25	15–20	10–15
50	15–20	10–15	5–10

Although these figures only consider road traffic as a primary source of noise and only apply to areas of London during the early 1960s, this method of assessing the sound protection required by a building could be applied to other situations. It would then be necessary to take site measurements before design decisions could be taken. Ways of achieving various degrees of noise reduction will be discussed in Chapter 4.

Of course, structureborne sound must equally be considered as a potential external noise source and, just as in the case of airborne sound, it must receive attention at the early design stages. The major external sources of structureborne sound are road and rail traffic (including underground railways); aircraft (particularly jet noise and sonic boom); wind effects; earthquakes; blasting (as from adjacent quarrying operations); external equipment (such as pile-drivers or road-drills). Useful design information is given by Skipp (1966), Norris *et al.* (1959), Sutherland (1968), Hertz and Rubinstein (1969) and Johns (1968, 1970). Although it is possible to estimate probable general design loads for various inputs such loads should, if possible, be measured or estimated for the particular building environment.

2.4.5. NOISE CRITERIA IN THE FUTURE

A report issued in 1975 by the Noise Advisory Council entitled *Noise Units* discusses the limitations of noise indices—the difficulty of accounting for the wide range of human response, the limited hours of validity specified for various criteria, the problem of mixed sources experienced simultaneously. Neither the L_{10} (18 hour) nor the NNI indices take into account the variation in background sound level. The Noise and Number Index excludes the noise from ground-running aircraft engines. The Corrected Noise Level was devised for prediction of complaints but complaints do not necessarily reflect the degrees of annoyance. A unified noise scale needs to take into account all these factors. From the evidence around L_{eq} appears to be a strong candidate for an international unified noise scale but the use of L_{NP} needs more consideration.

Gierke in Stephens (1975) shows a correlation based on measurements made in the USA between day-night average sound level, $L_{d,n}$ and population density ρ (people per square mile) as

$$\bar{L}_{d,n} = 10 \lg \rho + 22 \quad (dB)$$

Beginning with the basic description for environmental noise given as

$$L_{eq} = 10 \lg \left\{ \frac{1}{(t_2 - t_1)} \int_{t_1}^{t_2} 10^{L_A(t)/10} \, dt \right.$$

for sound levels L_A in dBA occurring over a time period $(t_2 - t_1)$. Gierke derives a day-night weighted $L_{eq(24)}$ which has been termed a day-night sound level $L_{d,n}$. The need to distinguish between daytime (07.00 to 22.00 hours) and night-time (22.00 to 07.00 hours) is based on the fact that many surveys have shown that noise is more disturbing and more annoying at night-time than during daytime. The importance of undisturbed sleep is discussed earlier in this chapter (see Section 2.4.2). Although the hearing threshold is increased during sleep the background sound levels are at least 10 dB lower during the night. Gierke gives the day-night sound level as

$$L_{d,n} = 10 \lg \left\{ \frac{1}{24} [(15)(10^{L_d/10}) + (9)(10^{(L_n+10)/10})] \right\}$$

where the night-time sound level L_n carries a penalty of 10 dB over the daytime sound level L_d. In quiet environments $L_{d,n} < 55$ dB and $L_d - L_n \triangleq 10$, whereas in noisy ones $L_{d,n} > 65$ dB and $L_d - L_n \triangleq 4$. $L_{d,n}$ can be expressed in terms of other indices such as noise exposure forecast and composite noise rating. As an approximate rule of thumb

$$L_{d,n} \triangleq CNEL \triangleq NEF + 35 \triangleq CNR - 35$$

At the present time it is not clear if the $L_{d,n}$ is suitable for use in the United Kingdom. The USA Environmental Protection Agency published in 1974 the guidelines given on page 151 for a noise abatement programme (see Gierke in Stephens (1975)).

The difficulties of attempting to predict the feelings of people exposed to any stress and the shortcomings of present criteria have been debated. A more complete picture of annoyance based on sensory perception has been assembled (see Figs. 2.8 and 2.44), but the quantitative interrelationships between the many parameters are not yet known although some research is now beginning to fill in the many gaps in the

USA GUIDELINES FOR ENVIRONMENTAL NOISE (STEPHENS, 1975)

Effect	Level	Area
Hearing Loss	$L_{eq(8)} \leqslant 75$ dB	Occupational and educational settings
	$L_{eq(24)} \leqslant 70$ dB	All other areas
Outdoor activity interference and annoyance	$L_{d,n} \leqslant 55$ dB	Outdoors in residential areas and farms and other outdoor areas where people spend widely varying amounts of time and other places in which quiet is a basis for use.
	$L_{eq(8)} \leqslant 55$ dB	Outdoor areas where people spend limited amounts of time such as school yards, playgrounds, etc.
Indoor activity interference and annoyance	$L_{d,n} \leqslant 45$ dB	Indoor residential areas
	$L_{eq(24)} \leqslant 45$ dB	Other indoor areas with human activities such as schools, etc.

F

These recommendations may be compared with those below issued by Danish authorities and reported by Ingerslev in Stephens (1975).

Location	Satisfactory environment $L_{eq(24)} \leqslant$		Unsatisfactory environment $L_{eq(24)} \geqslant$	
	Road Traffic	Air Traffic	Road Traffic	Air Traffic
Weekend area. Recreational area in countryside.	40 dB(A)	80 PNdB	50 dB(A)	90 PNdB
Residential area. Hospitals. Recreational area in urban zone.	45	85	55	95
Hotels, churches, theatres, blocks of offices.	50	90	60	100
Activities with a low inside noise level. Small shops.	55	90	65	100
Department stores, supermarkets.	65	95	75	105
Activities with a high inside noise level.	70	100	80	110

state of knowledge. Of course, more knowledge raises more questions. *Annoyance* has been defined as a feeling of displeasure associated with any stimulus in the environment believed by an individual or a group to be adversely affecting them; this view has also been expressed by Borsky (1964). *Distraction* has been differentiated from the all-embracing word annoyance and confined to involuntary attention switching. Kryter (1970) uses the term *perceived noisiness* which he defines as the subjective impression of the unwantedness of a not unexpected, non-pain or fear-producing sound as part of one's environment. In other words the stress is particularised to the case of noise, and rating scales have been empirically derived to express the degree of perceived noisiness. These scales, based on social surveys and acoustical measurements, have tended to take the general form:

$$\text{Percentage annoyed} = k_1(S) + k_2(N) + k_3,$$

where $k_1(S)$ is a stimulus function in terms of a mean energy level or peak-background level difference, $k_2(N)$ is a function of the number of events and k_3 is a constant. These scales have been derived for communities of people, the personal differences in response are masked by the regression analysis constants k_1, k_2 and k_3. It is important that a deeper knowledge of individual response is searched for if we wish to have a thorough understanding about stress. Moreira and Bryan (1972), Bryan and Tempest (1973) and Bryan (1973b, 1974) have asked the question: are all people the same in their response to noise and is it scientifically correct to consider the population as homogeneous as far as noise annoyance is concerned? Work showing the range of susceptibility to noise has been mentioned already (see Fig. 2.9). Bryan in these articles comments that there is growing evidence to suggest that there may be at least two distinct types of reaction to noise—that of the adaptor and that of the noise-sensitive individual who represents 20–30% of the population. The former can become used to living and working in quite high levels of noise, whilst the latter is incapable of either without being subjected to severe mental strain. The point has also been reiterated throughout this chapter that no adequate measure of mental strain exists, and until strain can be assessed the assumption that adaptation relieves stress effects may be false. Schultz (1972) states that no physical measurement of noise has yet been developed which will yield correlations with individual responses higher than about 0·45. But hope may be gleamed from recent American Work (Tractor Project 253-004, Final Report Volumes I, II, Document No. T-70-Au-7454-U Austin Texas, 1970), and the research of Moreira and Bryan (1972) already discussed relating noise annoyance to personality. The American study relates individual response to aircraft noise and social predictor variables; noise exposure, rated only fourth in order of importance, fear of crashing and distance from the airport being higher priority order factors.

Bryan and his collaborators prefer to collect data using field studies, whereas Rice (1973) has shown that provided experiments are carefully designed useful indicators of weakness in community annoyance criteria can be studied in the laboratory. Certainly basic measurements on the fundamental effects of stress on the brain and the nervous system may be carried out under laboratory conditions. Rice's work, however, is not of a neurophysiological nature; various taped noises are presented to subjects, and the reliability and validity of the experiments are ensured by using statistical experimental methods. One important result of this work to date is the evolution of a *goodness factor* defined by the ratio of

$$\frac{\delta_s}{\delta_p} = \frac{\text{standard deviation of unit values at subjective equality levels}}{\text{standard deviation of unit values of the noise set}}$$

This concept has suggested that for traffic noise the noise pollution level L_{NP} ranks better than the mean energy level L_{eq}, used in other parts of Europe and also in America. The choice of L_{10} levels for specifying standards on traffic noise is also vindicated.

Grandjean (1974) has also carried out laboratory-simulated traffic noise investigations. A preliminary report by Voigt *et al.* (1974) concluded that equivalent noise level L_{eq} and the cumulative L_{10} noise level are the most appropriate parameters to predict the effect of traffic noise on people, but they established an *annoyance index* defined in terms of L_1, and L_{99} levels as

$$\text{AI} = 0.29\, L_1 + 0.27\, L_{99} - 19,$$

which may in the future prove to be the best predictor. There are some contradictions between the work of Voigt *et al.* reported at the Ergonomics Research Society 1974 Annual Conference and that of Rice described about no doubt due to differences in the statistical experimental methods used by each. Many keys need to be found to unlock doors as yet unknown.

2.5. The Effect of Noise on Human Performance

The influence of noise (both its presence and its absence) on chess tournaments has not as yet been made the subject of scientific study. But I would suggest that if there is a chess-playing student who wishes to benefit the world of chess he might do worse than to devote himself to a study of the subject for his PhD. It is patently clear that to play chess properly, especially when international tournaments are concerned, it is vital to have as complete a silence as possible.

Human nature being what it is, as long as you have more than one person in the playing room there exists the danger of a whispered conversation that sometimes seems to swell to a roar to the sensitive ears of the hapless chess-master. Bobby Fischer, in particular, blessed or cursed with a sense of hearing that might be useful in detecting the existence of submarines at great depth, has often objected to people speaking whilst he is playing.

Fischer's objection to audience noise at Reykjavik was not confined to their snoring. The above-mentioned student would be well advised to take that remarkable occasion as the chief source for his data and evidence concerning this sonorous subject.

There was the terrifying sound, rather like that of a horde of Valkyries, made by the unwrapping of the paper surrounding boiled sweets by some 300 children. There was the curious sound resembling a series of small explosions created by the entry into the auditorium from the buffet of various spectators and this despite the fact that the doors through which they passed were manned by stewards who prevented too frequent or noisy a use of the entrances and exits.

Though it is odd that Fischer's hyper-sensitivity to noise increased as he approached world status there is no doubt that he has done much to raise the standards of chess events in the matter of good conditions for play.

HARRY GOLOMBEK: (*The Times*, 13 *September* 1975).

McNair (1972) contends that in the absence of any hypothesis as to why the three states of thermoneutrality, thermal comfort, and working efficiency should be different it appears to be reasonable and realistic to assume that they are the same. Wilkinson (1974) argues against this on the basis that what is comfortable depends on the type of task, as the basic definition of comfort given at the commencement of Section 2 suggests, and on the individual interpretation of comfort in a given situation. Reliable individual differences, states Wilkinson (1974) will only be revealed by repeated measures of a given effect upon the same people and under the same conditions. When asked to assess conditions individuals may differ in what they mean by words such as comfort, annoyance or pleasant so that verbal measures as well as objective ones have a range of meaning for a given sample of people. When an observer experiences a range of conditions he will tend to place the centre point of the rating scale at the centre of the range of conditions sampled. Performance and comfort are related. Comfort conditions depend on the task in hand as well as the individual.

The arousal level concept and its rôle in the stochastic dominance interpretation was described in section 2.1. The arousal hypothesis has been used to explain with some, but not entire, success the effects of environmental stress on man working in various task situations, and there is now even some physiological evidence suggesting that it is a reasonable hypothesis. The effect of noise on human behavior has received more attention than thermal stress by experimental psychologists, notably Broadbent and his co-workers at the British Medical Research Council's Applied Psychology Research Unit in Cambridge (see Broadbent, 1958a).

What empirical evidence is available to explain the behavioural response of man to noise? We do not propose to deal with temporary or permanent deafness, which begins to occur in environments exceeding a level of about 75–85 dBA and is unfortunately all too common in industry today mainly because of inadequate legislation; reference should be made to Burns (1968) as a starting point for further information and references on the subject.

The early work of Broadbent (1954, 1958a, 1960, 1965) in England and Jerison (1957) in Ohio showed that performance of complex tasks, which may be defined as those having high rates of information input, began to deteriorate after about 0·5 to 1 h in 100 dB environments. But what does the word performance mean? Errors made, speed of working and the response time were measured for each subject who had to observe random changes of dial pointers placed around him, or perhaps had a five-choice task matching holes on an electrical board with those appearing visually and in a random sequence. The number of errors increased, but speed of working and response time remained unaffected. This contrasts with the effect of another type of stress, sleeplessness, which slows down performance and increases the response time of the individual but does not affect the error score (Broadbent, 1966). Later experiments by Hockey (1969) at Cambridge and McGrath at the Human Factors Centre in Los Angeles have shown that noise (64–95 dB) can *improve* performance of simple work tasks (i.e. those with a low rate of information imput).

Comparatively few studies have been carried out on the interactions of stresses such as various combinations of noise, heat and light acting together on a person. Grether *et al.* (1971, 1972) describes the possibilities when two stresses act upon a person simultaneously. They may result in

* no effect
* a simple additive effect
* a synergistic effect (i.e. more than additive)
* an antagonistic effect

If two stresses applied independently produce, say, an increase in error rate, then it is likely that any safety margin in the task will, if not consumed by one stress, be consumed by them both (Broadbent, 1971). These two stresses would then be acting synergistically. Grether *et al.* (1971, 1972) were unable to demonstrate any additive effects in heat, noise and vibration interaction studies. Wilkinson (1969) makes some further analytical comment on the work of Viteles and Smith (1946). At 31·1°C, 50% relative humidity performance of naval ratings carrying out various tasks (serial reaction, tracking, coding and inspection tasks) was better in 90 dB than in 80 dB noise, but at 36·1°C, 50% relative humidity, the reverse was true. Thus, in the first case the noise seems to be counteracting the effects of the heat but in the second case their effects are synergistic. A current study on noise and heat interaction is being conducted by McK. Nicholl (1974).

Other investigations have been made to find out how various stresses interact with one another. Wilkinson (1958, 1964, 1966) and Corcoran (1962) have shown that incentives can partly or wholly cancel out the effect of sleeplessness but when combined with heat stress only give minor increases in performance that remain almost independent of the temperature. The affect of adding noise is quite different. For an inspection of vigilance type task with a low rate of input information, noise helps to maintain a high arousal level and can even cancel out the effect of sleeplessness, but if the information inputs occur at a high rate the addition of noise can be detrimental.

Poulton *et al.* (1974) show that when heat is combined with a loss of sleep the arousal level is not reduced as much as when sleep loss is acting as a stressor. This interaction is similar to that between noise and sleep. Noise, however, can produce some effects different from those of heat. For instance, in tracking tasks subjects remain on target longer in a noisy condition than in a very warm condition (Hockey, 1970a; Poulton *et al.*, 1974). Working on the classical five-choice task man makes more errors in noise and heat than in normal controlled environments, but in noise the average rate of work remains unchanged whereas heat slows down the rate of work. The reliability and sensitivity of stress interaction experiments depend on the type of task and also the duration of the task besides the statistical experimental design (Poulton, 1965, 1970). Some of the conditions used by Pepler (1959), Hockey (1970b) and Poulton *et al.* (1974) are summarised in Table 2.40.

Colquhoun and Edwards (1975) investigated the interaction of noise with alcohol on a task of sustained attention. 18 subjects carried out a 30 min 5-choice serial response test 6 times under all possible combinations of alcohol (at zero, 'low' and 'high' concentrations in the bloodstream) with quiet (70 dB) and continuous 'white' noise (100 dB) conditions. Speed of responding was significantly decreased by noise, but not

TABLE 2.40. SOME CONDITIONS FOR STRESS INTERACTION EXPERIMENTS

Researcher	Task	Subjects	Temperature	Noise	Sleep deprivation
Pepler (1959)	Tracking with peripheral lights Five-choice (Leonard, 1959)	Naval enlisted men	38°C dry-bulb 32°C wet-bulb	←—————————————————→ See Corcoran (1962), Wilkinson (1963)	
Hockey (1970b)	Tracking with peripheral lights				
Poulton et al. (1974)	Tracking with peripheral lights Five-choice (30 min) Wilkinson auditory vigilance (30 min)	Twelve naval enlisted men 18–35 years old	38°C dry-bulb 32°C wet-bulb	70 dB white noise 1000 Hz tone for 0·37–0·5 s on head-phone every 2 s on white noise background of 70 dB	One night before test

by alcohol; there was no interaction between the two stressors in this aspect of performance. However, a clear interaction was observed in accuracy, the error rate increasing with alcohol under quiet conditions, and decreasing under noise conditions, in such a way that the separate effects of the two stressors were effectively cancelled out when administered in combination. The results are interpreted as supporting the view that noise functions as an 'arouser' and that alcohol, in its physiological rôle as a cortical depressant, acts in the opposite direction.

Bryan (1973a) describes experiments in which he attempted to discover how noise affects the mental performance of students completing intelligence tests over a 30 min period in a room having a background noise level of 80–90 dBA. Two groups each of twenty students partook in the experiments. One group (A) was tested in the quiet first and then in the noise, and the other group (B) in the noise first and then the quiet. Each student completed two intelligence tests in the quiet and one in the noise. It was found that noise had no significant effect on the *mean* performance. However, the distribution of the IQ scores for the subjects working in the quiet was a normal one, whereas that for those working in the noise was skewed towards lower values indicating that there were fewer higher scores when tests were undertaken in the noisy climate (Fig. 2.55). The change in the IQ score due to noise is shown in Fig. 2.55b; the correlation between change in score and IQ was about 0·4, significant at the 1–2% level. Students with high IQ scores tended to show a deterioration in their performance when working in the noisy environment, whereas the students with low IQ scores improved their performance in the noisier climate. The extent of the change is such that it makes about a 10% difference to the score at either end of the IQ scale.

An explanation of these results can be ventured in terms of arousal level theory. The high IQ scoring person is working at his optimum arousal level; too much noise

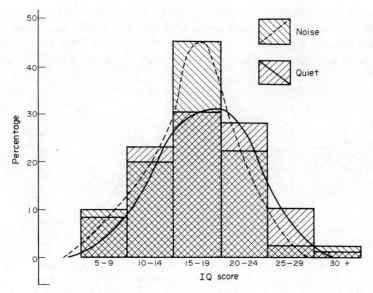

(a) The distribution of IQ scores for both groups A and B in the quiet and in the noise. The two distributions are different. That in the quiet is normal whilst that in noise is skewed to low values. The noise has reduced the level and the number of the peak IQ score.

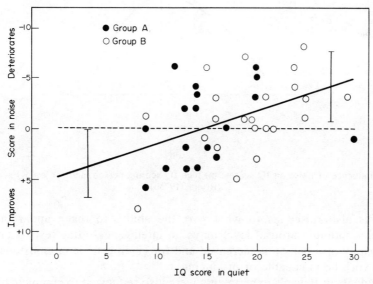

(b) The effect of noise is to reduce the scores of subjects with high IQ and to increase the scores of subjects with low IQ.

FIG. 2.55. Effect of noise on IQ scores (Bryan, 1973b).

increases this level beyond the optimum and overloads the sensory system with the result that less work is done and more mistakes are made. For a low IQ scoring person the increased noise level increases the arousal level to the individual's optimum level and more work with fewer errors is done (Fig. 2.56a and b). The

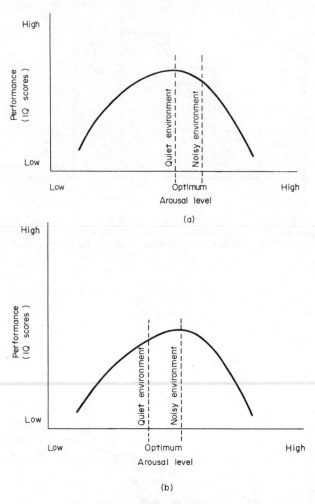

FIG. 2.56. Influence of noise on IQ scores: (a) high IQ scoring person, and (b) low IQ scoring person (Bryan, (1973b).

question is also raised as to what part the ability to reach and remain at an individual's optimum arousal level plays an intelligence-rating tests. Certainly the work capacity of a person is increased and for given genetic circumstances will be the best work he is capable of producing.

The students in Bryan's experiments were also requested to complete a questionnaire in which they were asked about their attitude to the noise and the effect they thought it had on their performance; they also completed an Eysenck personality inventory: 40% of the group found the noise annoying but this did not correlate with performance; 90% of the group thought that the noise had slowed them down or caused them to make more mistakes, but this was not found to be true; 75% of the students stated they preferred to study in noise they described as "quiet" or "noticeable". These experiments did not show any differences in performance between extroverts and introverts.

More research is needed to establish the effect of noise on the performance of mental tasks, but this work provides many interesting points for debate and further investigation.

What conclusions can be made from this work?

(a) *The type of task is important*—noise can be beneficial and act as a stimulant for simple tasks which easily become boring and have little motivation, whereas for complex tasks noise is detrimental. This may be summed up in the Yerkes–Dodson law shown diagrammatically in Fig. 2.57 and from which it can be seen that there is

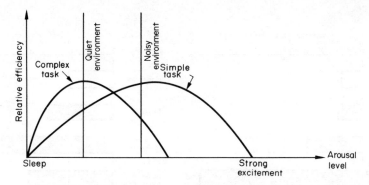

FIG. 2.57. Yerkes–Dodson law.

an optimum arousal level which is lower for complex tasks than simple ones; above the optimum level, concentration and attention is lacking due to anxiety caused by an overloading in the sensory system (corresponding to Fig. 2.7b), and, below the optimum, boredom and lack of motivation result in a lack of environmental awareness (i.e. a stiff response to the environment again as shown in Fig. 2.7). The optimum arousal level occurs when concentration and attention on the central stimulus is unimpaired by peripheral stimuli (see section 2.3).

(b) *Various stresses* like heat, noise, sleeplessness interact and compensate one another if any pair are at opposing ends of the arousal continuum, e.g. sleeplessness and noise, boredom and noise, alcohol and boredom (Fig. 2.58b–d), whereas stress pairs at the same end of the arousal continuum, e.g. lack of sleep and boredom (Fig. 2.58a) reinforce one another, but the effect on the individual depends on their arousal level at the time the stresses are imposed.

(c) *Stresses affect performance in different ways*—noise can increase errors, sleeplessness can slow down the rate of working; again the polarity of the arousal spectrum suggests that suboptimal arousal states are more likely to result in slower rather than inaccurate working, whereas stresses which raise the arousal level above the optimum will result in faulty judgement because the sensory system becomes overloaded.

Many of the experiments that have been referred to used relatively high noise levels. It may be argued that people adapt to their environment, but what long-term effects does stress have on human behaviour? It would be interesting to see, by making physiological measurements, if the neuron activity levels in the central nervous system increase as the stresses of life become more abundant, various and intense. Above about 90 dBA, noise affects the human system involuntarily, but response to noise below this level depends much more on the type of individual. Some general trends have been noted: so-called professional people tend to be more easily bothered by noise than people working in non-professional grades; people who have difficulties in making personal adjustments to life find environmental stress more troublesome to cope with than a "happy-go-lucky" individual; there are people who are more susceptible to sounds than others (see Fig. 2.9). Johansson (1970) has shown that intelligence interacts with the effect of noise on performance, and Arvidsson et al. (1965) concludes that personality is a factor in complaints of noise disturbance.

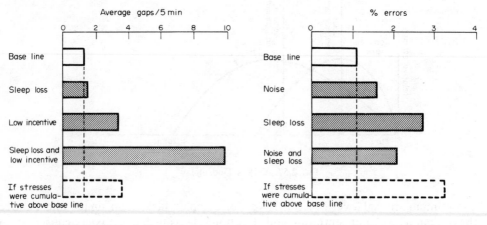

(a) Effect of lack of sleep and boredom
(b) Effect of noise and lack of sleep

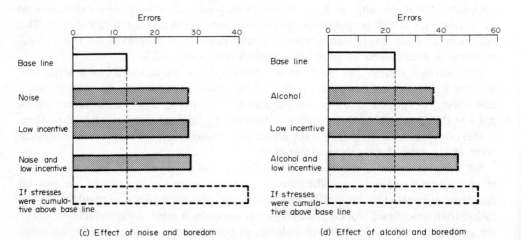

(c) Effect of noise and boredom
(d) Effect of alcohol and boredom

FIG. 2.58. Effects of various stress combinations on performance (Wilkinson, 1972).

TABLE 2.41. INTERACTION OF NOISE AND PERFORMANCE

Research topic	Source	Remarks
Effect of noise on repetitive productivity	Laird (1933)	Complex steady noise above 512 Hz affected production more than pure tones, muscular strain, increased urine output, humming in ears, after exposure to 80–90 dB. Varying complex noise had worst effect on production. Four male subjects carried out repetitive task requiring close eye and hand co-ordination for $4\frac{1}{2}$ h
	Weston and Adams (1935)	The productivity of 10 textile workers wearing ear defenders was higher than that of 10 workers not wearing them (Fig. 2.59)
	de Almida (1950)	Absenteeism in electric punch-card work room reduced when noise level reduced
	Kryter (1950)	Noise does not appear to reduce non-auditory work activity; can improve performance by masking distracting auditory signals
	Felton and Spencer (1957)	Ego involvement in high status occupation affects concern about noise
	Broadbent (1960)	Noise reduced from 99 to 89 dB in a photographic processing factory resulted in fewer broken film rolls and equipment shut downs
	Stikar and Hlavac (1963)	Noise caused decrement in typing performance
	Kovrlgin and Mikeyer (1965)	Increasing noise level (78–95 dB) increased postal letter sorting errors
Effect of music on productivity	Ayres (1911)	Lively music stimulated athletes to respond with greater effort even though they were near exhaustion in a six-day cycle race
	Wyatt and Langdon (1935)	No consistent relationship between music and work
	Humes (1941)	Slow–fast music resulted in fewer errors by semi-skilled labour assembling radio tubes
	McGehee and Gardner (1949)	Eighty-four-minute music selections presented on some days over a five-week period. Complex tasks not affected but employees believed that they had benefited.
	Freeman and Neidt (1959)	Familiar background music had no significant effect on the memory of the contents of a film being watched at the same time.
	Houston and Jones (1967)	Familiar loud noise facilitated performance of subjects carrying out Stroop Colour Word test
	Kryter (1970)	Concludes that it is difficult to quantify the beneficial effects of music in industry. Effects may be transitory and related to temporary changes in worker morale, may be or may be not beneficial, effects are maybe small compared to task or motivation factors

TABLE 2.41 (*Continued*).

Research topic	Source	Remarks
	Croome (1974)	No measured evidence but many firms in the hosiery industry have said their employees prefer some music on for a few hours in a working day
Effects of noise on mental output of school children	Butcher (1938)	Ten-minute arithmetic tests with intermittent noise, continuous noise, continuous musical note, records of music; children unaware of tests. Work distraction was *least* for intermittent noise and *most* for records; errors decreased as distraction increases (due to increased effort to counteract distraction?)
Effect of background music on children	Mitchell (1949)	Silent reading test: music aided pupils (12 years old) with IQ > 100 but adversely affected those with an IQ below this
	Hall (1952)	Reading comprehension: background music aided all pupils (14–15 years old); greatest advantage for those with under average IQ
Effect of noise on subtraction task carried out of naval ratings	Broadbent (1959b)	One group worked in 70 dB for two sessions; another in 100 dB on one day and 70 dB on another; third group had one session in 70 dB followed by one in 100 dB on next day. Time to solve problems increased more for group working in noisier climate; there is an after effect of noise which affects future performance. Deterioration in performance most marked for extroverts
Influence of noise on discrimination tasks	Sanders (1961)	Forty air force ratings exposed to (a) steady white noise at a level of 70 dB, (b) varying noise levels 65–90 dB (average 75 dB) presented as 16 tones (85–1360 Hz in equal steps of 85 Hz, equal loudness); concluded that unpredictable noise affects performance more than monotonous noise (also see Broadbent (1953))
Effect of noise on vigilance tasks	Tarrière and Wisner (1962)	Thirty-five subjects working on a 1·5 h vigilance task in 35 dB, 90 dB (car noise), 90 dB (speech and music); decrease in vigilance performance in all cases but in varying degrees
	Smith (1947) McBain (1961) Corcoran (1963b) Kirk and Hecht (1963) Oltman (1964) Berlyne *et al.* (1966) Weinstein and Mckenzie (1966) Schwartz (1967) Houston (1968)	Performance of various vigilance, or monotonous, tasks was found to improve with the introduction of sound
Effect of noise on complex tasks	Smith (1951) Broadbent (1954) Bindra (1959) Plutchik (1959)	Introduction of arousing stimuli was detrimental to performance of various complex tasks

TABLE 2.41 (*Continued*).

Research topic	Source	Remarks
	Teichner *et al.* (1963) Woodhead (1966) Boggs and Simon (1968)	
	Smith (1961) Park and Payne (1963) Konz (1964) Brown *et al.* (1965) Hoffman (1966) Carlson (1967) Slater (1968)	Effect of noise found to be insignificant
Effect of train noise on primary school children doing mental work	Nagatsuka (1964) Maruyama (1964)	Disturbance was high when levels, due to trains, rose to 58–62 dBA but much less at 54 dBA
Influence of train noise upon school children	Izumiyama (1964)	At 62 dBA articulation index was < 60%; at 50 dBA it rose to 80%
Effect of noise from Orly Airport on primary and secondary school nearby	Guinot (1965)	Less of 50% teaching time; classes calmer and teaching less tired when flights postponed during foggy weather, serious long-term effects, e.g. loss of ability to concentrate, more impatience
Environmental conditions in offices	Manning (1967)	Levels of 55–65 dBA evaluated as "quiet" in land-scaped areas
Eighteen male students subjected to mathematics and reading comprehension under sound levels of 20–25 dB, 45–55 dB, 75–85 dB	Lehmann *et al.* (1965)	Higher marks obtained under 20–25 dB condition, a little difference obtained for reading com-prehension under 20–25 dB and 45–55 dB condi-tions. Work was also slower with louder sound levels
Reactions of teachers and secondary school children	Bruckmeyer and Lang (1968)	Questionnaires showed that 60% of the teachers and 50% of the children found sound levels of 50–55 dBA disturbing
	Vulkan and Stephenson (1970)	Sound level of 45 dBA advocated as criterion for classrooms
Reading comprehension for 13-year-old children under levels of 45–55 dBA, 55–75 dBA, 75–90 dBA	Slater (1968)	Noise found to have no effect on performance
Acoustical conditions in landscaped offices	Croome (1969a) Schreiber (1971)	Levels of < 45 dBA evaluated as too quiet, 50–55 dBA found to be satisfactory
Effects of noise on mental and motor performance[*]	Kryter 1970 (Chapter 11)	General review
Effect of unfamiliar and unpredictable noise on school children. Part of a series of experiments to investigate effect of information content of	Wyon (1970)	Short bursts of white noise 55–78 dBA of random duration, sequence and intermittency but sound pressure level below variable noise background of each of four groups (two control groups) to-talling 110 children 12 years old working for 80 min; observations made through one-way mir-

TABLE 2.41 (*Continued*).

Research topic	Source	Remarks
variable noise on performance		rors. Significant noise decrements due to noise. In a self-paced numerical task children worked more slowly but more accurately, possibly because noise interrupted them and had to start again. In a test of recognition memory, children exposed to noise less confident, less consistent in their answers
Acoustical conditions in lecture rooms	Croome (1971), Croome and Brook (1974)	Subjective evaluations showed that the range 42–47 dBA classified as quiet and satisfactory
Effects of intermittent noise on performance of serial decoding task and on physiological response	Conrad (1973)	Sixteen university students worked in periodic and aperiodic 93 dBA broadband noise. Studies indicated negligible decremental effect of noise on short-term memory task performance. Physiological variables recorded were finger photoplethysmographic blood volume pulse amplitude, pulse rate and forearm electromyogram; subjective response measured by noise annoyance questionnaire. Finger-blood volume pulse response increased (i.e. increased intensity of effort) in all the noise conditions and was higher for subjects that were more highly annoyed than other less annoyed
Effect of 90 dBA sound levels on forty students carrying out IQ tests	Bryan (1973a)	Decreased performance of high IQ scoring students; increased performance of low IQ scoring students
Noise and personality	Broadbent (1957)	Neurotic persons find noise move annoying.
	Bakan (1959)	Secondary task improved extroverts' primary task performance more than for introverts'.
	Cohen *et al.* (1966)	Anxious subjects performed worse than nonanxious ones during high noise
	Davies and Hockey (1966)	A 95 dB noise was beneficial to extroverts in a visual cancellation task, but not to introverts
	Day (1967)	Less anxious subjects paid more attention to complex material and converse of this for more anxious subjects
	Cohen (1968)	Steelworkers and other industrial workers showed more social conflict at home and at work, chronic fatigue and neuroticism
The effects of noise stress on blood glucose level and skilled performance.	Simpson *et al.* (1974)	Thirty-two male subjects were randomly assigned to one of four groups. Groups A and B were preloaded with glucose, Groups C and D were not. Groups A and D carried out a pursuit rotor task under 50 dBA (non-stressful), white noise. Groups B and C carried out the same task under 80 dBA (stressful) white noise. Blood samples were taken from each subject and blood glucose levels as well as time on target were measured. Statistical

TABLE 2.41 (*Continued*).

Research topic	Source	Remarks
		analysis of the results showed that noise stress impaired performance and that pre-loading with glucose attenuated the impairment. This change in performance was accompanied by a reduction in the high blood glucose level caused by pre-loading. Pre-loading in the non-stressful condition impaired performance without an associated fall in the high blood glucose level. A suggestion is made that blood glucose levels may be used as indicant of stress.
The effects of noise and time of day upon age differences in performance at two checking tasks.	Davies, (1975)	Forty older (age range 65–72 yr) and 40 younger subjects (age range 18–31 yr) performed a 15 min cancellation task and a 20 min version of the Continuous Performance Test (CPT). They worked either in noise (95 dBA) or quiet (70 dBA) and either in the morning (10 a.m.) or the afternoon (2 p.m.). In the cancellation task older subjects worked significantly more slowly and significantly less accurately than younger ones. However, noise significantly improved the rate of work for older subjects, but not younger ones, without impairing accuracy. Older subjects, but not younger ones, also were significantly more accurate in the afternoon. The performance of older subjects on the CPT was significantly inferior to that of younger subjects. Afternoon testing again improved the performance of older subjects to a greater extent. Some explanation is offered by the evidence suggesting that older subjects have a lower arousal level characteristic compared to younger ones.

Table 2.41 is an attempt to gather some facts which will lead to satisfactory criteria, and yet how conflicting some of the findings are. Probable satisfactory sound levels are 50–55 dBA for landscaped offices, 40–45 dBA for classrooms, 35–40 dBA for lecture rooms and 30–35 dBA in examination rooms; but notice the comments referring to non-quantifiable factors such as lack of concentration and increased impatience, besides straightforward communication interference. The need for feedback data on environmental criteria is illustrated by the classroom acoustical requirements. On the basis of speech interference level 55 dBA is advocated by many as a criterion, but the Inner London Education Authority have found that in their experience a 10% sound level of 45 dBA will permit more concentration and less distraction (Vulkan and Stephenson, 1970).

Mehrabian and Russell in their 1974 treatise *An Approach to Environmental Psychology* (published by the Massachusetts Institute of Technology) suggest that performances can be expressed as

$$p = k_1(P) - (k_2I + k_3I_e + k_4A)^2,$$

where P is the pleasure dimension, I is the information rate of the task, I_e is the information rate of the environment in which the task is performed and A is the

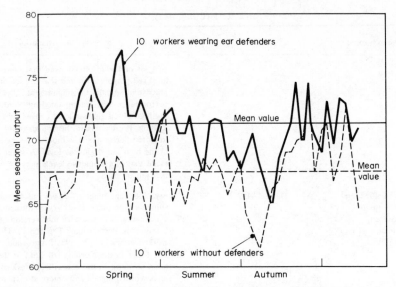

FIG. 2.59. Effect of reducing the noise level by 10–15 dB on the work output of textile workers (Weston and Adams, 1935).

arousal level of the individual. Subjective scales for assessing P, I, I_e and A are given. Intuitively and by everyday observation this dependence of task performance on pleasurable work and environment, which in degree will depend on the individual, makes good sense. More empirical evidence is needed to define this function more explicitly for all kinds of work, environment and groups of people.

2.6. Noise and Speech Communication

2.6.1. CHARACTERISTICS OF SPEECH

There are good reasons for giving particular attention to the effects of noise on speech communication. Much evidence now exists suggesting that the information content of noise has an influence on the probability of distraction occurring. Speech therefore is a primary distractive noise source. Language as a means of communication between human beings is a fundamental ingredient of any culture and as such is common in varying degrees to any job or occupation.

In spaces where speech is the prime mode of communication it is important that the reverberant conditions, the background sound levels, the positions of speaker and listeners allow the speech to be clearly understood, i.e. the words spoken must be intelligible to all the listeners who may have a wide range of "hearing efficiency" and for all speakers with their wide range of speaking styles and voice quality. There is some evidence, too, showing the importance of gestures which act as auditory cues, besides instinctive lip-reading at close inter-communication distances.

Very early on in the history of acoustics the human voice was being carefully studied. Originally much was taken for granted, as always seems to be the case, until such time as a full understanding of the mechanics and techniques of voice production became desirable. Such a time came with the development of communication systems. Their rapid improvement and worldwide acceptance created a need for further improvements, new techniques, higher standards, better signal-to-noise ratios and thus better intelligibility. In the search for greater intelligibility in the conservations, sometimes dimly heard across the telephones of America, it was realised that intelligibility could not be easily measured and quantified. Not only did intelligibility need defining and measuring, but also the human voice and its many qualities and characteristics needed studying throughout its complete spectrum.

Beranek (1954) found that the human voice (male) has, on average, a power spectrum that peaks at about 500 Hz and a spectrum (in octave bands) that drops off above 1000 Hz at a rate of about 8 dB per octave. The effective range is contained within the octave bands between 250 and 4000 Hz, speech energy below 200 and above 7000 Hz contributing almost nothing to speech intelligibility. In each frequency band speech has a dynamic range of about 30 dB; the peak values lie about 12 dB above the long-time r.m.s. levels. At high frequencies the voice becomes very directional. Each syllable of speech lasts about 0·125 s and the average interval between syllables is about 0·1 s. The voice shows great variability between people and in use.

French and Steinberg (1947) state that one of the distinguishing characteristics of speech is its dynamic nature. Conversation at the rate of 200 words per minute, corresponding to about four syllables and ten speech sounds per second, is not unusual. During the brief period that a sound lasts, the intensity builds up rapidly, remains comparatively constant for a while, then decays rapidly. The various sounds differ from each other in their build-up and decay characteristics, in length, in total intensity and in the distribution of the intensity with frequency. With the vowel sounds the intensity is carried largely by the harmonics of the fundamental frequency of the voice and tends to be concentrated in one or more distinct frequency regions, each sound having its own characteristic regions of prominence. The consonant sounds, as a group, having components of higher frequency and lower intensity than the vowel sounds. In addition, the intensity tends to be scattered continuously over the frequency region characteristic of each sound. Thus, when the elementary sounds are combined in sequence to form syllables, words and phrases, there is a continuous succession of rapid variations in intensity, not only in particular frequency regions, but also along the frequency scale. The interpretation of speech received by the ear depends upon the perception and recognition of these constantly shifting patterns. Vowel sounds are not as critical to speech intelligibility as the consonant sounds. In some ways it is unfortunate that the consonant sounds are so weak and, therefore, are easily masked by noise. Because they are also composed of mainly high frequencies they are easily distorted by absorption and transmission loss.

With these initial studies on the physics of the human voice, several research projects were started in order to arrive at the measurement of speech intelligibility. Fletcher and Steinberg (1930) devised a method of *articulation* testing using word lists. The method consisted of several "judges" listening to specially constructed lists of monosyllabic words (sometimes meaningless words, phrases or sentences being used). These lists aimed at reproducing the vowel are consonant sounds in the proportion that they occur in the language being used. They would be spoken at a fixed rate of about three per second. The fractions of the vowels and indicated consonants that were understood correctly were determined separately, and the percentage syllable articulation was calculated from:

$$PA = 100\{1 - (1 - V_w C_w^2)^{0.9}\},$$

where PA is the percentage articulation (for syllables), V_w is the fraction of vowels correctly heard and C_w fraction of designated consonants correctly heard.

The researchers of the Bell Telephone Company did much work on the formulation of the theories of speech intelligibility in the 1930s. Beranek (1954) expanded these theories and did much to relate the new concepts of intelligibility, articulation and feelings of privacy. However, much of the modern work and thought was initiated by the postulates and subsequent tests performed by French and Steinberg (1947). They saw that tests using speakers, listeners and word lists were not independent in their results from the skill and experience of the testers. More importantly they noted the fact that syllable articulation, in common with other subjective measures, is not an additive measure of the importance of the contributions made by the speech components in the different frequency regions. Stated differently, the articulation observed with a given frequency band of speech is not equal to the sum of the articulations observed when the given band is subdivided into narrower bands, which are then individually tested. They decided that, for the purpose of establishing relations between the intelligence-carrying capacity of the components of speech and their frequency and intensity, a more fundamental index, free of the above defects, was needed. Such an index, called the *articulation index*, could be derived from the results of articulation tests. The magnitude of this index was taken to vary between zero and unity, the former applying when the received speech was completely unintelligible, the latter to the condition of best intelligibility.

Thus the articulation index was based on the concept that any narrow band of speech frequencies of a given intensity will carry a contribution to the total index which is independent of the other bands with which it is associated and that the total contribution of all bands is the sum of the contributions of the separate bands. French and Steinberg (1947) made the note that this independence was not absolutely true; the contribution of a band may be modified somewhat by masking produced by intense speech in neighbouring bands. Therefore, letting ΔA represent the articulation index of any narrow band of speech frequencies and n the number of narrow bands into which the total band is subdivided for computational purposes, the articulation index A of the total band reaching the listener is

$$A = \sum_{1}^{n} \Delta A.$$

The value of ΔA, which is carried by any narrow frequency band, varies all the way from zero to a maximum value ΔA_m as the absolute levels of speech and noise in the ear are independently varied over wide ranges. Letting W represent the fractional part of ΔA_m, which is contributed by a band with a particular combination of speech and noise, the value of articulation index for that band is given by

$$\Delta A = W \, \Delta A_m.$$

Hence

$$A = \sum_{1}^{n} W \, \Delta A_m.$$

The establishment of relations for computing A thus involved two main steps:

(i) the determination of the increments of frequency which give equal values of ΔA_m throughout the frequency range;
(ii) the determination of relationships between W and the levels of speech and noise in the ear.

Eventually they decided on twenty frequency bands between the limits of 250 and 7000 Hz. Each of these makes a 5% contribution to the articulation index when all bands are at their optimum levels. Beranek (1947) later revised these frequency bands and their limits, and it was from this revision that the modern standard was developed by Kryter (1962) and later by Working Group S3-36 of the American National Standards Institute, for their *Standard Methods for the Calculation of the Articulation Index* (ANSI S3.5—1969).

Later, more detailed work correlated modern, standardised, articulation tests to the articulation index as shown in Fig. 2.60. At the same time, however, further ideas were being formulated about the possibility of there being a link between speech

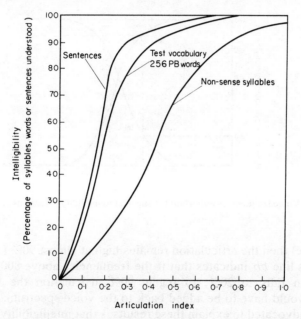

FIG. 2.60. Relationship between speech intelligibility and articulation index.

intelligibility and speech privacy by Cavanaugh *et al.* (1962). They said that: "It is a matter of simple observation that even though the background noise level around us is well above our threshold of hearing, we almost invariably become accustomed to it, accept it and, most of the time, are altogether unaware of it. On the other hand, we sometimes express strong dissatisfaction at intruding speech whose r.m.s. levels are no greater than those of the background noise."

Speech intelligibility was known to be determined not by the level of the speech but rather by the ratio of speech to noise. Their experimental work confirmed the assumed relationship between privacy and intelligibility, and Cavanaugh *et al.* (1962) were able to state that the most critical 10% of his test subjects began to feel a lack of privacy when the articulation index reached 0·05. They then went on to develop their findings in the context of offices, but these were invariably of the cellular type. Their calculations and applications concentrated on transmission loss situations between neighbouring offices separated by walls or partitions.

The final stage in the development of the articulation index as a measure of speech privacy was the work done by Kryter (1962). He established firm methods for its calculation and had these methods ratified as American standards. Since this work the development has tended along the lines of improvement upon these standards. The application of the articulation index as a design tool is the focus of attention for modern-day researchers.

Other experiments described by Stephens and Bate (1966) have demonstrated the effect of voice frequency content on articulation and hence intelligibility. Consider Fig. 2.61 which shows that if low frequencies are filtered out of the voice (e.g.

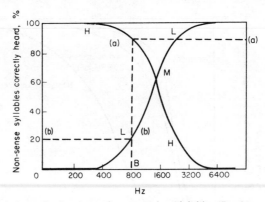

FIG. 2.61. Voice frequencies important for speech intelligibility (Stephens and Bate, 1966).

0–800 Hz range) then the articulation remains high at almost 90% for this frequency range, whereas line *bb* indicates that if the frequencies above 800 Hz are removed the articulation falls to 20%; to regain 90% articulation the frequency range 800–2300 Hz would have to be added back to the voice spectrum. One explanation that has been advocated to explain these results is that intelligibility depends on high frequencies, the low frequencies acting as a carrier wave.

Other factors affecting speech intelligibility are now given:

* at a signal-to-noise ratio of −18 dB consonants cannot be distinguished from one another (Fig. 2.62);
* speech with a very weak, or a very high level of vocal effort is not intelligible as speech in the range 50–80 dB (at 1 m from talker) for a constant signal-to-noise ratio;
* effect of other voices interfering with speech is shown in Fig. 2.63;

Consonants in front of vowel /a/

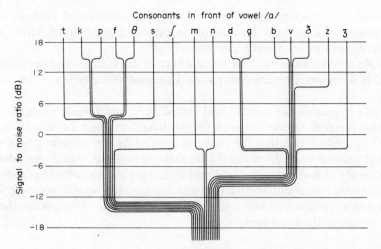

FIG. 2.62. Masking of consonants by randomly white noise (Miller and Nicely, 1955).

psychological attention switching can occur at lower sound levels than those shown;

* binaural listening aids intelligibility of speech because phase and intensity differences between the speech and noise signals entering two ears give localisation cues;
* it already has been stated that consonants hold most of the meaning in speech and these parts of the speech wave are often less intense than those containing the vowel sounds; peak clipping by less than 6 dB can trim the vowel sounds and, after amplification, leave the intensities of the consonant and vowel sounds in balance;
* reverberation can cause the decay of one sound to blur the energy growth of a succeeding sound;
* there are large variations in level and character between the speech acoustical spectra of people;

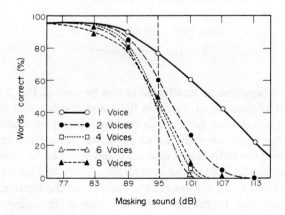

FIG. 2.63. Word intelligibility as a function of the intensity of different number of masking voices; the level of the desired speech was held constant at 94 dB (Miller, 1947).

* a number of psychological factors not well understood such as attention switching due to high meaningful information content of visual or aural sources also effect speech communication;
* people tend to increase their vocal effort as the background noise increases; this is sometimes referred to as the Lombard voice reflex.

Sound power spectra for normal and raised male and female voices have been measured by Fletcher (1953), Beranek (1954), Kingsbury and Taylor (1970) and are shown in Fig. 2.64; notice that the most recent results for women school teachers suggest that normal voice levels of 75 dB may be realised, which are some 10 dB above the results of earlier work. Field measurements by the author have shown that some men and women unconsciously raise their voice sound pressure level by 5–10 dBA when speaking on the telephone and that large variations in the speech spectrum do occur between people.

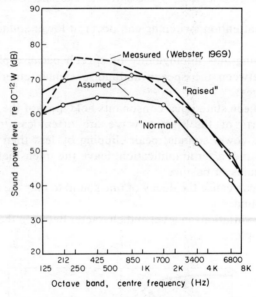

FIG. 2.64. Voice spectra.

The effect of reverberation on intelligibility can be seen in Fig. 2.65. The intensity curve for monosyllabic words of duration 0·3 s and spoken at a rate of one every 0·4 s follows the reverberation time curve but decaying from it after each duration time (Fig. 2.65a); the total intensity curve is shown dotted and the peaks give an indication of how well the individual syllables stand out. The corresponding loudness curves can be seen in Fig. 2.65b; the dotted lines show the effect of reducing the reverberation time from 4 to 2 s. Figure 2.65c shows that the audience can reduce the possible interference due to reverberation so that in this case the dotted line, representing the loudness of the third syllable, stands distinct from the decay curve for the first syllable when the hall is full but not when it is less than half full.

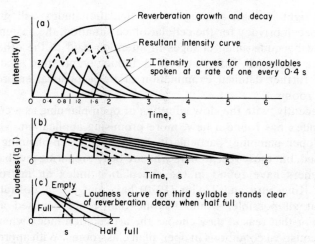

FIG. 2.65. Effect of reverberation on sound (Stephens and Bate, 1966).

Conclusions that can be drawn from these factors are as follows.

(a) Speaking slowly enables the energy peaks for each speech sound to stand out well above the decaying reverberant field for the previous spoken sounds; as the reverberation time of a space increased, it is necessary to speak more slowly.

(b) Design spaces for an optimum (note: *not* minimum) reverberation time; low enough to allow a peaky speech sound intensity curve to develop but not too low to make the speaker feel the strain of raising the voice level. Reverberation provides auditory feedback to speakers and listeners which is valuable physiologically because the ear can integrate sound patterns from balanced direct-reverberant sound fields easier than from direct ones for the purposes of interpreting the sensory inputs into a continuous thread of thought; and psychologically, because people in enclosed spaces intuitively expect some reverberation which forms a contrast impression with the free field mainly experienced outside in Nature; and, finally, the sense of sound envelopment seems to add other qualities to the sound which are not at this time properly understood.

(c) Basic factors like voice quality, hearing efficiency and manners of presentation will vary considerably, making it difficult to give quantitative steps for designing a space for speech communication.

Thanks to the history of its development the articulation index has long been used as a quality measure in the design of electroacoustical communication systems. An articulation index of 0·7 is regarded by telephone engineers as the lower limit for satisfactory intelligibility. Its application to the acoustical design of cellular offices has been ably demonstrated and tested by Cavanaugh et al. (1962), but this usage has not met with wide acceptance. The privacy criteria for cellular offices are rarely related to the articulation index and generally concentrate on the partition noise reduction index and the possibility of flanking transmission. In this attitude designers may well be right in that, although the articulation index will give an efficient measure of speech privacy for the cellular office situation, the design values are too easily affected by quite small changes in attention paid to the other criteria. The inclusion of a cross-talk attenuator in a linking ventilation system can easily preclude any further detailed examination of the acoustical environment in two neighbouring rooms.

It is only recently, with the slow adoption of open-planning as a concept, that the articulation index has found a new, more promising application. The problems of privacy and open-planning, particularly landscaped open-planning, have already been mentioned, but those researchers who are busy preparing feedback information for the designers have found in the articulation index an instrument of great potential. As Kingsbury and Taylor (1970) have put it: "A desirable method for analysis, if at all possible, would be one that could be used also for design prediction." For that reason they choose the articulation index when attempting to analyse the acoustical conditions in open-plan classrooms. With appropriate inputs it can be used for either analysis or design prediction.

Following this publication, Yerges and Bollinger (1972) have done more work on the acoustics of open-plan educational facilities. They devised a design method for the application of the articulation index to the layout of classes within open-plan schools. In principle their method could be applied, with equal success, to the layout of work stations in the *Bürolandschaft* situation. However, in practice it may well prove too detailed for an office design. Maybe their concept of a "privacy contour" could be applied to a typical floor layout with just a few work stations. It is difficult to see, though, how one can accurately design the whole by looking closely at one small and unique section. Undoubtedly, their method will find better applications in the situation for which it was designed. The shape of their contour allows for speaker orientation, which is justified in a school, with its static conditions, but cannot be applied to the dynamic situation found in an office.

Hegvold (1971) has specifically concentrated on the acoustical design of land-scaped offices using the articulation index. He has again laid out a design procedure for determining privacy conditions, but this falls short of real requirements. He makes no allowances for discrete frequencies within the background noise spectrum, and has tried to oversimplify the case by merely using single-figure dBA readings.

Hegvold's choice of limits on the articulation index for both privacy and communication are realistic, however. For privacy it is suggested that the articulation index should be <0.2, whereas for speech communication their value needs to be >0.7 (Fig. 2.66).

Kryter (1962), Pirn (1971), Farrell (1971) and Lewis (1973b) have all discussed the articulation index and how it is affected by the alteration of several variables. In particular, the last three authors mentioned above have discussed these variables in relation to the landscaped office situation. Kryter (1962) has done exploratory work on the effects of different room reverberation times and has also looked into the resultant articulation index change when, for a given communication system, be it natural vocal or electroacoustic, the listener has visual contact with the speaker's face. As one might expect, the addition of visual contact improves communication in most marginal situations. Taken to its limits, the additive effects of visual cues become lip-reading, but even gestures can communicate great meaning. Modern

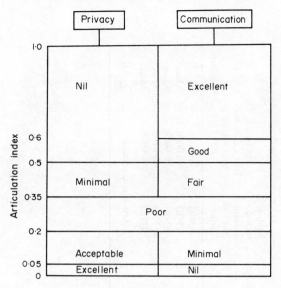

FIG. 2.66. Values of articulation index suitable for privacy and communication (Lewis, 1973a).

anthropologists and zoologists are only just beginning to appreciate the importance of man's common link with animals in the use of gestures.

But most of Kryter's work was concentrated into the development of standard methods for the calculation of the articulation index. In the particular situation of the landscaped office one cannot be quite so precise as to the effect on the articulation index of the two previous variables. Later works have tended to accept that a worker who does not wish to be disturbed will avoid visual contact with others who may be speaking. However, there is room for much further study in this area, as one cannot account for those situations in which a speaker wishes to avoid giving away any visual cues or gestures that may detract from his confidential privacy. Up till now the cure has always been to provide visual barriers like movable screens.

Kryter (1969) has developed a method for the calculation of the articulation index. The steps are shown in Table 2.42.

Thus the articulation index summed over the octave band centre frequencies is 250–4000 Hz will be

$$\sum_{f=250}^{f=4000} \text{AI}_f = \left\{ L_w + 10\lg\left(\frac{4}{R} + \frac{Q}{4\pi r^2}\right) + 12 - \text{NR} \right\} \sum_{f=250}^{f=4000} F. \tag{2.15}$$

In terms of L_{10} and L_{90} levels

$$\sum_{f=250}^{f=4000} \text{AI}_f = \left\{ L_w + 10\lg\left(\frac{4}{R} + \frac{Q}{4\pi r^2}\right) + (L_{10} - L_{90}) - L_{90} \right\} \sum_{f=250}^{f=4000} F. \tag{2.16}$$

The symbols R, Q and r have their usual meaning; NR (or L_{90}) is the design noise rating sound pressure level remembering that activities in an average lecture room or quiet office can contribute 5 dBA or more over the background sound level due to airconditioning systems, lighting and other services; L_{10}–L_{90} is the peak allowance. AI_f will be a minimum when $L_w = L_{90}$, together with a low peak allowance.

TABLE 2.42. CALCULATION OF THE ARTICULATION INDEX

Octave band centre frequency (Hz)	Voice spectrum power level, L_ω (10^{-12} W) (dB)	Sound pressure level, $L_p = L_w + 10\lg\left(\dfrac{4}{R} + \dfrac{Q}{4\pi r^2}\right)$ (see equation (10.31))	Peak allowance or $(L_{10}-L_{90})$ (dB)	Ambient sound pressure level or L_{90} (dB)	Articulation index conversion factor $F^{(a)}$	Resultant
250	77	$+L_{P1}$	$+12$	$-(NR)$	$\times 0\cdot0024$	AI_{250}
500	75	$+L_{P2}$	$+12$	$-(NR)$	$\times 0\cdot0048$	AI_{500}
1000	72	$+L_{P3}$	$+12$	$-(NR)$	$\times 0\cdot0074$	AI_{1000}
2000	65	$+L_{P4}$	$+12$	$-(NR)$	$\times 0\cdot0108$	AI_{2000}
4000	57	$+L_{P5}$	$+12$	$-(NR)$	$\times 0\cdot0078$	AI_{4000}
					Articulation index	$\displaystyle\sum_{250}^{4000} AI_f$

$^{(a)}$Note: $\Sigma F = 0\cdot0024 + 0\cdot0048 + 0\cdot0074 + 0\cdot0108 + 0\cdot0078$.
$\Sigma F = 0\cdot0332$.

Articulation index $\displaystyle\sum_{250}^{4000} AI_f = (L_{P1} + 12 - NR)0\cdot0024$ or $\{L_{P1} + (L_{10} - 2L_{90})\}0\cdot0024$.

$(L_{P2} + 12 - NR)0\cdot0048$ or $\{L_{P2} + (L_{10} - 2L_{90})\}0\cdot0048$.

$(L_{P3} + 12 - NR)0\cdot0074$ or $\{L_{P3} + (L_{10} - 2L_{90})\}0\cdot0074$.

$(L_{P4} + 12 - NR)0\cdot0108$ or $\{L_{P4} + (L_{10} - 2L_{90})\}0\cdot0108$.

$(L_{P5} + 12 - NR)0\cdot0078$ or $\{L_{P5} + (L_{10} - 2L_{90})\}0\cdot0078$.

Substituting Sabine's reverberation time formula in equation (2.15) or (2.16) the articulation index takes the form

$$\sum_{f=250}^{f=4000} \mathrm{AI}_f = \left\{ L_w + 10 \lg\left(\frac{25T(1-\bar{\alpha})}{V} + \frac{Q}{4\pi r^2}\right) + 12 - \mathrm{NR} \right\} \sum_{f=250}^{f=4000} F. \qquad (2.17)$$

Notice how low values of AI_f, as required for privacy, occur when $\bar{\alpha}$ approaches 1 or when the difference $(L_w - \mathrm{NR})$ is low (i.e. use broadband making sound to increase NR relative to L_w). The advantage of using the articulation index is that the interdependence of the various factors can be appreciated and related directly to the primary acoustical objective of achieving intelligible communication between speaker and listener and privacy from other people in adjacent areas. In accordance with the findings of Kingsbury and Taylor (1970) for the classroom situation, a minimum articulation index of 0·7 should be designed to achieve good speech communication, whereas for the converse condition of providing privacy a maximum figure of 0·4 is recommended but in practice a design value of $\mathrm{AI} = 0·2$ is preferable.

Insertion of the articulation frequency components in equation (2.15) will enable the distance r between speaker and unintended listeners or, alternatively, the necessary increase in broadband background sound level, NR+, to be calculated which will ensure that the speech sound patterns are not intelligible.

As the speaker-to-listener distance increases so will the articulation index decrease. With an assumed attenuation rate of between 4 and 6 dB per doubling of distance, changes of between 0·15 and 0·20 can be expected. If the attenuation rate exceeds the free field rate then these changes will be still greater. However, it must be remembered that only the close field of 4–6 m radius from a work station should be considered. Greater distances than this will not only fall well within the 3 dB per doubling of distance "plateau", but will also bring many other noise sources into contention. Very few landscaped offices can economically support each work station in its own 6 m diameter circle. Spatial attenuation rates are important to the articulation index, but they should be considered in the main as the most important controlling factor of the background noise level, thereby affecting privacy indirectly by control of the signal-to-noise ratio.

Speaker-to-receiver orientation has also been considered in some depth (see Fig. 10.5). As the speaker turns from the receiver, the received speech level decreases (assuming no exceptional reflections of focusing effects). Considering the receiver as a non-directional point (which, surprisingly, considering the shape of the ear, can be assumed correct for our purpose), each 90 degree change in speaker orientation results in a decrease of articulation index of approximately 0·15. So a maximum decrease of 0·3 may be obtained when the change in orientation is 180 degrees, i.e. when the speaker's back is to the receiver. At worst this would span the range from acceptable to zero privacy. For the design of an office, though, one should always assume the worst case of orientation, i.e. with speaker facing the listener. It is impossible in the practicable situation to restrict people from moving or turning. This is not to say that nothing can be gained from correct desk layouts, with thought given to speaker orientation. Certainly the average situation can be improved, but it should be looked upon as a possible bonus, not a permanent acoustical feature.

Speech effort and background noise level are undoubtedly the two most important variables in relation to the articulation index. It is becoming increasingly obvious that these two factors should be studied together in much greater depth. Generally it is felt that the landscaped office environment has the effect of reducing conversational levels. Maybe one feels subconsciously too open, and this lack of psychological protection creates a sensation of vulnerability. Undoubtedly it all ties back to the original search for privacy, be it visual, acoustical, physical, psychological or a mixture of all four. However, there are other possible points to be noted here remembering that sometimes people unconsciously raise their voice by 5–10 dB when speaking on the telephone, but not enough is known yet about this phenomenon. Individual voice levels and spectra vary considerably, and, with a potential range of 30 dB, voice-level changes with a histrionic speaker could cause the articulation index to fluctuate by as much as 0·6, thereby spanning the range from confidential privacy to good communication. To a certain extent speech training and education could be an essential part of a landscaped office worker's curriculum. Reference has been made already to the Lombard voice reflex, and initial studies (Lane, 1971) suggest that, for every 2 dB increase (or decrease) in background noise, the vocal effort is increased (or decreased) by 1 dB up to a maximum level at which the vocal chords physically become strained.

As the articulation index is a frequency-weighted measure it is not easy to specify the effect of a single-figure change in background noise level. Each case would have to be worked out on its own merits unless one assumed a linear change throughout the frequency bands. In the latter case it is justifiable to say that a change in the background level of 3 dB causes a change in the articulation index of 0·1. A reduction in level of 6 dB is therefore sufficient to transform a condition of acceptable privacy (AI = 0·15) to one of fair communication (AI = 0·35). The importance of the background noise level to the articulation index cannot be overemphasized, which is why writers like Farrell (1971) and Cowell and Motteram (1973) have tried to stress the significance to designers of the new concept of electronic *sound conditioning*. With complete control over background noise levels one has almost complete control over the privacy conditions within a space. The limiting factors are obviously the maximum levels tolerable as ambient noise of visual cues and the situation where one has perfect communication.

Cross-talk arises when conversation in one room is transferred to another via the ductwork and hence limiting the degree of privacy (Fig. 2.67). The speaker in room 1 acts as a sound source which is transmitted via the ductwork system to the unintended listener in room 2.

Thus for intelligibility of speech source L_w, in room 1

$$\sum_{f=250}^{f=4000} (AI_f), \quad \geqslant 0 \cdot 7.$$

For privacy in room 2, L_{w_1} must be unintelligible. However, L_{w_1} has been transmitted down a duct of length L having an attenuation of n dB unit length, and a sound generation (which will help to mask L_{w_1}) of g dB per unit length and emitted through a duct terminal area S; thus for speech source L_{w_1},

$$\sum_{f=250}^{f=4000} (AI_f)_2 \leqslant 0 \cdot 2,$$

$$\frac{0 \cdot 2}{\sum F} = L_{w_1} - (n + g)L + 10 \lg S + 12 - NR_2 + 10 \lg \left(\frac{4}{R} + \frac{Q}{4\pi r^2}\right)_2.$$

Note the peak allowance of 12 refers to L_{w_1}. Strictly speaking a duct end reflection should be included in n but neglecting it gives a margin for error. The work of Miller and Nicely (1955) was shown in Fig. 2.62 and suggests that $L_{w_1} - NR_2$ should be 18 dB if consonants are to be indistinguishable from one another. Hence, using $\Sigma F = 0 \cdot 0332$ (see Table 2.42) and summing over all frequencies,

$$(n + g)L = 30 - \frac{0 \cdot 2}{0 \cdot 0332} + 10 \lg S + 10 \lg \left(\frac{4}{R} + \frac{Q}{4\pi r^2}\right)_2.$$

The effective attenuation n, or masking g, can be ascertained.

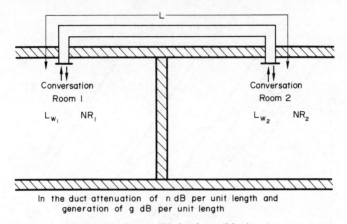

In the duct attenuation of n dB per unit length and generation of g dB per unit length

FIG. 2.67. Cross-talk via airconditioning duct.

The principal advantage of using the articulation index for estimation of cross-talk attenuator requirements is that all the relevant design factors are included—background noise; speech information content; attenuation requirements; effect of duct length, terminal and the room; masking noise requirements can also be estimated if required (see Waller, 1969).

2.6.2. NOISE LEVEL CRITERIA FOR SPEECH COMMUNICATION

Although the advantages of using the articulation index have been stressed, the index is not yet widely used in practice. Bowman (1973) has pointed out some of the limitations of the index. Beranek (1971, p. 586) gives a simplified procedure for speech privacy design. What follows is a brief summary of noise criteria for speech communication that are used in practice.

A useful criterion has been advocated by Webster (1969). He proposed the use of Fig. 2.68 as a method of specifying the noise level in a building in order to permit *preferred octave speech interference level (PSIL)* which is the average sound pressure level in the octavebands 500, 1000 and 2000 Hz,

or
$$PSIL = dBA - 7 dB$$

or $$PSIL = PNdB - 20\,dB$$
or $$PSIL = SIL + 3\,dB$$

where SIL is the *speech interference level* conceived by Beranek (1947a, b) and defined as the average sound pressure level in the octave band of 600–1200, 1200–2400 and 2400–4800 Hz. Figure 2.68 shows how the distance between speaker and listener may be estimated for satisfactory speech communication in various background noise conditions. In a survey by Croome (1969a) of the acoustical environment in landscaped offices a level of 50–55 dBA (43–48 PSILdB) was found to be satisfactory; referring to Fig. 2.68 this corresponds to an intercommunication distance of 3–4 m for people using normal voice levels; this happens to correspond to a distance which has been found to give good informal social groupings in open-plan offices.

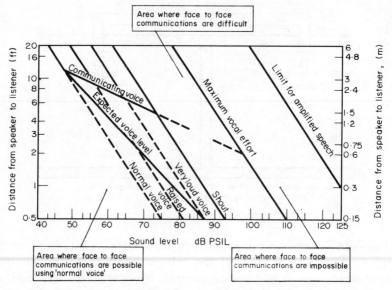

FIG. 2.68. Voice sound levels needed to communicate over various distances.

Current design practice is to use the noise rating curves (i.e. NR curves) shown in Fig. 2.69 as a method of specifying the sound level in a building in order to permit intelligible speech communication.

They were obtained empirically for a small sample of people by Kosten and van Os (1962) using broadband sources, and in principle they are similar to the noise criterion (NC) curves proposed by Beranek (1956). Each curve is classified by a number corresponding to the SIL; if the loudness level (phons) in a space exceeds the NR number by more than 22 units, then speech communication will be difficult. The relationship between articulation index, speech and background sound pressure levels is shown in Fig. 2.70, and it can be seen that a difference of 22 dB between the latter two levels will ensure an articulation index of 0·8. Recommended noise ratings for various kinds of buildings are shown in Table 2.43. Remember that the resultant sound level in a space is due to external sources, activities within and around the space, besides the noise originating from the airconditioning system. In setting the

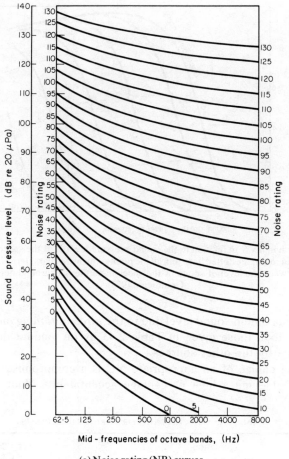

(a) Noise rating (NR) curves.

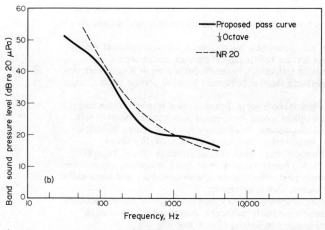

(b) Proposed noise pass curve compared with NR criterion curve (Clark and Petrusewicz, 1970–1).

Fig. 2.69. Noise criteria curves.

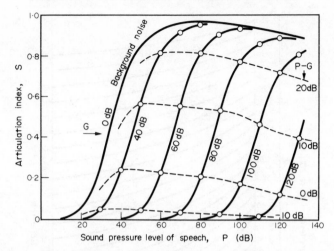

FIG. 2.70. Articulation index S in relation to the sound-pressure level P of the speech and to the level of background noise G in room; the dotted lines join all the points which are characterised by differences of 20, 10, 0 and −10 dB between the level of sound of the speech and the background noise (i.e. $P − G$) (after Grandjean, 1973).

criterion for the maximum sound level due to the *airconditioning only* the NR figure should be lower than those given in Table 2.43 by an amount dependent on the contribution made by the other sources.

In practice, the choice of an appropriate NC is a compromise between system economics and achieving a high percentage acceptability. Young (1964) demon-

TABLE 2.43. NR CRITERIA FOR VARIOUS BUILDINGS

Situation	NR criterion
Concert halls, opera houses, studies for sound reproduction, live theatres (< 500 seats)	20
Bedrooms in private homes, live theatres (> 500 seats), cathedrals and large churches, television studios, large conference and lecture rooms (< 50 people)	25
Living rooms in private homes, board rooms, top management offices, conference and lecture rooms (20–50 people), multipurpose halls, churches (medium and small), libraries, bedrooms in hotels, banqueting rooms, operating theatres, cinemas, hospital private rooms, large courtrooms	30
Public rooms in hotels, ballrooms, hospital open wards, middle management and small offices, small conference and lecture rooms (> 50 people), school classrooms, small courtrooms, museums, libraries, banking halls, small restaurants, cocktail bars, quality shops	35
Toilets and washrooms, large open offices, drawing offices, reception areas, offices, halls, corridors, lobbies in hotel, hospitals, laboratories, recreation rooms, post offices, large restaurants, bars and night clubs, departmental stores, shops, gymnasia	40
Kitchens in hotels, hospitals, laundry rooms, computer rooms, accounting machine rooms cafeteria, canteens, supermarkets, swimming pools, covered garages in hotels, offices, bowling alleys	45
Landscaped offices	35–45
Workshops	50–55

strated that the subjective data obtained by Beranek correlated just as well with noise levels expressed in dBA as with the NC rating. It is certainly easier to refer to and to measure a single-figure value, but when noise control systems are being designed it is essential to have a knowledge of the sound spectrum. For noise sources having a spectral shape that is similar to the NR curves it can be taken that the NR rating is equivalent to (dBA − 6).

The NR curves are mainly concerned with the loudness of noise over all frequencies whereas annoyance to noise depends not only on this attribute, but also the presence of tones at particular pitches, the number of tones, their duration and intermittency and their information content. If a sound is known to have a significant tone 5 dB is added to the NR number in an attempt to mask it (see Table 2.45). The validity of this design procedure was checked by the author by carrying out experiments on eight different classes of students exposed to noise, which included 80 Hz (at amplitudes of 67 or 73 dB), 160 Hz (at amplitudes of 55 dB or 64 dB), 315 Hz (at amplitudes of 59 dB or 65 dB) or 630 Hz (at amplitudes of 50 or 54 dB) tones during the duration of a one hour lecture. A Wilcoxon matched-pair signed rank test was used to discover the significance of the thermal, visual or acoustical environment on annoyance and distraction. The 80 Hz and 160 Hz tones did not cause any significant reaction on the distraction scale whereas the higher frequencies were significant. A Kruskal Wallis one-way analysis of variance showed that pitch had a more pronounced distraction effect than intensity of the tone but the higher intensities were more distracting than the lower ones, in fact the 315 Hz at 65 dB was intolerable and was switched off after 35 minutes. It is suggested that the NR corrections applied to lecture rooms should be + 10 dB, + 5 dB, and − 5 dB in the frequency ranges 63–125 Hz, 125–250 Hz and > 250 Hz respectively.

The Wilson Committee (1963) recommended the L_{10} sound levels shown in Table 2.44 which should not be exceeded for more than 10% of the time in the buildings classified.

TABLE 2.44. RECOMMENDED L_{10} VALUES (WILSON, 1963)

	dBA (L_{10} values)	
New factories	50	
Old factories not typical of the area	55	
Old factories in keeping with the area	60	
General areas in schools, offices and hospitals where speech intelligibility is important	55	
	Day-time	Night-time
Houses in country areas	40	30
Houses in suburban areas but away from main traffic routes	45	35
Houses in busy urban areas	50	35

Allowances to these recommendations and those shown in Table 2.43 should be made as Table 2.45 shows.

A survey of noise problems that occur in over 400 households by Clark and Petrusewicz (1970–1) has led them to advocate some modified NC curves. Other results from the survey are described in more detail in Chapters 3 and 4. The present method of designing the noise requirements for a bedroom would be to select NR 25 (see Table 2.43 and Fig. 2.69a) and apply a correction of − 5 (see Table 2.45) if a pure

TABLE 2.45. CORRECTION TO NR OR dBA RECOMMENDATION

Character of noise:	
Pure tone (i.e. whistle, hiss	
screech or noticeable humming	
noise)[a]	−5
Impulsive noise[a] (i.e. bangs, thumps,	
hammering, clanks, hammering or riveting)	−5
Time:	
Weekdays 8 a.m. to 6 p.m.	+5
Evening up to 10 p.m.	0
Weekends	0
Night time 10 p.m. to 7 a.m.	−5
District (adopt one correction only):	
Rural (residential)	−5
Suburban or urban, no road traffic	0
Residential urban	+5
Urban with light industry or main roads	+10
General industrial area	+15
Heavy industrial area	+20
Intermittent noise	
(use to determine intermittent limiting level	
but not applicable during evening	
or night):	
Noise occurring for 15 min/h	+5
5 min/h	+10
1 min/h	+15
1 min/half-day	+20

[a]Powell (1971) suggests, on the basis of his experimental work, that 15 dBA is a better allowance to gain a 95% acceptability rating for impact noise; using repetition rates of 5–100 impacts per second, Powell showed that the stated allowance of 5 dBA in Table 2.45 only predicts the mean loudness level.

tone is easily perceptible. Using a five-point acceptable–totally unacceptable scale, Clark and Petrusewicz (1970–1) explored the subjective effects of noises commonly encountered in houses, especially pump noise, and found that this standard NR25-5 was quite unacceptable in the households in their survey.

Accepting the assumption that a discrete frequency is just masked when the energy it contains is equal to the energy contained in the critical bandwidth B centred on the discrete frequency, then the level to which the pure tone must be reduced to be masked is calculated from

$$\text{power spectrum level per Hz} = \text{broadband sound pressure level} - 10 \lg B.$$

This study showed that it was not the overall sound level which was so important as the level necessary to ensure that any discrete tones are masked.

The new proposed criterion is shown in Fig. 2.69b and compared with NR 20. The greatest differences between the new proposal and the traditional criterion occur below 100 Hz and in the 250–700 Hz frequency range. The latter frequencies are due to the noise transmitted through the water and the piping to the radiators and cause the most trouble. Noise in the frequency range 1000–4000 Hz at the pump due to air in suspension and water turbulence tends to be damped by the system before it reaches the radiator.

We have already seen that the reaction to noise depends on a host of factors. There are the physical attributes, displayed in the temporal and spatial sound patterns, and the psychophysiological factors such as the arousal level of the individual, sensitivity of the hearing system, type of work task—these, and probably other unknown factors, form an intricate array from which we evaluate our environment. In the future it can be expected that environmental design recommendations will include allowances for these factors as our knowledge about them extends.

2.6.3. LANDSCAPED OFFICES

Some general features of landscaped offices have already been discussed in the early part of this chapter. The acoustical characteristics of these offices may perhaps best be illustrated by using the European field surveys made by Hoogendoorn (1973) and Nemecek and Grandjean (1973a, b). The latter carried out objective and subjective measurements in fifteen Swiss landscaped offices. Five hundred and nineteen employers (80% men, 75% in the 20–40-year-old age group) were asked individually a series of questions about their experiences and opinions of landscaped offices. The range of physical noise climates in the offices is shown in Fig. 2.71 from

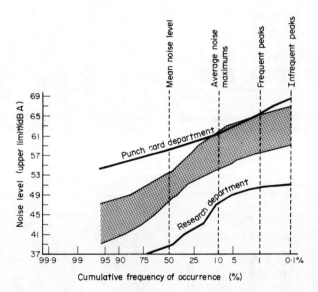

FIG. 2.71. Cumulative frequency curves of noise level in 15 landscaped offices (Nemecek and Grandjean, 1973a, b).

which it can be seen that the mean noise level (L_{50}) varied from 48–53 dBA, whereas the frequent peak levels (L_1) ranged from 58–65 dBA. In contrast to the work of Hay (1973), no correlation was found between the degree of noise disturbance and the L_1, L_{10} or L_{50} levels.

Before this work, German landscaped office acoustical design was guided by the work of Zeller (1964) and Gottschalk (1968), both of whom recommended an L_{50} level of 52–55 dBA, whereas the German Society for the Protection of Work proposed 50–55 dBA; Zeller also suggested that no peaks should exceed 60–65 dBA. Boyce (1974) reports average levels of 54 dBA and peak levels of 62 dBA, but concludes that using the level of noise above as a criterion as an indicator of noise disturbance is of limited usefulness because it is the information content and the unpredictability of the stimulus which are also important.

An analysis of the subjective responses from the study by Nemecek and Grandjean (1973a, b) is given in Fig. 2.72a–d.

Figure 2.72a shows that:

* 35% of those questioned were very much disturbed by noise;
* 45% were slightly disturbed;
* 20% were not disturbed at all;
* office clerks and clerical assistants were a little less disturbed by noise than directors and graduates;
* men were more bothered by the office noise than women.

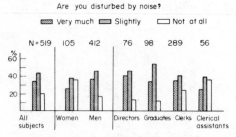

(a) Frequency distribution of answers concerning noise disturbances; distribution between the social groups are significant with $p < 0.01$ (chi^2-test).

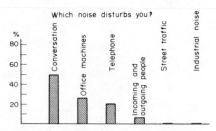

(b) Answers to the question concerning noise sources; 411 people questioned whether much or slightly disturbed; 762 multiple answers = 100%.

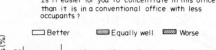
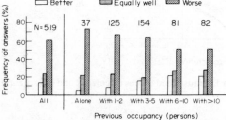

(c) Frequency distribution of assessments on concentration possibilities for all 519 people questioned and five groups with various previous occupancies; the differences between the five groups are significant with $p < 0.05$ (chi^2-test).

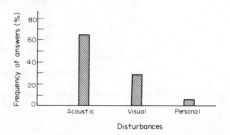

(d) Frequency of data on distraction causes; 448 multiple answers = 100%; those 313 people who complained of worse concentration in landscaped offices were questioned.

FIG. 2.72. Subjective responses to noise in landscaped offices (Nemecek and Grandjean, 1973b).

Figure 2.72b shows clearly that conversations were the most bothersome noise source, being mentioned by 46% of the respondents, while office machines and telephones were only mentioned on 25% and 19% of occasions; external noise was not a problem. Many subjects reported that it was the content of the conversation rather than the loudness that was disturbing.

Figure 2.72c gives data relating to a comparison between people's concentration in landscaped offices and their previous offices. The results show distinctly that 61% of subjects found it more difficult to concentrate in a landscaped office than in a small office, whereas only 25% found no difference in their ability to concentrate, and 14% found it easier to concentrate in a landscaped office. The differences were most marked for people that had worked alone or in offices accommodating up to five people.

Figure 2.72d shows that people were more often disturbed by noise than by visual causes.

Boyce (1974) assessed and compared the noise climate in conventional and landscaped offices. No significant differences were found between them for disturbance assessed from judgements marked on an often–seldom–never scale. By the time people had settled down in the new landscaped office, age and sex differences were also not significant. However, draughtsmen and managerial groups were more disturbed by noise than clerks and secretarial staff; this point was also found to be true in Nemecek's and Grandjean's study. Table 2.46 shows the sources of noise that were found to be disturbing. Except in one case all primary sources of noise disturbance were internal, people talking and the telephone. In the landscaped office, road noise is not mentioned, but the range of disturbing internal noise sources increases to include airconditioning, typing and office machines.

The provision of space per person is an important factor in office design. Too little produces tensions in the individual due to overcrowding, and too much is plainly uneconomical. Past work suggests $8 \cdot 5$–14 m^2 per person is satisfactory. Boyce (1974) gives evidence that between $8 \cdot 5$–$12 \cdot 6 \text{ m}^2$ per person 11% of the people in his study felt overcrowded, but at $7 \cdot 4 \text{ m}^2$ per person overcrowding becomes a very serious problem, 41% of subjects experiencing this disadvantage.

Nemecek and Grandjean (1973a, b) also produced some correlations between noise effects and room-design aspects; these are shown in Table 2.47 together with references to other data.

The influence of density of occupation on sound level has been measured by Hoogendoorn (1973). The results illustrated in Fig. 2.73 show that when the occupational density is doubled there is a 6 dB increase in the mean and the peak sound levels. There is a tendency for people to raise their voices when the room they are in is more densely occupied (i.e. the Lombard voice reflex), and this may account for part of this large increase in sound level. It has been suggested that too little acoustical continuity occurs when the occupational density is very low, say at about 15 m^2 per person. Several sources of information indicate that a gross floor area of about 10 m^2 per workplace is preferable.

Hoogendoorn (1973) reveals some interesting facts in a comparison survey he conducted at the landscaped offices of two firms, Volvo at Göthenburg in Sweden and Otto Maier at Ravensburg in Germany. The principal similar and dissimilar features are shown in Table 2.48. Contrast is provided by the ceiling structure and

TABLE 2.46. PERCENTAGE FREQUENCY OF OCCURRENCE OF DISTURBING NOISE SOURCES IN OFFICES (BOYCE, 1974)

Conventional offices				Landscaped office	
1	2	3	4		
Pre-war heavy six-storey office block with central light well; in city centre near main road	1950 lightweight highly glazed six-storey slab office block; in city centre near main road	Pre-war converted suburban houses	As 2 but two-storey in rundown industrial area	Short-term responses	Responses made after a year of occupation
Road noise 77	Telephone 70	Telephone 61	Telephone 91	Telephone 67	Telephone 75
Telephone 51	People talking 61	People talking 45	People talking 75	People talking 55	People talking 63
People talking 40	Road noise 57		Road noise 25	Airconditioning 34	Airconditioning 49
	Office machines 20			Typing 28	Office machines 35
				Office machines 21	Typing 33

TABLE 2.47. INFLUENCE OF ROOM DESIGN ASPECTS ON NOISE ASSESSMENT

Room-design aspects	Noise climate	
	Better	Worse
Number of workplaces:		
Nemecek and Grandjean (1973a, b)	52–120	20–50
Gottschalk (1968)	> 80	
Zeller (1964)	30–50 minimum	
Gross floor area:		
Nemecek and Grandjean (1973a, b)	475–1355 m²	252–445 m²
Gottschalk (1968)	> 1000 m²	
Zeller (1964)	300–500 m² (minimum)	
Height of room:		
Nemecek and Grandjean (1973a, b)	2·5–2·7 m	2·96–3·30 m
Zeller (1964)	2·8 m (maximum)	
Gross floor area per workplace:	9·0–11·2 m² per workplace	
Zeller (1964)	8·0 m² per workplace (minimum)	
Hoogendoorn (1973)	—	15.0 m² per person gives too little acoustical continuity

finish. The use of a highly absorbent ceiling trims the noise peaks, but notice that there is no appreciable difference in the average sound levels (Fig. 2.74a–c).

Although to a more limited degree than for music, spaces designed for speech communication require a balance of direct and reverberant sound, the reverberant field provides feedback to the auditory sensory system; besides, in enclosed spaces there is an "expectancy" of reverberant sound based on contrast patterns between the free field of Nature and the sound envelopment in buildings experienced by an

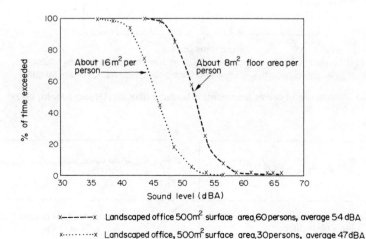

x——x Landscaped office 500m² surface area, 60 persons, average 54 dBA

x········x Landscaped office, 500m² surface area, 30 persons, average 47 dBA

FIG. 2.73. Effect of occupational density on sound level (Hoogendoorn, 1973).

TABLE 2.48. COMPARISON OF TWO LANDSCAPED OFFICES
(HOOGENDOORN, 1973)

Characteristics	Volvo	Otto Maier
Similar aspects:		
Floor surface area	900 m²	1200 m²
Occupation density	11 m² per person	10 m² per person
Activities	General clerical work	
Room finishes:		
10 mm carpet	Similar	
Absorbent screens	One to every four persons	
Average sound level		
(L_{50}) (see Fig. 2.74)	57 dBA	56 dBA
Different aspects:		
Ceiling finish	100 mm thick	Flat absorbent
(see Fig. 2.74a)	wood slabs,	ceiling
	void, egg-	
	crate grid	
Peak sound levels	64 dBA	58 dBA
(L_1) (see Fig.		
2.74b and c)		
Transient peak index	High	Low

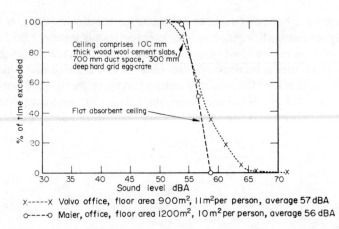

x-----x Volvo office, floor area 900m², 11m² per person, average 57 dBA
o----o Maier, office, floor area 1200m², 10m² per person, average 56 dBA

(a) Comparison of ceiling treatments in landscaped offices (Hoogendoorn, 1973).

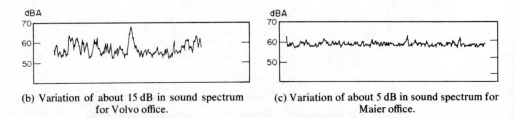

(b) Variation of about 15 dB in sound spectrum
for Volvo office.

(c) Variation of about 5 dB in sound spectrum for
Maier office.

FIG. 2.74. Comparison of two landscaped offices with different ceiling treatments (Hoogendoorn, 1973).

individual from an early age, i.e. buildings should be designed to have spatial responsiveness. An acoustically dead space cannot satisfy this condition. An extreme case is an anechoic room which gives a feeling of "strangeness" or some form of sensory deprivation, whereas in practice a dead space will have $\bar{\alpha} \simeq 0\cdot4$, and field investigations have shown that this gives a hushed, unnatural environment (see Croome, 1969a). On the other hand, too much reverberation will result in a climate which is too noisy.

Landscaped offices are distinguished by a high width (or breadth) to height ratio, and so the walls play a negligible part in the sound distribution. The floors are usually carpeted and so, together with furniture and people, the sound absorption of the space is fixed, the remaining variable being the ceiling. The effect of multiple sources and their sound distribution in low-height rooms has been given some attention in section 10.5. Here a different approach will be made in the light of current research.

The acoustical design of landscaped offices is not straightforward for a number of reasons. Workplaces are adjacent to one another so that there is the possibility of conversations, telephones, typewriters and general background noise (e.g. shutting of the file drawers) being distracting either by direct sound transmission or by ceiling reflections between workplaces. Measurements by Schultz (1971a), West (1973a), Aldersey-Williams (1973a), Day (1973), Hoogendoorn (1973), Peutz (1973) and Croome and Brook (1974) have shown that the spatial attenuation rate varies in any given office and can exceed the direct free field condition of 6 dB per doubling with distance; the reverberation time also varies throughout the office. It seems that the intermittency of the numerous noise sources (i.e. conversations, telephones and office machinery) produces a non-diffuse sound field rendering the use of the classical acoustical equations invalid; the reverberation time equation (derived in Chapter 10) for instance takes the form

$$T(x, y, z, t) = \frac{0\cdot16V(x, y, z)}{A(x, y, z)},$$

where $T(x, y, z, t)$ is a spatial–temporal distribution function of reverberation time; the $V(x, y, z)$ and $A(x, y, z)$, functions suggest that the office should not be considered as one large volume, but as a composite number of volumes each having its own absorption characteristic. It is suggested that these volumes should be categorised by the resultant effect of the sound sources in each volume, e.g. secretaries with typewriters could form one group with its particular sound climate, office machines another, jobs needing a lower sound level for concentration another, and so on.

Peutz (1973) has found that the attenuation of sound pressure level per doubling of distance is a function of floor area S, height of the room h and the measured reverberation time T_m; this relationship is shown in Fig. 2.75 and can be expressed in SI units as

$$\Delta = 7\cdot821 \left(\frac{S}{hT_m^2}\right)^{0\cdot01776}. \tag{2.18}$$

Notice that shape of the space, denoted by the floor area-to-height ratio, is partly governing the sound distribution; in Chapter 10 a height-to-width ratio is recommended to obtain the optimum sound distribution in concert halls.

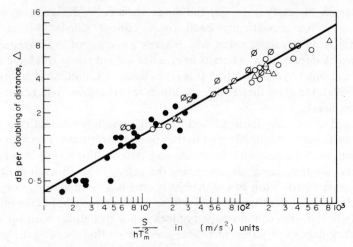

FIG. 2.75. The attenuation per doubling of distance Δ variation with the floor area S over the height h times the square of the measured reverberation time T_m (Peutz, 1973).

The total sound pressure in a space will be shown to be the sum of the direct and reverberant field contributions (see section 10.4),

$$p^2 = \frac{\rho c W}{4 \pi r^2} + \frac{4 \rho c W}{R}.$$

This assumes that the directivity factor for the source is unity. The direct sound pressure level varies with distance according to the well-known law (see Chapter 10)

$$L_p = L_w - 20 \lg r - 10 \lg 4\pi.$$

Between two points r_1 and $2r_1$ the spatial attenuation rate is

$$\Delta = (L_{p_1} - L_{p_2}) = 6 \text{ dB per doubling of distance.}$$

The reverberant sound pressure expression has no distance term in it. However, we have seen from Peutz (1973) that his measurements in landscaped offices show that the spatial variation in reverberation time and the room geometry must be taken into account. Suppose that the spatial function $\phi(r)$ represents an attenuation parameter so that total sound pressure in a space is given by

$$p^2 = \left(\frac{\rho c W}{4 \pi r^2} + \frac{4 \rho c W}{R} \right) \frac{1}{\phi(r)}. \tag{2.19}$$

As the receiver moves away from the source the sound power contributions from the image sources (Day, 1973) become important. Thus assume a sound power function $W/\phi(r)$ such that at low values of r, $W/\phi(r) \to W$.

The sound pressure level can now be expressed as

$$L_p = L_w - 10 \lg \phi(r) + 10 \lg \left(\frac{1}{4 \pi r^2} + \frac{4}{R} \right).$$

Between two points r_1 and $2r_1$ the sound pressure level difference, or the spatial attenuation rate, is

$$\Delta = 10\lg\left\{\frac{\phi(2r_1)}{\phi(r_1)}\right\} + 10\lg\left\{\frac{\left(\dfrac{1}{4\pi r_1^2}+\dfrac{4}{R}\right)}{\left(\dfrac{1}{16\pi r_1^2}+\dfrac{4}{R}\right)}\right\}.$$

Using the expression due to Peutz (1973),

$$10\lg\left\{\frac{\phi(2r_1)}{\phi(r_1)}\right\} = 7\cdot821\left(\frac{S}{hT_m^2}\right)^{0\cdot01776} - 10\lg\left\{\frac{\left(\dfrac{1}{4\pi r_1^2}+\dfrac{4}{R}\right)}{\left(\dfrac{1}{16\pi r_1^2}+\dfrac{4}{R}\right)}\right\}.$$

A general pattern of sound decay is shown in Fig. 2.76; three distinct regions can be seen. Near the source of noise Δ is about 6 dB per doubling of distance; this decreases to 3 dB per doubling of distance at distances of about $0\cdot5h$ to $3h$ (h is the room height in metres) from the source; beyond a distance of about $3h$ from the source Δ depends on the amount of absorption and room dimensions, but the attenuation rate may exceed 6 dB per doubling of distance. This pattern can be recognised in some of the predicted and measured attenuation curves shown in Fig. 2.76a–e. The fluctuations in the attenuation tend to be less at higher frequencies. Notice that the predicted curves (Fig. 2.77a and b) compare favourably with measurements shown in Fig. 2.77d and e up to 10–15 m from the source; the discrepancy in the 20–30 m region is attributed to the fact that the theoretical predictions have not included the effect of the walls. West (1973a) also made attenuation measurements in offices having zero and half furniture densities; his results were very similar to those shown in Fig. 2.77d for a normal furniture density.

Using the empirical data shown in Fig. 2.76 in equation (2.18) three conditions

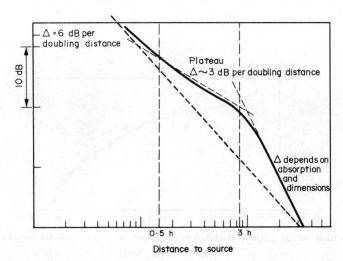

FIG. 2.76. The relative sound pressure level versus the distance to the source (in relation to height h) in a landscaped office (Peutz, 1973).

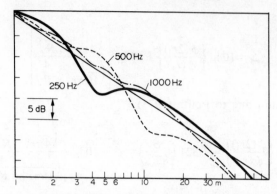

(a) Predicted attenuation curves for an office with a perfectly absorbing ceiling and a carpeted floor (source height 1·16 m) (West, 1973a).

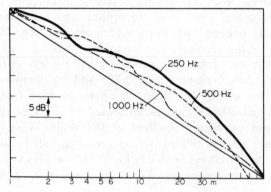

(b) Predicted attenuation curves for an office with an "acoustimetal" suspended ceiling (perforated tiles backed with a 25 mm layer of Stillite in a polythene bag) having a 1·9 m airspace (West, 1973a).

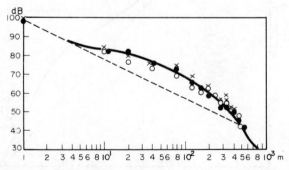

(c) The attenuation versus distance in different directions in a very large landscaped office of about 7000 m² (Peutz, 1973).

FIG. 2.77. Typical sound attenuation curves in landscaped offices.

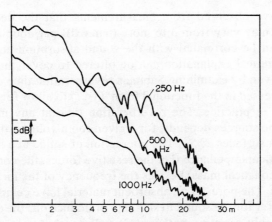

(d) Measured attenuation in the Kew Office with normal furniture density, diagonal traverse (West, 1973a).

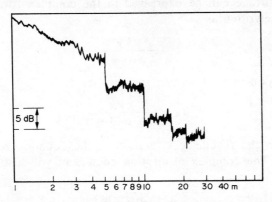

(e) Measured attenuation in the Kew Office with hard screens placed at 5 m intervals for 1000 Hz octave band source (West, 1973a).

FIG. 2.77. (*Continued*).

apply:

(a) near the source where $r < 0.5h$ (in metres) Δ is 6 dB per doubling of distance, hence

$$7.821 \left(\frac{S}{hT_m^2}\right)^{0.01776} = 6;$$

(b) at distances of $0.5h < r \leqslant 3h$, Δ is 3 dB per doubling of distance and

$$7.821 \left(\frac{S}{hT_m^2}\right)^{0.01776} = 3;$$

(c) when $r > 3h$ then $\Delta > 6$ dB per doubling of distance,

$$7.821 \left(\frac{S}{hT_m^2}\right)^{0.01776} > 6.$$

Until now we have observed from measurements that the decay of sound in a landscaped office may vary from 3 to more than 6 dB per doubling of distance and the decay rate can be correlated with the sound absorption and geometry of the office. What theoretical explanation can be offered to cover these issues?

A first hint is given by examining Sabine's reverberation time equation which has already been expressed in the functional form $T(r, t)$ to allow for the variations with distance r found in practice. The reverberation time at any instant t, at a point distance r from the sources depends, for a given room volume, on the absorption of sound at the room surfaces, $\Sigma\, Sa$. The interaction of sound at a surface depends on the surface acoustical impedance (i.e. the resistive forces, the compressibility of the surface and the acoustical mass) besides the frequency of the incident sound and its angle of incidence. The pores of an absorbent material have cellular spaces the air in which forms a buffer for the sound to sink into or bounce off; the entry to the cellular space offers a highly resistive pathway. It will be shown in Chapter 6 that the acoustical impedance has resistive and reactive components so that the absorption coefficient at a surface can be expressed in the complex form $\alpha + j\psi$. Sabine's equation can be rewritten as

$$T = \frac{kV}{\Sigma\, S(\alpha + j\psi)},$$

giving a real component of T as

$$T_R = \frac{kV\alpha}{\Sigma\, S(\alpha^2 + \psi^2)}.$$

Notice that when the reactance is zero then the equation reverts back to the traditional form. The complex absorption coefficient will also modify the room constant R to

$$R = \frac{S(\alpha + j\psi)}{1 - (\alpha + j\psi)}.$$

Day (1973) has explained the spatial attenuation rate in terms of broadband interference between image sources. As the receiver moves from the source, a line of images form the auditory view and the sound power at the receiver is that due to the direct pathway from the real source plus the contribution from the image sources (Fig. 2.78). In practice the situation is complicated further by the presence of not just

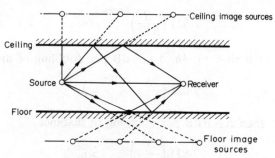

FIG. 2.78. Sound image formation.

one but a multitude of sources, some of which operate intermittently. Strictly speaking, equation (2.19) should consider a multitude of complex sound power sources which interfere producing very high spatial attenuation rates far from the source; instead another approach using the attenuation function $\phi(r)$ has been developed here for ease in the mathematical treatment of the problem, and also because the research of Peutz has established the nature of such a function. The problem may be complicated further because speech sources can defy physical measurements that attempt to assess privacy; attention switching (e.g. cocktail party effect) may occur with noises having a high information content.

Rewriting equation (2.19) in a complete form gives:

$$\frac{p^2}{\rho c} = \frac{1}{4\pi}\left\{\frac{W_1}{r_1^2\phi(r_1)} + \cdots \frac{W_n}{r_n^2\phi(r_n)}\right\} + \frac{4\{1-(\bar{\alpha}+j\psi)\}}{S(\bar{\alpha}+j\psi)}\frac{\Sigma(W + \cdots W_n)}{\phi(r)}$$

$$= \frac{1}{4\pi}\left\{\frac{W_1}{r_1^2\phi(r_1)} + \cdots \frac{W_n}{r_n^2\phi(r_n)}\right\} + \frac{4}{S}\frac{\Sigma(W + \cdots W_n)}{\phi(r)}\left(\frac{\alpha}{\alpha^2+\psi^2}\right)$$

$$+ \frac{4}{S}\frac{\Sigma(W_1 + \cdots W_n)}{\phi(r)}\left(\frac{j\psi}{\alpha^2+\psi^2}\right), \qquad (2.20)$$

where

$$10\lg\left\{\frac{\phi(r_n)}{\phi(r_{n-1})}\right\} = 7{\cdot}821\left(\frac{S}{hT_m^2}\right)^{0{\cdot}01776} - 10\lg\left\{\frac{\left(\frac{1}{4\pi r_n^2}+\frac{4}{R}\right)}{\left(\frac{1}{4\pi r_{n-1}^2}+\frac{4}{R}\right)}\right\}.$$

Day (1973) expresses the complex pressure amplitude contributed by each image in the form

$$\bar{p}_n = \frac{p_0}{r_n}|\tilde{R}_c|^{n_c}|\tilde{R}_f|^{n_f}\exp(-jkr_n)\exp j(\psi_c n_c + \psi_f n_f),$$

where R_c and R_f are the amplitudes of the complex pressure reflection coefficients of the ceiling and floor respectively, r_n is the distance from image to receiver, n is the number of reflections forming the image, k is the wave number (i.e. $2\pi/\lambda$ or $2\pi f/c$), and ψ_c and ψ_f are the arguments of the complex pressure reflection coefficients. The complex pressure coefficients are functions of the surface acoustical impedance, the angle of incidence, the frequency of sound and the image order number.

In considering the need for privacy, or good speech communication, use can again be made of the articulation index (see Croome and Brook, 1974), although the form given in equation (2.17) needs to be modified to allow for the spatial attenuation function $\phi(r)$ and the complex absorption coefficients, thus

$$\sum_{250}^{4000}\frac{\text{AI}_f}{\Sigma F} = L_w - 10\lg\phi(r) + 10\lg\left[\frac{4\{1-(\alpha+j\psi)\}}{S(\alpha+j\psi)} + \frac{Q}{4\pi r^2}\right] + (L_{10}-L_{90}) - L_{90}. \qquad (2.21)$$

Two conditions apply. Within the communication radius of $r = 4$ m then $\Sigma\,\text{AI}_f > 0{\cdot}7$, whereas for privacy at $r > 4$ m, $\Sigma\,\text{AI}_f < 0{\cdot}15$. (Note that from Table 2.42 $\Sigma F = 0{\cdot}0332$). Equation (2.21) can be solved using these two conditions giving the design values for α or, alternatively, for the required background sound level.

In calculating the articulation index the basic equation used to correct the idealised voice power level spectrum and to convert it into a sound pressure level spectrum is given by

$$L_p = L_w + 10 \lg \left(\frac{Q}{4\pi r^2} + \frac{4}{R} \right),$$

where L_w is the sound power level (dB re 10^{-12} W); L_p is the sound pressure level (dB re $20\,\mu$Pa); Q is the directivity factor, here taken as unity, to allow for the worst case of speaker-to-listener orientation, that is, face-to-face; r is the communication distance from speaker to listener here taken as $4\cdot0$ m as for an average landscaped office situation; and R is the room constant and the term $(4/R)$ is sometimes known as the *room effect*.

Obviously, if the articulation index is to be used as a design tool it is important that the effects of certain variations to parameters are well known and understood.

In the study by Croome and Brook (1974) the term $[Q/(4\pi r^2)]$ was found to be a factor of about 10 larger than the room effect term at $r = 2$ m, and even at $r = 8$ m it was of the same order as the $(4/R)$ term. This study did not, however, consider a complex value for α. Thus, maintaining a constant $Q = 1\cdot0$ leaves the possible variation of r, the speaker-to-listener distance.

Difficulty remains in assessing the attenuation function $\phi(r)$. Measurements by Aldersey-Williams (1973b) show that the required spatial attenuation rate Δ is related to the total sound absorption per unit area of floor A by the empirical laws shown in Fig. 2.79.

$$\Delta = 4\cdot5A + 2 \text{ for large offices } (A > 1) \text{ with screens,} \tag{2.22a}$$

$$\Delta = 3A + 3 \text{ for large offices } (A < 1) \text{ without screens,} \tag{2.22b}$$

$$\Delta = 6\cdot5A + 3 \text{ for small offices without screens.} \tag{2.22c}$$

It is difficult to compare the work of Aldersey-Williams (1973b) with that of Peutz

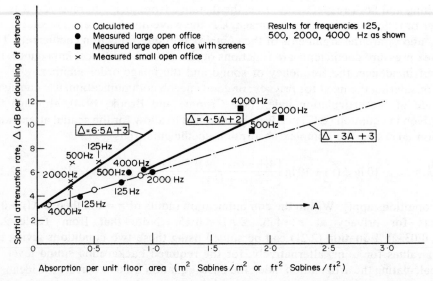

FIG. 2.79. Sound reduction with distance plotted against absorption/unit area (Aldersey-Williams, 1973b).

(1973) because in the latter case the spatial attenuation rate is expressed in terms of the measured reverberation time T_m rather than the room absorption A_T (see equation (2.18)). It is not valid to use the Sabine equation $T_m = (0{\cdot}16V)/A_T$ in low-height rooms having a non-diffuse sound field in cases where the complex absorption coefficient has to be considered and where the absorption is increased non-linearly with respect to the space dimensions by use of screens. The non-linear nature of the spatial attenuation rate for spaces with screens is shown clearly later in Fig. 2.82. Baines (1973) makes the point that reverberation time measurements in landscaped offices do not seem to have any meaning.

Further research is required to ascertain spatial attenuation rate function $\phi(r)$. Table 2.49 summarises some data for the landscaped office studied by Croome

TABLE 2.49. COMPARISON OF SPATIAL ATTENUATION RATES

Parameter	Frequency (Hz)				
	250	500	1000	2000	4000
Reverberation time (s):					
Measured values					
(Croome and Brook, 1974)	0·95	0·90	0·98	1·10	1·10
Calculated values using:					
(a) Sabine equation	1·05	0·63	0·45	0·43	0·42
(b) Eyring equation	0·98	0·55	0·35	0·34	0·39
Spatial attenuation rate (dB):					
Measured 1–2 m	+6·5	+6·0	+8·0	+9·0	+12·3
2–4 m	−1·0	−3·5	−3·0	+5·5	+6·0
4–8 m	+2·5	+7·5	+6·0	+4·5	−2·5
8–16 m	−2·5	+4·0	−2·0	−3·0	+3·0
Calculated values (dB per distance doubling) using:					
(a) Peutz formula (equation (2.18))	+8·70	8·70	8·70	8·65	8·65
(b) Aldersey-Williams formula (equations (2.22a, b, c))	+5·03	7·03	9·03	9·36	9·52

and Brook (1974). From this work, the following conclusions can be made:

(a) classical formulae due to Sabine and Eyring cannot be used to calculate the reverberation time at frequencies above 250 Hz due to the non-diffuse nature of the sound field;

(b) spatial attenuation rates exceeding 6 dB per doubling of distance can be expected in landscaped offices;

(c) a multitude of noise sources are often encountered in landscaped offices which produces a very non-uniform sound field; it is thus difficult to describe the spatial attenuation rate in such cases by an orderly behaved mathematical function, and average values become meaningless;

(d) the idea of treating the office as an entity when carrying out the acoustical design may be inappropriate; it is possible that each work station should be dealt with individually.

Absorption materials help to control the reverberant sound field (see Chapter 4 section 4.6). If they are to be effective then the distance between the noise source and the receiver must be greater than the free field radius r_d, which is found by equating the free field and reverberant sound field energies thus:

$$\frac{WQ}{4\pi r_d^2 \phi(r)} = \frac{4W}{R\phi(r)}.$$

Therefore

$$r_d = \sqrt{\frac{RQ}{16\pi}}.$$

If a source of sound (e.g. a person talking) is separated by a distance greater than r_d from the receiver (e.g. unintended listener), then the reverberant field can be controlled by absorption materials at the surrounding surfaces. Notice that if the directivity index Q is unity, then

$$r_d = 0 \cdot 14\sqrt{R}.$$

This is very similar to the expression for r_d to be derived in Chapter 4, but is distinguished by the complex sound absorption coefficient implicit in the room constant.

How, then, can the necessary sound absorption for a landscaped office be calculated? Ideally the ceiling of a landscaped office should reflect sound at near normal incidence ($\theta = 90°$) but absorb rays that are incident at an angle given by

$$\tan \theta \leqslant \frac{(h-1)}{(r/2)} \quad \text{(Fig. 2.80)},$$

where h and r are in metres; it is assumed that the average ear level is approximately 1 m from the floor for a seated person.

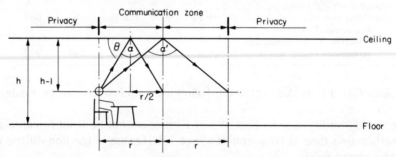

FIG. 2.80. Ceiling reflections in landscaped offices. Ceiling to reflect all sound in range of angles α to α' back into communication zone: ceiling to absorb sound at, in the limit, all angles $< \tan \theta \leqslant (h-1)/(r/2)$ (h and r in metres).

Besides room surfaces and people, the ceiling and the use of movable absorbent screens in the space play a key rôle in controlling the total absorption necessary to satisfy a criterion given by equation (2.22a, b, c). In practice, screens do not seem to affect the total absorption too much but are really more effective acoustically in screening particular noise sources. The interplay between the feelings of visual and acoustical privacy are not understood. Screens may also assist the aesthetic purpose (e.g. colour); the breaking up of large areas of open space in the vertical plane; the provision of visual privacy is important and also may act as a source of reverberant sound for a particular work station if they have a reflective inside-facing surface. In practice they are also used as pinboards and working group dividers. Care should be taken when selecting any materials that the random incident absorption coefficient (i.e. *not* normal incidence coefficient) is obtained; for ceilings, select materials having high values of $\bar{\alpha}$ for grazing incidence where possible. There can be as high as a 20% variation in the absorption coefficient between that measured at normal incidence and that at grazing incidence. Hoogendoorn (1973) suggests that in his experience a flat absorbent ceiling needs to have an absorption coefficient of 0·8–0·9 and claims that a value of 0·5 is too low. The ventilation and light fittings should not occupy more than 10–15% of the ceiling area unless an absorbent coating is applied.

At the present time materials which absorb sound at preferential angles of incidence are not commercially available in general. An alternative is to use a composite ceiling having a combination of reflective and absorptive elements. An egg-crate ceiling, for example, can have absorbent vertical sides but reflective horizontal surfaces. A V-form of egg-crate ceiling offers more advantages because it can accommodate the environmental services with a consequent saving in space. Sculpted ceilings scatter sound in many directions and help to ensure that no reflections predominate.

Schultz (1971a) recommends the following approach to designing the ceiling and screen absorption requirements in landscaped offices. Referring to Fig. 2.81 the

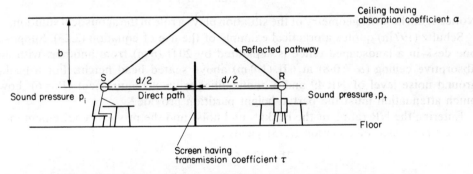

FIG. 2.81. Sound pathways between adjacent working places in landscaped offices.

acoustical energy emitted from the source is

$$\epsilon_s = \frac{p_i^2}{\rho c},$$

whereas that energy arriving at R will have two components, that directly transmitted through the screen and that reflected from the ceiling, thus

$$\epsilon_R = \frac{p_i^2}{d^2 \rho c} + \frac{(1-\alpha)p_i^2}{2^2(b^2 + d^2/4)\rho c}.$$

The attenuation of the screen is $\Delta_s = -10 \lg (\epsilon_R/\epsilon_s)$.

$$\Delta_s = -10 \lg \left\{ \frac{p_i^2}{p_i^2 d^2} + \frac{(1-\alpha)}{2^2(b^2 + d^2/4)} \right\},$$

but the transmission coefficient for the screen in $p_i^2/p_i^2 = \tau$.

Therefore

$$\Delta_s = -10 \lg \left\{ \frac{\tau}{d^2} + \frac{(1-\alpha)}{(4b^2 + d^2)} \right\}.$$

The attenuation of the screen must ensure that the difference between the speech and the background sound levels is sufficient to provide privacy at R. Schultz (1971a) suggests that a suitable design goal is

$$\Delta_s = 71 - L_{90} + P,$$

where L_{90} is the background noise rating for the space and P is a privacy requirement parameter ($P = 0$ for normal privacy and $P = 6$ for confidential privacy). Hence

$$71 - L_{90} + P \leqslant -10 \lg \left\{ \frac{\tau}{d^2} + \frac{(1-\alpha)}{(4b^2 + d^2)} \right\}. \tag{2.23}$$

(i) When $\tau = 0$ the attenuation is $10 \lg \left\{ \frac{(4b^2 + d^2)}{(1-\alpha)} \right\}$ giving an infinite attenuation when $\alpha = 1$, or $10 \lg (4b^2 + d^2)$ when $\alpha = 0$.

(ii) When $\tau = 1$ the attenuation is $-10 \lg \left\{ \frac{1}{d^2} + \frac{(1-\alpha)}{(4b^2 + d^2)} \right\}$ giving an attenuation of $20 \lg d$ when $\alpha = 1$, or a value of $10 \lg \left\{ \frac{d^2(4b^2 + d^2)}{2(2b^2 + d^2)} \right\}$, when $\alpha = 0$.

Notice again how the geometry of the situation plays a rôle in the acoustical solution.

Schultz (1971a) quotes a practical example of the use of equation (2.23). Suppose one desk in a landscaped office is separated by 20 ft (6 m) from another, with an absorptive ceiling ($\alpha = 0.8$) at 6 ft (1.8 m) above seated head height. For a background noise level of NR 49 and a confidential privacy requirement ($P = 6$), how much attenuation must the partial-height partition provide?

Entering the NR rating of the background noise and the privacy requirement into the left-hand side of equation (2.23) yields

$$71 - 49 + 6 = 28 \text{ dB}.$$

The four quantities in the right-hand side of the formula must, therefore, be chosen so that this attenuation term Δ is at least 28 dB. Suppose the partial-height barrier has an effective attenuation (for $d = 20$ ft and $b = 6$ ft) of 6 dB, that is $\tau = 0.25$. Then

$$\Delta_S = -10 \lg \left\{ \frac{0.25}{400} + \frac{(1-0.8)}{400 + 144} \right\},$$

$$\Delta_S = -10 \lg \left(\frac{1}{1600} + \frac{1}{2720} \right) = 30 \text{ dB},$$

$$\Delta_S = \left(\begin{array}{c} \text{direct} \\ \text{path over} \\ \text{barrier} \end{array} \right) + \left(\begin{array}{c} \text{ceiling} \\ \text{path} \end{array} \right).$$

Thus the requirement is satisfied. If the ceiling reflection were not present this attenuation would increase to 32 dB.

Some experimental results obtained by Aldersey-Williams (1973b) and Peutz (1973) and showing the effect of screens have been shown in Fig. 2.82.

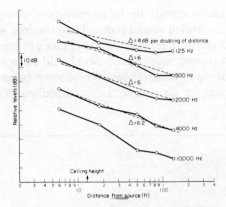

(a) Spatial attenuation rate in an open office of floor area 12 000 ft² (Aldersey-Williams, 1973b).

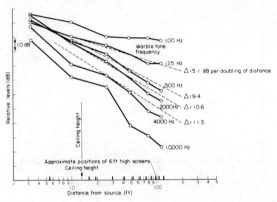

(b) Spatial attenuation rate in the same open office as (a) but including absorbent screens.

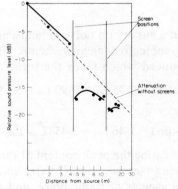

(c) The effect of two screens each 1·6 m high, placed in the line of measurement, on the spatial sound attenuation rate (Peutz, 1973).

FIG. 2.82. Spatial attenuation rates in landscaped offices.

Reference has been made already to the possibility of controlling the background noise level by using a controlled sound input. The principal advantage is that it can provide different sound environments at various places in the office. Beranek (1971) has suggested the background noise spectrum should lie within the shaded region shown on Fig. 2.83 for open-plan offices. Noise may be produced electronically (Waller, 1969) or by using an air source, i.e. air-handling diffuser.

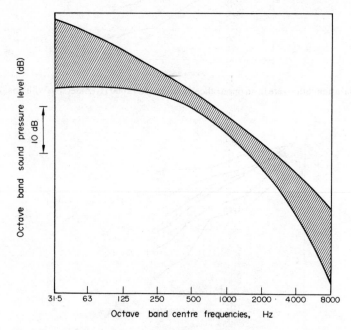

FIG. 2.83. Suggested shape of background noise spectrum for landscaped architectural spaces; if the noise spectrum is exactly that of the upper edge of the shaded area, add 7 dB to the level for the 1000 Hz band to obtain L_A, the A-weighted level for the noise in dBA; for a noise with the shape of the lower edge of the shaded area, $L_A = L_{1000\,Hz} + 6$ dBA (Beranek, 1971).

Hay (1973) has carried out a survey on noise in landscaped offices and attempted to correlate objective and subjective measurements. From her results a linear regression equation was deduced which takes the form:

$$\% \text{ dissatisfied} = 2\cdot45\,L_{10} - 0\cdot99\,L_{90} - 57\cdot5$$

or

$$\% \text{ dissatisfied} = 1\cdot46\,L_{90} + 2\cdot45\,(L_{10}-L_{90}) - 57\cdot5.$$

Notice the $(L_{10}-L_{90})$ term indicating the peak content of the noise; a similar term was found in the traffic noise index (see equation (2.9a)), which attempted to correlate dissatisfaction with physical aspects of traffic noise, and also in the noise pollution index (see equations (2.13) and (2.14)).

Another useful aspect of this research was described by Hay (1972) (and in more detail by Hay and Kemp (1972)); the results of this are summarised in Table 2.50.

TABLE 2.50. THE RELATIONSHIP BETWEEN THE PERCENTAGE OF PEOPLE WHO NOTICED AIRCONDITIONING NOISE, AND THE NOISE RATING VALUES FOR THE OFFICE AND FOR THE AIRCONDITIONING NOISE (HAY AND KEMP 1972)

Office NR	35	37·5	40	42·5	45	47·5	50
Airconditioning NR 25	9	8	7	6	4	3	2
Airconditioning NR 27·5	10	9	8	7	6	5	4
Airconditioning NR 30	12	11	10	9	8	7	6
Airconditioning NR 32·5	14	13	12	11	10	9	7
Airconditioning NR 35	15	14	13	12	11	10	9
Airconditioning NR 37·5	17	16	15	14	13	12	11
Airconditioning NR 40	19	18	17	16	15	14	13
Airconditioning NR 42·5	21	19	18	17	16	15	14
Airconditioning NR 45	22	21	20	19	18	17	16

Hay (1972) comments on these results:

In specifying noise design goals for a landscaped office, a balance between airconditioning noise and the office activity noise should be sought, for the acceptibility of the airconditioning noise does not depend upon its absolute level and frequency content, but rather on its relationship to the other noises involved. This shows the designer the importance of finding out the expected level and frequency content of the office activity noise, at the feasibility stage of design. In particular, if the background office noise is expected to be NR 45, then the design value of the airconditioning should be NR 40, in order to limit the percentage of people aware of noise from the airconditioning to 15 per cent.

It is worth writing Table 2.50 in a slightly different form as in Table 2.51.

TABLE 2.51

Difference (dB) Office NR – airconditioning NR	Percentage of people noticing airconditioning noise
+25·0	2
+22·5	3–4
+20·0	4–6
+17·5	6–7
+15·0	7–9
+12·5	8–11
+10·0	9–13
+7·5	10–14
+5·0	12–16
+2·5	14–17
+0·0	15–18
−2·5	17–19
−5·0	19–20
−7·5	21
−10·0	22

From this it can be seen that 1 in 5 people notices the airconditioning noise when it is 5 dB above the background level, but even when the airconditioning noise is 12·5 dB below the background level about 10% of people still notice it. For non-airconditioned buildings people are less sensitive to internal noise sources such as ventilation systems because open windows permit external noise to enter and mask the internal sources. The nature of the airconditioning spectrum also affects the subjective response: systems exhibiting peaks in the spectrum will be more noticeable generally.

2.6.4. OPEN-PLAN SCHOOLS

National Swedish Building Report 50: 1969 contains a draft version of a climatological standard for traditionally planned schools. The following comments are based on Swedish experience of the more recent open-plan schools and are described by Antoni *et al.* (1972).

The new schools are based on working groups of varying size and on a form of collaboration between teachers and pupils which ignores the concept of class boundaries. It thus follows that as far as premises are concerned there is less need for dividers between different working units. The key requirement is, instead, flexibility and scope for group work. The merits of daylighting and direct visual contact with the outdoors must be weighed against other factors pertaining to work and the environment. Consequently, buildings have become deeper, sizes of premises vary and links between room units are more open. The administrative unit, however, continues to be a class of 25–30 pupils.

The acoustic planning of group study areas is based on the criterion that each group must be afforded maximum scope for following their own discussions while ensuring that the amount of disturbance caused by other groups is minimized. Both factors depend upon the level of background noise which should be between 40 and 50 dBA. Groups containing more than ten people and activities which are sensitive to noise (e.g. language instruction) should be accommodated in separate rooms. Case studies show that group study areas should be furnished taking into account the nature of different activities and their capacity for transmitting and receiving noise. No more than 0·25 persons per m^2 of gross area should be present in these spaces at any one time in view of the risk of the different groups disturbing each other. A 75% level of use, which probably seldom will be exceeded, corresponds to 0·35 persons per m^2, which can thus serve as a rule when allotting dimensions. Sound absorbent materials should be used as much as possible, in particular on ceilings. In smaller areas for group study, and in areas of unsuitable shape, sound absorbent materials will also be needed on walls.

It is at present difficult to draw up strict regulations governing the amount and degree of acoustic separation to be provided between units of basic space. This stems from the fact that the need for this separation has not been evaluated in the light of the significance of flexibility and team work in certain forms of teaching.

CHAPTER 3

NOISE SOURCES IN BUILDINGS

Ordinarily there is not a close connection between the flow of air in a room and its acoustical properties, although it has been frequently suggested that thus the sound may be carried effectively to different parts. On the other hand, while the motion of the air is of minor importance, the distribution of temperature is of more importance, and it is on reliable record that serious acoustical difficulty has arisen from abrupt differences of temperature in an auditorium. Finally, transmission of disturbing noises through the ventilation ducts, perhaps theoretically a side issue, is practically a legitimate and necessary part of the subject.

(W. C. SABINE, *Collected Papers on Acoustics*,
Dover, 1964.)

3.1. General Aspects

Noise within buildings may originate from:

(i) the mechanical engineering services (e.g. heating, ventilation and airconditioning systems, fluid supply the drainage systems);
(ii) the electrical services (e.g. lighting);
(iii) the circulation services (e.g. lifts);
(iv) the communication services (e.g. telephone);
(v) the process machines (e.g. forging machines, typewriters);
(vi) the people (e.g. impact noise from walking, slamming doors, movement of
chairs, cross-talk between one space and another);
(vii) the external sources (e.g. traffic, aircraft, machinery).

A limited amount of acoustical data is available on these sources, and it is suggested that manufacturers are consulted or acoustical measurements are taken on existing similar installations when insufficient data is available to design the acoustical environment. This especially applies to categories (ii) to (iv); in cases (v) to (vii) a field assessment is inevitable.

The acoustical environment in the occupied space is the resultant of the noise arriving to the space from the engineering services, from adjacent areas by direct or indirect transmission and cross-talk, from the external environment and from noise generation within the space.

Figure 3.1 shows a general mechanical services layout in a building. The plant-room may be within or outside the building housing the people. Noise emitted

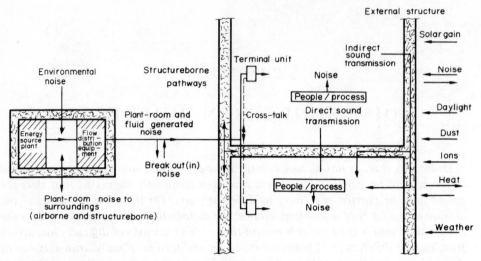

FIG. 3.1. Building, environment, and services noise pathways.

in the plant-room will be:

(i) conducted to the occupied spaces by the fluid distributing the energy or by flanking transmission in the distribution network;
(ii) transmitted through the plant-room structure to adjacent areas inside and/or outside the building.

Further, noise generation may occur in the distribution system if the fluid velocities are too high and/or at the terminal units.

Although the mechanisms of noise control will be discussed in Chapter 4 and some aspects of control of structureborne sound in Chapter 9, reference will be made to the control of the noise sources as they arise throughout this chapter. Noise sources and their control share the same breath. It is the manufacturer's responsibility to produce quiet equipment. The building environmental engineer and the architect need to work together on the noise implications of their building design. Noise control procedure has a systematic pattern:

(i) design good aerodynamic and hydrodynamic distribution systems;
(ii) select plant and equipment carefully considering the noise implications;
(iii) locate noisy areas away from critical areas wherever possible;
(iv) use suitable building materials and structures for control of noise penetration;
(v) lastly, use sound control equipment where necessary only after the above issues have been exhausted.

3.2. Sound Generation in Airflow Systems

The general wave equation for matter will be shown to be (see equation (8.19))

$$\frac{\partial^2 \phi}{\partial t^2} - c^2 \nabla^2 \phi = 0. \tag{3.1}$$

This describes the motion of a wave moving with a velocity c through a medium in which the particle displacement, the particle pressure, particle velocity or density are represented by the parameter ϕ. One of the assumptions made in deriving the general wave equation is that the propagation is taking place in a steady, source-free field, i.e. the interaction with other sound sources is neglected, and it is assumed that no fluctuations take place in the medium (i.e. the term "fluctuation" is used to cover macro-effects, like turbulent flow, not micro-disturbances such as the random particle movements due to Brownian motion). In airconditioning the sound field cannot be described by this equation because the airflow field is not steady. Turbulent eddies and vortices may be formed by flow separation in bends, by discontinuities, such as valves, dampers, guide vanes, fan blades, heater or cooler battery coils, or by turbulent behaviour in the boundary layer at duct walls or equipment surfaces.

3.2.1. SOUND GENERATION MECHANISMS

The model that has been used to develop the theory of aerodynamic sound generation is that of a stress system in the fluid acting on fluid elements causing them to condense or dilate, and thus the elements act as acoustic radiators. In essence, the model is analogous to that of a solid body being set into vibration by a periodic force, but since it is the fluid elements which are acting as acoustic radiators the term "pseudo-sound" is used to distinguish the different nature of the source. The stresses are due to fluctuations in the pressure and velocity fields, shear forces in the flow, the viscous nature of the fluid and any external body forces.

The simplest mechanism occurs when a fluid flow periodically fluctuates. A periodically fluctuating pressure field in a fluid can be imagined as a pulsating spherical source, referred to as a *monopole*, which radiates sound in a similar manner to its material counterpart. An example in practice occurs when fan blades periodically displace volumes of air.

When turbulent flow interacts with a moving or stationary surface, by turbulence shedding or impact on the surface, the force exerted by the flow may be resolved into "drag" and "lift" components, and the sound generated is represented by a *dipole*, the axis of which is aligned with that of the fluctuating force. The dipole can be thought of as two monopoles separated by a small distance and pulsating 180° out of phase with each other. Most of the sound generation mechanisms in airconditioning systems can be represented by sources having dipole character, e.g. fan noise, impact of airflow on bends, duct walls or valves. Gordon (1969) has shown that the sound power radiated W from a dipole source is given by

$$W \propto \frac{\tilde{\Delta p}^2 \epsilon^2 D^3}{h \, \rho c}, \tag{3.2}$$

where $\tilde{\Delta p}$ is the fluctuating pressure differential in the region of the discontinuity, ϵ is the area taken up by discontinuity, D is the duct diameter, h is the duct wall thickness and ρc is the acoustic impedance of the air.

If the dipole source distribution is random the sound spectrum is of a broadband nature (i.e. vortex noise), whereas if it is a well-ordered distribution, a discrete sound spectrum results (e.g. rotational noise in fans).

Many cases arise in airflow systems where turbulent mixing occurs without the presence of surfaces offered by duct walls or other solid surfaces (e.g. induction airstream mixing, boundary layer turbulence). For this case Lighthill has proposed that a *quadrupole* (i.e. a combination of two dipoles aligned in opposite or parallel direction) represents a model of the acoustic source. It can be shown that the general wave equation is modified due to pseudo-sound sources and takes the form

$$\frac{\partial^2 \phi}{\partial t^2} - c^2 \nabla^2 \phi = \frac{\partial Q}{\partial t} - \frac{\partial F_i}{\partial x_i} \frac{\partial^2 T_{ij}}{\partial x_i \partial x_j},$$ (3.3)

where $\partial Q/\partial t$ is the rate of introduction of mass into the system, represented by monopole sources (e.g. air displacement by fan blades); $\partial F_i/\partial x_i$ is the fluctuating external force field represented by dipole sources (e.g. fan noise); T_{ij} is the stress tensor represented by quadrupolar source distributions and accounts for fluctuating pressure and momentum fields, viscous forces and the acoustic radiation pressure (see Lighthill, 1952, Ffowcs–Williams in Stephens, 1975).

3.3. Sound Generation in Waterflow Systems

Just as in airflow systems, sound is transmitted from equipment such as cooling towers, airwashers, boilers, pumps and refrigeration plant through the liquid (i.e. heating or colling water, liquid refrigerant) itself or through the pipe walls over long distances. It may be radiated directly to the surroundings (i.e. airborne sound), or may be transmitted through supports or contact with the building structure (i.e. structureborne sound) where the energy can be reradiated from the surface as airborne sound.

Sound may be generated within the liquid as described in the previous section, but, in addition, two other phenomena can be responsible for sound emission— *waterhammer* and *cavitation*. Waterhammer arises from sudden rises in pressure which cause the distribution system to be shock excited; valves which create sharp pressure rises should be avoided. Thermal expansion and contraction stresses in the pipework can also create waterhammer. Cavitation can best be understood by studying Fig. 3.2. When the vapour pressure and the surface tension in the bubble acting together exceed the hydrostatic pressure surrounding the bubble, then rupture occurs. Many such bubbles, due to air and other gases in the liquid, behaving in a similar manner cause intense sound to be generated due to shock waves formed at rupture and also due to the beating of bubbles, which are in violent motion, against the pipe wall. Of considerable interest to scientists has been the weak emission of light that can accompany the bubble rupture, an effect which has been named *sonoluminescence* and suggests a way of measuring cavitation. Grein (1974) discusses cavitation in detail.

Figure 3.3a–d shows the sound levels that can be expected in various situations. Conclusions that can be made from this work are:

(a) the sound pressure level is inversely related to the pipe diameter;
(b) the shape of the fittings is more important than the number of fittings used;
(c) the cumulative effect of a number of fittings is not additive;
(d) corroded pipes have been found to give higher sound emissions than clean ones.

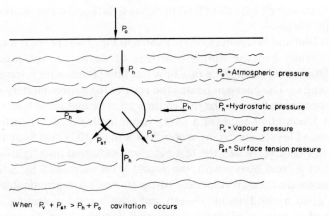

When $P_v + P_{st} > P_h + P_o$ cavitation occurs

FIG. 3.2a. Cavitation process.

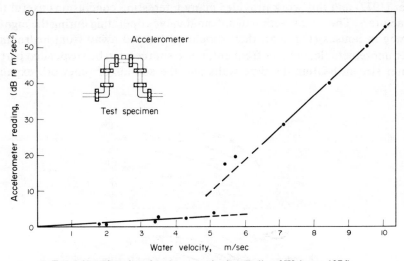

FIG. 3.2b. Pipe vibration due to cavitation (Ball and Webster, 1976).

Figure 3.2b presents results obtained by Ball and Webster (1976) for vibration due to cavitation in a double offset arrangement. Cavitation is established at a water velocity of about 5 to 6 m/s, whereas in straight pipes it is not detected at water velocities of up to 14 m/s. Experiments with water velocities up to 7 m/s using one elbow or two elbows at various spacings showed no evidence of cavitation. The water temperature was 15 °C in these tests.

Kristensen (1964) and Kamber (1968) give the relationship between sound pressure level and water flow velocity for $\frac{1}{2}''$, and $1\frac{1}{2}''$ pipes as shown in Fig. 3.4a. These measurements suggest that higher water velocities (2–4 m/s) would be acceptable in many situations; higher velocities have the advantages of providing larger pressures for circulation, smaller pipe sizes and also assist in clearing air pockets, although a compromise has to be met between these benefits and the higher pump power necessary. Endejann (1975) claims that American field experience shows that water velocities of about 2·4 m/s (8 ft/s) can be used in commercial areas, but should be a maximum of 1·2 m/s (4 ft/s) in houses, flats and quiet areas of buildings to avoid noise problems.

The work of Ball and Webster (1976) is shown in Fig. 3.4b. There is a scatter of about 11 dB in the results because different fittings of similar design produce different amounts of sound energy at a given rate of water flow; the scatter is larger at low water velocities (< 5 m/s). The curve indicates the results for approximately 4 m of straight 5 mm bore copper pipe fixed to the test room wall. The test room had a reverberation time of about 5 seconds hence for a living room having a reverberation time of 0.5 seconds the sound levels due to the water flow will be 10 lg (5/0.5) or 10 dB less than those shown in Fig. 3.4b. When the test pipe was mounted on a light-weight backing panel instead of a rigid heavy wall, the sound level increased by 3 to 5 dBA. This research suggests that water velocities of up to 3 m/s are practicable with suitable fittings, with good hydrodynamic design and with a rigid mounting wall.

The design of valves to minimise noise emission has been discussed by Göselle (1959, 1967, 1968), Göselle and Voitsberger (1970), Rückward (1970) and Schuder in Crocker (1972) and this work provides many interesting conclusions useful to valve manufacturers. The on–off action of solenoid valves operating during the night can be disturbing in houses (Fig. 3.5); they should be located away from bedrooms.

Other unpredictable, but less frequent, noise sources can be suspended particles in the water stream beating the pipe walls and the sounds of pipework expansion.

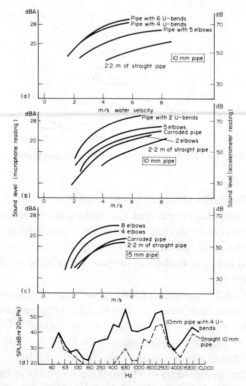

FIG. 3.3. Noise in water networks (Henderson, 1959).

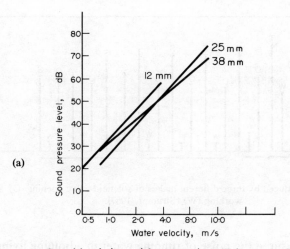

(a) Sound levels in straight water pipes (Kristensen, 1964).

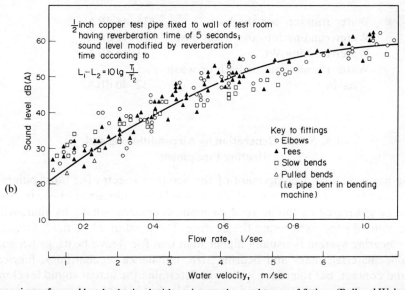

(b) Comparison of sound levels obtained with various makes and types of fittings (Ball and Webster, 1976).

FIG. 3.4 Sound levels emitted from water distribution systems.

Grandjean (1973a) states that: "The most important sources of noise in kitchens, bathrooms and toilets are the noises associated with water installations, and the highest of these is the sound of water flushing."

Kurtze (1964) measured these values:

Water running into the bath	75 dBA
WC—flushing	78 dBA
WC—cistern re-filling	70 dBA.

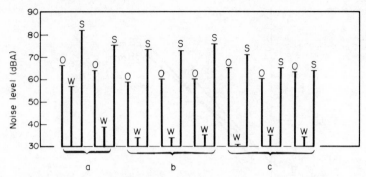

FIG. 3.5. Sound levels produced by three different makes of solenoid valve opening (O), shutting (S) and working (W) (Strumpf, 1967).

Kamber (1968) measured the noise of running water in adjoining living rooms and obtained the following values:

Water running into the bath	35–44 dBA
Water running into the wash basin	26–42 dBA
Water-flushing WC	33–40 dBA
Water running out of bath or wash basin	32–40 dBA.

3.4. Sound Generation by Airconditioning and Heating Equipment

Bearing noise may be a component of the acoustic spectra for fans, pumps and motors. Sleeve bearings support the shaft on an oil film maintained by shaft rotation. Noise can be generated by the oil feed mechanism or occasionally by shaft-whirl in which the journal processes round the bearing. The level of white noise from a ball and roller bearing system is usually higher than that for sleeve bearings because of the metallic material, mass and bearing size, imbalance, eccentricity, lubrication method and contact, but tolerance wear will determine the actual sound level in any bearing system. Ball and roller bearings are noisier but help to make a more compact machine and are cheaper than sleeve bearings. Spring mis-match assemblies have become common practice in an attempt to reduce ball-bearing noise.

Out-of-balance forces are another source of vibration in rotating machinery. These may arise from an improperly designed supporting structure or may be inherent in the machine part themselves. Wear, dirt impingement and improper maintenance let out-of-balance forces develop.

3.4.1. FANS

The components in a fan sound spectrum can be classified as rotational, broad-band, vortex and mechanical noise.

(a) *Rotational noise*

This arises from:

* periodic lift forces due to the interaction between the velocity fields of
  stationary (e.g. scroll or volute outer casing) and moving (e.g. impeller)
  surface and
* the pulsation of air as the impeller displaces air in its pathway.

In each case a discrete frequency, the blade passage frequency f, is emitted
together with possible harmonics, where

$$f = nN \text{ (Hz)} \tag{3.4}$$

for a fan with number of fan blades n and an impeller speed in revolutions per
second N.

A velocity profile for the air is shown in Fig. 3.6a in which the maximum can be
seen to occur between the fan blades and the minimum at the blade edges. It is these
velocity profiles passing through the cut-off section which produce rotational sound;
the cut-off is the narrowest dimension between the impeller and the fan housing. The
frequency and amplitude of the sound energy emitted depends upon the cut-off
distance Δr (Fig. 3.6b) and also upon the velocity profile shown in Fig. 3.6a, smoother
profiles generate less sound; fans with a high number of impeller blades have flatter
velocity profiles than those with a fewer number. The work of Leidel (1969)
illustrated by Fig. 3.6b, shows how changing the cut-off width from 2 to 40 mm
results in the rotational noise level being decreased by as much as about 20 dB. In the
case of double-inlet centrifugal fans, noise can be reduced by displacing the blades
of the two rotors by half a blade division. Rotational noise increases in level as the
fan speed increases, to approximately a fourth power law (Finkelstein, 1974).

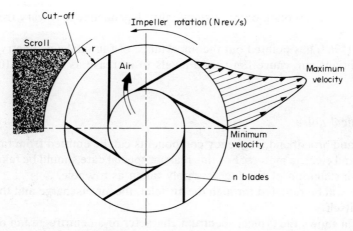

(a) Components of rotational noise. Fundamental frequency $f = nN$. Rotational noise level $\propto N^4$ approximately.

FIG. 3.6. Rotational noise in fans (Finkelstein, 1974).

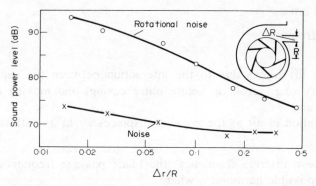

(b) Influence of cut-off size on rotational noise (Leîdel, 1969).

Fig. 3.6. (*Continued*).

(b) *Broadband and vortex noise*

This originates from the fluctuations in the turbulent air patterns across the fan blades and in the velocity field of the fan intake air and vortex shedding from the edges of the fan blades. Vortex noise have a frequency characteristic determined by the Strouhal number S, where

$$f = S\frac{v}{d} \tag{3.5}$$

for an air flow of velocity v, and having a characteristic dimension d; S usually lies in the range of 0·15 to 0·20 (see section 3.4.2). The resulting sound spectrum due to the turbulence fluctuations and the vortex components usually has a broadband character with a maximum level at the frequency given by equation (3.5). Research over the years has shown that the turbulence power sound level increases as the sixth power law of velocity (i.e. $L_w \propto v^6$), but for air velocities encountered in low- and high-pressure airconditioning systems, this dependence may vary from $L_w \propto v^5$ to $L_w \propto v^7$.

Sharland (1969) has pointed out the importance of fan intake geometry; turbulent intake conditions can cause fan sound levels to increase as much as 10 dB.

(c) *Mechanical noise*

Discrete and broadband frequency components can be emitted from fan bearings, belt drives and electric motors. For this reason, special care should be taken over the design and installation of the fan assembly taken as a whole.

Fan noise will be radiated through the air inlet, the air discharge and the vibrating fan casing itself.

Figure 3.7a shows the typical spectrum character of a centrifugal fan operating at 310 and 620 rev/min as measured by the author. At the lower speed the discrete noise components due to rotational noise and mechanical sources predominate, whereas at the higher speed the broadband aerodynamic noise masks most of the discrete noise

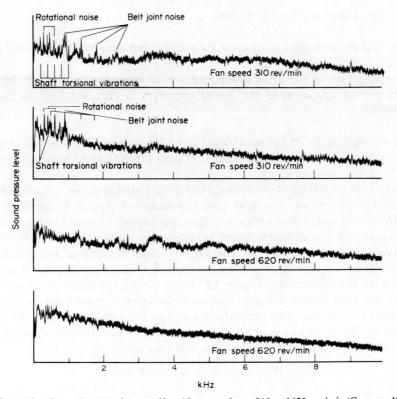

(a) Narrow band sound spectra for centrifugal fans running at 310 and 620 rev/min (Croome, 1969)

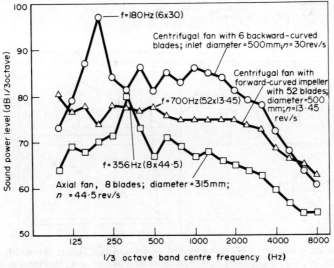

(b) Typical noise spectra of fans: rotational sound fundamental frequency $f = nN$ (Hz); 1st harmonic $f_1 = 2f$; 2nd harmonic $f_2 = 3f$ (Finkelstein, 1974).

FIG. 3.7. Centrifugal fan sound spectra.

components. These spectra show that:

(a) the higher harmonics of discrete frequency sound may be transmitted through the system;
(b) the selection of the drive components is as important as that of the fan;
(c) the nature and level of the sound spectrum is very sensitive to fan speed change.

Some measurements of fan spectra made by Finkelstein (1974) are shown in Fig. 3.7b. Notice that when the blade number is high the rotational noise component is absent from the spectrum.

Fan selection should be made considering the aerodynamic and acoustic specification, besides the more practical considerations of space requirements. The aerodynamic and sound performance of fans are related, as would be expected from the discussion on aerodynamic sound generation in section 3.2; work by Bommes (1968) and Finkelstein (1974) has carried our understanding of this further. Eck (1973) also quotes evidence supporting the view that the lowest sound levels occur at the maximum fan efficiencies. Figure 3.8 shows the vector diagrams for centrifugal and axial flow fans. Bommes argues that the aerodynamic sound from the fan may be considered in two parts—that generated by airflow through the impeller W_1 and that generated by the air passing over struts, guide vanes, or past the cut-off in the case of centrifugal fans W_2. In general, the total acoustic power is

$$W = W_1 + W_2 = k_1 w^{n_1} + k_2 c^{n_2}, \qquad (3.6)$$

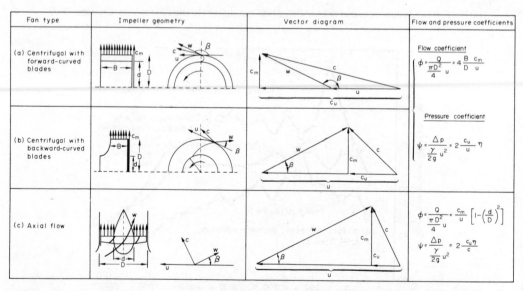

FIG. 3.8. Vector diagrams for fans. A = arbitrary proportionality factor; B = width of impeller; c = absolute air velocity leaving impeller; c_m = vertical component of c; c_u = tangential component of c; D = outside diameter of impeller; d = hub diameter of axial flow fans or inlet diameter of centrifugal fans; g = acceleration due to gravity; Δp = total pressure difference; Q = air volume flowrate; u = tip speed; w = velocity of air leaving impeller relative to impeller; β = angle between u and w vectors; γ = ratio of specific heat capacities at constant pressure and constant volume, 1.4; η = fan efficiency.

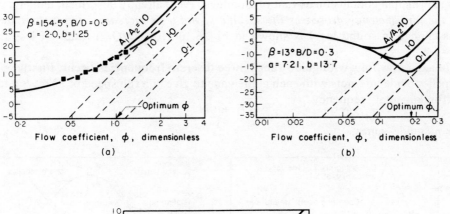

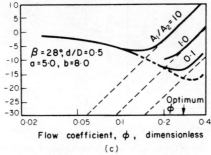

FIG. 3.9. Predicted change in sound level with change in operating point for (a) centrifugal fan with forward curved blades, (b) centrifugal fan with backward curved blades, (c) axial flow fan (Bommes, 1968).

where w is the velocity of the air leaving the impeller taken relative to the impeller, c is the absolute velocity of the air leaving the impeller, k_1 and k_2 are constants which depend on the fan type and the fan size, and n_1 and n_2 are indices which depend on the fan type and the magnitude of the air velocity. Assuming $n_1 = n_2 = 5$, Bommes derived relative sound level contours for three different types of fans with varying flow conditions; these are shown in Fig. 3.9. It can be seen that axial and backward-curved bladed centrifugal fans show a distinct minimum sound level near the optimum air flow condition; this minimum does not occur for forward-curved bladed centrifugal fans. The sound levels of the axial flow fan are a little higher than that of the backward-curve centrifugal fan; in practice this is partly compensated for by the extra low frequency (< 300 Hz) attenuation required by centrifugal fans although less is required at higher frequencies (> 300 Hz).

Finkelstein (1974) develops an expression for the overall sound power level in terms of the airflow rate Q (m³/s), the fan total pressure p (Pa) and the specific sound power level L_s (dB), thus:

$$L_w = L_s + 10 \lg Q + 20 \lg p, \qquad (3.7)$$

where L_s is a constant which depends upon the type of fan besides its operation speed and efficiency. Empirical values of the specific sound power level are shown in Fig. 3.10e. The research of Finkelstein (1974) illustrated in Fig. 3.10a–c reaches the same principal conclusion as that of Bommes (1968) already described, namely, that *as the fan efficiency decreases the specific sound level increases*; the magnitude of this change in sound level is shown in Fig. 3.10d for (Q/Q_{opt}) ratios of less than 0·6–1·4.

In axial flow fans with guide vanes three discrete frequencies occur. Firstly, each impeller blade interacts with each guide vane to give at a fan speed of N rev/s a tone of fundamental frequency

$$f_1 = n_g N$$

for n_g guide vanes.

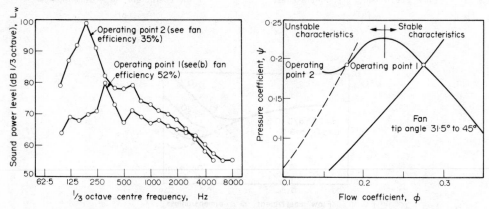

(a) Sound power level of a 315 mm diameter axial flow fan in the range of stable and unstable characteristics (in range of unstable characteristics $L_w = 100$ dB; $\eta = 35\%$; in range of stable characteristics $L_w = 84$ dB; $\eta = 52\%$).

(b) Position of working point on ϕ/ψ diagram; see (a) for significance of operating point selection on sound level.

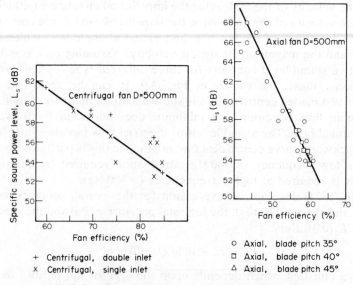

+ Centrifugal, double inlet
× Centrifugal, single inlet

○ Axial, blade pitch 35°
□ Axial, blade pitch 40°
△ Axial, blade pitch 45°

(c) Variation of specific sound power level with efficiency.

FIG. 3.10. Fan efficiency and noise output (Finkelstein, 1974).

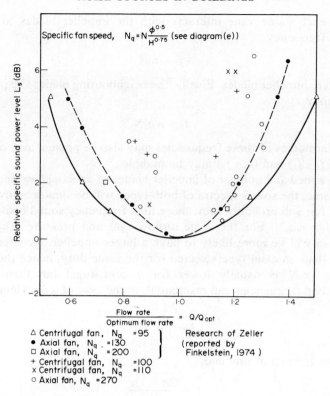

(d) Variation of specific sound power level as a function of the relative flow rate.

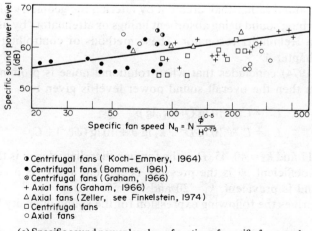

(e) Specific sound power level as a function of specific fan speed.

FIG. 3.10. (*Continued*).

Secondly, each guide vane interacts with the impeller blades to give another fundamental frequency

$$f_2 = n_i N$$

for a fan with n_i impeller blades. Finally, the neighbouring blades slip past each other and a frequency

$$f_3 = n_i n_g N$$

can occur. Harmonics of these frequencies may also be present and, of course, beat frequencies $(f_3 - f_1)$ and $(f_3 - f_2)$ may be audible.

If the size, speed and number of impeller blades of an axial flow and a centrifugal fan are the same, the sound spectra of both types will be similar at low frequencies, but the axial fan will probably show more high-frequency sound content due to the component $f_3 = n_i n_g N$. For the same airflow rate and pressure requirements the centrifugal fan will be more likely to have a larger impeller diameter and run at a lower speed than an axial type selected for the same duty, hence the fundamental frequency $f_2 = n_i N$ is usually lower for a centrifugal fan than for its axial counterpart. Fan frequencies can resonate if, in the case of a duct length l closed at both ends,

$$f = \frac{nc}{2l} \quad \text{for} \quad n = 1, 2, 3, \ldots,$$

or, if the duct is open at one end,

$$f = \frac{(2n - 1)c}{4l},$$

where c is the velocity of sound. This is in principle just the same as the production of resonant frequencies in organ pipes or in the airways of woodwind and brass instruments. Musicians use mutes to alter the tone quality of their instruments (i.e. by harmonic damping); they produce various frequencies by opening and closing various lengths of their instruments by valves or keys. In airconditioning, designers wish to subdue sound rather than create it by altering the geometry of the ductwork system by absorbing sound using absorbent linings or attenuators by using resonance filters such as a Helmholtz resonator. These methods of controlling sound will be discussed in Chapter 4.

Finkelstein (1974) concludes that when rotational noise is prominent in the fan sound spectrum then the overall sound power level is given by

$$L_w = L_s + 10 \lg Q + k_1 \lg p \tag{3.8a}$$

or $\qquad L_w = L_s + 10 \lg D^2 + k_2 \lg v + 10 \lg (\phi \psi^2) + C, \tag{3.8b}$

where $k_1 = 15$–17 and $k_2 = 40$–45, D is the fan impeller diameter, v is the air velocity, ϕ is the flow coefficient, ψ is the pressure coefficient and C is a constant. When broadband sound is prevalent, $k_1 = 20$ and $k_2 = 50$.

Eck (1973) derives the following expression for the sound intensity level produced by a fan:

$$L_I = 10 \lg \left\{ K\phi\psi \left(\frac{1}{\eta} - 1 \right) \frac{\rho}{2} v^{3+n} \right\} \tag{3.8c}$$

where ϕ is the flow coefficient, ψ is the pressure coefficient, η is the fan efficiency, v is the average air velocity of the airflow in the fan, K is a constant and the power n varies from 1·5 to 4; both n and K are determined by measurement.

The frequency curve of a fan may be drawn using the relative frequency spectrums which have been averaged from experimental results shown in Fig. 3.11a–c; at each centre frequency these relative levels are substracted from the overall sound power levels. Recent measurements of sound power levels and relative levels have been made by Finkelstein (1974) and are compared with other people's

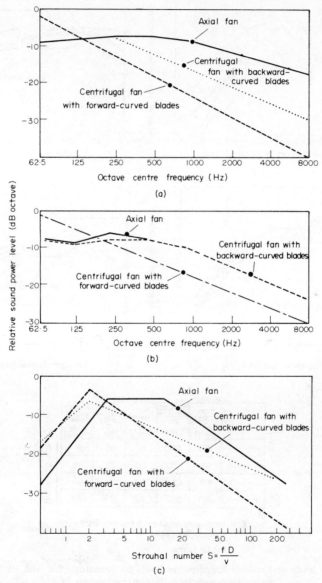

FIG. 3.11. Relative sound power level for centrifugal and axial flow fans as determined by various research workers: (a) Allen (1957) and Beranek (1960); (b) Bommes (1961); (c) Wilkström (1964).

research in Figs. 3.12a–c. Wilkström (1964) suggests that it is better to plot the relative power spectra against the non-dimensional frequency $f_m D / v$; this is a more valid method if comparisons with other work are to be made because third-octave centre frequencies f_m are more sensitive than octave band frequencies and display the discrete tones, and also fan-tip speed is included, which is a parameter used by designers of airconditioning systems.

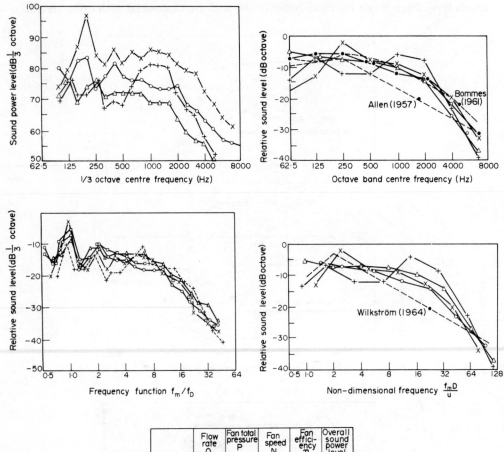

	Flow rate Q (m³/s)	Fan total pressure P (N/m²)	Fan speed N (rev/s)	Fan efficiency η (%)	Overall sound power level L_W (dB)	
o——o	108	880	30	85	91	Single inlet
△——△	98	170	20	70	85	Single inlet
x——x	317	620	33	74	99	Double inlet
+——+	125	570	22	80	89	Double inlet

f_m = third–octave centre frequency (Hz)
D = impeller diameter
u = fan tip speed

(a) Sound power levels and relative sound power levels for centrifugal fan with six backward-curved blades; inlet diameter = 500 mm (Finkelstein, 1974).

FIG. 3.12. Fan sound power measurements.

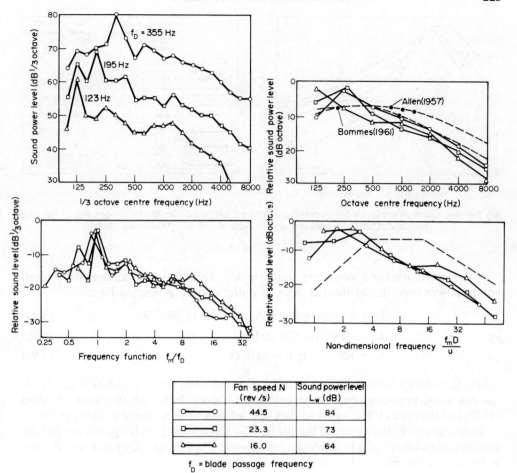

	Fan speed N (rev /s)	Sound power level L_w (dB)
○——○	44.5	84
□——□	23.3	73
△——△	16.0	64

f_D = blade passage frequency

(b) Sound power levels and relative sound power levels for an axial fan of 315 mm diameter (Finkelstein, 1974).

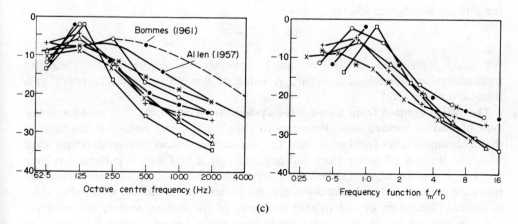

(c)

FIG. 3.12. (*Continued*).

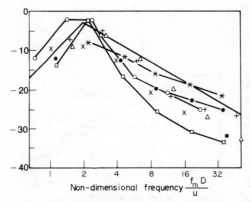

	Fan diameter D (mm)	Number of blades n	Fan speed N (rev/s)	Sound power level L_w (dB)
x———x	500	12	18·7	96
+———+	501	12	13·0	98
o———o	505	6	27·5	96
□———□	636	6	17·3	97
△———△	555	12	12·3	93
•———•	500	7	17·7	96
———	500	20	17·2	95

(c) Relative sound power levels for a centrifugal fan with backward-curved blades operating at fan total pressure of 500 Pa and delivering air at a rate of 1.95 m³/s (Wilkström, 1964).

FIG. 3.12. (*Continued*).

If the manufacturer's measured sound levels are not available, then the overall sound power level L_w at the fan inlet of outlet can be estimated from

$$L_w = 67 + 10 \lg P + 10 \lg p_s \quad \text{(dB re } 10^{-12}\text{ W)} \tag{3.9a}$$

or
$$L_w = 40 + 10 \lg Q + 20 \lg p_s \tag{3.9b}$$

or
$$L_w = 105 + 20 \lg P - 10 \lg Q, \tag{3.9c}$$

where Q is the air volume flow rate (m³/s), P is the rated motor power (kW) and p_s is the fan static pressure (Pa). These calculated values, based on the work of Allen (1957) and Beranek (1960), apply to fans operating at their optimum working point.

Baumann (1974) shows that if the blades of a fan are irregularly spaced around the impeller, rotational noise is displaced into the very low-frequency range and the fan noise is less disturbing. A parameter θ is defined by

$$\theta = \frac{\Sigma \, \Delta \theta_j}{z}$$

for a blade number z, where

$$\Delta \theta_j = \frac{\alpha_{\text{opt}} - \alpha_j}{\alpha_{\text{opt}}}$$

for $\alpha_{\text{opt}} = 360/z$ and α_j is the angle between successive blades; impeller circumference subtends $\Sigma \, \alpha_j = 360°$. A value of $\theta = 0·9$ is shown to result in a minimum noise level.

The sound emitted from a belt-drive system depends on the belt speed and the type of friction surface used. Referring to Figs. 3.13a–c and 3.14a–d it can be seen that a chrome leather friction surface, for instance, produces an overall sound level which is about 6 dB lower than that produced by a belt with a patterned rubber surface. From the many modern belts available, endless (i.e. jointless) multi-ply types are recommended. Needless to say, the pulleys of the fan and motor should be balanced. Depending on the overall tolerance of the driving system (i.e. pulleys, belts, motor and fan shafts), variations in the angular velocity of the fan shaft will occur, the so-called *drehfehler* phenomenon; out-of-balance forces, belt slips and variations in input power accentuate and make this innate feature irregular.

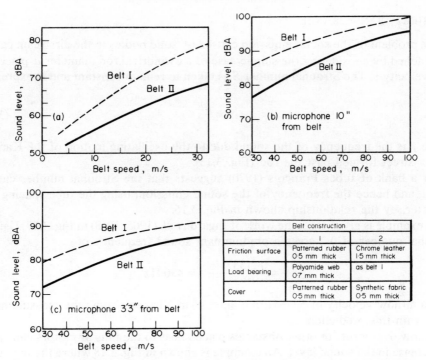

FIG. 3.13. Sound level emitted from two fan belts as a function of belt speed (Tope, 1971).

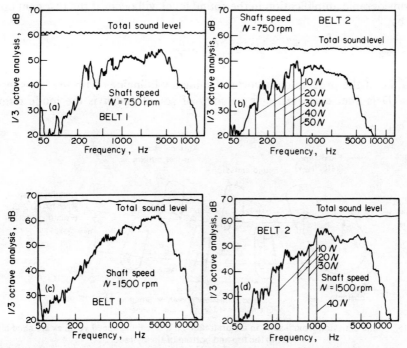

FIG. 3.14. Sound spectra of fan belt noise at shaft speeds of 750 and 1500 rpm (Tope, 1971).

3.4.2. HEAT TRANSFER COILS

The problem of the sound emission caused by solid bodies in the airstream can be approached by considering the simple case of a cylindrical rod diameter d placed in flow velocity v. The Strouhal number S is taken to remain constant and is expressed by

$$S = \frac{fd}{v},$$ (3.10)

where f is the frequency of the sound due to the oscillating motion of the Karman vortex street formed past the rod (Fig. 3.15).

For a bank of tubes François (1970) suggests that the Strouhal number can be found, and hence the frequency of the sound emission, using the tube spacing and diameter, by the relationship shown in Fig. 3.16.

An example is given from the work of Ingard *et al.* (1965, 1968) in Fig. 3.17. Using a Strouhal number of 0·2 gives a predominant sound frequency of

$$f = \frac{0 \cdot 2(5375/60)}{(0 \cdot 5/12)} = 430 \text{ Hz}$$

for an airflow velocity of 5375 ft/min across $\frac{1}{2}$ in. diameter rods; the measurements agree with this prediction.

Airflow over struts or other obstacles placed near to the fans or motors can cause an increase in the sound level. An example is shown in Fig. 3.18 where the top curve reveals a higher sound pressure level for a fan where the airflow is directed over some locking pins belonging to the fan housing; the lower curve gives the acoustic performance for an unimpeded flow.

A considerable amplification in the sound level will occur if the resonant condition given by

$$\frac{Sv}{D} = n\frac{c}{2D}$$ (3.11)

occurs, i.e. if the predominant vortex frequency coincides with the transverse duct modes (D is duct diameter, c is velocity of sound and n is the mode number).

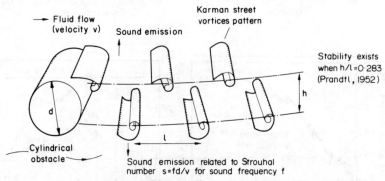

FIG. 3.15. Sound emission from a solid in an airstream. At a threshold speed vortices are shed alternately from the top and bottom of the cylinder.

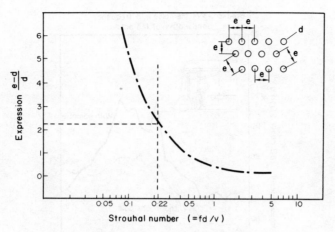

FIG. 3.16. Sound emission from a bank of tubes (François, 1970).

3.4.3. PUMPS

Noise from pumps is due to mechanical sources (originating mainly in the bearings) from electrical equipment, water turbulence and cavitation; other factors such as inadequate anchorage, insufficient bearing lubricant, deterioration, corrosion and wear of the pump parts, make the pump vibrations become more noticeable. The pump impeller becomes unbalanced by the build up of dirt and sediment on it, or

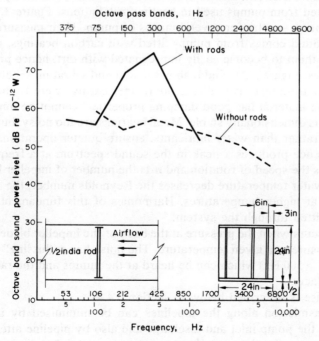

FIG. 3.17. Sound emission from $\frac{1}{2}$ in. diameter rods in an airstream velocity of 5375 ft/min as measured by Ingard et al. (1968).

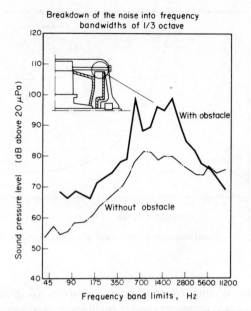

FIG. 3.18. Effect of obstacles in airflow near fan blades on sound emission. Fan is driven by a 120 kW motor with a speed of 1500 rpm (François, 1970).

wear, producing uneven working forces on the impeller shaft and blades.

Clark and Petrusewicz (1970-1) have carried out a detailed research programme on the sound emitted from pumps used in heating installations. Figure 3.19a–d shows sound and vibration spectra for several makes of pump. Their measurements show that the least sound comes from pumps fitted with carbon bearings, although this material allows them to become easily impregnated with dirt, hence producing out-of-balance forces. Figure 3.20a and b shows sound and vibration spectra for pumps with different bearing materials. PTFE fluorosint bearings were found to give the best results. The material has good damping properties, cannot be impregnated by dirty water and is dimensionally stable. Metal bearings are too noisy, but if they have to be used, oil rather than water lubricants, ensure quieter operation.

Water turbulence produces a peak in the sound spectrum at a frequency of Nn (Hz) where N is the speed of rotation and n is the number of impeller blades. Since an increase in water temperature decreases the Reynolds number, the sound output also decreases at higher temperatures. Harmonics of this fundamental frequency may be transmitted through the system.

Cavitation occurs when the pressure at the back of the impeller blade drops below the vapour pressure at a given temperature. The noise has a "hissing" or "spitting" type of sound (> 500 Hz) which can be heard at the pumps and the radiators; local boiling takes place.

Electrical noise is described below in section 3.4.4.

Flanking transmission along the pipelines can be minimised by using flexible connections on the pump inlet and discharge and also by pipeline attenuators (Fig. 3.21). Pumps need vibration isolation at support points. To eliminate the blade passage frequency the flexible connections should be at least five times the pipe

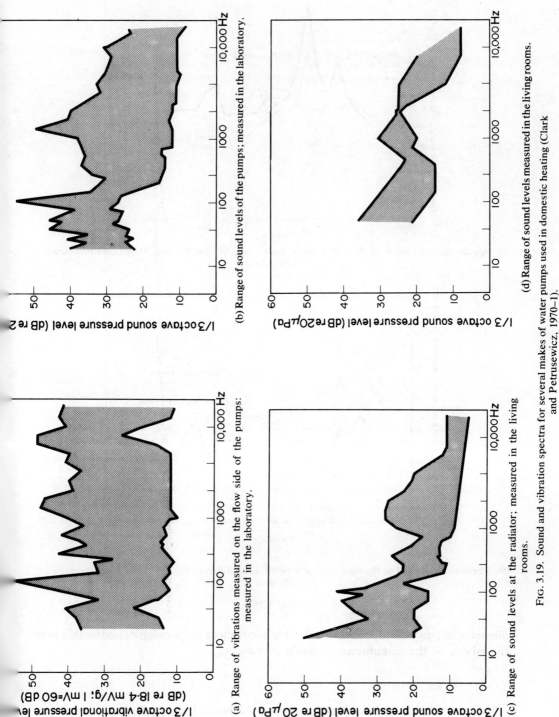

(a) Range of vibrations measured on the flow side of the pumps: measured in the laboratory.

1/3 octave vibrational pressure lev (dB re 18·4 mV/g; 1 mV=60 dB)

(b) Range of sound levels of the pumps; measured in the laboratory.

1/3 octave sound pressure level (dB re 2

(c) Range of sound levels at the radiator; measured in the living rooms.

1/3 octave sound pressure level (dB re 20 μPa)

(d) Range of sound levels measured in the living rooms.

1/3 octave sound pressure level (dB re 20 μPa)

Fig. 3.19. Sound and vibration spectra for several makes of water pumps used in domestic heating (Clark and Petrusewicz, 1970–1).

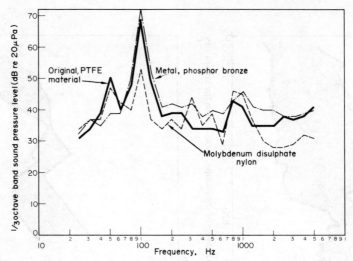

(a) Measurements of noise of the pump motor itself using different bearings (Clark and Petrusewicz, 1970–1).

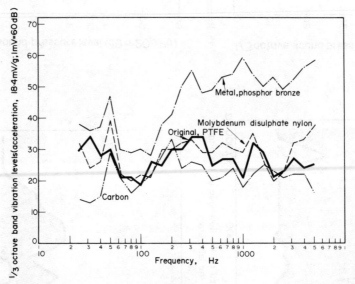

(b) Vibrations measured at the flow side of a pump using different bearings (Clark and Petrusewicz, 1970–1).

FIG. 3.20.

diameter in length. Purpose-made flexible connectors are available, and in this case the advice of the manufacturer should be sought.

3.4.4. ELECTRIC MOTORS

Noise in rotating electrical machinery has mechanical, aerodynamic and magnetic origins.

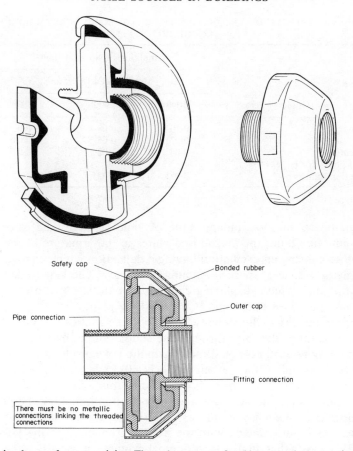

Safety cap   Bonded rubber

Outer cap

Pipe connection

Fitting connection

There must be no metallic
connections linking the threaded
connections

FIG. 3.21. Noise damper for water piping. These dampers are fitted between the water pipe and the fitting (e.g. pressure flush, angle valve, bath-tap system) and reduce the transmission of structural and water-borne noise quite considerably (with acknowledgements to Metzeler Ltd., Peterborough, England).

Bearings and brushes contribute towards the mechanical sources. Brush noise is only really significant for high-speed d.c. machines; as will be seen from the work of Clark and Petrusewicz (1970–1), bearing noise can develop as ball bearings wear if any imbalance develops besides inherent bearing noise dependent on the bearing materials and their linkage with the other motor components.

High-speed electric machines emit high aerodynamics sound levels (see Burke *et al.*, 1969). Machine ventilation is a major source of airborne noise.

Magnetic forces acting on the cores of the stator and the rotor produce mechanical vibrations, especially when the natural frequencies of the exciting forces are near those of the machine structure. Yang (1973) discusses the power estimation of the magnetic sound sources and suggests the vibration limits shown in Table 3.1.

Fans and pumps in airconditioning and heating plants are usually driven by induction motors of the squirrel-cage type. There is only a limited amount of data at present which defines clearly the sources of noise in electric motors, and this has been summarised succinctly by Clark and Petrusewicz (1970–1).

TABLE 3.1. VIBRATION LIMITS FOR ROTATING ELECTRICAL MACHINES
(YANG, 1973)

Grade	Amplitude of vibration (mm)		
	Motor speed (rev/min)		
	1000	2000	3000
Maximum	0·10	0·08	0·06
Good	0·07	0·05	0·04
Excellent	0·04	0·02	0·01
Typical limits for large induction motors	0·025	0·025	0·02

Most single-phase motors, independent of pole number and speed, emit a 100–120 Hz induction hum; the magnetised plates in the armature of the same sign repel each other causing an oscillation having a deflection which changes sign each half period giving a tone at twice the mains frequency. The amplitude of the hum depends on the load. Phase displacing condensers can be used to ensure that the maximum amplitude does not occur at the working point.

Fluctuations in the airgap flux density bring about magnetically induced vibrations during the movement of the rotor through one stator slot pitch and consequently an imbalance force rocks each pole on its seating in the yoke producing a slot frequency usually between 500 and 1000 Hz given by $S_R(N/60)$ for a rotor speed of N rev/min and a number of rotor bars S_R. Methods of reducing this source of noise include slot skewing, short-circuiting the pole tips and the pole body, making the pole arc less defined, altering the yoke mass or stiffness and selecting a running speed so that the slot frequency does not coincide with the natural frequency of the yoke.

King (1965) suggests that, firstly, the principal magnetic noise arises from the rotor bar frequency and the magnetisation by the stator winding each half-cycle. This produces sideband frequencies of $\{S_R(N/60) - 2f\}$ and $\{S_R(N/60) + 2f\}$, where S_R is the number of rotor bars, N is the rotor speed in rev/min and f is the mains frequency. Secondly, noise is generated due to the stator magnetisation component $2f$.

Sound is radiated inefficiently if the wave velocity v, relative to the velocity of sound in air c, is low, i.e. if $v/c < 1$. For a motor having a rotor radius R, number of stator slots S_s and a number of poles p, the wave velocities are $\dfrac{2\pi RNS_R}{(S_s - S_R)60}$ for the slot frequency and $\dfrac{2\pi RNS_R}{(S_s - S_R \pm 2p)(60 \mp 2f)}$ for the lower and upper sideband frequencies.

The choice of stator and rotor slot numbers is an important consideration at the motor design stage if the sound emission is to be low. Slot frequency harmonics may also be reduced by the use of magnetic wedges in open slots, and the lower harmonics may be curbed by suitable choices of winding pitch and slot skew, but using generous airgaps is the most effective way of limiting harmonic fluxes (Pomfret, 1973).

Clearly we rely on manufacturers to produce quiet electric motors. Beyond this the building environmental engineer needs to select the driving machinery for the environmental and utility building services systems with due attention to the noise emission, to use sound reduction enclosures and sound attenuators where necessary.

Pomfret (1973) gives typical no-load noise ratings as shown in Table 3.2.

Carter (1932), Ridley (1963), Walker and Kerruish (1960) and Pomfret (1973) (Table 3.3) suggest ways of defining the actual noise output providing useful guidance for designers of electric motors.

TABLE 3.2. TYPICAL NO-LOAD RATINGS FOR MOTORS (POMFRET, 1973)

Type	Output (kW)	Test speed (rev/min)	Enclosure	Measured Noise rating (dB)	Upper Limit of normal NRN[a] (dB)
Squirrel-cage motors	670	900	SP[d]	71	85
	1120	495	CACW	65	90
	1200	1485	CACA	86	92
	1650	990	CACW	72	90
	1750	495	PV	83	90
	1750	246	CACW	65	90
	5100	1490	DP	90	95
	5100	1490	DP[b]	73	95
	1500	2980	CACA	89	92
Variable-speed a.c. motors	16	2250	PV	56	80
	29	1600	SP	66	80
	67	1500	SP	78	80
	127	720	DP	75	80
	220	1270	SP	72	85
	490	1630	SP	78	90
	1230	1045	NEMA II	79	92
DC motors	19	575	TE	50	70
	33	1470	SP	62	75
	67	2000	EV[c]	65	85
	193	1750	FV[c]	71	85
	127	1750	FV	64	85
	470	600	SP[c]	77	85

[a]An upward tolerance of 3 in the noise rating number (NRN) is allowed.
[b]Fitted with air inlet and outlet attenuators.
[c]Forced ventilated.
[d]The abbreviations refer to various enclosure types:

CACW   Totally enclosed machine with internal cooling air passing through a water-cooled heat exchanger.

CACA   Totally enclosed machine with internal cooling air flowing through an air to air heat exchanger.

PV   Pipe ventilated. Clean cooling air drawn into the motor through a duct and discharged into the atmosphere.

DP   Drip-proof. Cooling air drawn into the motor and discharged into the atmosphere; water drips to be prevented from entering the motor.

SP   Splashproof otherwise as DP

TE   Totally enclosed motor.

EV   Enclosed ventilated machine.

FV   Force ventilated machine.

NEMA II   Is a special enclosure which permits cooling air to be drawn through the motor via a cowl which prevents driving rain, snow, leaves, etc., from entering the machine. It refers to an American specification issued by the National Electrical Manufacturers Association.

TABLE 3.3. UPPER LIMIT OF NORMAL RATING ON NO-LOAD (POMFRET, 1973, APPENDIX C OF BEAMA 225/1969)

Rating (kW)		NRN[a]		
		Rated Speed		
Above	Up to	3000–1501 (rev/min)	1500–1001 (rev/min)	1000 and below (rev/min)
	2·5	60	60	55
2·5	6·3	70	65	60
6·3	16	75	70	65
16	40	80	75	70
40	100	85	80	75
100	250	85	85	80
250	630	90	90	85
630	1100	90	90	85
1100	2500	92	92	90
2500	6300	95	95	90
6300	16000	97	97	92

[a]See Table 3.2.

3.4.5. COMPRESSORS

Control of compressor noise affords a good example which illustrates the interface of architecture and environmental engineering in building design. The basic questions to be answered by the building environmental engineer and the architect are:

* What type of refrigeration system is best suited to the building under consideration?
* Where should the equipment be located?
* How much space will the plant require?

Mechanical or vapour compression refrigeration is distinguished by its use of reciprocating, open centrifugal, hermetic centrifugal or rotary screw compressors, all of which are electrically driven; whereas heat energises steam or high temperature, hot-water absorption, steam or gas turbine centrifugal, and direct gas-fired absorption refrigeration systems. Absorption machines have a reputation for being quieter than electrically driven centrifugal chillers, but the choice depends upon many other factors than noise output. In any case, the heat generation plant necessary for the absorption system can emit sound levels equivalent to those measured for hermetic centrifugal chillers (see Sessler, 1973).

Sessler (1973) notes the following features of absorption refrigeration which are made in contrast to those for vapour compression systems:

* use in buildings with high heating load requirements or where 24 h operation is needed;
* capital cost of plant is more expensive;

* less efficient than vapour compression systems hence more heat rejection plant needed;
* plant installations use more floor space;
* ventilation requirements more critical;
* sound levels from *total* plant is similar to that of electrically driven compressors, but latter has more high frequency noise.

Noise control of a refrigeration system assumes the patterns of decision-making already referred to several times throughout this book—careful selection of equipment; co-ordinated planning by the architect and the building environmental engineer with regard to location of the equipment and the choice of building materials, the use of airborne and structureborne sound control devices. This assumes good building construction (i.e. the building composite structure should not *leak* sound or form bridge pathways for sound).

Refrigeration plant can be placed at basement level, at an intermediate floor level, at roof level or it may be installed in a power-house which is located away from the main building. The pros and cons of these choices are discussed in Chapter 4. As regards chiller location the acoustical penalties may be as severe at basement level as at roof level although the latter will clearly introduce more critical structural considerations. Chillers at basement level will have high condenser water-pumping costs.

After selection and location of plant has been decided, noise control is achieved further by enclosing the machines, spot noise control or by treating the plant-room itself as an enclosure. The latter is often, but not always, preferable because many plant items raise noise problems besides refrigeration equipment. Plant which has some particular problem will, however, need individual treatment; this should be ascertained at the early design stage by the building environmental engineer and the manufacturer. Until comparatively recently there has been a scarcity of acoustical data about refrigeration machines. That the rotor, the drive and the impeller cause noise generation is well established.

The noise radiated by a centrifugal compressor is characterised by the compressor blade passage frequency. Acoustical data from ten reciprocating compressors is shown in Fig. 3.22a. The compressors referred to were rated at 15–150 hp, had 4–12 cylinders and were direct driven at 1200/1750 rev/min or belt-driven at a slower speed. One-third octave band analysis has revealed that pure tones can be expected to occur at the piston stroke rate (i.e. rev/s × no. of cylinders) and its harmonics (Fig. 3.22b).

Blazier (1972) has described a sound field survey made on 38 centrifugal chillers both hermetic and open-drive types, and 14 chillers using reciprocating compressors; this equipment was designed by five major American manufacturers. The sound pressure level measurements are plotted in octave bands in Fig. 3.23a–d.

Generalised noise spectra for use in establishing sound control requirements are shown in Fig. 3.23e; the spectral shapes are different for hermetic and reciprocating machines. Notice that there appears to be no significant correlation between the sound emitted and the machine duty, although the variations in machine design may obscure any such correlation. When the chillers are operating on light load the noise level is higher than that measured at full load especially in the lower and mid

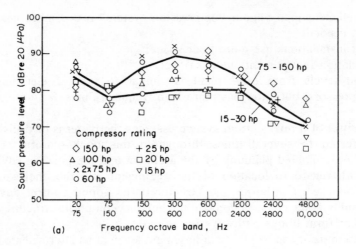

(a)

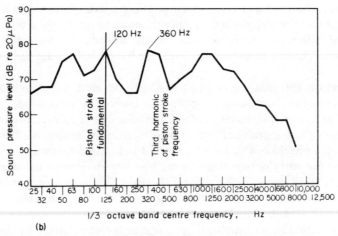

(b)

FIG. 3.22. Sound characteristics for reciprocating refrigeration compressors: (a) data from 10 installations, (b) a 4 cylinder, 1800 rpm compressor (Hoover, 1962).

frequencies. Open-drive centrifugal chillers have a sound level about 5 dBA higher than that of the hermetic type.

The quieter two-thirds of the data shown in Fig. 3.23 is summarised in Fig. 3.24a–c. The *minimum* airborne sound reduction required for a floor between the plant-room and an adjacent private office or conference room is shown. Building design solutions for plant-rooms will be discussed in section 4.5.1.

Sessler (1973) shows data that display marked differences when compared with Blazier's work (Fig. 3.25a). These arise primarily because of the lack until recently of a suitable standard measurement method which would allow direct comparison of data from different manufacturers; this difficulty has been remedied by Standard 575, *Method of Measuring Machinery Sound within Equipment Rooms* issued by the Airconditioning and Refrigeration Institute of America.

Sessler (1973) makes the following points:

(a) compressors do not radiate sound equally from all surfaces (Fig. 3.25b), and if it is proposed to site a chiller within, say, about 6 m of a wall, this should be taken into account because these differences are in the same order as the pressure doubling allowance of 3 dB;

(b) low-load compressor operation results in higher sound levels than high load (Fig. 3.25b and c), a point already made by Blazier (1972) (Fig. 3.23b); these variations may be only 0–1 dB in lower and upper octave bands, but may be as high as 6–8 dB in mid-frequency bands especially near the impeller blade passage frequency and its lower harmonics due to backflow from the condenser into the compressor;

(c) particular makes and models of compressor display characteristic noise spectra (Fig. 3.25d);

(d) the scroll cut-off in a compressor needs careful design; burrs on the cut-off can cause very high noise levels; sound baffles installed in the crossover line between the first and second stage of the compressor, and also in the line from here to the condenser, can absorb much of the compressor noise (Fig. 3.25e); diffuser plates can minimise noise also;

(e) the type and amount of refrigeration can also affect the noise levels.

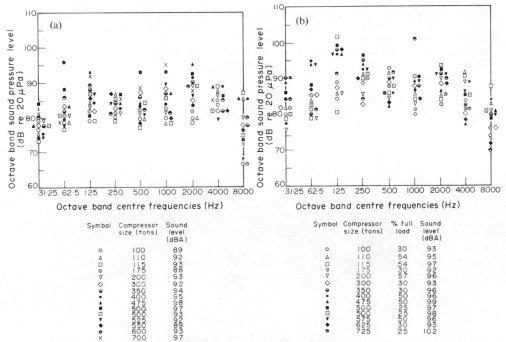

(a) Noise levels of 15 *hermetic centrifugal chillers* operating at or near design *full load*. Weighted sound level in dBA is also indicated for each machine.

(b) Noise levels of 14 *hermetic centrifugal chillers* operating at significantly less than design *full load*. Corresponding weighted sound levels in dBA are also shown.

FIG. 3.23. A survey of sound measurements taken at 3 ft (1 m) from chillers (Blazier, 1972).

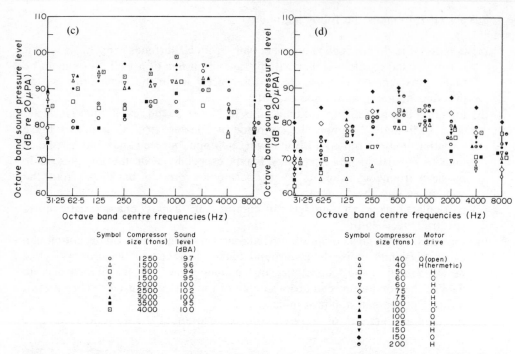

(c) Noise levels of 9 *open-drive centrifugal chillers* operating at approximately full-load conditions. Both direct-drive and geared machines are represented in sample.

(d) Octave band noise levels of 14 chillers using *reciprocating compressors of both hermetic and open type* on full load. Large data scatter is due to wide range of variables in machines designed for the same field service.

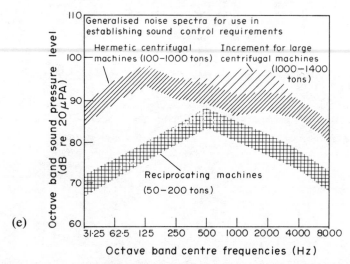

(e) Design curves shown are based on the upper 50 percentile noise levels found in the field survey of chiller equipment. They are recommended for use in establishing architectural noise isolation aims and in preparation of limit specifications for equipment.

FIG. 3.23. (*Continued*).

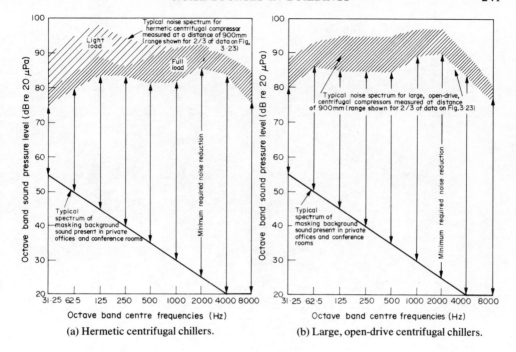

(a) Hermetic centrifugal chillers.

(b) Large, open-drive centrifugal chillers.

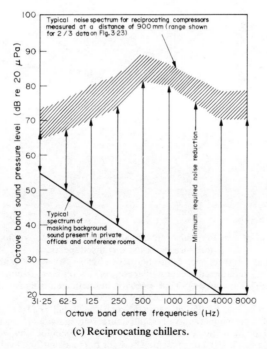

(c) Reciprocating chillers.

FIG. 3.24. Summary of two thirds of the acoustic data shown in Fig. 3.23 for both light and full-load operation. Noise control requirements for a typical building are shown (Blazier, 1972).

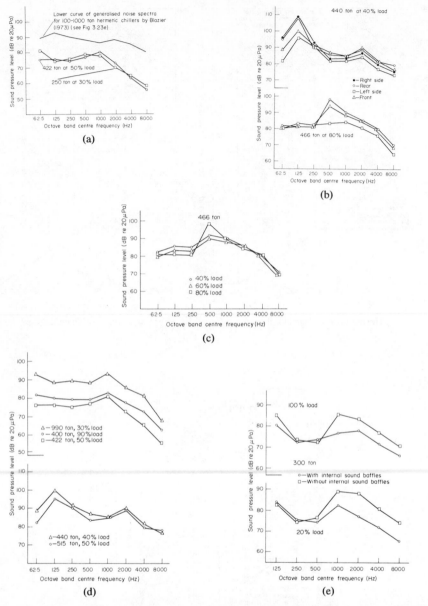

(a) Sound levels at 1 m of two hermetic centrifugal chillers (1·5 m above floor); note relationship of measured level to generalised spectra.

(b) Sound levels taken at 1 m from each major hermetic chiller surface (1·5 m above floor); positions are referenced to each chiller control panel as being front.

(c) Sound levels at 1 m (1·5 m above floor) of a 466 ton hermetic centrifugal chiller showing variation of levels with load.

(d) Sound levels at 1 m of two groups of hermetic centrifugal chillers (1·5 m above floor); each group consists of chillers of same model family of manufacture; note individual "signatures" for each group.

(e) Sound levels of 300 ton hermetic centrifugal chiller taken by manufacturer, with and without baffles in the compressor lines; data presented both for full load and 20% load conditions.

FIG. 3.25. Comparative acoustic performance of chillers (Sessler, 1973).

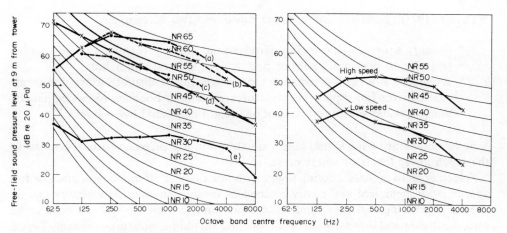

(a) Cooling tower sound levels; types (a) to (d) have same cooling duty:

(b) Two-speed cooling tower.

(a) induced draught axial fan contra-flow;
(b) cross-draught axial;
(c) induced draught mixed flow fan contra-flow;
(d) forced draught centrifugal fan contra-flow;
(e) water noise from small induced draught contra-flow cooling tower.

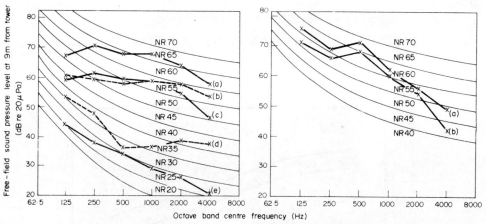

(c) Noise reduction methods;

(a) axial flow fan;
(b) as (a) with straight-through silencer;
(c) axial flow at three-quarter speed;
(d) as (a) with pod-type silencer;
(e) centrifugal fan.

(d) Noise reduction using a fibrous baffle: (a) tower with no sound insulation; (b) tower with fibrous baffle around shell and screened louvres.

FIG. 3.26. Cooling tower noise (Taylor, 1973).

3.4.6. COOLING TOWERS

Taylor (1973) classifies cooling tower noise into two categories:

* *primary sources*—due to fan aerodynamic noise, mechanical noise sources (e.g. due to bearings, chain or belt drives), electric motor hum;
* *secondary sources*—originating from water noise, discrete tones due to air passing over fan-supporting structures, rattling or booming of casing panels, unusual water pressure distributions (e.g. badly sited ball valves).

Some comparative fan sound levels are shown in Fig. 3.26a and are in contrast to the much lower levels of water noise.

Selecting an over-sized cooling tower with low rates of air movement and low fan speeds is an inefficient and costly method of noise reduction. Attenuators can be used remembering that they contribute a pressure loss on the system, need to be fitted upstream and downstream of the fan and should be moisture-resistant. Taylor (1973) advocates the use of centrifugal fans for quiet cooling tower operation and also the use of a two-speed arrangement in locations where night running may disturb residents nearby (see Figs 3.26a, b and c). The low speed derates the tower, but this is compensated for by the lower wet-bulb temperatures which occur during night hours. Notice the relatively small effect on noise reduction shown in Fig. 3.26d of using a fibrous baffle.

Dyer and Miller (1959) have investigated the nature of cooling tower noise, and their work suggested that in general the spectrum is that as shown in Fig. 3.27; the low-frequency energy arises from the fan, and the high frequency is due to splashing of the water.

The overall sound power level can be estimated from

$$L_w = 115 + 10 \lg P \, (\text{dB re } 10^{-12} \, \text{W}), \tag{3.12}$$

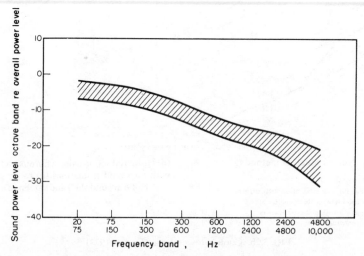

FIG. 3.27. Relative sound power level generalised spectrum for induced draught cooling towers (Dyer and Miller, 1959).

where P is the total rated fan horsepower. The sound pressure level at a distance r metres from the cooling tower for each octave band can be calculated from Fig. 3.27 and the expression

$$L_p = L_w - 10 \lg (1 \cdot 2\, r^2)\ \text{dB}. \tag{3.13}$$

3.4.7. BOILERS

The boiler and chimney can be considered to be a Helmholtz resonator excited by the combustion process. Losch (1970) gives the resonant frequency as

$$f = \frac{c}{2} \sqrt{\frac{A}{VH}}, \tag{3.14}$$

where A is the chimney cross-sectional area (m^2), H_0 is the chimney height (m), r is the chimney radius at the boiler outlet (m), $H = H_0\{1 + (\pi r/2)\}$, V is the volume of combustion chamber, $c = = c_0\sqrt{1 + (\theta/273)} =$ velocity of sound at temperature $\theta\,^{\circ}\text{C}$ and $c_0 =$ velocity of sound at 273 K.

Airborne sound is conducted from the boiler to the chimney and then to the surroundings, besides being radiated from the boiler shell and conducted as structureborne sound to the foundations. Sound levels tend to depend on the type of burner, very little variation occurring for different types of fuel. The sound level at 1 m from the fuel burner, based on the work of Locher and Nassenstein (1971), may be taken as

$$L = 12 \cdot 5 \lg q + 20 (\text{dBA}), \tag{3.15}$$

where q is the thermal output in W. The accuracy of this level will be about ± 5 dBA. For a boiler exceeding an output of 10^4 W (i.e. a sound level of 70 dBA) great care should be taken in locating the plant-room with respect to critical noise areas and to using a high mass (> 500 kg/m^2) or floating plant-room structure. For a boiler-house with n boilers each having a rating of q watts the level would be

$$L = 12 \cdot 5 \lg q + 10 \lg n + 20\ (\text{dBA}) \tag{3.16}$$

3.4.8. COMBUSTION EQUIPMENT NOISE

In reviewing combustion noise Cummings (1973) categorises two types:

(a) *turbulent combustion noise*, arising from random turbulent fluctuations in the flame reaction zone;

(b) *periodic oscillation noise*, due to inherent system instabilities which may occur in control systems or from pulsations in air and fuel supplies; this source of noise is less common, whereas (a) is always present.

The problem of noise arises in all types of fuel-burning systems although liquid and gaseous fuels tend to burn more noisily than solid fuel. Ancillary equipment such as draught fans, pumps and motors will also contribute towards the noise.

The origins of *turbulent combustion noise* are in the interaction between the energy conversion of the flame and the combined aerodynamic and acoustic characteristics of the installation. A normalised spectrum of the combustion noise taken from the work of Smith and Kilham (1963) is shown in Fig. 3.28 and is

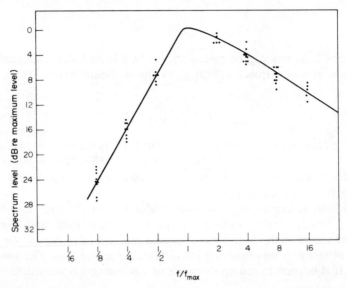

FIG. 3.28. Normalised spectrum of combustion noise (Smith and Kilham, 1963).

applicable for a premixed gaseous flame. Spectra for diffusion flames display similar characteristics but show a marked variation in shape for different fuels. Briffa *et al.* (1973) have shown that the source of sound in the combustion process coincides with a position between the inner and outer flame cones (Fig. 3.29), and that the preferred direction of noise emission from premixed flames lies at an angle in the range 40–80° (Fig. 3.30). Convection and refraction determine the directivity pattern. Diffusion flames exhibit similar directivity characteristics as premixed flames at low frequencies; at high frequencies the preferred direction of sound emission varies with the type of fuel.

The noise output from premixed flames depends on the fuel flow velocity v, the burn velocity v_b and the burner diameter d. Briffa *et al.* (1973) offer the following empirical expressions for the sound pressure level where v and d are combined in the Reynolds Number, Re:

$$L_p = -73.5 + 25.7 \lg v_b + 23.4 Re, \text{ for lean mixtures,} \tag{3.17a}$$

$$L_p = -71.6 + 6.5 \lg v_b + 32.0 Re, \text{ for rich mixtures,} \tag{3.17b}$$

$$L_p = -55.3 + 7.7 \lg v_b + 27.2 Re, \text{ for lean and rich mixtures.} \tag{3.17c}$$

The fuel used was an ethylene–air mixture with burner diameters ranging from 2 to 13·4 mm, burn velocities were 0·22–0·85 m/s, cold gas velocities (used in the calculation of the Reynolds number) were 2·5–95 m/s. Clearly a reduction in the sound pressure level occurs when either v_b or Re are reduced. Turbulence in the

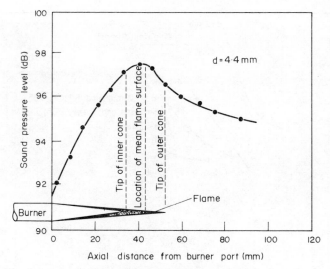

FIG. 3.29. Variation of sound pressure level with axial distance along the flame for a burner diameter of 4·4 mm (Briffa *et al.*, 1973).

air–fuel stream can be decreased by ensuring complete mixing of the fuel with the air and by elimination of any eddy flows before mixing.

Typical values of the thermo-acoustic efficiency for premixed and diffusion flames are shown in Fig. 3.31. In general, premixed flames tend to be noisier than diffusion ones. The thermo-acoustic efficiency η increases with thermal output q, and $d\eta/dq$ varies inversely as the burner diameter.

Hurle (1968) has extended the work of Thomas and Williams (1966) to show that the sound generation mechanism for a turbulent premixed flame emitted from a given burner diameter is

$$p_i(r, t) = \frac{\bar{\rho}}{4\pi r} \left\{ \frac{d}{dt} (E - 1) Q_i \right\}_{t-\tau},$$

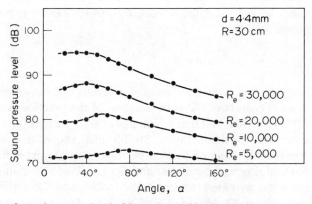

FIG. 3.30. Variation of sound pressure level with angular position for a burner diameter of 4·4 mm, an axial flame distance of 30 cm and a range of Reynolds Numbers (Briffa *et al.*, 1973).

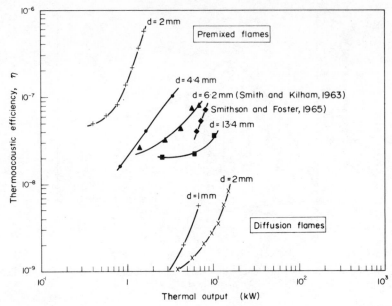

FIG. 3.31. Comparison of the thermo-acoustic efficiencies of premixed and diffusion flames (Briffa *et al.*, 1973).

when p_i is sound pressure of the ith flame element situated a distance r from the microphone, $\bar{\rho}$ is the density of the ambient gas, E is the volumetric expansion ratio, Q_i is the volume flow rate of the fuel–air mixture and τ is the time for the sound to pass the distance r from the flame to the microphone. This derivation assumes that a turbulent flame can be represented by a monopole sound source distribution evolving heat and gas. If the flame dimensions are appreciably less than the wavelength of the sound radiated, then the total sound emission over the whole combustion zone will be

$$p(r, t) = \sum p_i(r, t) = \frac{\bar{\rho}}{4\pi r} (E - 1) \left(\frac{dQ}{dt}\right)_{t-\tau}.$$

This work has been applied to liquid fuel as well as gas fuel sprays.

Some of the most recent theoretical work by Strahle (1973) expresses the total emitted sound power as

$$P = \frac{1}{4\pi c\bar{\rho}} \int_V \overline{\rho_{tt}\rho_{tt}} \, V^* dV,$$

where V^* is the volume over which the second derivative of the density fluctuations at the flame front ρ_{tt}, is correlated; V is the volume of the reaction zone. This agrees with the work of Ribner (1964) and that which has already been referred to, by Hurle, on sound emission from turbulent flows. Strahle also shows that his theory is in agreement with the results obtained from experiments on premixed flames by Smith and Kilham (1963) and those of Smithson and Foster (1965) by using two turbulent flame models, called the wrinkled laminar flame model, and the distributed reaction zone model, from which he obtained scaling laws for sound power and flow parameters in premixed turbulent gas flames. Thus the sound power P may be represented by

$$P \propto U^{4-3q}S^{r+3q}l^{2+r}, \tag{3.18}$$

where U is the mean flow speed, S is the laminar flame speed, l is a characteristic burner dimension and q and r are empirical exponents. If q and r are suitably chosen, the empirical expression of Smith and Kilham (1963) results:

$$P \propto U^2 S^2 l^2. \tag{3.19}$$

Cummings (1973) obtains a formula

$$P \propto S^2(E-1)^2 u^2 l^4 L^{-2}, \tag{3.20}$$

where E is the volume expansion ratio of the reactants, u is the typical r.m.s. turbulent velocity component and L is a typical turbulence length scale; again, if $L \propto l$ and $u \propto U$, then the Smith and Kilham formula is obtained.

There is a conflict between the requirements for good combustion and those for low noise output. Good flame stability needs high secondary flows in the burner, and intense combustion demands high turbulence, and hence high flow velocities. Both of these requirements increase the sound emission. Cummings (1973) makes the point that for quieter operation it is better to have a series of small flames instead of a single flame.

Periodic oscillation noise gives rise to discrete tones in the combustion sound spectrum which are due to control or flow instabilities and system resonances. Variations in the heat release rate can couple with acoustic pressure oscillations in the combustion chamber. Vortices in the combustion chamber may be shed at discontinuities and can couple with the resonant frequency modes of the chamber; besides this the vortices are also sensitive to pressure fluctuations in the space. Cummings (1973) considers that most combustion oscillations are of the feedback type and describes recent research which aims to increase our understanding of this source of noise.

Referring to Fig. 3.32 transfer functions are defined as:

Flame transfer function $\qquad G = \dfrac{Q_3}{Q_4}.$

Combustion chamber
transfer function $\qquad Z = \dfrac{P}{Q_3 + Q_{ext}}.$

Burner transfer function $\qquad H = \dfrac{Q_4}{P}.$

From these functions conditions for self-excited oscillations are obtained. If there is no external excitation, $Q_{ext} = 0$ and $ZHG = 1$.

Since these functions are complex quantities conditions for neutral stability are

$$|ZHG| = 1, \tag{3.21}$$

$$\arg(ZHG) = 0, \tag{3.22}$$

and for stable operation $|ZHG| < 1$ at any frequency where $\arg(ZHG) = 0$.

The control of burner noise should begin at the equipment design stage with careful attention being given to the geometry and the dynamic performance of the

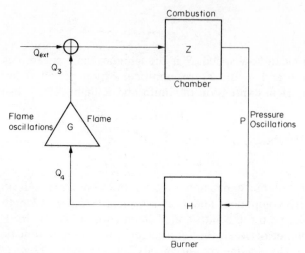

FIG. 3.32. Feedback elements of a combustion system: Q's denote volume flow rates; P denotes pressure (Cummings, 1973).

combustion system. General guidance on the suppression of noise from burners is given by Sutherland (1969), Gupta (1973), Briffa *et al.* (1973) and Cummings (1973).

The principal aerodynamic noise components occur at the fuel injector arising from the shear layer between the high-velocity fuel jet and the low-velocity air which it entrains. In order to lower the sound emission the pressure drop across the injector should be as low as is compatible with good combustion; multi-hole injectors help by reducing the Reynolds number. The turbulence level of the mixture emerging from the burner ports can also be lowered by designing them to be deep and small in diameter. This helps to produce a homogeneous laminar mixture stream; there should be, however, sufficient turbulence to ensure thorough mixing of fuel and air. A Reynolds number of < 2000 is recommended for the port fuel–air stream.

Another way to reduce turbulence is to reduce the fuel and combustion airflow rates, but this is limited by the amount of energy required from the fuel and also by the air supply needed to ensure efficient combustion. Gupta (1973) reports that if the fuel is injected in two directions radially as well as axially, a reduction of about 4 dB in the burner sound level can be achieved. Roberts and Leventhall (1975) consider methods of reducing noise in gas burners involving streamlining of mixture flow.

Helmholtz resonators and quarter-wave tubes are required to filter out discrete noise components, otherwise appreciable reductions in the broadband sound level may be obtained by using sound absorption materials on the primary and secondary airways. Other remedial measures include sealing the combustion chamber, making an acoustical enclosure for the unit and attenuating the noise emitted from ancillary equipment. If draught stabilisers are used they should be fitted with an acoustical baffle.

Alteration of the supply pipework configuration can eliminate standing waves that may be present upstream of the burner. Modification of the combustion chamber geometry by re-positioning of a refractory wall, or by altering the gasflow pathway, will change the sound energy spectrum.

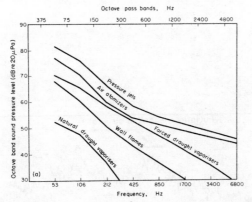

(a) Average domestic oil burner noise curves (Kennedy, 1966).

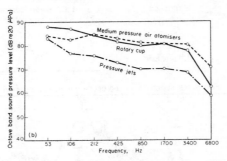

(b) Average industrial oil burner noise curves (Briffa and Kennedy, 1965).

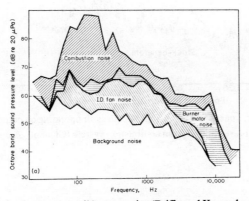

(c) Analysis of rotary cup oil burner noise (Briffa and Kennedy, 1966–7).

FIG. 3.33. Sound spectra for various types of oil burners.

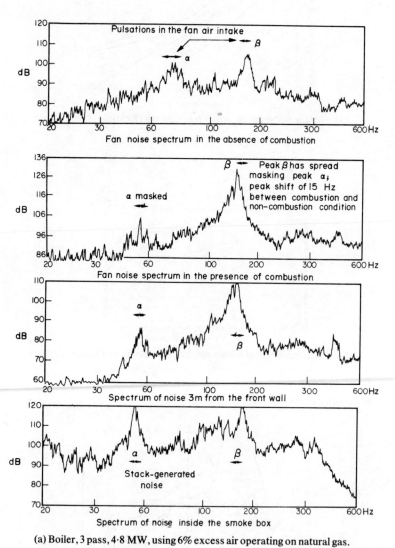

(a) Boiler, 3 pass, 4·8 MW, using 6% excess air operating on natural gas.

FIG. 3.34. Spectra of noise emitted from combustion system (Ghamah–Zadeh and Briffa, 1969).

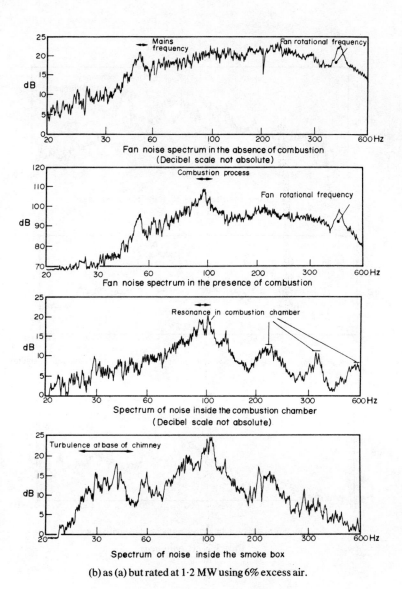

Fan noise spectrum in the absence of combustion
(Decibel scale not absolute)

Fan noise spectrum in the presence of combustion

Spectrum of noise inside the combustion chamber
(Decibel scale not absolute)

Spectrum of noise inside the smoke box

(b) as (a) but rated at 1·2 MW using 6% excess air.

FIG. 3.34. (*Continued*).

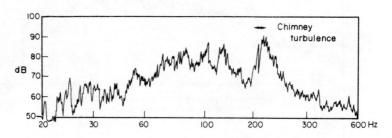

Spectrum of noise 20 m from the top of chimney

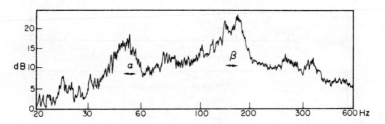

Spectrum of noise 12 m from the top of chimney (decibel scale not absolute)

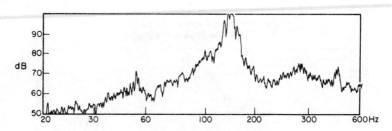

Spectrum of background noise with combustion in progress

FIG. 3.34. (*Continued*).

Research and development of an ultrasonic oil burner unit has been carried out for several years. The principle is that a piezoelectric crystal sets the atomising head into longitudinal vibration at a frequency of between 10–100 kHz depending on the droplet size required. The oil is fed through a velocity transformer on the vibrating head and is atomised. Besides achieving quiet operation a number of other advantages accrue—very short ignition times; no high fuel pressures developed; the atomiser is self-cleaning and has no moving parts; atomising efficiency is easily controlled by regulating the current to the piezoelectric crystal and is consistent over a wide fuel flow range; turndown ratios of 50:1 are possible (Korn, 1971).

Our knowledge of combustion noise is in its infancy, and this section has attempted to serve as an introduction to the subject. It is clear from the papers presented at the Second AGA-IGT Conference held in Atlanta, Georgia, USA, on 5–7 June 1972, that rapid progress is likely to occur in the next few years.

To conclude, Figs. 3.33 and 3.34 show some acoustic spectra of combustion noise taken from the work of Briffa and Kennedy (1966–7), Kennedy (1965) and Ghamah-Zadeh and Briffa (1969) which illustrate some of the features that have been discussed.

3.4.9. CHIMNEYS

Noise has been observed to originate from the chimney in some installations. This is due to draught fans, combustion noise and/or aerodynamic excitation of the chimney structure due to wind. Aerodynamic stabilisers should be fitted near the top of chimneys situated in exposed areas in order to break down the large Karman vortices formed on the windward side into smaller scale eddies having shorter decay times.

The sound level in chimneys exhausting fuel gases from a boiler rated at q watts may be estimated from

$$L = 10 \lg q + 45 \,(\text{dBA}). \tag{3.23a}$$

The octave-band sound pressure levels may be assessed from these dBA levels using spectrum corrections given in Book B of the *IHVE Guide 1970*. Again, the accuracy is in the order of ± 5 dBA. The sound pressure level at a distance metres from the chimney outlet may be estimated from

$$L_{P,r} = L_w - 10 \lg 4\pi r^2 + D, \tag{3.23b}$$

where L_w is the sound level at $r = 0$, i.e. $L_w = L_P$ (at $r = 0$) $+ 10 \lg A$, for a chimney of cross-sectional area of A (m^2) and directivity index D (see Chapter 10, equation 10.16).

3.4.10. STEAM AND GAS TURBINES

The work of Iredale (1973) should be referred to for guidance in designing noise control for this type of plant.

3.5. Sound Generation Along the Air-Distribution Network

The level of aerodynamic sound increases with an increase in the airflow velocity due to the greater strength of fluctuations in the pressure field. Stewart (1969) has compared the airflow generated noise for various duct fittings (Table 3.4). He suggests that a limited extrapolation to other sizes and velocities may be obtained using the expression for overall sound power level given as

$$L_w = K + 10 \lg\left(\frac{A}{A_0}\right) + 55 \lg\left(\frac{v}{v_0}\right), \tag{3.24}$$

where K is given in Table 3.4; A and v are the duct cross-sectional area ($A_0 = 300\,\text{mm} \times 300\,\text{mm}$ or 12 in. by 12 in.) and mean airflow velocity respectively ($v_0 = 10\,\text{m/s}$ or 2000 ft/min). The following sections will review the work of Ingard, Oppenheim and Hirschorn on sound generation in various duct elements.

TABLE 3.4

Fitting	K
Straight duct	38
3:1 contraction (to 12 in. by 12 in.*)	47
90° bends:	
Radiused	48 (estimated)
Mitred with turning vanes	56
Mitred without turning vanes	57
Duct with 90° tee (5% draw-off)	55
Dampers:	
Open	44
15° closed	53
45° closed	65

*300 mm × 300 mm.

3.5.1. STRAIGHT DUCTS

Figure 3.35a shows an octave band overall power spectrum for air passing through a 600 mm × 600 mm (24 in. by 24 in.) mild steel duct at various velocities. The following conclusions can be made:

* a change of slope occurs in the spectrum when the fundamental transverse mode is propagated at about 270 Hz;
* an increase of about 18 dB occurs in the overall power level per doubling of air velocity;
* the measured trends compare favourably with equation (3.24).

German data is presented in Fig. 3.35b. The loudness level, L phons, can be related to the fan speed, N (rpm), by the approximate relationship

$$L = 54\left(\frac{N}{2160} + 1\right).$$

See Chapter 5, equation 5.2 for similar work on centrifugal fans.

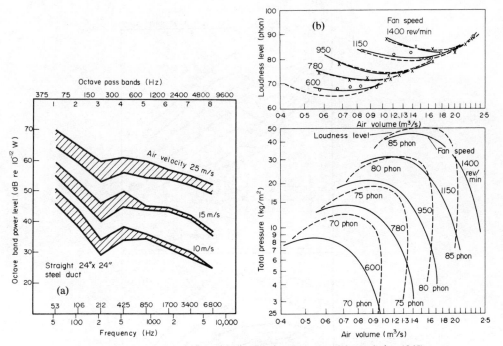

(a) Sound power level variation with air velocity (Ingard and Oppenheim, 1965).
(b) Loudness level as a function of quantity of air and of the characteristic of a Pollrich axial flow fan of 450 mm diameter according to measurements by Zeller reported in Eck (1973).

FIG. 3.35. Sound levels related to airflow rates.

3.5.2. BUTTERFLY DAMPERS

The sound generation spectra for a 588 mm ($23\frac{1}{2}$ in.) by 588 mm ($23\frac{1}{2}$ in.) by 3 mm ($\frac{1}{8}$ in.) steel damper installed in a 600 mm × 600 mm (24 in. by 24 in.) duct are shown in Figs. 3.36a–c. When fully open, the spectrum is similar to that for an empty duct, but as the damper is closed the airflow pattern changes; pressure field perturbations appear with a consequent rise in sound level and variation in spectrum shape.

3.5.3. TRANSITION PIECES

Data is given for abrupt and gradual transition pieces in Fig. 3.37a–d. The conclusions that can be made are:

(a) abrupt transition pieces emit more sound than gradual ones;
(b) the slope of the spectra is modified when the fundamental and higher order transverse modes appear;
(c) the higher generated sound levels occur when large changes occur in the acoustic impedance;
(d) Soroka (1970) has measured the sound levels generated by air discharged from various duct geometries at velocities up to 20 m/s. The measurements were

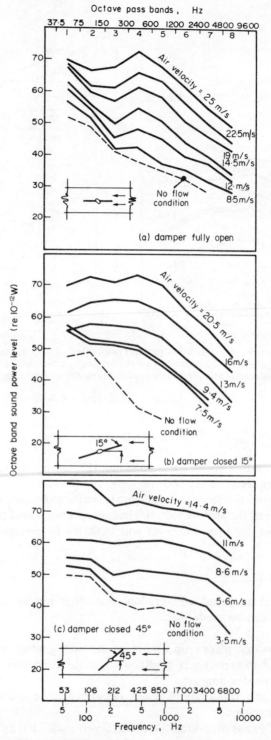

FIG. 3.36. Influence of damper position on airflow sound levels (Ingard and Oppenheim, 1965).

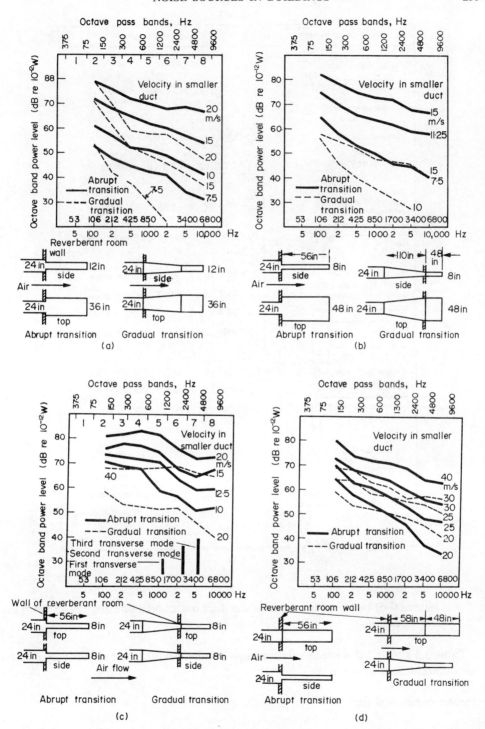

Fig. 3.37. Influence of abrupt and gradual transitions on airflow sound levels (Ingard and Oppenheim, 1965).

TABLE 3.5

Duct arrangement	Sound level (dBA)
Circular (diameter d) 45° 0·61m Measurement point M	$L_p = 65\cdot5\ \lg_{10} v + 12\ \lg_{10} d - 200\cdot6$
Square (side a)	$L_p = 59\cdot6\ \lg_{10} v + 3\cdot4\ \lg_{10} a - 172\cdot2$
Rectangular (aspect ratio, $r \geqslant 1$)	$L_p = 59\cdot6\ \lg_{10} v + 3\cdot4\ \lg_{10} a + 3\cdot8\ \lg_{10} r - 172\cdot2$
Circular duct with short radius 90° elbow at termination 45° 0·61 m M	$L_p = 73\cdot5\ \lg_{10} v - 203\cdot2$ (100–200 mm elbows) $L_p = 51\cdot5\ \lg_{10} v - 136\cdot4$ (250 mm elbows)
Circular ducts with short radius 90° elbows and 1·5–2 m of ducting 1·5–2 m 45° 0·61m M	$L_p = 70\cdot1\ \lg_{10} v - 201\cdot3$ (100 mm elbow) $L_p = 75\cdot7\ \lg_{10} v - 214\cdot7$ (150 mm elbow) $L_p = 64\cdot4\ \lg_{10} v - 177\cdot5$ (200 mm elbow) $L_p = 56\cdot8\ \lg_{10} v - 151\cdot7$ (250 mm elbow)
Circular ducts with long radius 90° bends d $r = 2d$ 45° 0·61 m M	$L_p = 66\cdot9\ \lg_{10} v - 197\cdot4$ (100 mm elbow) $L_p = 62\cdot5\ \lg_{10} v - 183\cdot0$ (150, 200mm elbows) $L_p = 32\cdot6\ \lg_{10} v - 83\cdot3$ (250 mm elbows)
Circular ducts with long radius 90° bends and 1·5–2 m of ducting d $r = 2d$ 1·5–2 m 45° 0·61m M	$L_p = 69\cdot7\ \lg_{10} v - 209\cdot0$ (100 mm elbow) $L_p = 72\cdot4\ \lg_{10} v - 215\cdot5$ (150 mm elbow) $L_p = 93\cdot2\ \lg_{10} v - 277\cdot4$ (200 mm elbow) $L_p = 69\cdot1\ \lg_{10} v - 193\cdot0$ (250 mm elbow)

made at 0·61 m from the centre of the duct outlet on a line angling 45° with the flow direction.

Table 3.5 shows the various relationships obtained with an air velocity of v (m/s) for different duct geometries.

Soroka concluded that:

(i) circular ducts may produce 20% higher sound levels than those for square ducts when airflows are in the order of 5 m/s, but at 15 m/s there is hardly any difference; the levels produced from rectangular ducts with aspect ratio r are increased by a factor of $3 \cdot 8 \lg r$ over those for square ducts;

(ii) one fitting in the ductwork system may increase the sound level of the system by between 30–100% depending on the size and form of the fitting;

(iii) duct lengths fitted after the fitting help to produce a lower system sound level;

(iv) easy radius 90° bends cause lower airborne generated sound levels than 90° elbows; properly designed bends can produce an attenuating effect which exceeds any generation level;

(v) at lower air velocities larger fittings generate higher sound levels than smaller ones, but for air velocities greater than about 10 m/s smaller fittings produce more sound; this trend was not apparent when a length of straight ducting followed the bend, i.e. sound generation increased as the bend size increased irrespective of flow velocity.

3.5.4. GENERAL CONCLUSIONS

The sound generation from any duct element depends upon:

* *the size of the duct element*: for a given air velocity the sound generated will increase with the size of the element volume which acts as an acoustic radiator;

* *large changes in acoustic impedances between one duct element and another*: highest sound generation levels occur with large changes in the duct dimensions at transitions or branch off-takes, i.e. when large changes occur in the airflow pattern; if several well-designed turning vanes are introduced at a bend they do not alter the generation sound level significantly although some back scatter of the low-frequency radiation (< 500 Hz) occurs (Fig. 3.38);

* *duct element geometry*: changes often occur in the sound generation spectral frequency fall-off when the duct cross dimension $D \geqslant nc/2f$; duct elements with high aspect ratios give a greater probability that the sound will be scattered into the higher order transverse modes; abrupt changes in duct element dimensions produce higher sound levels than gradual ones due to the greater turbulence intensity.

The sound generated in airflow systems is only partly dependent upon pressure drop. This is not unexpected when it is considered that acoustical efficiencies are normally in the order of 10^{-4}; very little of the energy lost is converted into sound radiation. In any case, the origin of pseudo-sound is the transient behaviour of the fluid pressure fields rather than steady-state pressure differentials.

Holmes (1973b) concludes that noise generated by airflow around bends fitted with turning vanes has been shown in his experiments to be dependent upon the vane pressure loss and the total vane surface area; the design of the vane may be important but there is insufficient evidence available to be conclusive. The octave band sound power level may be predicted from the equation

$$L_w = K + K_v + K_A + K_c + K_n + K_s \text{ (dB re } 10^{-12} \text{ W)}, \qquad (3.25)$$

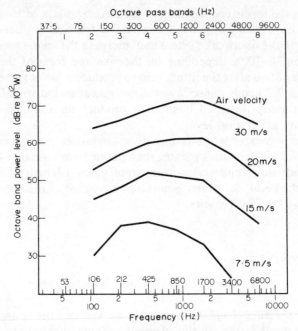

FIG. 3.38. Sound power spectra for airflow through a 90° elbow with seven equi-spaced turning vanes in a 200 mm × 200 mm (8″ × 8″) duct (Ingard and Oppenheim 1968).

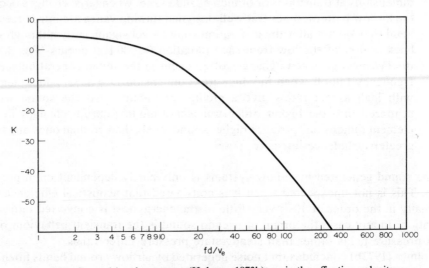

FIG. 3.39. General bend spectrum (Holmes, 1973b). v_e is the effective velocity =

$$\frac{\text{volume flow rate (m}^3\text{/s)}}{\text{duct area (m}^2\text{)} \times \text{blockage factor}}.$$

$$= \frac{Q}{A_D B_e}$$

(values of B_e are given in Holmes, 1973b); f is the frequency; d is the duct height.

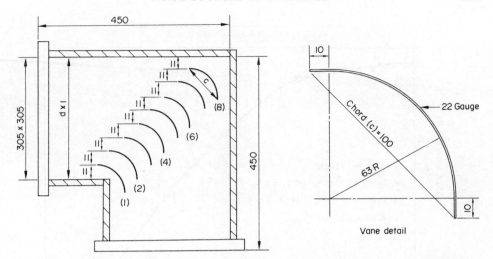

FIG. 3.40. Bend and vane detail (Holmes, 1973b): (1) refers to position of single vane; all dimensions in millimetres.

where K is the general bend spectrum (Fig. 3.39), K_v is the velocity correction, K_A is the duct area correction, K_c is the vane chord length correction, K_n is the vane number and K_s is the spectrum correction.

Values of K_v, K_A, K_c, K_n, K_s are tabulated in Holmes (1973b); the bend and vane designs used in these experiments are shown in Fig. 3.40. From Fig. 3.41b it can be seen that bends with vanes radiate more sound power than those without them, particularly in the frequency range 250–1000 Hz. Figure 3.41b and c shows the general band spectrum factor and also comparisons with other work. Note that at

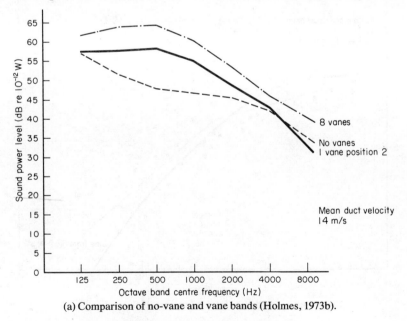

(a) Comparison of no-vane and vane bands (Holmes, 1973b).

FIG. 3.41. Sound emission from bends.

$$L_w = 10 \lg K + 10 \lg A_D + 10 \lg C + 10 \lg n + 50 \lg v_e$$

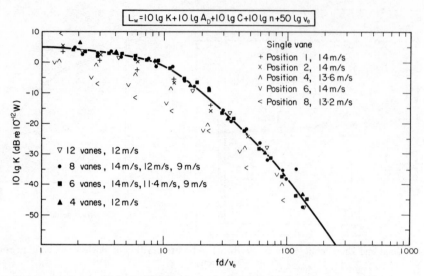

(b) Correlation of test data (Holmes, 1973b); A_D is the duct area (m²); C is the vane chord (m); n is the number of vanes; $v_e = Q/(A_D B_e)$ (see Fig. 3.39).

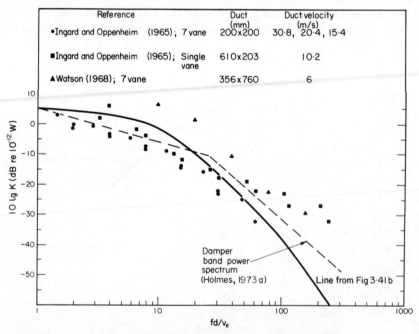

(c) Comparison with data from references (Holmes, 1973b).

FIG. 3.41. (*Continued*).

positions 8 and 6 on Fig. 3.41b there is some discrepancy; Holmes (1973b) suggests that this is to be expected as the vanes were quite close to the outer edge of the bend allowing strong separation to occur at the inner edge, thus energy losses became located on the wall of the bend rather than at the vane itself. This work is important because Holmes (1973b) gives a method of estimating the noise levels generated by bends that are fitted with vanes.

3.6. Air Terminal Devices

The category of terminal devices considered here are those connected to an air distribution system, i.e. grilles, diffusers, induction units, dual and single duct units and reheat units; these are used for controlling and/or distributing air in spaces. The acoustical situation for an air terminal device is:

(i) there is a flow of acoustical energy through the distribution system, arising from the central station plant and the system; the terminal may attenuate the forward propagation of this sound by absorption or reflection; this attenuation is the insertion loss of the terminal;
(ii) airflow through the terminal generates noise.

The insertion loss is a function of the form of the terminal and any acoustical treatment provided; it is best determined by test. Airflow-generated noise can be expected to be dipole in character, its magnitude is indicated by equation (3.2) or, as Curle (1955) has shown in terms of bulk flow parameters, it is approximately

$$\text{sound power} \propto \rho v^6 c^3 L^2 g(Re),$$

where ρ is the fluid density, v is the fluid velocity, c is the velocity of sound, L is the dimension of surface and, $g(Re)$ is a Reynolds number function.

For nearly all our systems ρ and c can be taken as constant and an area term A can be substituted for L^2. Hence

$$\text{sound power} \propto v^6 Ag(Re)$$
$$= Kv^6 Ag(Re). \tag{3.26}$$

The proportionality constant K takes account of such factors as turbulence intensity, surface effects and acoustical radiation parameters. For similar terminals and flow conditions a simplified expression for sound generated by airflow is

$$\text{sound power} = Kv^6 A, \tag{3.27}$$
$$\text{sound power level} = 10 \lg (Kv^6 A)$$
$$= K^1 + 60 \lg v + 10 \lg A. \tag{3.28}$$

This is very similar to equation (3.24). The coefficients of the logarithmic factors have been found experimentally to vary, and this accounts for the slight difference in the equations. One reason for this is the neglect of the Reynolds function $g(Re)$, which is a function of both v and A.

The frequency spectrum of the sound power can be expected to be related to the frequency spectrum of the decaying turbulent eddies and the acoustical radiation

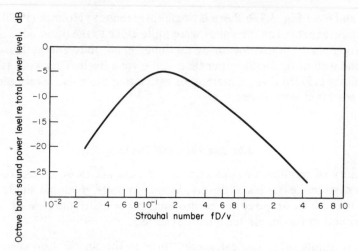

FIG. 3.42. Sound emission from vaned bends. Total sound power level $= 10 \lg \left[\dfrac{A \rho v^8}{2c^5} \right] - 40$ dB re 10^{-12} W

where parameters refer to joint outlet.

impedances of the terminal. The correlation of the sound power spectrum for a jet of air is shown in Fig. 3.42.

3.6.1. DIFFUSERS AND GRILLES

Chaddock (1957) found experimentally that the total sound power level of circular ceiling diffusers could be expressed by the equation

$$\text{sound power level} = 32 + 13 \lg A + 60 \lg v, \qquad (3.29)$$

where A is the minimum flow area in neck (m^2) and v is the velocity at minimum flow area (m/s).

Data on the frequency distribution of the sound power level relative to that predicted by equation (3.29) are shown in Fig. 3.43 for several experimental investigations, including Chaddock (1957) and Hardy (1963) for circular ceiling diffusers and Marvet (1959) and Helies and Cadiergues (1965, 1967) for grilles.

It is apparent that not only are there differences between diffusers and grilles but there are differences within each type. It follows that individual manufacturers' data are necessary rather than predictions from generalised formulae.

The adjustment of vanes, cones or dampers can also modify the sound power spectrum; an example of the latter is given by Woods and Smith (1969).

Holmes (1973a) has shown that the noise generated by airflow over grilles and dampers is dependent upon the pressure loss across them. Using three double deflection grilles, the sixth power law equation (3.28) was verified and hence established that the noise source has a dipole character (Fig. 3.44a). Hardy (1963) has shown that generalised power spectra can be obtained if the spectrum power level (sound power in a unit frequency band) is plotted against the Strouhal number, as shown, for instance, in the generalised jet spectrum depicted in Fig. 3.42, and in this

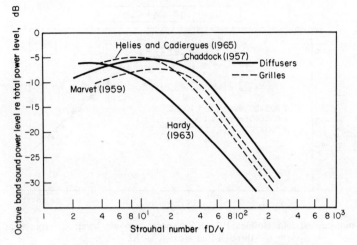

FIG. 3.43. Sound power characteristics for grilles and differences.

case equation (3.28) can be written as

$$10 \lg S = 10 \lg K + 10 \lg A + 10 \lg l + 50 \lg v \text{ (dB re } 10^{-12} \text{W),} \qquad (3.30)$$

where $10 \lg S$ is the spectrum power level, i.e. equals octave band power level minus $10 \lg$ (centre frequency bandwidth), and l is a representative grille length; K, A and v are as for equation (3.28). If $K' = Kl$, then the experimental data obtained by Holmes (1973a) is shown in Fig. 3.44b and compared with other data in Fig. 3.44c. The ordinate of Fig. 3.44c is given by a modified form of equation (3.30); thus

$$10 \lg K = 10 \lg S - 10 \lg (B_e A_D) - 50 \lg (v_D / B_e) - 10 \lg d + 30 \text{ (dB re } 10^{-12} \text{W),} \qquad (3.31)$$

where B_e is a blockage factor defined as the ratio of free area to duct area (values are given by Holmes (1973a)), A_D is the area of the test duct (m²), v_D is the mean velocity in the test duct (m/s) and d is the thickness of the grille vanes (mm).

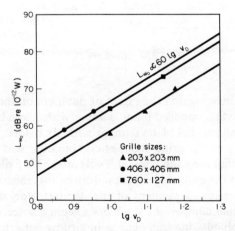

(a) Grille: overall sound power as a function of duct velocity (Holmes, 1973a).

FIG. 3.44. Sound power characteristics of grilles.

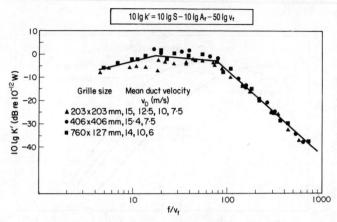

(b) Correlation of grille test data (Holmes, 1973a); S = spectrum power level; A_f = free area; v_f = velocity through free area; $K^1 = Kl$.

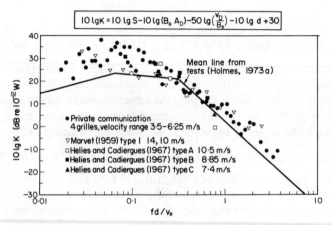

(c) Comparison of grille test data (Holmes, 1973a); B_e = blockage factor; A_D = duct area (m²); v_D = duct velocity (m/s); d = grille vane thickness (mm); $v_e = \dfrac{\text{air flow rate (m}^3\text{/s)}}{A_D B_e}$.

FIG. 3.44. (*Continued*).

Holmes (1973a) also investigated the effect of dampers on the grille sound power spectrum using a multi-vane opposed blade damper with seven blades, a single-plate damper and a multi-vane parallel blade damper with six blades. The results showed that the spectrum shape of the grille–damper combination was very similar to that for the grille alone, but that the sound power levels are about 5 dB higher (Fig. 3.45a).

Figure 3.45b confirms the expected dipole nature of the noise source and also that the overall sound power is independent of the damper type and setting, although notice that the single-plate damper set at 45° does seem quieter, and Holmes (1973a) suggests that this is probably due to a change in airflow pattern at the high angle of incidence. From the graphical plot shown in Fig. 3.45c it can be concluded that the assumptions made for this work (i.e. treating the damper as a simple duct constric-

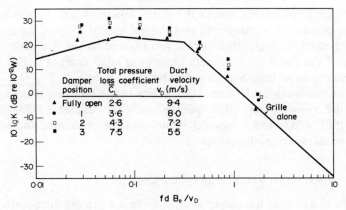

(a) Grille and damper test results (Holmes, 1973a).

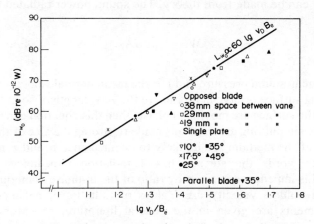

(b) Damper: overall sound power as a function of effective velocity (Holmes, 1973a).

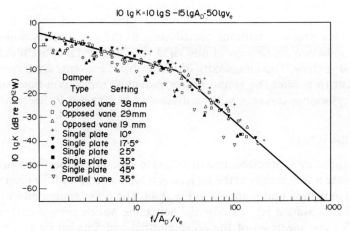

(c) Damper: correlation of test data (Holmes, 1973a).

FIG. 3.45. Influence of grille damper control on sound emission.

tion; constant static pressures across the ducts upstream and downstream of the damper; static pressure is recoverable and the noise is of a dipole nature generated by the interaction of the high turbulence levels created by the damper on the blades and the walls of the duct) are reasonable except at large damper angles and for the parallel vane/damper at high Strouhal numbers.

The value of this work by Holmes (1973a) lies in the fact that good agreement is found with the research of other people such as Marvet (1959), Helies and Cadiergues (1965, 1967) and a method is outlined for estimating the noise output from grille and damper combinations.

3.6.2. SOUND PROPAGATED FROM VIBRATING SURFACES

Is it possible to estimate the sound emitted from a vibrating duct surface, a fan or an induction unit casing? In the absence of empirical data it is difficult to verify some predictions that can be made from theory. The sound power radiated by a vibrating surface of area A is

$$W = \int_A \bar{p}\bar{v}\, dA, \tag{3.32}$$

where $\bar{p}$ is the mean sound pressure and $\bar{v}$ is the mean normal velocity of surface.

If the vibrating surface is large in relation to the wavelength of sound being excited and all parts of the surface are moving in phase, then this equation can be solved.

In practice the simplifying assumptions made regarding the size of the surface and the boundaries of the medium are unlikely to occur, hence p and v need not be in phase. In other words the surface has a radiation impedance; the resistive component of this impedance is given by ratio of the in-phase components of sound pressure p and volume velocity vA. Values of radiation impedances for simple source arrangements are given in the general literature, see Morse (1948) and Beranek (1954).

If the vibrating surface is not rigid, then flexural waves can be set in motion, i.e. all points on the surface are not moving in phase. With out-of-phase movements of adjacent areas of the surface, fluid can flow from the areas of positive pressure to those of negative pressure without contributing to the net sound volume velocity. The acoustic radiation efficiencies of non-rigid vibrating surfaces are much less than that for a rigid surface. The complexity of the problem prevents any simple analysis being sufficient to predict the sound power radiated from a non-rigid surface.

Control of vibrating surfaces will be discussed in Chapter 9, Section 9.6.2.

3.6.3. INDUCTION UNITS

These units provide appreciable insertion loss to sound from the distribution system and also generate noise, mainly at the air jets. A typical example of the insertion loss of a commercial induction unit is shown in Fig. 3.46. The noise generated by air jets is a function of the maximum jet velocity and hence the nozzle pressure. However, the casing modifies the spectrum of the noise emitted and data on jet noise cannot be directly applied; for this reason manufacturers' data should be used. The sound power output can also be modified by dampers controlling the secondary flow.

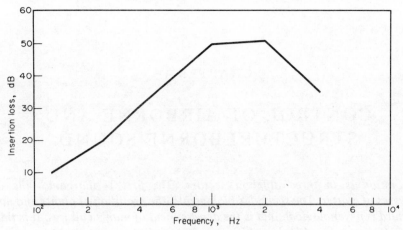

FIG. 3.46. Insertion loss for an induction unit.

3.6.4. HIGH-PRESSURE TERMINAL EQUIPMENT

These generate noise in the flow-mixing control sections; these are often followed by an attenuating section or sometimes recommendations are given by manufacturers on adding a length of lined duct at outlet. They also provide insertion loss to the system noise.

Generated noise can reach occupied rooms through the outlet or through the casing. Casing-radiated noise is of considerable importance if the unit is situated in a different area from the terminal, e.g. in the ceiling space of the room below. Some units are connected to the distribution system with a length of flexible duct; this may provide very little sound insulation to noise generated within the unit and radiated through the inlet connection. An example of the influence of flexible duct connections on the casing radiation from a dual duct controller is shown in Fig. 3.47.

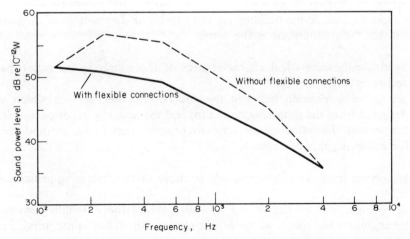

FIG. 3.47. Influence of flexible duct connections on the casing sound radiation from a dual duct controller.

CHAPTER 4

CONTROL OF AIRBORNE AND STRUCTUREBORNE SOUND

Evolution is in three different sectors. The first is inorganic—the cosmic processes of matter. The second is biological—the evolution of plants and animals. The third is psychological, and is the development of man's cultures. It is this third stage which is now critical, and if we are to survive as a species it can only be by replacing nature's controls by our own, not only birth control but control of the whole environment.

(JULIAN HUXLEY)

4.1. Basic Principles of Sound Control

There is a threefold problem: to protect the building from external noise sources; to ensure that the building being designed is not a noise nuisance to people in buildings and spaces nearby; to control the amount of noise generated within the building.

Increasingly, the building environmental engineer and the architect are being requested to advise on all aspects of the problem. The degree of external protection and internal control required depends on the use of the building. For example, broadcasting studios and concert halls will require a stringent acoustical criterion (i.e. NR 15–25); in a factory, however, the permitted noise levels will be much higher (i.e. NR 55–85)—in practice often much higher than audiologists would recommend. In designing the indoor noise climate care is also needed to ensure that people in adjacent spaces *outside* the building are not aggrieved. The pattern of diagnosis and treatment for these situations is the same:

(a) ascertain the acoustical characteristics of the source, propagation pathway and receiver;
(b) provide a mis-match between the source and the receiver which will be designed from the data obtained in (a) and the acoustic criterion specified for the project; the following sections aim to review briefly the methods available for achieving this mis-match.

Noise control involves employing one or more of the following principles:

(a) designing a source to achieve a set acoustical tolerance: building environmental engineers have little to say in this matter, which lies in the province of the equipment manufacturers, but they can exercise discretion in selecting fans,

272

compressors and machinery besides designing efficient aerodynamic and hydrodynamic energy distribution systems;

(b) providing a large separation between the source and the receiver, making the most benefit of natural attenuation; this involves planning of building sites and also the layout of spaces within buildings;

(c) enforcing natural attenuation along the source–receiver pathway by one or a combination of:
   (i) using a high mass structure to reflect the energy before it reaches the receiver (see section 11.4);
   (ii) installing an absorption attenuator to dissipate the energy via friction over a broadband frequency range;
   (iii) employing interference attenuators to reduce the sound level at specific frequencies by phase cancellation;
   (iv) using resonance attenuators, which may be tuned to provide attenuation over various frequency bandwidths by varying the mass, stiffness and damping characteristics of the attenuator cells;
   (v) providing materials or isolators with high isolation efficiency characteristics (i.e. low transmissibility) to minimise the propagation of structureborne sound (see Chapter 9);

(d) the acoustical conditions in the receiving space may, in addition, be altered by:
   (i) using absorbent materials applied to various surfaces to reduce the reverberant energy level;
   (ii) use of soft-floor finishes to minimise impact noise;
   (iii) introducing background noise;
   (iv) isolating the room structure from that of adjacent spaces;

(e) in industrial situations the employment of restricted work schedules and/or of ear-defenders by workers will be necessary in acoustical environments equal to or exceeding noise levels of 80–85 dBA (see Burns, 1968);

(f) legislation—usually out of the province of the environmental engineer, but refer to the book *The Law of Noise* published by the Noise Abatement Society in 1969 for further information.

It can be seen that noise control requires a lot of common sense as well as technical expertise. In what follows it will be useful to check some of the fundamental issues that will be discussed in Chapters 9 and 11, and to these add a pragmatic view.

4.2. Sound Insulation

Sound insulation is largely an interplay between mass, stiffness and damping—the basic factors that will be described in Chapter 6. In Chapter 11 we shall see that the mass law is affected by resonance at lower frequencies and coincidence at higher ones. The zone in between can be extended to avoid these ineffective insulation regions by:

(a) using a less stiff panel, hence lowering the resonant frequency $\{f_r \propto \sqrt{(k/m)}\}$ but raising the critical frequency $\{f_c \propto (1/h)\sqrt{\rho/Y}\}$;

(b) decreasing the panel thickness h, hence raising f_c but resulting in a reduced panel mass;

(c) increasing mass hence lowering f_r and raising f_c; in practice, cost usually limits the amount of structural mass;

(d) increasing the amount of damping;

(e) increasing the stiffness and decreasing the coincidence frequency to a very low value.

Good sound insulation is usually the best combination of low stiffness, high mass and high damping possible in a given situation for a given cost; manufacturers have the problem of providing high impedance mis-matches between noise sources and people at low cost.

In order to give the mass law some practical perspective, Fig. 4.1 shows some typical structures and their sound insulation values plotted on a theoretical mass law curve using data from Doelle (1972). In general, it can be seen that the mass law serves as a reasonable guide for estimating the sound insulation of a structure, but notice where deviations occur and these are summarised in Fig. 4.2. Humphreys and Melluish (1971) give cost and performance data for walls and floors.

* Cavities cause an increase in sound insulation (Fig. 4.2b and d) so that cavity walls such as structure 4, or partition structures 12–17, on Fig. 4.1, have sound reduction index values lying well above the mass law curve.

* Flanking paths can cause a serious deterioration in sound insulation! For example, structure 3 on Fig. 4.1 shows the ill effect of wire ties in a cavity wall.

* A number of other variables contribute towards good sound insulation and these are summarised on Fig. 4.2c, examples of which are illustrated by various structures in Fig. 4.1. Briefly, use leaves of different thicknesses to mis-match coincidence frequencies; use absorbent materials in cavities to prevent a build-up of reverberant sound; use resilient connections at suspension points or at bridging pathways; use wide spacings for studs in partitions and stagger studs if possible; eliminate sound leaks by the use of perimeter caulking or acoustic sealants.

* Good quality materials and installation are essential.

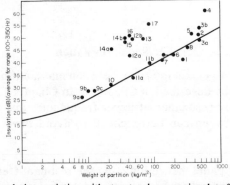

FIG. 4.1. Sound insulation variation with structural mass using data from Doelle (1972).

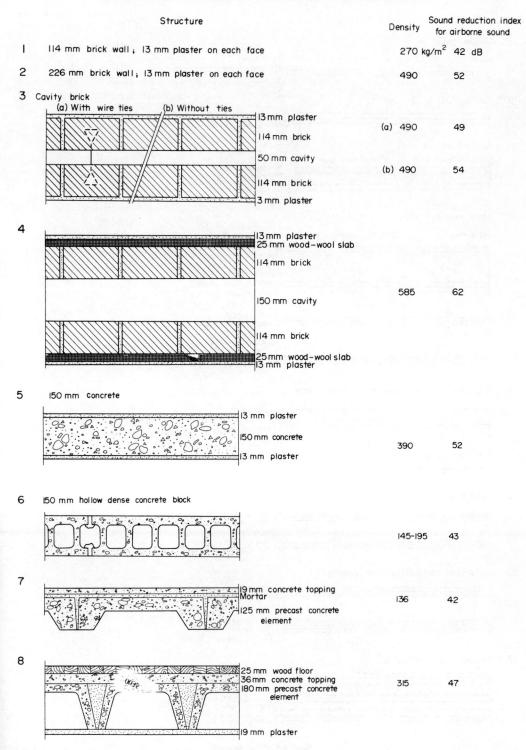

	Structure	Density	Sound reduction index for airborne sound
1	114 mm brick wall; 13 mm plaster on each face	270 kg/m²	42 dB
2	226 mm brick wall; 13 mm plaster on each face	490	52
3	Cavity brick (a) With wire ties (b) Without ties 13 mm plaster 114 mm brick 50 mm cavity 114 mm brick 3 mm plaster	(a) 490 (b) 490	49 54
4	13 mm plaster 25 mm wood–wool slab 114 mm brick 150 mm cavity 114 mm brick 25 mm wood–wool slab 13 mm plaster	585	62
5	150 mm concrete 13 mm plaster 150 mm concrete 13 mm plaster	390	52
6	150 mm hollow dense concrete block	145–195	43
7	19 mm concrete topping Mortar 125 mm precast concrete element	136	42
8	25 mm wood floor 36 mm concrete topping 180 mm precast concrete element 19 mm plaster	315	47

FIG. 4.1. (*Continued*).

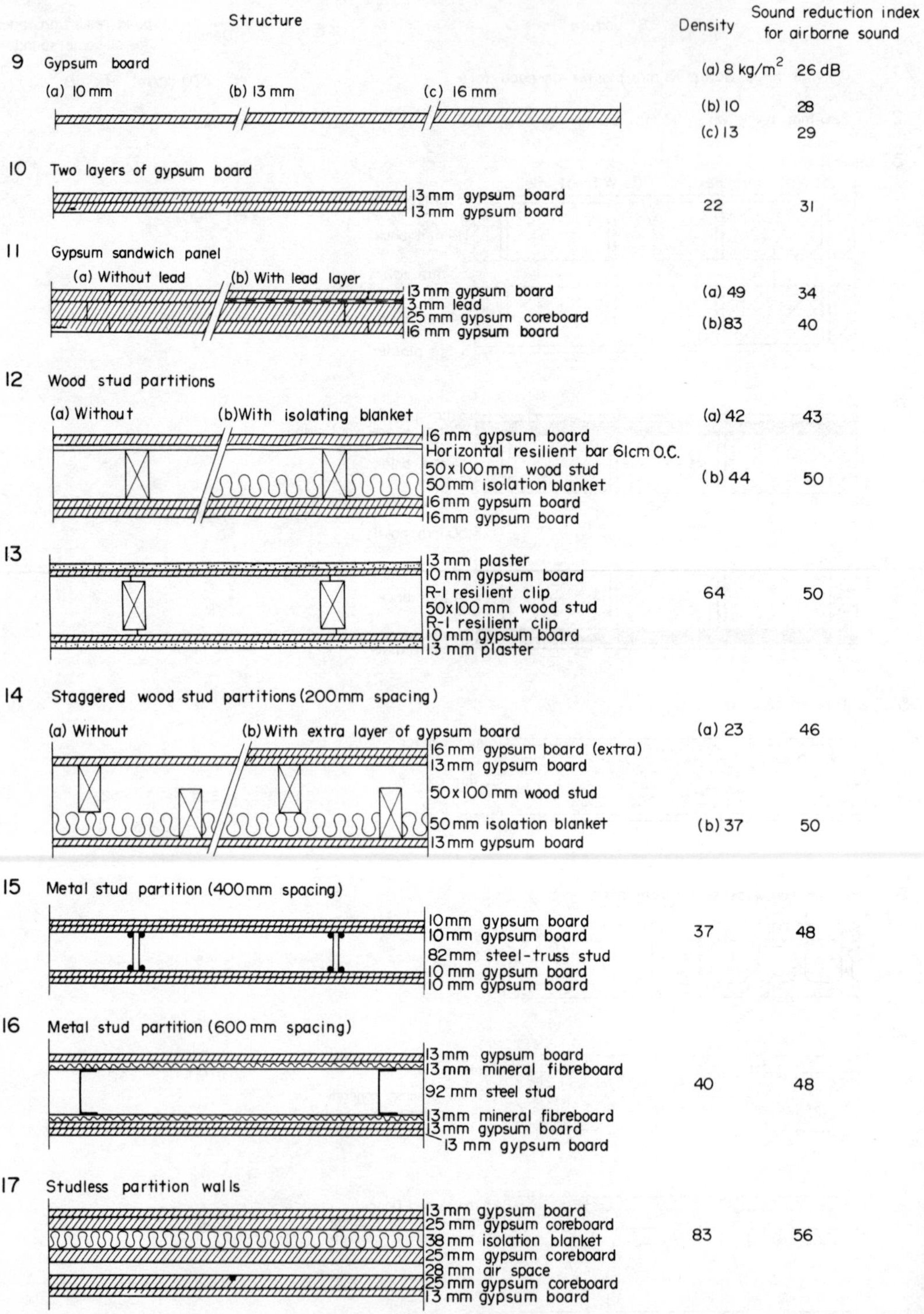

Structure	Density	Sound reduction index for airborne sound
9 Gypsum board	(a) 8 kg/m^2	26 dB
(a) 10 mm (b) 13 mm (c) 16 mm	(b) 10	28
	(c) 13	29
10 Two layers of gypsum board		
13 mm gypsum board / 13 mm gypsum board	22	31
11 Gypsum sandwich panel		
(a) Without lead (b) With lead layer	(a) 49	34
13 mm gypsum board / 3 mm lead / 25 mm gypsum coreboard / 16 mm gypsum board	(b) 83	40
12 Wood stud partitions		
(a) Without (b) With isolating blanket	(a) 42	43
16 mm gypsum board / Horizontal resilient bar 61cm O.C. / 50 x 100 mm wood stud / 50 mm isolation blanket / 16 mm gypsum board / 16 mm gypsum board	(b) 44	50
13		
13 mm plaster / 10 mm gypsum board / R-1 resilient clip / 50x100 mm wood stud / R-1 resilient clip / 10 mm gypsum board / 13 mm plaster	64	50
14 Staggered wood stud partitions (200mm spacing)		
(a) Without (b) With extra layer of gypsum board	(a) 23	46
16 mm gypsum board (extra) / 13 mm gypsum board / 50 x 100 mm wood stud / 50 mm isolation blanket / 13 mm gypsum board	(b) 37	50
15 Metal stud partition (400mm spacing)		
10mm gypsum board / 10 mm gypsum board / 82mm steel-truss stud / 10 mm gypsum board / 10 mm gypsum board	37	48
16 Metal stud partition (600 mm spacing)		
13 mm gypsum board / 13 mm mineral fibreboard / 92 mm steel stud / 13 mm mineral fibreboard / 13 mm gypsum board / 13 mm gypsum board	40	48
17 Studless partition walls		
13 mm gypsum board / 25 mm gypsum coreboard / 38 mm isolation blanket / 25 mm gypsum coreboard / 28 mm air space / 25 mm gypsum coreboard / 13 mm gypsum board	83	56

FIG. 4.1. (*Continued*).

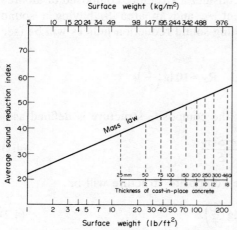

(a) The approximate average sound reduction index of solid single-leaf partitions can be estimated from this mass law.

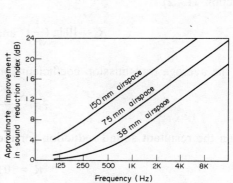

(b) Approximate improvement in the sound reduction index of multiple partitions with various airspaces over single-leaf partitions of the same total weight.

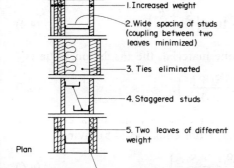

Sound insulation increases with:

1. Increased weight
2. Wide spacing of studs (coupling between two leaves minimized)
3. Ties eliminated
4. Staggered studs
5. Two leaves of different weight
6. Resilient attachments
7. Maximum separation between two leaves
8. Isolation blanket in airspace
9. Perimeter caulking

(c) Some layouts of multiple partitions with details contributing to increased sound insulation.

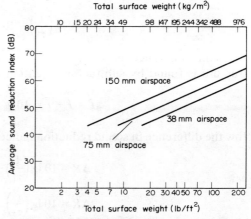

(d) Approximate average sound reduction index of multiple partitions with different airspaces.

FIG. 4.2. Effect of other variants, other than mass, on sound insulation of structures using data from Doelle (1972).

Since many building structures are composite ones it is useful to be able to assess the reduction of insulation due to the introduction of a material having a lower insulation value than the basic structure. Consider a structure composed of an area S_1 having a sound transmission coefficient τ_1, and an area S_2 having a sound transmission coefficient τ_2. The corresponding sound reduction indices will be (see section 11.4.2)

$$R_1 = 10 \lg \left(\frac{1}{\tau_1}\right) \quad \text{and} \quad R_2 = 10 \lg \left(\frac{1}{\tau_2}\right).$$

If the average transmission coefficient for the composite structure is defined as

$$\bar{\tau} = \frac{S_1 \tau_1 + S_2 \tau_2}{S_1 + S_2}, \tag{4.1}$$

then the resultant sound reduction for the composite structure will be

$$\bar{R} = 10 \lg \left(\frac{1}{\bar{\tau}}\right). \tag{4.2}$$

Using equation (11.24) in Chapter 11, the sound insulation for a composite structure is

$$\bar{I} = \bar{R} - 10 \lg \left(\frac{S_1 + S_2}{A}\right) \tag{4.3}$$

and inserting equations (4.1) and (4.2)

$$\bar{I} = 10 \lg \left(\frac{S_1 + S_2}{S_1 \tau_1 + S_2 \tau_2}\right) - 10 \lg \left(\frac{S_1 + S_2}{A}\right). \tag{4.4}$$

For a structure area $(S_1 + S_2)$, composed of one material, the insulation value is

$$= 10 \lg \left(\frac{1}{\tau_1}\right) - 10 \lg \left(\frac{S_1 + S_2}{A}\right); \tag{4.5}$$

hence the difference in sound insulation brought about by having a composite structure is found by subtracting equation (4.4) from equation (4.5) giving

$$\Delta I = I - \bar{I} = 10 \lg \left\{\frac{S_1 \tau_1 + S_2 \tau_2}{\tau_1 (S_1 + S_2)}\right\}. \tag{4.6}$$

Now the difference in sound reduction indices will be, assuming $\tau_1 < \tau_2$ (or $R_1 > R_2$),

$$\Delta R = 10 \lg \left(\frac{1}{\tau_1}\right) - 10 \lg \left(\frac{1}{\tau_2}\right),$$

$$\Delta R = 10 \lg \left(\frac{\tau_2}{\tau_1}\right).$$

Therefore
$$\frac{\tau_2}{\tau_1} = \text{antilg} \left(\frac{\Delta R}{10}\right), \tag{4.7}$$

and if the proportion of materials used is expressed by the ratio

$$P = \frac{S_2}{(S_1 + S_2)} \tag{4.8}$$

then by inserting expressions (4.7) and (4.8) into equation (4.6) gives the change in insulation as

$$\Delta I = 10 \lg \left[1 - P \left\{ 1 - \text{antilg} \left(\frac{\Delta R}{10} \right) \right\} \right].$$ (4.9a)

Equation (4.9a) is plotted in Fig. 4.3a, which is a useful design aid. When the element

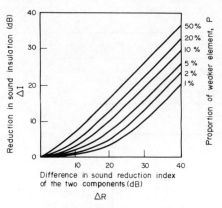

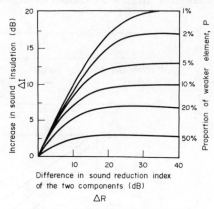

(a) Reduction in sound insulation of a good component by inclusion of a weaker one.

(b) Increase in sound insulation of a weak component by inclusion in an area having a higher insulation value.

FIG. 4.3. Sound reduction index of component structures; these graphs are useful for assessing the effect of glazing on the sound insulation of the building façade.

added to the structure has a higher sound reduction, then the structure, i.e. $\tau_1 > \tau_2$ (or $R_1 < R_2$), then Fig. 4.3b results, the equation for which is

$$\Delta I = 10 \lg \left[1 - P \left\{ 1 - \frac{1}{\text{antilg} \left(\dfrac{\Delta R}{10} \right)} \right\} \right].$$ (4.9b)

Example

A wall 5 by 3 m has an average sound reduction index of 45 dB. What will be the effect of installing a door 1 by 2 m and weighting 28 kg/m² into the wall? Calculate the further loss of insulation due to a 25 mm crack all around the door.

	R	τ	S	τS
Wall	45 dB	$3 \cdot 16 \times 10^{-5}$	15 m²	$4 \cdot 74 \times 10^{-4}$ m²
Door	31 dB	$7 \cdot 95 \times 10^{-4}$	2 m²	$15 \cdot 90 \times 10^{-4}$ m²
	(from mass law, Fig. 4.1)			

Hence

$$\bar{\tau}(S_1 + S_2) = 20 \cdot 64 \times 10^{-4}.$$

Therefore

$$\bar{\tau} = 1 \cdot 216 \times 10^{-4}$$

and the effective sound reduction index will be

$$R \simeq 39 \text{ dB}.$$

This may be checked either using Fig. 4.3a or by using equation (4.9a) and is seen to be in close agreement for this case, i.e. the wall insulation has been reduced by 6–8 dB. Repeating this procedure for the airgap and using the transmission coefficients for airgaps as shown in Fig. 4.4:

	R	τ	S	τS
Wall and door	39 dB	$1 \cdot 216 \times 10^{-4}$	17 m^2	$20 \cdot 64 \times 10^{-4} \text{ m}^2$
Airgap	—	$0 \cdot 65$ (see Fig. 4.3 at 400 Hz)	$0 \cdot 144 \text{ m}^2$	$0 \cdot 0937 \text{ m}^2$

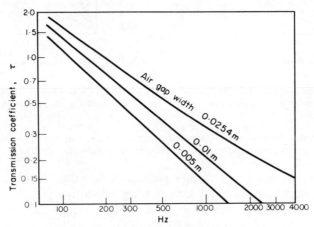

FIG. 4.4. Sound transmission coefficients of air gaps in a wall 50 mm thick (Ingerslev and Nielsen, 1944b).

Hence $\bar{\tau} = 5 \cdot 57 \times 10^{-3}$ resulting in a further insulation reduction of about 16·5 dB which, again, may be checked either with Fig. 4.3a or by using equation (4.9a).

4.3. Urban Planning and Noise Control

4.3.1. THE DISTANCE OF BUILDINGS FROM NOISE SOURCES

Noise emanating from an external source will decay as the sound energy spreads out at a rate depending on the nature of the source (geometry and acoustic excitation), the pressure and nature of boundary surfaces and weather conditions. Work by the Building Research Establishment and also by Rathé (1966) has shown that the attenuation of traffic noise with distance takes the form illustrated in Fig. 4.5a and b. Levels occurring at only 1% of the time almost follow the classical inverse square law; these are mainly composed of the discrete frequency components which tend to behave like point sources. Background noise levels are composed of multitudinous contributions, and their decay deviates markedly from the inverse square law. Measurements made in the near field of an acoustic source (about 10 m for traffic noise) are generally unreliable for reasons to be discussed after equation 10.10, Chapter 10.

Some extra useful information about the data presented in Fig. 4.5b is worth

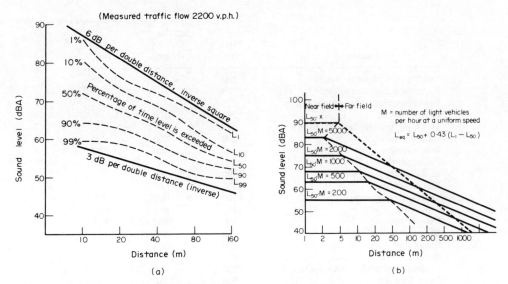

FIG. 4.5. Variation in traffic sound levels and distance for various traffic densities based on work by (a) Langdon and Scholes (1968), and (b) Rathé (1966).

noting (see Grandjean, 1973b). In the near field the sound pressure level can be calculated from

$$L_p = 9 + 20 \lg M.$$

As the traffic density (M vehicles per hour) increases, the extent of the near field x is reduced according to the law

$$x = \frac{10\,000}{M} \text{ (in metres).}$$

The positioning of a building with respect to other noise sources usually requires a field investigation at the planning stage. Reichow (1963) has made the suggestion shown in Table 4.1 for minimum distances between various industries and residential area to prevent noise penetration and air pollution.

Since this time the noise climate has increased. Grandjean's (1973b) review of all the work concerned with setting noise limits for buildings and regional development in Europe suggests that houses, schools and hospitals should be located at least 300 m away from main roadways. If this limit cannot be met, very special care is needed to provide a building structure with a very high acoustical performance.

4.3.2. NOISE REDUCTION BY BARRIERS

Barriers placed between the sound source and the receiver produce an acoustic shielding effect due to sound diffraction and also due to sound reflection and

TABLE 4.1

	Recommended minimum distances between industrial buildings and residential areas		
	Pollutant		
Industry	Noise and vibration (m)	Smoke, soot, dust (m)	Smell (m)
Sawing and planting works	200	150	—
Paper-bag manufacturers	200	—	—
Bakery equipment manufacturers	200	—	—
Chain makers	200	120	—
Dairy equipment manufacturers	200	—	—
Canning factory	200	100	400
Iron foundry	200	400	100
Boiler works	500	—	—
Mill construction works	700	200	—
Shipbuilding	900	600	—
Traction equipment	1100	700	200
Tin works	1100	600	600
Copper mill	500	1100	600
Bronze works	1100	600	600
Sugar mill	500	500	1650
Rubber works	200	1100	1650
Combustion engine manufacturers	2200	600	400
Blast furnace	500	2200	1100
Cellulose and paper mill	200	600	2200
Mineral oil works	300	1100	2200
Hydrogenating works	500	1200	2200
Ship propellor works	700	400	—
Iron construction works	800	400	—

absorption at the surface. Maekawa (1968) measured the diffraction of a thin, rigid semi-infinite screen in free space and obtained the results shown in Fig. 4.6 which give the sound attenuation for various Fresnel zone numbers. The Fresnel zone number N for sound of wavelength λ is given by

$$N = \frac{2}{\lambda}\delta, \qquad (4.10)$$

where δ is the path difference between the acoustic source and receiver and can be expressed in terms of the barrier height if necessary.

Differences between the theoretical and the actual performance of a noise barrier in practice occur due to the wind and air turbulence, the ground effect, air absorption at high frequencies and temperature gradients. Upwind from a source the sound waves are refracted upwards creating sound shadows, whereas downwind from a source the sound is deflected downwards towards the ground. The effect of temperature gradients on atmospheric propagation will be discussed in section 8.7 of Chapter 8. Turbulence tends to scatter sound and limit any attenuation due to wind sound shadows for instance. The ground effect can cause large attenuation in excess of those predicted by the inverse square law; it arises from the interference of the sound wave fronts emitted from the real source and its image. The magnitude of the ground effect depends on the nature of the ground; source and receiver proximity to

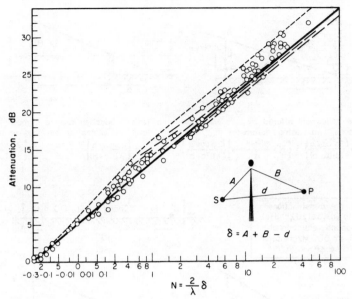

FIG. 4.6. Sound attenuation of barrier in free space (Maekawa, 1968, and in Stephens, 1975).

the ground; and the distance between source and receiver. Scholes *et al.* (1971) carried out experiments using a barrier made of 50 mm wood sheets, cement rendered on one side and weighing 25 kg/m^2; the propagation of sound from a point source was measured for barrier heights of 1·8, 3·0, 3·7 and 4·9 m. The conclusions drawn from these experiments were that, in general, Maekawa's work gives reasonable predictions about barrier performance except when the ground effect predominates; wind has different effects on screened and unscreened propagation; wind gives most advantages at higher frequencies, lower receiver heights and greater source–barrier–receiver distances—the converse not being true; higher barriers result in lower sound levels at the receiver except at frequencies where the ground effect is important.

In practice, noise reduction is often achieved by embankments, cuttings or roadway mounds which have a finite size and appreciable thickness. Maekawa's work (1968 and in Stephens, 1975) still forms a good basis for the calculation of the sound reduction, but account must be taken of transmission loss and any absorption occurring at surfaces or due to atmospheric conditions. Rathé (1969) has simplified Maekawa's approach.

Figure 4.7, taken from Rettinger (1959), shows four different constructions which give the same amount of sound reduction. Some German constructions have used sound absorbent cells on the surface of the overhang or extension wall shown in Fig. 4.7f and g.

4.3.3. LANDSCAPING

Belts of woods and plantations can offer significant sound reduction but only if used in belt widths of not less than about 50 m. Figure 4.8 displays sound reduction data which has been amassed from several sources for conifer plantations. For the most effective results plantations should be dense and made up of lofty trees.

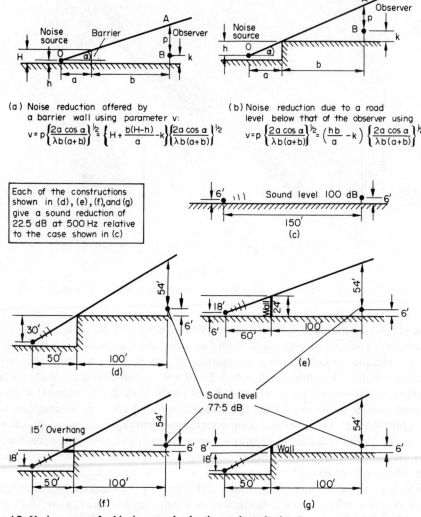

(a) Noise reduction offered by a barrier wall using parameter v:

$$v = p \left\{ \frac{2a \cos \alpha}{\lambda b (a+b)} \right\}^{\frac{1}{2}} = \left\{ H + \frac{b(H-h)}{a} - k \right\} \left\{ \frac{2a \cos \alpha}{\lambda b (a+b)} \right\}^{\frac{1}{2}}$$

(b) Noise reduction due to a road level below that of the observer using

$$v = p \left\{ \frac{2a \cos \alpha}{\lambda b (a+b)} \right\}^{\frac{1}{2}} = \left(\frac{hb}{a} - k \right) \left\{ \frac{2a \cos \alpha}{\lambda b (a+b)} \right\}^{\frac{1}{2}}$$

Each of the constructions shown in (d), (e), (f), and (g) give a sound reduction of 22.5 dB at 500 Hz relative to the case shown in (c)

FIG. 4.7. Various ways of achieving sound reduction evaluated using the parameter v (Rettinger, 1959).

4.3.4. TOWN PLANNING

For new town developments, sufficient information is available which should prevent serious noise contamination occurring. A number of measures in addition to those discussed in preceding sections can be adopted such as:

(a) *traffic management*: existing traffic may be re-routed and new routes designed to skirt housing developments;

(b) *industrial premises*: new housing developments should be restricted near industrial premises and acoustic field surveys conducted to ascertain the magnitude of the problem; conversely, new industrial developments should be confined to industrial estates;

(c) *airports*: the planning of new airports has received much national interest in

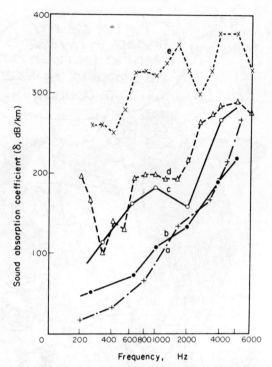

F<small>IG.</small> 4.8. Sound absorption in thick woods according to various measurements. (a) Wiener and Keast (1959); (b) measurements in deciduous woods by Meister and Ruhrberg (1959); (c) measurements in thick evergreen woods by Meister and Ruhrberg (1959); (d) measurements in evergreen woods over 33–50 m stretches by Embleton (1963); (e) as (d) but over 50–70 m stretches.

the last decade. The effect of aircraft noise on the population is becoming better understood. Figure 2.53 showed the percentage of people affected by various noise and number index (NNI) levels; the implications of this on the design of airports is discussed by Richards (1969) and Ollerhead (1973);

(d) *orientation of buildings*: tall buildings may sometimes be used to shelter low-height buildings; housing estates may be positioned end-on to main traffic routes so that walls containing windows are not directly facing the noise source.

4.3.5. TALL BUILDINGS

The Greater London Council Noise Study established that height does not appear to be a major factor in reducing average noise levels. Discrete noises (e.g. vehicle horns and brake squeal) tend to show a decrease with increase in building height because they act as free-field point sources, whereas streams of traffic behave as longitudinal sources. Some factors worth taking into account are:

(a) podiums can have a marked effect on floors immediately above;
(b) tall buildings bordering streets can cause a high noise level due to the reverberant field set up by the building surfaces;
(c) in high-rise flat and office blocks a difference of more than 15 dB may occur

Even the Open Golf Championship needs environmental planning (*Daily Express*, 14 July 1973).

between the front and the back of the building especially if there is a courtyard
or traffic-free square at the back;

(d) at higher levels any attenuation of sound due to ground absorption disappears.

Grandjean (1973b) gives the data shown in Table 4.2 based on German and
Swedish experience.

4.4. The Rôle of the Building Structure in the Protection of Buildings
from External Sources of Noise

A building is a climatic moderator. The structure of a building must protect the
occupants from the ingress of atmospheric pollutants and inclement weather
conditions; it must permit tolerable thermal, solar, day lighting and noise levels

TABLE 4.2. INCREASE OF EXTERNAL NOISE LEVEL WITH BUILD-
ING HEIGHT (GRANDJEAN, 1973b)

Distance from building façade to centre of nearest traffic lane (m)	L_{50}	
	Ground floor (dBA)	Higher floor (dBA)
25	68	68
50	64	65
100	59	61
200	53	58
400	45	54

	Building height (dB)			
	7 m	10 m	16 m	22 m
50 m	+3	+4	+4	+4
100 m	+2	+3	+6	+7

Add these values to ground floor sound level

inside the building; in addition, it must possess aesthetic and the necessary structural qualities. Our immediate concern is the acoustic pathway between the inner and outer environment of a building, but it must be remembered that the selection of a building structure is a *gesamt* process involving the optimisation of several factors.

The architectural trend during the last decade has been towards lightweight buildings with large glazing areas. One problem that has arisen from this has been overheating. Table 4.3, which is based on research carried out by the Building Research Establishment, shows the maximum glass areas recommended in order to keep the internal temperature below 24°C in buildings of light and heavyweight construction.

This work shows that buildings situated in noisy areas should have less glazing than those in quiet surroundings where solar heat gain is the guiding criterion. Of course, the problem of overheating can be solved by using external or internal sun controls preferably in an airconditioned building, but the main point is that the selection of building fabrics is an optimisation process involving several factors.

Clearly, the building design team should *jointly* consider the selection of the building envelope so that the environmental, as well as the structural and aesthetic qualities, are not overlooked. Before selecting a suitable sound reduction index for the structure, steps must be taken to limit the noise reaching the building.

In any building the weakest sound barriers are the glazing areas and doorways.

TABLE 4.3

Building construction	External noise environment	Maximum glass area expressed as a percentage of the external wall area
Heavy	Quiet	50
	Noisy	25
Light	Quiet	20
	Noisy	15

Secondary sound transmission pathways also exist depending on the type of building (e.g. fresh-air intakes or exhaust outlets in airconditioned office blocks, chimneys in houses) and the quality of the construction (e.g. air leakage at the joints of the building members). Figure 4.1 showed the mass law for a typical range of building structures. As expected from classical sound transmission theory (see Chapter 11), the transmission loss increases with the mass of the structure, the qualification being that the structure must have a low resonant frequency range and also a high flexural wave frequency to avoid coincidence effects.

The architectural trend in recent years has been towards using extensive glazing areas in commercial buildings. In order to limit the ingress of dust and limit the noise transmission, sealed buildings with airconditioning have become common. Glazing controls the amount of natural daylight in spaces, but if the windows are single panes, and openable, large heat losses occur; solar gains also present a problem. Experience has shown that windows have a psychological function—to give the building user contact with the outside world—so that elimination of glazing is not the answer. Table 4.4 compares the insulation values of single- and double-glazing which is part of a 220 mm brick wall.

TABLE 4.4. BRITISH STANDARD CODE OF PRACTICE CP3/1960

Percentage of glazing in 220 mm wall	Single window (openable) (dB)	Double window (sealed, 4 mm glass with 200 mm air space and absorbent lined reveals) (dB)
100	20	40
75	21	41
50	23	43
33	25	44
25	26	45
10	30	47
0	Value of 220 mm wall	50

The insulation values for single windows vary by 10 dB over the range of 10–100% in glazing percentage areas and by 7 dB over this range for double windows. If 10% or more of the walls have windows that are wide open, the sound reduction for the composite wall will not be more than about 10 dB as can be seen from Table 4.5 (and by using equation (4.9)), which also gives a comparison of various window constructions in open and closed positions.

The use of sealed window units means that mechanical ventilation is necessary. Special window-ventilator units have been developed for use in houses (Fig. 4.9). For intermittent noise sources (e.g. aircraft) an automatic window has been installed in some buildings; the opening of the window is controlled by the current flow in a microphone circuit. Opening lights which permit airflow but attenuate sound by continuous reflection between the overlapping leaves give a cheaper solution to the problem especially in houses (Fig. 4.10).

Summarising, the principal factors which should be considered by the architect and the environmental engineer with regard to the acoustical performance of windows are:

(a) *weight*: the variation of the mean sound insulation with glass thickness is shown in Fig. 4.11; in practice 12 mm glass will be the maximum thickness

TABLE 4.5. BRITISH STANDARD CODE OF PRACTICE CP3/1960

Type	Description	Approximate sound insulation (dB)
Single windows	Wide open	5
	Slightly open	10–15
	Closed but openable	18–20
	Sealed (24–30 oz glass)[a]	23–25
	Sealed (6·35 mm plate)[a]	27
	Sealed (9·53 mm plate)[a]	30
Double windows	Ventilated	15–20
	Closed but openable with 200 mm airspace and absorbent lined reveals	30–33
	Sealed (4 mm glass, 200 mm airspace lined)	40
	Sealed (6·35 mm plate glass, 200 mm airspace lined)	42

[a]Note 32 oz glass is equivalent to 4 mm thick plate glass; 6.35 mm = $\frac{1}{4}$ in. plate; 9.53 mm = $\frac{3}{8}$ in. plate glass.

FIG. 4.9. A well-known problem after having double glazing is how to provide an efficient air inlet without at the same time permitting the ingress of noise. The Silentair Dove shown is an acoustical ventilation unit which introduces up to 140 m³/h of fresh air into the room. Models are available which enable the air stream to be tempered during the winter months. They have been found to be particularly useful around airports, motorways and other areas where double glazing has to be installed for acoustic rather than for thermal reasons (with acknowledgements to Sound Attenuators Ltd.).

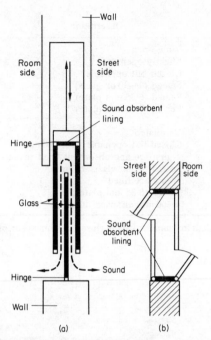

FIG. 4.10. Two types of baffled opening windows.

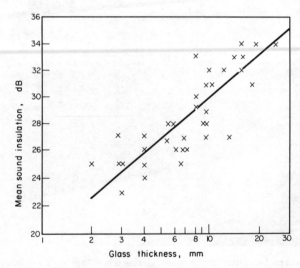

FIG. 4.11. Variation of sound insulation with glass thickness for random incident sound.

considered, and cost often means that a 4–6 mm thickness will be the likely outcome since an increase in the order of only 2 dB is likely;

(b) *cavity width*: 100 mm airspace is an absolute minimum for some gain in insulation at medium and high frequencies, whereas 200–300 mm is a desirable design objective; some measurements are shown in Figs. 4.12a and b (for optimum heat transfer keep the width to at least 15 mm);

(c) *absorbent linings*: the airspace forms a resonant cavity; a 25 mm lining will increase the mean insulation value by about 2 dB due to absorption of the reverberant field energy in the cavity;

(d) *damping*: the vibration of the panes can be minimised by using foamed rubber gaskets (Figs. 4.13a and b); improvements of the insulation in the critical region can be obtained by using laminated glass with resilient interlayers (Fig. 4.14), by using different glass thickness for each pane to avoid coincidence frequency matching (Fig. 4.15) or by using non-parallel panes (Fig. 4.16);

(e) *mechanical separation*: panes should be isolated from one another and the opening in the structure housing the frames (separate frames are preferable to one common frame) should not be bridged (again see Fig. 4.13a and b);

(f) *sealing*: this is one of the most important factors in the acoustical design of windows and has been described in detail by Lewis (1971); gaps most commonly occur between the opening frame and the subframe, but they also appear between the glass and the opening frame, the fixed frame and the subframe, the subframe and the wall opening, and maybe due to poor tolerances, poor workmanship, thermal stresses or age deterioration;

(g) *size and shape of glass*: as the size of a pane decreases, the transmission loss in the coincidence region is improved, whereas the resonance frequency is increased; little difference in performance occurs in the mass law region;

(h) *edge fixing*: this has already been referred to under the heading of damping; sound incident upon the pane causes it to bend, the bending wave is reflected at the edge to form a flexural wave which at and above the critical frequency causes a large increase in the sound radiated from the pane; if the edges of the glass are clamped between hardwood beads, are suspended in mastic, or are

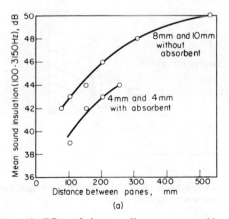 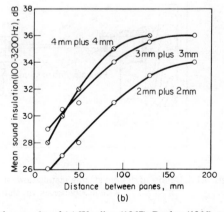

FIG. 4.12. Effect of air space distance on sound insulation results of (a) Woolley (1967), Bazley (1966) and (b) Brandt (1954).

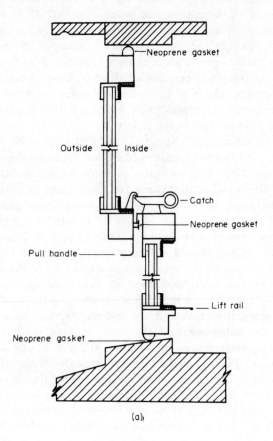

(a)

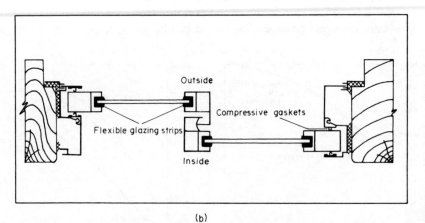

(b)

FIG. 4.13. Details of double glazing installations sealed opening lights (a) double-glazed double-hung sash with flexible seals, (b) sliding window with flexible seals.

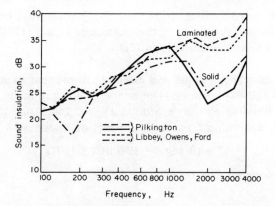

FIG. 4.14. Advantage of using laminated glass (two 3.2 mm glasses with 1 mm interlayer); increased stiffness reduces the effect of coincidence above 1500 Hz (Pilkington, 1970).

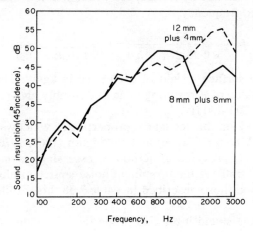

FIG. 4.15. The probability of coincidence is reduced by using different glass thicknesses in double (Oosting, 1967).

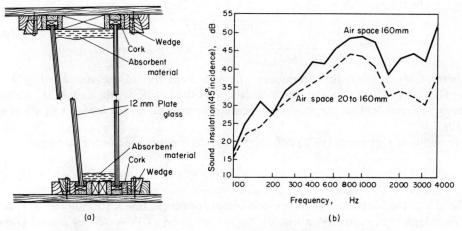

FIG. 4.16. Effect of wedge shaped airgap on sound insulation (Oosting, 1967).

clamped between neoprene strips, then the flexural wave energy is attenuated by absorption and reflection at the mounting; Lewis (1971) and Utley (1969) have studied this problem (Fig. 4.17f).

A number of practical problems and attempts to prevent them are illustrated in Fig. 4.17 which is self-explanatory. The condensation problems caused by glazing are dealt with in *Building Research Station Digest*, **140**, April 1972.

The mean insulation value I for a glass panel of thickness h mm is given by

$$I = 10 \cdot 5 \lg h + 19 \cdot 3 \text{ dB} (\pm 3 \text{ dB}) \tag{4.11}$$

and the critical frequency can be estimated from

$$f_c = \frac{12\,000}{h}. \tag{4.12}$$

For a double-glazing unit having an airspace width of d mm and pane thickness h_1, h_2 mm, the resonance frequency is

$$f_0 = \frac{1200}{\sqrt{0 \cdot 25\ d(h_1 + h_2)}} = \frac{1200}{\sqrt{0 \cdot 1\ d(m_1 + m_2)}}, \tag{4.13}$$

m_1 and m_2 being the surface weight of each pane in kg/m². Thermal type double-glazing has a small value of d resulting in an undesirably high resonance frequency usually in the range 200–400 Hz.

An extensive survey on the acoustical properties of windows has been made by Lewis (1970, 1971), Marsh (1971a–c) and Pilkington (1964, 1970).

De Lange (1968) has provided a method of assessing glazing requirements for buildings where road traffic is the prominent noise source producing external noise climates comparable to those described in Table 2.38; his results are shown in Fig. 4.18.

The reduction in the sound insulation properties of windows due to cracks and airgaps has been treated in great depth by Gomperts (1972). Empirical evidence is given which, in general, supports the theoretical derivation of the sound transmission coefficient for a crack given as

$$\tau = \frac{mk \cos^2 (ke)}{2n^2 \dfrac{\sin^2 k(L + 2e)}{\cos^2 (ke)} + k^2 \{1 + \cos k(L + 2e) \cos (kL)\}}, \tag{4.14a}$$

where $k = (2\pi f b / c)$ for frequency f, and velocity of sound c; $L = d/b$; $e = (1/\pi) \ln (4 \cdot 5/k)$; b is the crack width; d is the crack depth; and m, n are the numbers dependent on the nature of incident sound and on the location of slit (Fig. 4.19).

In addition, Gomperts (1972) defines, firstly, the damping factor for an airgap as

$$D = \left\{ 1 + \frac{6}{\pi} \frac{\nu}{cb} L(L + 2e) \right\}^{-2}, \tag{4.14b}$$

where ν is the kinematic viscosity coefficient for air ($\nu = 1 \cdot 56 \times 10^{-5}$ m²/s at 20°C and 0·76 m Hg), and, secondly, a *window factor* W given as $W = l/S$ for a total airgap length l in a façade containing a total window area S; l and S must be expressed in

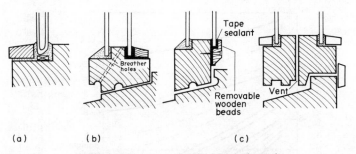

(a) Typical factory-sealed double-glazing units.
(b) Typical glazed *in situ* double-glazing. Provide one 6 mm breather hole per 0.5 m² of window and plug with glass fibre; paint or varnish wood exposed to airspace.
(c) Typical double windows, coupled type.

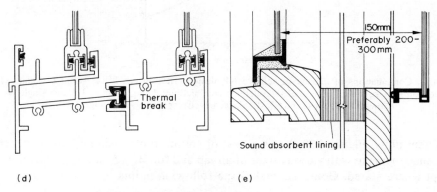

(d) Horizontal sliding type metal.
(e) Dual glazed window for sound insulation.

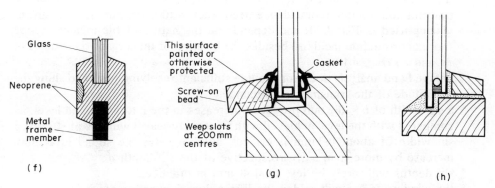

(f) Flexible mounting of glass in neoprene gasket giving resonance damping.
(g) Drained glazing system to prevent water remaining contained in unit.
(h) Typical stepped unit for frames that are too small or unsuitable for enlarging to fit standard units.

Fig. 4.17. Practical aspects of glazing systems as illustrated in *Building Research Station Digest*, **140**, April 1972.

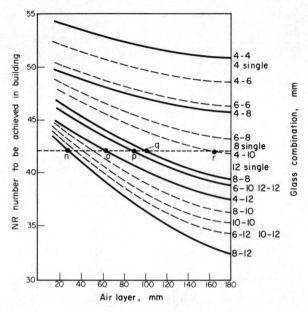

FIG. 4.18. Combinations of glazing to achieve NR design criterion e.g. NR42 could be achieved by combination *n*, *o*, *p*, *q*, *r* or any other intermediate combination; 4 mm, 8 mm and 12 mm single glazing allows NR 50, NR 42 and NR 40 to be achieved respectively (de Lange, 1969).

consistent units. Figure 4.20 shows a host of sound transmission curves for 12 mm thick single-glazing with various sizes of airgap and for $W = 5$, and $m = 8$ with $n = 1$ except where stated. Gomperts makes the following points:

* without damping, the minima of the transmission curves are independent of the crack widths;
* with damping these minima are strongly dependent on the crack width;
* the sound transmission minima occur in the upper part of the coincidence transmission region (coincidence frequency = 1000 Hz for 12 mm single-glazing used in Fig. 4.20) and depends on the nature of the incident sound (normal or random incident) besides the location of the airgaps (i.e. values of m and n) (Fig. 4.19);
* octave band analysis does not provide sufficient resolving power to show the magnitude of the transmission minima;
* a slit width of $b < 0.2$ mm produced increases in the internal sound level, as compared with the level achieved with a properly sealed window of < 2 dB; a slit width of about 0.3 mm is critical, as above this the sound level may increase by more than 3 dB irrespective of the slit depth d;
* slit depths will rarely be less than 40 mm in practice;
* the window factor will seldom be larger than 5 in practice;
* a difference of 2 dB in the sound level is not audible to most people.

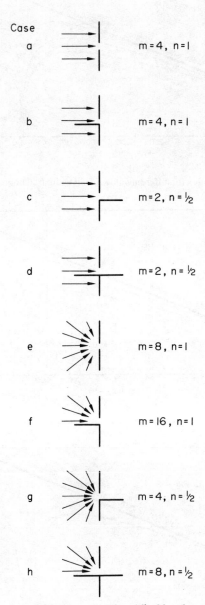

FIG. 4.19. Values for m and n in different cases; only applicable at low values of ke (Gomperts, 1972).

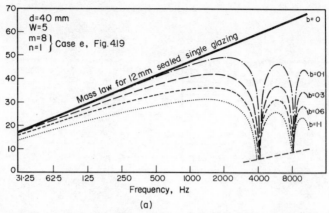

(a)

Resulting sound reduction index for different values of slit width b ; damping is neglected ($D = 1$).

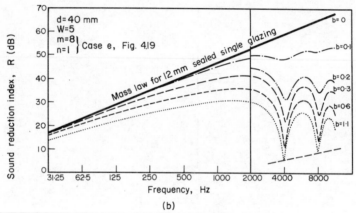

(b)

As Fig. (a) above but with damping taken into account.

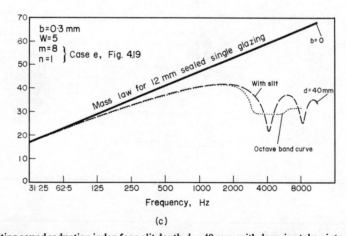

(c)

Resulting sound reduction index for a slit depth $d = 40$ mm, with damping taken into account.

FIG. 4.20. Influence of airgaps on the sound transmission of 12-mm-thick single glazing (Gomperts, 1972).

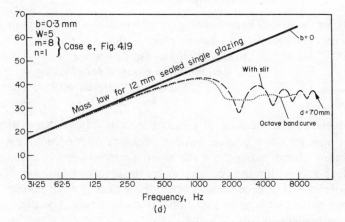

Resulting sound reduction index for a slit depth of $d = 70$ mm, with damping.

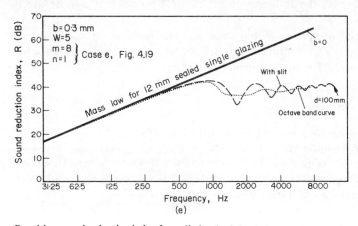

Resulting sound reduction index for a slit depth of $d = 100$ mm, with damping.

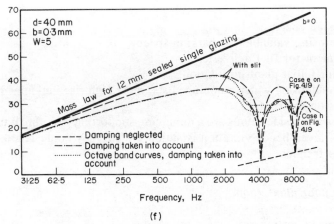

Comparison of cases (e) and (h) shown on Fig. 4.19.

FIG. 4.20. (*Continued*).

4.5. Noise Control within Buildings

.... noise [in dwellings] had catastrophic effects not only on nervous fatigue but also on social relations between families and within one family.

(P. CHOMBART DE LAUWE, *Proceedings of the First International Council for Building Research (CIB) Congress*, Rotterdam 1959, Elsevier, 1961.)

A general layout for a mechanical services system was shown in Fig. 3.1. The plant-room is a major noise source, but high velocity airconditioning systems can generate airborne sound along the distribution system at levels comparable with those from the plant. Within the occupied space the resultant sound level consists of contributions from lighting, airconditioning, the external environment and many miscellaneous sources such as impact noise from footsteps, opening and shutting doors, office equipment, lift motors and so on.

In this chapter the following noise control procedures will be discussed:

* layout of spaces within buildings and location of noise sources;
* the use of structural components to isolate one noisy area from another;
* the use of sound control equipment.

4.5.1. PLANT-ROOMS

A plant-room may be situated at roof-level, basement-level or at an intermediate level in a building; a separate power house may be preferred. In some cases several plant-rooms are required. The answer to the question—Where should the plant-room be situated?—depends on several factors. Listed below are some of the considerations that need to be taken into account.

Basement plant-room

* low space costs;
* maybe using valuable car-parking space;
* convenient for fuel deliveries;
* easy for initial plant installation;
* high condensing water-pumping costs; most high-rise buildings have cooling towers on the roof;
* structural considerations are a minimum, supporting structure more rigid than at roof-level, hence vibration isolation system may be designed for a higher natural frequency.

Intermediate floor plant-room

* high space costs;
* providing plant-room ventilation is difficult;
* higher probability of influencing more spaces with noise.

Roof-level plant-room

* space costs lower than for intermediate floor locations but initial plant installation may be more difficult;
* fuel pumping costs higher than for basement location;
* structural considerations necessary;
* very high static heads may necessitate specially constructed plant, pipework and fittings.

Separate powerhouse

* higher building cost;
* may be an uneconomic use of land;
* environmental interference with main building a minimum if pollutants expelled correctly and if it does not spoil the view.

FIG. 4.21. Roof-level acoustic plenum. A louvred penthouse plenum is an accepted way of handling roof-level discharge outlets or intakes controlling the inflow and outflow of noise; they must be weather-resistant too. These penthouse plenums, the largest being 55 ft (17 m) long, consist of a series of acoustic louvres, joined together to form the four walls of the enclosure, with an acoustically treated pitched roof. The whole unit is then sealed to its concrete base to ensure maximum attenuation. The acoustic performance of these louvres, reaches a maximum in the 1 kHz octave band, giving a sound reduction of 18 dB (with acknowledgements to Sound Attenuators Ltd.).

With regard to noise the following general cardinal rules apply:

(a) design to limit the airborne and structureborne sound levels reaching other areas within, and also outside the buildings; this will necessitate the use of duct or pipeline attenuators and vibration isolators;

(b) use high sound-insulating structures for construction of the plant-room (Table 4.6);

TABLE 4.6

Insulation required between plant-room and adjacent spaces (dB)	Minimum weight of structure (kg/m²)
55	1000
50	500

(c) ensure that the plant-room is airtight (i.e. beware of the joints in the structure, and of the pipes or ducts passing through the plant-room structure); all necessary supply air inlets or exhaust air outlets should be provided with acoustical louvres or attenuators which will also have to filter out external sources (Fig. 4.21); gasketed or double doors are advocated (Fig. 4.22);

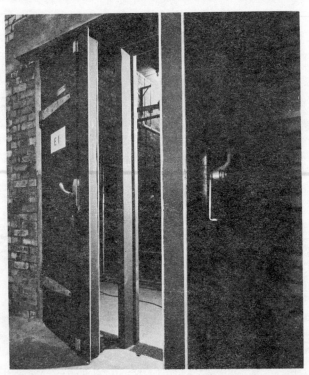

(a) A pair of 45 dB steel doors on the equipment access to an acoustics laboratory. This laboratory is used for testing components ultimately to be installed in nuclear reactors, and hence extremely high sound pressure levels are generated. In order to contain these high sound levels but at the same time permit the free movement of large components in and out of the laboratory, it was necessary to devise a special threshold.

FIG. 4.22. Sound-resistant and airtight doors (with acknowledgements to Sound Attenuators Ltd.).

(b) Photograph shows the way in which a sound attenuator 45 dB door was modified to eliminate the need for a raised threshold. The thickness of the door was increased by 4 in. (10 cm) at the bottom, and a cavity lined with fibreglass provided immediately under the door created a similar effect to a Helmholtz resonator. In this way it was possible to eliminate the airtight perimeter seal without degrading the performance of the door.

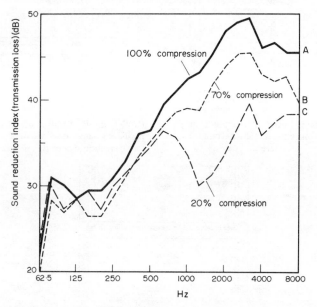

(c) Performance of 38 dB door under different pressures (A) 100% compression, (B) 70% compression, (C) 20% compression; all compressions are airtight. Air leakage is less than 3.5 m³/h with a pressure differential of 250 mm in the case of steel doors irrespective of whether the pressure is positive or negative.

FIG. 4.22. (*Continued*).

(d) locate the plant-room as far away from working areas as is practicable;

(e) plant-rooms installed at roof level should have floating floors (see section 4.5.3).

Each case has to be considered on its merits, and these guidelines will only serve as the absolute minimum conditions required.

As a practical example of plant-room design, Blazier (1972) describes three different construction techniques (Fig. 4.23). The first scheme has walls and a floor simply constructed of 150–200 mm-thick, solid, dense concrete; noise is transmitted directly through the floor and side walls of the plant-room to the space below. In the second scheme the addition of a 100 mm-thick floating floor isolated from the structural slab limits the direct transmission across this member but still allows direct conduction of sound via the side walls. For critical situations a third scheme is proposed in which an additional interior wall supported by the floating floor reduces the transmission through the side walls. The noise reduction obtained by field measurements is illustrated in Fig. 4.24a. Variations in construction detailing quality of workmanship and accuracy in measurement all limit the field performance data. Utilising these constructions for a plant-room housing a hermetic centrifugal compressor, it can be seen from Fig. 4.24b that the dense concrete slab alone is not sufficient to meet the noise design criteria for the receiving space adjacent. The floating floor construction meets the criterion only at frequencies below 63 Hz and above 500 Hz. Scheme III is satisfactory, although the design margin is only within the limits of experimental accuracy at 125 Hz.

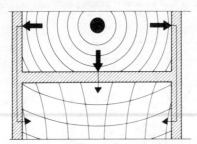

(a) Scheme I. Structure floor slab only generally consisting of 150–200 mm thick concrete.

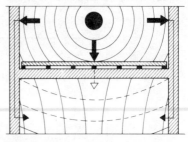

(b) Scheme II. Addition of 100 mm thick resiliently supported "floated" floor slab to structural floor slab of scheme I.

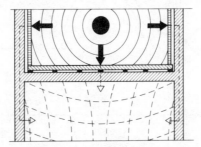

(c) Scheme III. Addition of inner walls bearing on floated slab of scheme II for control of sidewall flanking.

FIG. 4.23. Three typical construction techniques used in upper-floor plant-rooms where control of noise transmission to spaces below is of concern. Construction used depends on noise isolation requirements (Blazier, 1972).

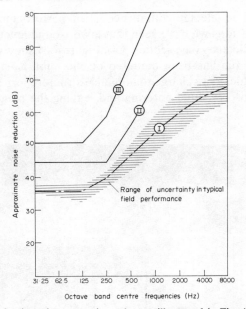

(a) Approximate noise reduction of construction schemes illustrated in Fig. 4.23. Range of uncertainty shown for scheme I is also applicable to schemes II and III and represents variation which can be expected in field performance. I, design curve for 150–200-mm-thick dense concrete structural floor; II, design curve for addition of 100-mm-thick floated floor slab with sidewall flanking; III, design curve for addition of 100-mm-thick floated floor slab with controlled sidewall flanking.

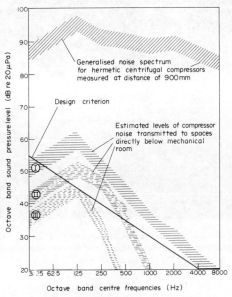

(b) Noise isolation performance of the three plant-room constructions of (a) when the source is an hermetic centrifugal compressor and the receiving space below is a private office or conference room. I, 150–200-mm-thick dense concrete structural floor; II, addition of 100-mm-thick floated floor slab with sidewall flanking; III, addition of 100-mm-thick floated with controlled sidewall flanking.

FIG. 4.24. Practical application of plant-room constructions shown in Fig. 4.23 (Blazier, 1972).

Equipment should be selected with its acoustical performance set as an important factor. In the case of fans we have seen that lower sound levels are associated with high aerodynamic efficiency (see section 3.4). Figure 4.25 shows a typical situation in which an axial flow fan has been mounted on vibration isolators and has flexible connections on the ducted discharge to minimise structureborne sound transmission. Acoustic splitters absorb sound entering and leaving the plant-room.

FIG. 4.25. Typical noise and vibration control measures for a 48 in. (122 cm) axial flow fan have been taken in this plant-room, e.g. vibration isolators under the fan, flexible connector between the fan and the duct and acoustic splitters on the air inlets to the area (with acknowledgements to Sound Attenuators Ltd.).

Sometimes acoustic enclosures are used to house equipment; the plant within them must be isolated from the structure to gain most benefit. The effective insulation for an enclosure is given by formula (11.24) derived in section 11.4.3 of Chapter 11,

$$I = R - 10 \lg\left(\frac{S}{A}\right),$$

where R is the sound reduction index, S is the total surface area of the enclosure and A is the total absorption of the receiving space; the absorption coefficient for the inside surfaces should be as high as possible. In Fig. 4.26 a diesel engine has been housed in an acoustic enclosure which resulted in the sound pressure level at 1 m from the engine being reduced from 110 to 97 dBA.

FIG. 4.26. Acoustic enclosure with one side removed. Rushton Paxman Ventura diesel engine shown develops 2000 bhp when running at 1500 rev/min and needed to meet a specification which called for a sound pressure level at 1 m to be reduced from 110 to 97 dBA. This was achieved using a standard modulator acoustic enclosure manufactured by Sound Attenuators Ltd. of Colchester. Because of the nature of the equipment it was necessary to provide easy access for maintenance, and for this purpose four sound-resistant and airtight doors were used. One further feature was the provision of air inlets for aspiration air which were also acoustically treated.

4.5.2. DISTRIBUTION OF SPACES WITHIN BUILDINGS

Wherever possible high noise level prone areas should be kept away from those that are required to be quiet. Having located the rooms, Parkin and Humphreys (1969) suggest the schemes shown in Fig. 4.27 for arriving at the degree of sound insulation required between various rooms in schools and offices; this scheme can be extended to other types of buildings, experience being the best criterion for interpreting the meaning of low, medium and high tolerance in terms of dBA levels for various kinds of room activity.

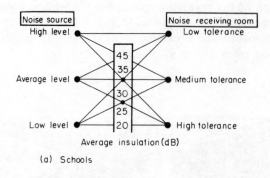

(a) Schools

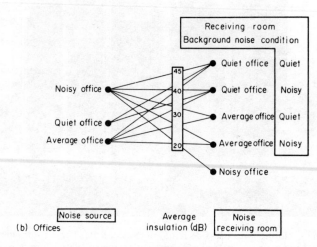

(b) Offices

Fig. 4.27. Suggested average sound insulation values for (a) schools and (b) offices (Parkin and Humphreys, 1969).

In Fig. 4.28 a lecture room in a modern building of a university having a ventilation system with a satisfactory acoustic performance has been positioned on the building side exposed to traffic, the wall of this side having a large amount of single-glazing; consequently this room is hardly used for its intended purpose because of the traffic noise disturbance.

It is paramount that the architect and the building environmental engineer plan the layout of the rooms in the building and select materials for the structure jointly, taking full account of the external and internal noise climates.

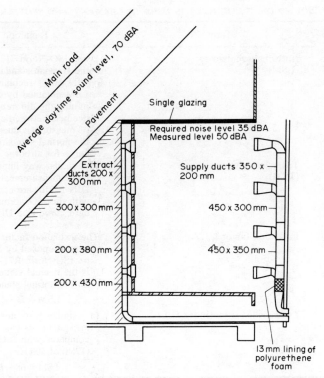

FIG. 4.28. A lecture room in a university situated near to a main road; traffic noise causes the sound level to rise to 50 dBA which is 15 dBA above the design criterion. Solution would be to use a different material or double-glazing to protect the room from traffic noise (Croome, 1969).

4.5.3. INTERNAL SOUND INSULATION AND ISOLATION

In section 2.4.4, a review was made of some of the surveys that have been carried out to ascertain what noises bother people. Many people, especially flat-dwellers and those living in terraced or semidetached houses, are even more troubled by noise when they are at home than when outside or at work. Neighbours' noise and traffic noise can be equally irritating. Hence the problem concerns the effectiveness of *external* walls and roofs, and *internal* walls, floors and ceilings as sound insulators. Bitter and van Weeren (1955) showed very clearly that staircase halls are frequently a source of noise complaints, and that the number of complaints of sound nuisance due to transmission through the floor may be twice as great as the number due to transmission through walls. He concluded that floors are the first item calling for improvement in sound insulation quality.

There remains a diversity of opinion as to what actual values of insulation to recommend, and before looking at the Building Regulations that have been issued on the matter in England and Wales it is worth looking at other European standards.

With grateful acknowledgement to Grandjean (1973a) for clear definitions in English, the terms shown in Table 4.7 are used in Europe.

TABLE 4.7. SOUND INSULATION TERMS USED IN GERMANY AND SWITZERLAND

Country	Term	Definition
Germany	Luftschallschutzmass (LSM)	Using ISO-Norm 717 (1968) the German standard DIN 4109[a] recommends LSM obtained by determining the difference in decibels between a curve of measured values and the normal standard ISO curve for airborne sound.
	Trittschallschutzmass (TSM)	In a similar way impact noise measurements (see DIN 52210) are compared with the ISO normal standard curve (see DIN 4109[a]).
Switzerland	Insulation factor (I_a)	The insulation factor has been proposed by SIA[b] based on ISO-Norm R717 (1968) and is the decibel value at 500 Hz on the normal standard curve $$LSM \simeq I_a - 52 \text{ dB.}$$
	Trittschallisolationsindex (I_i)	In a similar way, impact sound measurements (see SIA[b]) are compared with ISO-Norm R717 (1968) at 500 Hz $$TSM \triangleq 68 - I_i \text{ dB.}$$

[a]DIN 4109: 1962: *Schallschutz im Hochbau*.
[b]SIA: May 1968: Schweiz. Ingenieur- und Architekten-Verein, *Empfehlung für Schallschutz im Wohnungsbau*, Zürich.

In recent years the specifications laid out in Table 4.8 have appeared. Grandjean (1973a) comments that some of the recommendations are very inadequate.

It is worth remembering that Bitter and van Weeren (1955) concluded that no significant correlation exists between sounds heard and sound insulation at frequencies above 480 Hz. Although the sound reduction index increases as the sound frequency increases, the ear is more sensitive at high frequencies. There is a tendency to speak of single-figure insulation values; this figure should be referred to a frequency below 500 Hz and should compensate for the high-frequency sensitivity of the ear.

The Building Regulations issued in 1965 for England and Wales (Scotland has its own Building Standards Regulations issued in 1963) include mandatory requirements for sound insulation; these were revised in 1972 as Statutory Instruments 317. The degree of insulation is specified by a grading system which has the following classifications for airborne and impact sound.

Party-wall grade is based on the performance of a single-leaf brick wall plastered on both sides and weighing at least 415 kg/m² and aims to reduce the noise from neighbours to a level that is acceptable to the majority. This grade is very sensitive to flanking transmission pathway which must be minimised.

Grade I is the highest insulation that is practical vertically between flats. It is based on the performance of a concrete floating floor. Noise from neighbours should be no

TABLE 4.8. SPECIFICATIONS FOR SOUND INSULATION OF PARTY WALLS

Country	Sound reduction index R (dB)	Necessary density for a single wall (kg/m^2)	Thickness of brickwork, including plaster (cm)
Germany[a]	51	600	42
Sweden[b]	48	350	24
Austria:[c]			
Internal	40	100	6–8
Between two flats	48	350	24
Between two houses	53	800	56
Holland[d]	48–52	—	—

Switzerland[e] (1968)	Insulation factor			
	Minimum requirement		Enhanced requirement	
	I_a	I_i	I_a	I_i
Party walls; walls of entrance lobby	50	—	55	—
Ceilings between flats	50	65	55	55
Party walls and ceilings between flats and business premises, restaurants, workshops, etc.	60	50	65	45
Doors separating flat from entrance lobby	20	—	25	—
Doors separating flat from outdoors	—	—	25	—
Window or glazed door	20	—	30	—
United Kingdom	See text discussion following this table			

[a]DIN 4109, January 1959.
[b]Anvisningar till Byggnadsstadgan, December 1945.
[c]Oenorm B 8125, October 1949.
[d]Gezondheitsorganisatie TNO, October 1952.
[e]See footnote (b) to Table 4.7.

more of a nuisance than other disadvantages which tenants may associate with living in flats, whereas with *grade II* it will be considered to be the worst feature, although at least half of the tenants will probably not be seriously disturbed. At a level of insulation 8 dB below Grade II, neighbours' noise becomes intolerable.

Grade II is omitted from the 1972 Building Regulations.

The insulation should *not be less* than the values shown in Fig. 4.29a for airborne sound; the sound levels produced underneath a floor by a standard impact machine operated on the floor should *not exceed* the values shown in Fig. 4.29b in order to meet the impact grading (see *BRE Digest*, **187**, March 1976). Table 4.9, taken from *Building Research Station Digest*, **102** (1969), shows the grading for typical solid and cavity-wall constructions.

The 1972 edition of the Building Regulations (see Part G and Schedule 12) make no reference to grade II, but recommend party-wall grade for the *minimum* airborne sound insulation requirements of walls and grade I for floors (Fig. 4.29a); for impact sound insulation of floors the grade I curve in Fig. 4.29b represents the maximum

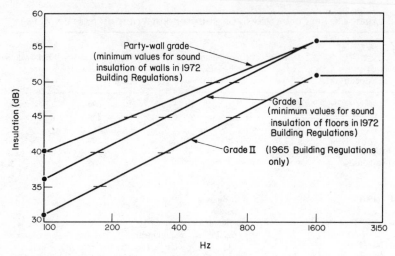

(a) Recommended airborne sound insulation grade curves for walls and floor.

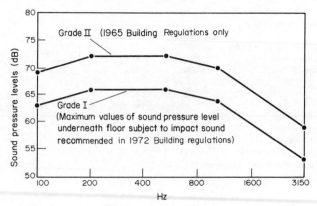

(b) Maximum sound levels under floor subjected to impact sound.

FIG. 4.29.

values of the sound pressure levels to be achieved underneath the floor which is being subjected to the vibration. Evidently, in light of experience it has been considered necessary to improve the sound insulation of buildings. Schedule 12 of the 1972 Building Regulations recommends constructions which will provide the necessary sound insulation, and this is reproduced here as Table 4.10.

It cannot be stressed sufficiently that the constructions in Table 4.10 will not provide adequate sound insulation unless the risks of flanking transmission and sound leaks are minimised. The resulting construction will only be considered as having met the party wall or grade I requirements if the aggregate deviations of the measured sound insulation and sound pressure levels over the given frequency range in Figs. 4.29a and b do not exceed 23 dB. The acoustic measurements should be made in accordance with BS 2750: 1956 (see Appendix I). The Building Regulations include some useful information on how to avoid flanking paths.

TABLE 4.9. TYPICAL SOLID AND CAVITY-WALL CONSTRUCTIONS TO SATISFY VARIOUS INSULATION GRADES (*Building Research Station Digest*, **102**, 1969)

Solid walls (plastered both sides except where stated)[a]	Minimum weight (including plaster[a]) (kg/m^2)	Grade (P = party wall)
Single-leaf brick wall (note: 225 mm brick wall weighs 490 kg/m^2)	415	P
175 mm concrete of density 2320 kg/m^3 (unplastered)		P
175 mm concrete of density 2080 kg/m^3 (plastered)		P
300 mm lightweight concrete of density 1200 kg/m^3		P
225 mm no-fines concrete of density 1600 kg/m^3		P
200 mm no-fines concrete of density 1600 kg/m^3		I (b)
Lightweight concrete (or equivalent)	220	II (b)
110 mm brick wall	220	II

Cavity walls with wire ties of butterfly pattern and plastered both sides	Minimum cavity width (mm)	Minimum weight (including plaster) (kg/m^2)	
Two leaves each consisting of:			
100 mm brick, block or dense concrete	50	415	P
Lightweight concrete with sound-absorbent surfaces facing cavity	50	300	P
Lightweight concrete with sound-absorbent surfaces facing cavity	75	250	P (b)
50 mm lightweight concrete of density 1280 kg/m^3	25	—	II (b)
100 mm hollow concrete blocks	50	—	II

[a]Plaster implies two-coat work at least 12 mm thick weighing not less than 24 kg/m^2.
[b]Grade II sound insulation is not now recommended for use in practice.

Of course, these Regulations are concerned with insulation between buildings owned or used by different groups or families of people; they do not refer to spaces subdivided within the same household or firm. It is worth noting the following regulations regarding walls separating habitable or non-habitable areas from refuse chutes:

... any wall which separates any habitable room in a dwelling from any refuse chute in the same building shall have an average mass (calculated over any portion of the wall measuring 1 metre square and including the mass of any plaster) of not less than 1320 kg/m^2.

Any wall which separates any part of a dwelling, other than a habitable room, from any refuse chute in the same building shall have an average mass (calculated over any portion of the wall measuring 1 metre square and including the mass of any plaster) of not less than 220 kg/m^2.

(Regulations G1, (2) and (3) of 1972 Building Regulations.)

Needless to say the selection of materials for building construction, where noise is an important aspect, still requires consideration of the structural, thermal, moisture and fire regulations besides those pertaining to sound.

Note that the standard method for rating the impact insulation value of floors and ceilings using a tapping machine and the International Standard Organisation grading curves has been much criticised by Fasold (1965), Schwirtz (1969) and Mariner and Hehman (1967). They reckon that the tapping machine does not realistically simulate the spectrum characteristics of footsteps and that greater emphasis needs to be laid on the high and low frequency components in order to provide acceptable acoustical conditions. Schultz (1971b) suggests that the loudness of impact sounds is not in itself a useful measure for estimating the isolation provided by floor–ceiling constructions and instead prefers to apply the basic principle of signal detection theory to footsteps,

TABLE 4.10. SOME BUILDING CONSTRUCTIONS WHICH WILL PROVIDE THE DEGREE OF SOUND INSULATION STIPULATED IN THE 1972 BUILDING REGULATIONS FOR ENGLAND AND WALES (SEE SCHEDULE 12 OF 1972 BUILDING REGULATIONS)

Part I: Walls providing resistance to the transmission of airborne sound (Fig. 4.29a)

Specification	Construction of wall
1	A solid wall consisting of:
	(a) bricks or blocks with plaster not less than 12·5 mm thick on at least one face; or
	(b) dense concrete cast *in situ* or panels of dense concrete having all joints solidly grouted in mortar; or
	(c) lightweight concrete with plaster not less than 12·5 mm thick on both faces of the wall;
	in each case the average mass of the wall (calculated over any portion of the wall measuring 1 metre square and including the mass of any plaster) being not less than 415 kg/m^2.
2	A wall having a cavity not less than 50 mm wide constructed of two leaves each consisting of bricks, blocks or dense concrete with plaster not less than 12·5 mm thick on both faces of the wall, and having any wall ties of the butterfly wire type, the average mass of the wall (calculated over any portion measuring 1 metre square and including the mass of the plaster) being not less than 415 kg/m^2.
3	A wall having a cavity not less than 75 mm wide constructed of two leaves each consisting of lightweight concrete with plaster not less than 12·5 mm thick on both faces of the wall and having any wall ties of the butterfly wire type, the average mass of the wall (calculated over any portion of the wall measuring 1 metre square and including the mass of the plaster) being not less than 250 kg/m^2.

Part II: Floors providing resistance to the transmission of airborne and impact sound (Fig. 4.29a and b)

Specification	Construction of floor
1	A floor consisting of:

(a) a solid concrete slab; or
(b) a slab of concrete beams and hollow infilling blocks of clay or concrete; or
(c) a slab of hollow concrete beams;

in each case having an average mass (calculated over any portion of the floor measuring 1 metre square and including the mass of any screed or ceiling plaster directly bonded to the slab but excluding the mass of any floating floor or suspended ceiling) of not less than 365 kg/m^2 and having either of the following laid upon it:

(i) rubber on sponge rubber underlay having a total thickness of not less than 4·5 mm; or
(ii) cork tiles not less than 8 mm thick.

2   A floor consisting of:

(a) a solid concrete slab; or
(b) a slab of concrete beams and hollow infilling blocks of clay or concrete; or
(c) a slab of hollow concrete beams;

in each case having an average mass (calculated over any portion of the floor measuring 1 metre square and including the mass of any screed or ceiling plaster directly bonded to the slab but excluding the mass of any floating floor or suspended ceiling) of not less than 220 kg/m^2 and having any of the following laid upon it:

(i) boarding nailed to battens so laid as to float upon a layer of glass fibre or mineral wool quilt, in either case capable of retaining its resilience under imposed loading; or
(ii) any covering directly applied to concrete or other cementitious screed, not less than 38 mm thick, so laid as to float upon a layer of glass fibre or mineral wool quilt, in either case capable of retaining its resilience under imposed loading; or
(iii) rubber on sponge rubber underlay having a total thickness of not less than 4·5 mm or cork tiles not less than 8 mm thick, in either case laid upon a dense airtight sealing layer upon lightweight screed, not less than 50 mm thick, of a density of not more than 1100 kg/m^3.

3   Boarding nailed to battens laid to float upon a layer of glass fibre or mineral wool quilt, in either case capable of retaining its resilience under imposed loading, the layer being draped over wooden joists, beneath which a ceiling of lath and plaster or of plasterboard, in either case not less than 19 mm thick, has been constructed, with pugging on the ceiling such that the combined mass of the ceiling and pugging is not less than 120 kg/m^2.

Part III: Floors providing resistance to the transmission of airborne sound only

Specification	Construction of floor
1	A floor consisting of a solid concrete slab having an average mass (calculated over any portion of the floor measuring 1 metre square and including the mass of any screed or ceiling plaster directly bonded to the slab but excluding the mass of any floating floor or suspended ceiling) of not less than 365 kg/m^2 and having any type of floor finish.

The sound insulation will only be achieved with good quality materials and with careful design and construction to eliminate flanking transmission.

comparing intruding sound with the background level,

$$d^{1} = \frac{S_{\text{r.m.s.}}}{B N_{\text{r.m.s.}}} \left(\frac{MW}{T}\right)^{1/2}, \tag{4.15}$$

where d^{1} is the detectability which can be directly related to annoyance; $S_{\text{r.m.s.}}$ is the r.m.s. sound intensity in bandwidth W; $N_{\text{r.m.s.}}$ is the r.m.s. background noise intensity in bandwidth W; M is the number of impacts available for potential audition in an event whose detectability is to be evaluated (Schultz states that a typical value is $M = 5$ for a person crossing an uncarpeted portion of an overhead room); T is the reverberation time corresponding to the decay slope of each impact as heard, and is effected by the reverberation of the structureborne sound, and the airborne sound in the receiving room; B is the rate of walking (impacts/sound); and W is the signal bandwidth (Hz) determined by the compliance of the heel or the floor covering.

Note that detectability is increased as absorption is added to the receiving room thus decreasing T. Also detectability depends upon signal bandwidth, not absolute frequency values as loudness does.

The trend towards lightweight construction has made it necessary to develop other ways of acquiring adequate sound insulation other than relying on mass. The degree of improvement depends on the amount of flanking transmission that occurs through wall-ties or studding connecting the leaves making up the composite structure, and through the surrounding walls, ceiling and floors. Continuous horizontal (or vertical) structural elements should include discontinuities at cross-junctions with vertical (or horizontal) ones. The *Building Research Station Digest*, **96** (1968) suggests that the minimum weight for each leaf of a party wall might be about 25 kg/m^2 exclusive of any framework. The construction of composite structures must take account of any movements due to heat or moisture which could cause joints to open up. A minimum cavity width of 225 mm seems to be reasonable in practice. This is in agreement with the work of Ford and Lord (1968) who have found by calculation and measurement that an average sound reduction index of 50 dB over the frequency range 100–3200 Hz is possible using a double skin having a combined thickness of 230 mm and a total weight of 59 kg/m^2 or, alternatively, a combined thickness of 80 mm with a total weight of 98 kg/m^2. Ford and Lord (1968) suggest the following relationships for estimating the average sound reduction index at about 400–500 Hz of a double-skin construction,

$$R = 20 \lg (md) + 34, \tag{4.16a}$$

where m is the surface density (kg/m^2) and d is the cavity width (m). Eck (1973) gives the sound reduction for a double wall as

$$R = 14 \lg (m_1 + m_2) + 24 \text{ to } 29. \tag{4.16b}$$

Absorption materials placed in the cavity decrease the intensity of the reverberant field in the cavity. Strategic positioning of the absorbent at the antinodes, where the particle velocity is a maximum, has been found to produce a negligible benefit (Ford, *et al.*, 1970). Damping materials stuck to the panels raise the coincidence frequency more effectively than the material in the cavity; in practice, application of damping materials externally and internally is recommended. The performance curves for a typical lightweight partition is shown in Fig. 4.30.

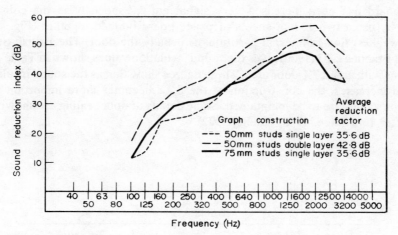

FIG. 4.30. Typical sound reduction index curve for a plaster partition (with acknowledgement to British Gypsum).

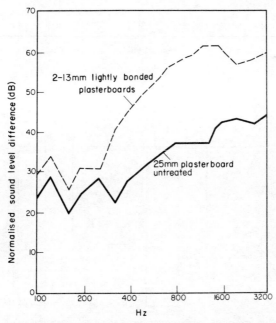

FIG. 4.31. Improved sound insulation using laminated structures. A single board exhibits coincidence because of its high bending stiffness whereas two sheets lightly bonded boards have a slightly higher stiffness and achieve a higher insulation in the coincidence region (Lord, 1969).

Increasing damping by the use of laminated structures has already been mentioned when discussing glazing. Lord (1969a) has shown how the sound insulation of a 25 mm plasterboard has been increased very significantly, especially in the coincidence region, by using two 13 mm sheets of plasterboard bonded together (Fig. 4.31).

The weakest link in internal partitions is usually the door. The British Standard Code of Practice CP3 (1960) gives the sound insulation values shown in Table 4.11 for partitions with about 7% door area. These figures show that as the sound insulation of the wall increases, the construction of the door becomes more important if good insulation values are to be maintained. Some details of door sealing are shown in Fig. 4.32 (after Lawrence, 1970; Doelle, 1972).

TABLE 4.11

	Insulation value of partition (dB)					
Construction	25	30	35	40	45	50
Door with large gaps around the edges	23	25	27	27	27	27
Light door with edge sealing treatment	24	28	30	32	32	32
Heavy door with edge sealing treatment	25	29	33	35	37	37
Double doors with sound lock	25	30	35	40	44	49

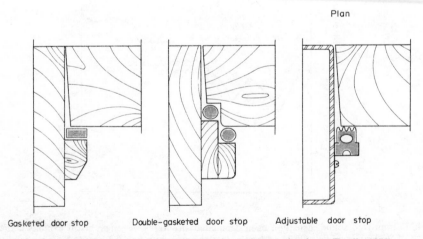

Plan

Gasketed door stop   Double-gasketed door stop   Adjustable door stop

(a) Door-stops used at the jambs and heads of sound-insulating doors (Doelle, 1972).

FIG. 4.32. Some details of door sealing to reduce airborne sound transmission (Doelle, 1972).

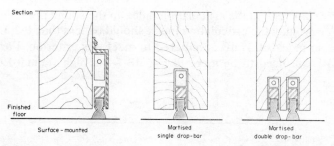

(b) Automatic drop-bar threshold closers used for sound-insulating doors (Doelle, 1972).

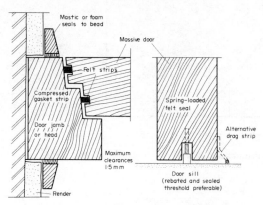

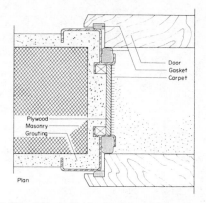

(c) Use of felt and drag strips (Lawrence, 1970).

(d) Sound-insulating double door with steel frame and perimeter absorption (Doelle, 1972).

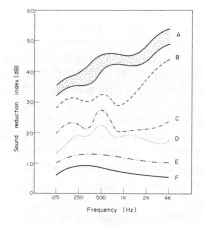

(e) Sound reduction index of different door constructions: A, sound-proof door (note spread of values); B, solid-core door gasketed; C, hollow-core door gasketed; D, ungasketed hollow-core door; E, louvred door (25% open); F, open door.

Brosio (1968) has related the articulation index to the sound reduction index (SRI) (Fig. 4.33). For privacy the articulation index should be less than 0·2, i.e. a minimum sound reduction index of 37 dB is required to meet this criterion. For management offices the SRI will need to be more like 45 dB for a high standard of privacy.

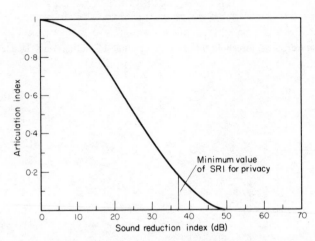

FIG. 4.33. Relationship between articulation index and sound reduction index (Brosio, 1968).

Gray *et al.* (1958) studied three groups of flats with regard to noise insulation and complaints of disturbance. Some of the results are summarised in Table 4.12. In general the 5 dB increment in the insulation from 45 to 50 dB is more useful than the 5 dB increment in the insulation from 40 to 45 dB.

Bitter and van Weeren (1955) surveyed 1200 people living in single flats in Rotterdam and The Hague. The work showed that brick floors were more effective than wooden ones in providing airborne and structureborne sound insulation. In a later study by Bitter and Horch (1958) an attempt was made to find the effect of different floor densities on disturbance from footsteps; some results are given in Table 4.13. A sample of 423 people stated that they were more bothered by music and radio talks than footsteps. A principal conclusion was that a solid floor should have a minimum density of 500 kg/m².

More recently van den Eijk and Bitter (1971) have carried out a survey of many flats in Holland in order to assess the annoyance from neighbours' footsteps. The results are summarised in Table 4.14 and Fig. 4.34. The results suggest that from the point of view of annoyance, rather than just hearing footsteps (often), a floating floor is no better than a solid floor of about 490 kg/m² but is preferable than a solid floor having a density of 300–410 kg/m².

Clark and Petrusewicz (1970–1) carried out noise surveys in more than 400 households. Many people were found to be uncomfortably aware of the central heating pump noise especially in the evening; noise coming from large radiator

TABLE 4.12. PENETRATION OF NOISE FROM FLATS ABOVE AND BELOW
(GRAY *et al.*, 1958)

	Percentages making these comments		
Average floor/ceiling sound insulation	Group I 50 dB	Group II 45 dB	Group III 40 dB
Comments			
Too much noise through ceiling	61	76	78
Too much noise from below	38	57	66
Conversation audible through ceiling	53	72	87
Conversation audible through floor	48	59	86
Talking is heard from flats on same level[a]	21	30	11
Radio or television above	70	80	90
Radio or television below[b]	64	59	91
Radio or television from flats on same level[a]	37	39	34
Radio or television above is disturbingly loud[b]	21	29	20
Radio or television below is disturbingly loud[b]	15	21	18
WC flushing heard[b]	80	87	84
WC flushing disturbingly loud[b]	13	28	8

[a]Floor or ceiling insulation does not contribute significantly to lateral sound insulation.

[b]Time of use, intermittency, type and volume of television or radio programmes will contribute towards large variations in subjective response beyond the normally expected personal differences.

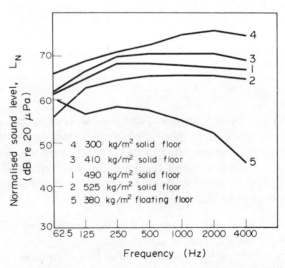

FIG. 4.34. Normalised impact sound levels in octave bands (van den Eijk. 1971).

TABLE 4.13. EXTENT OF PENETRATION BY AND DISTURBANCE FROM
FOOTSTEPS IN THE FLAT ABOVE (BITTER AND HORCH, 1958)

Density of floor (kg/m²)	Footsteps heard (%)	Footsteps disturbing (%)	Number questioned
500	14	3	399
400	33	8	423

TABLE 4.14. FLOOR INSULATION FROM FOOTSTEPS (EIJK AND BITTER, 1971)

Type of floor	Hearing and annoyance from neighbours' footsteps in % of dwellings visited					Number of dwellings visited
	Heard				Annoyed	
	Seldom	Sometimes	Often	Total		
Solid concrete 300 kg/m²	12	18	20	50	9	382
Solid concrete 410	2	16	15	33	8	423
Solid concrete 490	3	8	3	14	3	399
Solid concrete 525	13	14	7	34	5	584
Floating 380 (load bearing floor 300 kg/m² plus floating screed of 80 kg/m²)	14	18	7	39	5	478

surfaces in dining rooms or bedrooms, and noises from bathrooms and toilets can also be troublesome. Periodic firing of the boilers and external traffic noise were not a problem, although the pump cutting in and out did cause many people to wake up. Boiler noise was a nuisance when local boiling gave rise to high-pitched, high-intensity noise, similar to noise arising from turbulence and cavitation but at much higher intensity. Important factors are:

* *the noise spectrum*: many pump spectra showed peaks in the range 315–630 Hz which caused serious annoyance;
* *location of boiler, pump, and solenoid valves*; install as far away as possible from bedrooms and living rooms;
* *furnishings*: reverberant surroundings increase the likelihood of noise nuisance;

* *radiators or pipes fixed to hard finish wall surfaces* cause vibrations to be conducted to other rooms besides possibly giving rise to noise in the room they are located especially if it is reverberant; it will be suggested in Chapter 9 that all water pipes or air ducts should be isolated from building surfaces besides having attenuators in the pipeways or airways (e.g. see Fig. 3.21).

The lesson to be learnt from this survey is do not neglect any possible source of noise however minor it may appear to be.

Ward (1974) reports a survey of 262 kitchens in houses built to current building regulations. One of her conclusions was that greater attention needs to be paid to sound insulation. Table 4.15 shows that clearly noise is a problem. The sound level of various kitchen appliances were measured (Table 4.16). The mean external and internal sound levels are given in Table 4.17. More data on noise in kitchens is provided by Holmberg and Thylebring (1974).

TABLE 4.15. FEATURES CONSIDERED UNSATISFACTORY IN KITCHENS (WARD, 1974)

Public sector households	Private sector households
45% lack of adequate ventilation	44% volume of storage
45% noise in kitchen	33% difficulty in opening windows
43% condensation	32% area of work surfaces
33% difficulty in opening windows	31% lack of ventilation
31% volume of storage	30% poor lighting
25% position of doors	30% difficult access
19% area of work surfaces	28% noise
	26% condensation
	23% poor layout

TABLE 4.16. DOMESTIC APPLIANCES: NOISE LEVELS (dB) MEASURED BY WARD (1974)

	dB		dB
Moulinex liquidizer	94	Frigidaire spin dryer	38
Duchess potato peeler	87	Servis super twin tub	37
Rolls-Rapide twin tub	83⎫	Hoover spin dryer	73
Moulinex mixer BT2A	83⎭	Miele Dex dishwasher	73⎫
Hoovermatic twin tub	80	Hotpoint Supermatic	72⎭
Frigidaire twin tub	78⎫	Central heating sounds	68
Kenwood Chefette	78⎬	X-pelair	69
Electrolux 87 vacuum (cylinder)	78⎭	Midland Electric 36 fridge	55
Kenwood Chefette	76	English Electric Slimline fridge 40	54
Hoover Spin-a-Rinse	74	Hotpoint Iced Diamond	48

Various recommended floor constructions with their insulation gradings are shown in Fig. 4.35 (i.e. Table 1 as described in *Building Research Station Digest*, **103**); of

TABLE 4.17. WARD, 1974
Mean sound pressure levels of *external* noise sources

| | Noise a problem in kitchen | | Noise not a problem in kitchen | |
	Public sector (dBA)	Private sector (dBA)	Public sector (dBA)	Private sector (dBA)
Town				
Traffic	57	48	56	49
Aircraft	56	—	70	—
Trains	—	—	65	56
Industry	55	—	—	—
Construction	63	—	57	62
Children	52	49	56	58
Pets/animals	—	49	40	47
Radio/TV	—	—	—	—
Estate noise	50	—	53	47
Suburb				
Traffic	52	47	51	48
Aircraft	—	85	53	54
Trains	—	—	45	—
Industry	—	—	—	—
Construction	55	48	46	50
Children	55	58	52	44
Pets/animals	43	44	47	39
Radio/TV	—	—	—	—
Estate noise	46	48	47	54

Mean sound pressure levels of *internal* noise sources

| | Noise a problem in kitchen | | Noise not a problem in kitchen | |
	Public sector (dBA)	Private sector (dBA)	Public sector (dBA)	Private sector (dBA)
Town				
Traffic	50	—	43	44
Conversation	45	46	43	45
Domestic NFP[a]	40	—	—	—
Central heating	37	—	46	38
Children	66	—	50	35
Impact sounds	—	—	—	—
Radio/TV	58	52	48	—
House/estate noise	50	—	51	52
Suburb				
Traffic	43	—	45	—
Conversation	42	46	43	44
Domestic FP[a]	—	64	46	65
Domestic NFP[a]	40	48	60	44
Central heating	44	44	44	—
Children	44	—	42	36
Impact sounds	—	—	—	65
Radio/TV	53	—	53	—
House/estate noise	36	—	61	39

[a]FP = food preparation. NFP = non-food preparation.

particular note are the floating floors (items 2 and 3 in Fig. 4.35) which provide isolation from any other part of the building structure. The resilient layer is usually glass wool or mineral wool (e.g. 13 mm long fibre glass-wool paper faced on one side with an uncompressed density of 36 kg/m³, or resin-bonded glass wool, 25 mm thick and weighing 100 kg/m³, or even possibly a 13 mm expanded polystyrene board having a density of 15–25 kg/m³ with a 90% strain recovery after compression); cane fibreboards and hair-felts are not recommended because of their tendency to compact under continuous loading. The resilient layer should be turned up at all edges which abut walls, partitions or other parts of the structure, and there must be no continuity between it and the structural base.

CONSTRUCTION	Floor finish	Grade	
		Airborne	Impact
(1) Concrete floor  — The basic floor construction of items (1) to (5) is assumed to weigh not less than 220 kg/m² (including any integral screed and plaster finish) and may be dense or lightweight reinforced concrete, hollow concrete beams, or concrete beams with hollow clay block infilling	Hard	II	4 dB worse than grade II
	Medium	II	II
	Soft	II	I
(2) Concrete floor with floating screed 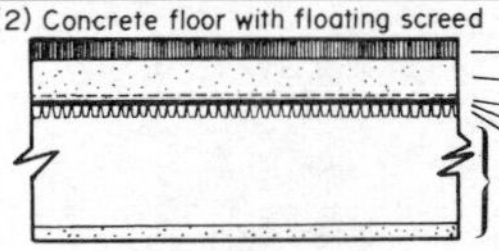 — Any floor finish — Screed (>65mm) — Wire mesh (20–50mm mesh) — Paper — Resilient layer (paper face upwards) — Not less than 220 kg/m²	Any	I	I
(3) Concrete floor with floating wood raft 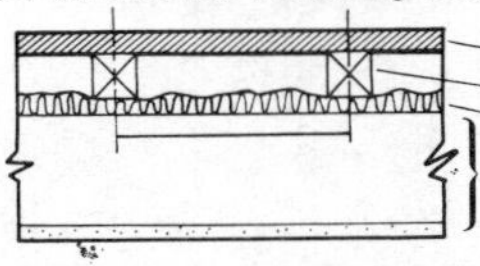 — Wood flooring — Battens (>50 x 50 mm cross-section) — Resilient layer (paper face downwards) — Not less than 220 kg/m²	Any	I	I
(4) Concrete floor with suspended ceiling 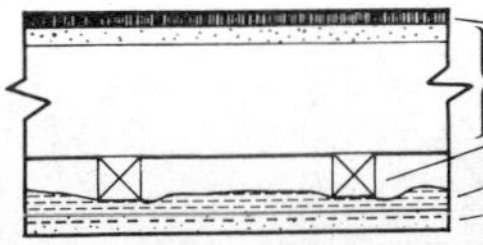 — Floor finish — Not less than 220 kg/m² — Battens wired to slab — Absorbent quilt — Heavy ceiling (>24 kg/m²) e.g. plaster on expanded metal lathing	Hard	I	2 dB worse than grade II
	Medium	I	II
	Soft	I	I
(5) Concrete floor with lightweight screed 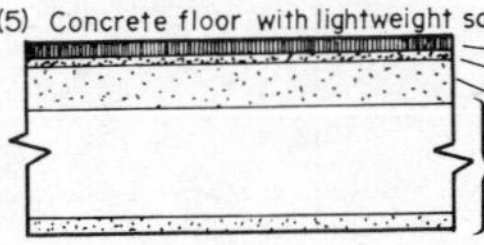 — Floor finish — Dense topping (see text) — 50mm lightweight screed (~1100 kg/m³) — Not less than 220 kg/m²	Hard	I	4 dB worse than grade II
	Soft	I	I
(6) Heavy-concrete floor 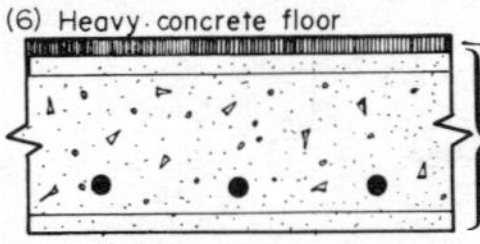 — Floor finish — Not less than 365 kg/m² including screed and plaster	Hard	I	4 dB worse than grade II
	Soft	I	I

FIG. 4.35. Sound insulation grading of concrete floors between flats (*Building Research Station Digest*, **103**, 1969).

A kinetic floating floor from America has recently become available in this country. The construction and high acoustic performance is shown in Fig. 4.36.

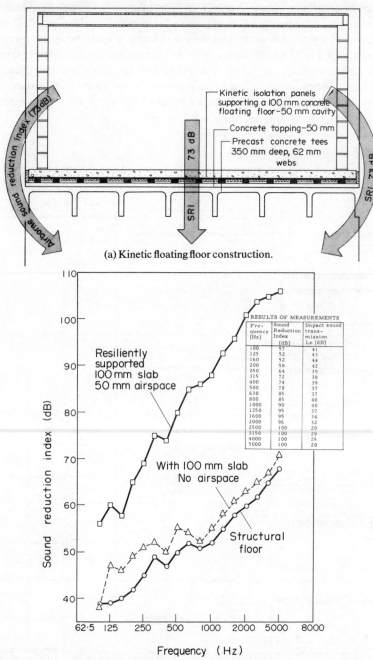

(a) Kinetic floating floor construction.

Kinetic isolation panels supporting a 100 mm concrete floating floor–50 mm cavity

Concrete topping–50 mm

Precast concrete tees 350 mm deep, 62 mm webs

Airborne sound reduction index (73 dB)

SRI 73 dB

SRI 73 dB

RESULTS OF MEASUREMENTS		
Fre-quency (Hz)	Sound Reduction Index (dB)	Impact sound trans-mission Ln (dB)
100	57	41
125	52	43
160	52	44
200	58	42
250	64	39
315	72	38
400	74	39
500	78	37
630	85	37
800	85	40
1000	90	40
1250	95	37
1600	95	34
2000	96	32
2500	100	29
3150	100	29
4000	100	25
5000	100	20

Resiliently supported 100 mm slab 50 mm airspace

With 100 mm slab No airspace

Structural floor

(b) Relative performances of a basic structural floor slab, a floor slab with an extra 100 mm of concrete cast on top, and the same structural floor with the 100 mm slab resiliently supported to provide a 50 mm airspace between the two masses. Notice the enormous increase in performance can be achieved by separating the two masses.

FIG. 4.36. Kinetic noise isolation systems (with acknowledgements to Sound Attenuators Ltd.).

Most wood-joist floor constructions are grade II or worse, but Fig. 4.37 shows floating heavy lath and plaster floors which may attain grade I when used in conjunction with thick walls (i.e. three or more walls *below* the floor must be at least one brick thick or of similar weight).

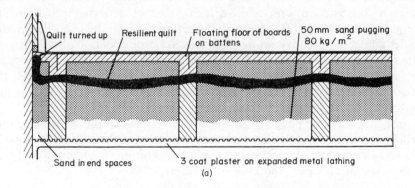

Sand in end spaces   3 coat plaster on expanded metal lathing
(a)

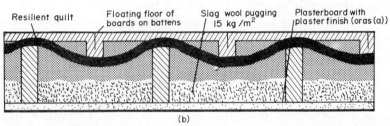

(b)

FIG. 4.37. Wood joist floor constructions to satisfy Grade I sound insulation shown in Fig. 4.29 (*Building Research Station Digest*, **103**, 1969) with (a) heavy pugging, and (b) light pugging.

Walker (1971) has carried out an extensive research programme on the sound insulation of wood-joist floors. His results are summarised in Fig. 4.38 and from them he drew these conclusions:

(a) the simpler forms of floors, construction types 1 and 3 as used in two-storey domestic houses, behave acoustically in an analogous manner to simple timber stud partitions; the depth of joist is of little significance, and the main factors controlling the insulation are the mass of the floor and the air leaks in the floorboards;

(b) improvements of about 8 dB average insulation to the standard house floor can be brought about by the use of either mineral wool pugging or resilient ceiling fixings;

(c) the use of either a floating raft floor or a resiliently mounted ceiling with mineral wool can increase the sound insulation of a standard house floor by as much as 12 dB average; both systems are acoustically equivalent;

(d) either a floating floor with a rigid ceiling, or a rigid floor with a resilient ceiling, behave acoustically rather like a double leaf partition and exceed the mass law value by about 10 dB;

(e) a floating floor with normal plasterboard ceiling pugged with mineral wool will not meet the standard required by the Scottish Building Regulations for party floors;

(f) combining a floating floor with a resilient ceiling and mineral wool pugging gives a construction which has a high value of acoustic insulation at all frequencies above 200 Hz; these constructions exceed the mass law value by 17 dB and generally give an average insulation of about 50 dB;

(g) mass loading the ceiling of an otherwise floating floor system with a suitable dry granular pugging can result in insulation values as high as 55 dB; these floors will meet the necessary grading standards when constructed so as not to be too rigidly linked to the flanking structure;

(h) if the heavy granular pugging is included in the sub-floor (i.e. not resting on the ceiling boards) it does not behave so usefully in dampening ceiling resonance but gives a more satisfactory structure in terms of fire resistance, likelihood of ceiling sag or collapse, and ease of prefabrication;

(i) if a suitably heavy and rigid plasterboard ceiling is used in conjunction with a floating floor dampened with one or more layers of plasterboard under the walking surface (total weight = 73 kg/m^2), heavy pugging is not necessary for the floor to meet the Scottish grading standards (Floor Construction 13);

(j) the most difficult criterion proved to be the low-frequency impact sound requirement for flats; subjective measurements have shown that this criterion is in fact fairly important in practice especially when dealing with floor weights of only 60 kg/m^2; it is thought that 65–70 kg/m^2 is about the minimum weight of floor which will, in the light of modern building technology, with any degree of certainty and within reasonable cost limits, meet the present British and international acoustic standards for floors in flats.

Note that resiliently mounted ceilings refer to ceiling boards screwed to 24 gauge Z-shaped metal channels and fixed across the underside of the joists.

The role of planning must not be forgotten. The layout of quiet and noisy areas within a house or hotel is just as important as the positioning and orientation of houses, hotels, offices or factories within an urban development plan.

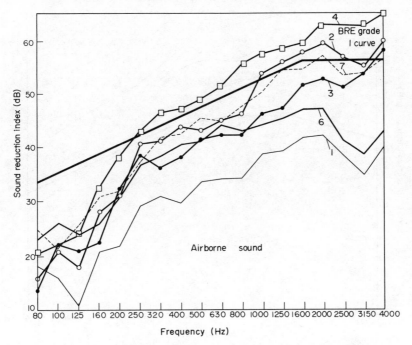

(a) Floor constructions *not* meeting the grade I standard for airborne sound (laboratory measurements).

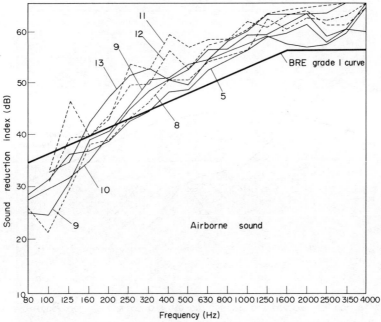

(b) Floor constructions meeting the grade I standard for airborne sound (laboratory measurements; 13 field-tested also).

FIG. 4.38. Airborne and impact sound insulation (laboratory measurements except where stated) of wood joist floors (Walker, 1971).

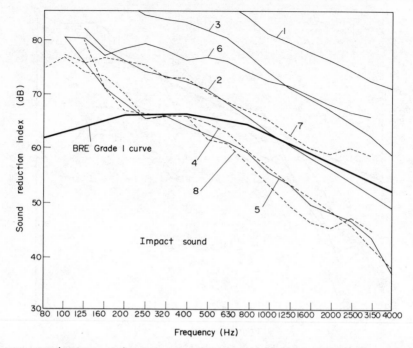

(c) Floor constructions *not* meeting the grade I standard for impact sound (laboratory measurements).

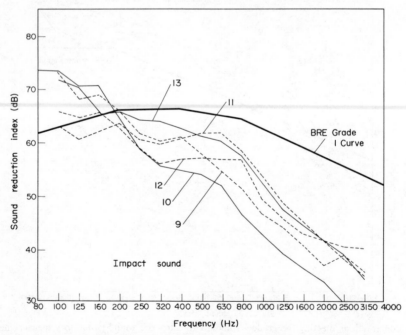

(d) Floor constructions meeting the grade I standard for impact sound (laboratory measurements; 13 field-tested also).

FIG. 4.38. (*Continued*).

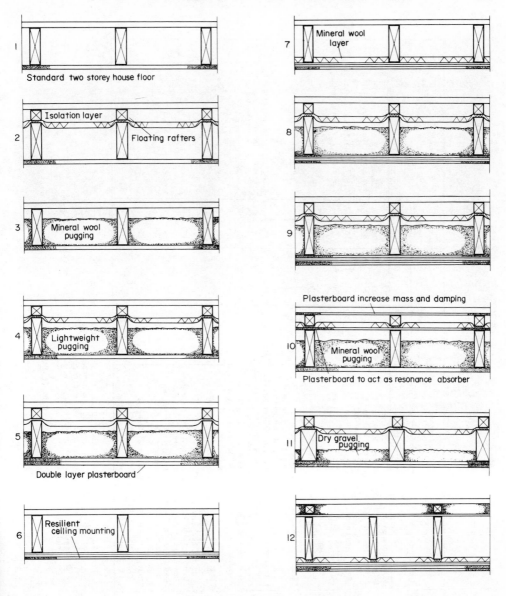

FIG. 4.38. (*Continued*).

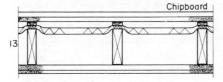

Construction No.	Type	Thickness (mm)	Weight (kg/m²)	Floor system	Plasterboard ceiling	Sound reduction index (dB) Measured 100–3200 Hz	Mass law	Impact insulation index Ii	Remarks
1	A	210	35	22 mm floorboards	1 layer 12·7 mm	31	32	88	1,2,3,4,6,7 unacceptable for grade 1 airborne sound
2	A	270	38	As above but floating	1 layer 12·7 mm	44	33	75	
3	A	260	50	22 mm floorboards	1 layer 12·7 mm but pugged MW*	40	34	85	5,8,9,10,11,12,13 acceptable for grade 1 airborne sound
4	A	270	52	22 mm floorboards floating	1 layer 12·7 mm pugged MW*	48	35	70	
5	A	283	62	22 mm floorboards floating	2 layers 12·7 mm pugged MW*	49	36	69	
6	A	226	37	22 mm floorboards	1 layer 12·7 mm resiliently mounted	38	33	78	1,2,3,4,5,6,7,8 unacceptable for grade 1 impact sound
7	A	226	39	22 mm floorboards	1 layer 12·7 mm resiliently mounted 25 mm MW in cavity	43	33	72	
8	A	290	65	Floating floor as above	2 layers 12·7 mm resiliently mounted and pugged MW	51	36	67	9,10,11,12,13 acceptable for grade 1 impact sound
9	A	300	72	Floating floor as above	2 layers 12·7 mm 1 layer 9.5 mm resiliently pugged MW	51	37	63	Laboratory grade 1
10	A	299	73	Floating floor 22 mm boards + 9.5 mm plasterboard	1 layer 19 mm pugged with MW*	49	37	63	Laboratory grade 1
11	C	215	108	Floating floor 22 mm chipboard	2 layers, 12·7 mm pugged with gravel (50 mm)	55	38	57	Laboratory grade 1
12	B	305	122	Semi floating 22 mm T & G sand-filled sub-floor 50 mm deep	2 layers, 12·7 mm resiliently + MW* mattress	53	40	58	Laboratory grade 1
13	A	282	68	Floating floor 22 mm chipboard and 19 mm gyproc plank	10 mm + 12·7 mm as double layer, no pugging	53	37	63	Laboratory and field grade 1

Type A: floors based on 175 mm x 50 mm joists at 400 mm centres.
B: 200 mm x 38 mm joists at 300 mm centres.
C: 150 mm x 75 mm joists at 300 mm centres.
* MW pugging implies granular mineral wood pellets laid about 100 mm deep.

FIG. 4.38. (Continued).

4.6. Noise Control Using Attenuation Equipment

4.6.1. ABSORPTION OF SOUND

Mention has already been made of the use of absorption materials for reducing the reverberant energy in the cavity of a lightweight structure. Can this be applied to a room with any benefit? The answer is yes, but application of such materials to room surfaces will only affect the reverberant sound field. Remembering that the energy in the direct sound field is $(WQ\rho c)/(4\pi r^2)$, and that in the reverberant field is $(4W\rho c/R)$ (see equations (10.12) and (10.30)), then equating these, the radius of the free field (in metres) is given by equation (10.34) now requoted as

$$r = 0.14\sqrt{RQ} \qquad (4.17)$$

which may be easily rearranged in terms of reverberation time and room volume if desired. Within the radius r the effect of increasing the absorption values of surrounding surfaces will be of no benefit, but beyond it they will decrease the sound level due to the reverberant field. Figure 4.39 gives a set of curves showing the reduction of the reverberant sound obtained by introducing material having an absorption coefficient α to a surface originally having an absorption value of α_0, the treated surface area forming $x\%$ of the whole surface area in the space.

From his experience, Sabine in Harris (1957b) suggests that the required amount of absorbent treatment to be applied to a room for satisfactory noise control can be judged from two working rules:

(a) the average absorption coefficient should be between 0·2 and 0·5; the former limit can be usually achieved by treating the entire ceiling, whereas to attain $\bar{\alpha} = 0.5$ usually requires half the wall area as well as the entire ceiling area to be treated;

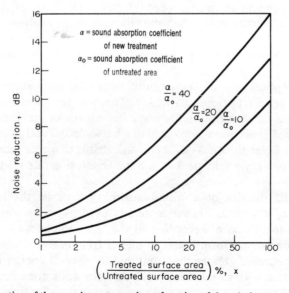

FIG. 4.39. Sound reduction of the reverberant sound as a function of the relative areas of absorbent treated and total surface areas.

(b) to produce a satisfactory improvement in an existing room, the total absorption after treatment should be between three and ten times the absorption before treatment if the ear is to perceive a distinct change.

Use rule (a) to check rule (b).

Absorption of sound relies on the dissipation of energy by viscous forces set up by the velocity gradient across the freely moving air in the centre of the pores and that which is almost stationary near the fibre surfaces. Isothermal conditions exist for low-frequency waves, whereas the rapidity of high-frequency waves allows no time for heat, arising from compressions and viscous energy dissipation, to flow to the fibre walls, and so adiabatic conditions apply. Besides the action of viscous friction, there is a substantial contribution to sound attenuation by damping due to the flow resistance of the pore network represented by

$$\Delta p = Rv^n, \tag{4.18}$$

where Δp is the pressure differential established due to the particle motion v of the sound wave in a network having a resistance R. The index n ranges from 1 at low frequencies to 2·5 at high frequencies. If R is much greater than the impedance of the sound wave ρc, then reflection will occur at the boundary, whereas if $R \ll \rho c$ there will be insufficient friction for any appreciable attenuation.

Absorption increases as the porosity (i.e. ratio of porous voids volumes to total volume) of the material is increased and as the flow resistance decreases. Clearly it is important that access to the pores should be made as easy as possible, and surface treatments, such as painting, which block up inlet apertures, should be avoided. Vibration of the fibre structures can make some additional contribution towards attenuation.

Sound-absorbing materials used in practice display no great variation in porosity. The ratio of voids to volume lies between 0·6 and 1. The flow resistance, however, is of the order of 0·1. Furthermore, the total flow resistance is always a function of the thickness of the layer. It is obvious from this that the correct choice of the layer thickness and the flow resistance is the criterion for the best possible attenuation of a given frequency.

A sound absorption rate of about unity when the frequency to be damped is achieved by a layer thickness equal to $\lambda/4$. The flow resistance should in this case amount to about $2\rho c$, where ρc is the wave impedance of air. From this frequency onwards the sound absorption rate oscillates between 0·9 and 1. At equivalent flow resistance, and a layer thickness of $\lambda/16$, the absorption rate amounts to approximately 0·5, and for a layer thickness of $\lambda/8$ the absorption rate is approximately 0·75 (Eck, 1973).

The greater the density of a sound-absorbing material the higher is its flow resistance. Amongst various fibrous materials used as sound absorbers, a density ranging from 20 to 120 kg/m³ according to the layer thickness is used.

Absorbent materials perform better at higher frequencies where the influence of flow resistance becomes greater than that due to viscous energy dissipation alone. Used in conjunction with membranes (such as dense foils, for example), which offer mass control, and airspaces offering stiffness control, the effectiveness of absorbents can be made more variable over a wide frequency range. Typical behaviour patterns are shown in Table 4.18.

TABLE 4.18

Type	Characteristic	Comments
Single thin porous screen Sound ρc ... d	α ... 1 ... 3 ... $\dfrac{4d}{\lambda}$	(a) Particle velocity and displacement are $\dfrac{\lambda}{4}$ out of phase, hence maximum absorption occurs where the particle velocity is a maximum i.e. $d = n\dfrac{\lambda}{4}$ $(n = 1,3,5.....)$ (b) Flow resistance of screen should match that of incident sound $R = \rho c$ for diffuse, random incidence $R = 2\rho c$ to $3\rho c$
Porous absorbers Airspace	α (b) (c) (a) Single layer $\dfrac{4d}{\lambda}$	(a) To develop a broadband range of absorption use successive layers of porous screens; cut-off occurs at about $\dfrac{d}{\lambda} \approx \dfrac{1}{4}$ (b) Use of airspace increases low frequency response (c) Shape of characteristic is similar for curtains, furniture, carpets, people and acoustic tiles
Membrane (or panel) absorber d(m) Membrane, m (kg/m^2)	m or d increasing α ... f Use of absorbent in airspace increeces mid/high frequency absorbption	A mass–spring system with a resonant frequency given by $$f = \dfrac{c}{2\pi}\sqrt{\dfrac{\sigma}{md}}$$ where σ = Poisson's ratio
Resonator absorbents (a) Simple Helmholtz resonator	1·0 α 0 ... 100 ... 1000 ... 5000	Resonant frequency given by $$f = \dfrac{c}{2\pi}\sqrt{\dfrac{a}{lV}}$$ where V is the volume of the cavity, a is the cross–sectional area and l the length of the cavity entry Side branches in airflow systems are a form of resonator Energy absorbed by friction in the neck Useful for filtering discrete tones
Perforated panel absorbents Area of hole a ... s ... d ... t	1·0 0·8 0·6 Absorption coefficient 0·4 0·2 0 b a c 125 250 500 1000 2000 4000 Frequency, Hz (a) Panel, 10% perforation area, over 25 mm mineral wool, over 25 mm airspace (b) As (a) but 50 mm mineral wool (c) Perforated plaster tile over 25 mm airspace	Application of Helmholtz resonator principle where resonant frequency given by $$f = \dfrac{c}{2\pi}\sqrt{\dfrac{a}{dts^2}}$$ Notice the gain in absorption at low frequencies by using absorbent material.

Notice that the addition of absorbent materials to a surface (e.g. acoustic tiles on a ceiling) has negligible influence on the insulation value of the surface. The reasoning is that such materials attract *more* sound energy to the surface than do reflecting materials, but since energy dissipation occurs in the pores, the resultant insulation value remains about the same. For appreciable reductions in sound transmission the absorbent would need to have a thickness comparable with the wavelength, which is not feasible in most circumstances (e.g. at 350 Hz the material thickness would be about 1 m!).

In selecting absorbent materials care should be taken to distinguish between the use of the *normal incidence* and *random incidence* absorption coefficients. Sabine's formula uses the latter and a series of calculations should be consistent and use the same one, obtaining corrections where necessary from the manufacturers of the material.

For perforated panels the absorption increases with frequency but there is an upper limit which for panels up to 5 mm thick is given by

$$f = \frac{3500\,d}{n},\tag{4.19}$$

where d is the perforation diameter in centimetres and n is the number of holes per unit length (i.e. holes per cm length). The sound absorption coefficient is about 0·9 at this frequency.

In Chapter 11 resonance formulae for double walls will be given (see equations (11.6a–c)). A layer vibrating at its resonant frequency is acting as a sound absorbing surface at that frequency, and by placing a sound absorbent material in the cavity the frequency range of this absorption by resonance can be extended. For a homogeneous panel, fixed securely but separated by an airspace from a second panel, the lowest natural frequency is

$$f = 0·103\,\frac{t}{b^2}c,\tag{4.20}$$

where t is the panel thickness (mm), b is the panel width (mm) and c is the velocity of sound.

The basic theory of a Helmholtz resonator will be referred to briefly in Chapter 9 (see equation (9.7)). It can be shown that the resonance frequency is given by

$$f = \frac{c}{2\pi}\sqrt{\frac{A}{Vt(t+2\Delta t)}}\tag{4.21a}$$

for a panel of area A, air volume V and panel thickness t; $2\Delta t$ is a correction factor (see Fig. 4.40a). Figure 4.40b shows a typical perforated resonator for which the resonant frequency is

$$f = \frac{c}{2\pi}\sqrt{\frac{A_2}{A_1(t+2\Delta t)b}\frac{1}{}},\tag{4.21b}$$

where $2\Delta t = \pi r/2$ for a perforation of radius r; the other symbols are given on Fig. 4.40b. Insertion of an absorbent material in the airspace broadens the frequency response of the absorption curve. Notice the marked effect displayed in Fig. 4.40d of changing the percentage open area (i.e. $(A_2/A_1)\times 100$).

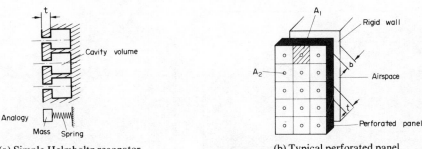

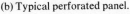

(a) Simple Helmholtz resonator.   (b) Typical perforated panel.

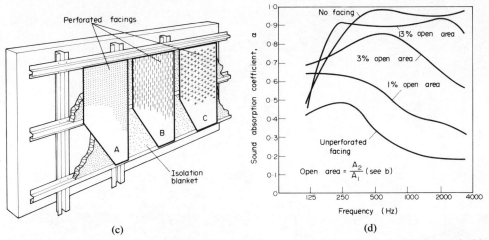

(c)   (d)

(c) Typical installation of a perforated panel resonator using various types of perforated facings and with an isolation blanket in the airspace: A, perforated board; B, slotted hardboard; C, perforated metal or plastic.

(d) Sound absorption of perforated panel resonators with isolation blanket in the airspace. The open area (sound transparency) of the perforated facing has a considerable effect on the absorption.

FIG. 4.40. Perforated panel resonators (Doelle, 1972).

4.6.2. APPLICATIONS OF SOUND ABSORPTION IN PRACTICE

Prefabricated absorption tiles in the form of perforated, unperforated, fissured or lay-in panels are available. They may be fixed to the surface using an adhesive, screws or nails, or may form part of a suspended ceiling system. The method of mounting can alter the absorption significantly (see Fig. 4.41a). Care must be taken in re-decorating, not only painting itself, but the method of painting can decrease the effectiveness of the absorbent as shown in Fig. 4.41b.

Sound absorption and aesthetics can march hand-in-hand quite happily. Architects often favour wood-slatted acoustical treatment for use as slit-resonator absorbers. Doelle (1972) shows some very effective examples of this in one of the lecture halls at the Université Laval in Quebec and in other Canadian buildings. Concrete blocks have also been used successfully as slit-resonator absorbers. Doelle (1972) recommends that the open area between the wood or concrete elements should be at least 35%.

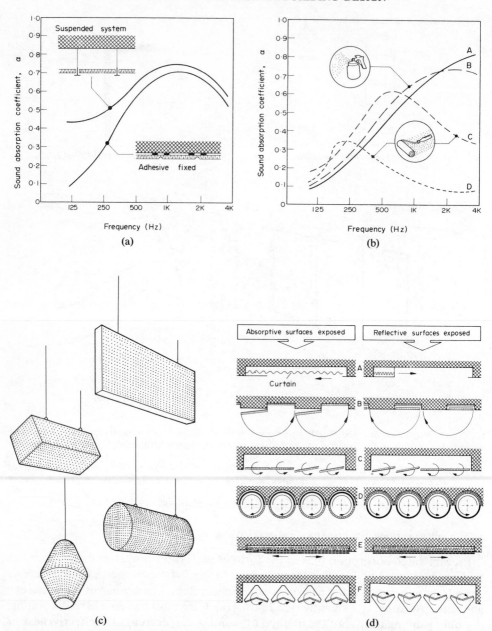

(a) Average absorption of typical acoustical tiles applied with adhesive or in suspension system.

(b) Effect of paint on porous prefabricated acoustical units: A, untreated surface; B, one coat of paint applied with spray-gun; C, one coat of paint applied with brush; D, two coats of paint applied with brush.

(c) Space absorbers can be suspended as individual units from the ceiling. They are used when the area of the room surfaces is not adequate for conventional acoustical treatment.

(d) Schematic illustration of variable absorbers which provide means for altering the absorption and thus the reverberation time: A, retractable curtain; B, hinged panels; C, rotatable panels; D, rotatable cylinders; E, sliding perforated panels; F, rotatable triangular elements.

FIG. 4.41. Application of sound absorption methods (Doelle, 1972).

Sometimes the surfaces and people in a space do not provide sufficient sound absorption. *Functional absorbers* can be suspended to provide the extra absorption needed (see Fig. 4.41c). They can be made visually interesting.

Another requirement in multi-purpose halls is to have a variable reverberation time without going to the expense of installing electronic assisted resonance or, alternatively, ambiophonic systems. Figure 4.41d illustrates some ways of doing this using adjustable constructions.

4.6.3. ATTENUATORS FOR AIRFLOW SYSTEMS

(a) Absorption Attenuators

As the name suggests, absorption attenuators use absorption materials whether lining the inside surface of the ducts or in purpose-made attenuator units.

Wells (1958) determined the sound reduction of a lined plenum chamber (see Fig. 4.42) as

$$R = -10 \lg \left[A \left\{ \frac{H}{2\pi d^3} + \frac{(1 - \bar{\alpha})}{S\bar{\alpha}} \right\} \right] \text{dB}, \tag{4.22}$$

where A is the area of the inlet, S is the total area lined with material having an average random incidence absorption coefficient $\bar{\alpha}$ and the other symbols are as described on Fig. 4.42. Plenum chambers should be designed so that each sound pathway has at least four reflections.

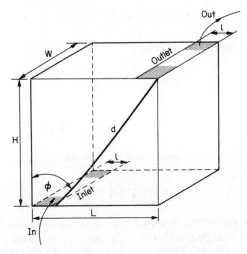

FIG. 4.42. Sound plenum attenuation unit (Wells, 1958).

A typical performance pattern for a splitter-type duct attenuator is shown in Fig. 4.43 from measurements made by Leskov (1970); the splitters were made of fibreglass having a density of 15 kg/m³, the fibres being 1–3 μ diameter and a flow resistance in the order of 10^4 rayls.

In the absence of specific acoustic performance data for a lined duct the following formula due to Sabine gives a reasonable estimate of the sound reduction R for a

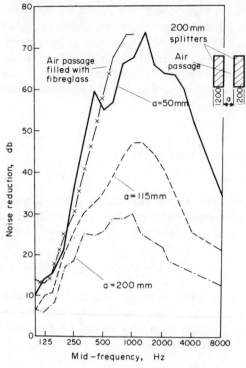

FIG. 4.43. Typical splitter-type duct attenuator (Leskov, 1970).

lined section under about 3 m in length:

$$R = \frac{P}{A} L \alpha^{1 \cdot 4} \text{(dB)},\qquad(4.23a)$$

where P is the perimeter of lined duct (m), A is the free cross-sectional area of duct (m^2), L is the length of lined duct (m) and α is the random incidence sound absorption coefficient.

German acousticians prefer generally to use the formula attributed to Cremer stated by Eck (1973) as

$$R = 1 \cdot 1 \frac{P}{A} L \left(\alpha + \frac{\alpha^2}{2} \right).\qquad(4.23b)$$

The advantage of designing the attenuator to have several narrow airways is evidently to increase the P/A ratio.

The thickness of the sound absorbent layer is governed by the lowest frequency to be absorbed. If a limited length of attenuating path is available for a specified sound absorption, the sound absorption coefficient should be as close as possible to one at the lowest frequency to be absorbed. The layer thickness as shown above must be approximately equal to $\lambda/4$. If a greater length is available for the attenuating path the layer thickness may also be thinner. This gives a smaller sound absorption coefficient for the lowest frequency to be absorbed. To achieve the same absorption as before the absorption path must then be longer.

Eck (1973) concludes that the sound absorbent layer thickness for obtaining the wall impedance must lie between $\lambda/4$ and $\lambda/5$ to suit the channel width. The layer must consist of very loose sound-absorbent substances, and for eliminating the longitudinal spread of sound energy transverse walls must be arranged on the inside of the layer at a distance of a few centimetres only. In this way we arrive practically at $\lambda/4$ long duct sections filled with absorbent materials which can be connected up laterally to the ducts to be attenuated.

By varying the layer thicknesses, by using a somewhat closer packing than called for in theory, the frequency range of the absorption can be enlarged even if the absolute limit of absorption is considerably increased thereby. This method of absorption offers advantages mainly for absorbing narrower frequency bands below 500 Hz provided that ample space is available for layer thicknesses of $\lambda/4$.

The advantages of the absorption attenuator are as described by Eck (1973) as:

   (i) wide-band absorption;
  (ii) raising of specified frequency ranges by corresponding build-up of sound absorbent layers;
 (iii) possibility of larger cross-sectional area and smaller pressure losses;
 (iv) simple and safe assembly.

(b) Interference Attenuators

Interference attenuators rely on the classical interference phenomenon whereby large reductions in sound energy are achieved at specific frequencies by sound waves having different phases interfering with one another. For the arrangement shown in Fig. 4.44 the sound reduction index is (see Morse, 1939; Harris, 1957a)

$$R = 10\lg\left\{1+\frac{(n^2-1)^2}{4n^2}\sin 2kl\right\}\mathrm{dB},\qquad(4.24)$$

where $n = a_2/a_1$, $k = 2\pi/\lambda$ and maximum reduction will occur at $kl = 1, 3, 5\ldots$.

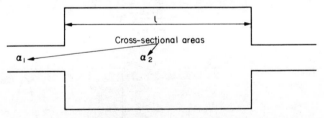

Fig. 4.44. Interference attenuator.

(c) Reflection Attenuators

Sound energy will be reflected at changes of the acoustical impedance in the system (e.g. at the end of an airflow system there is a large mismatch between the room and the duct acoustical impedance). The reduction in sound level is

$$R = 10\lg\left\{\frac{(n+1)^2}{4n}\right\},\qquad(4.25)$$

where n is the ratio of the smaller to the larger cross-sectional area. This arrangement is in effect a low-pass filter. The reflection attenuator is mainly suitable for absorbing low frequencies.

(d) Resonance Attenuators

This is similar to an absorption attenuator, but in this case the absorbent layer takes the form of a perforator or slotted resonator (as was shown, for example, in Fig. 4.40b).

A series of Helmholtz resonators fitted to a duct of cross-sectional area A will each provide a reduction at a frequency f (see Cremer, 1953; Beranek, 1960) given by

$$R = 10 \lg \left\{ 1 + \frac{(\alpha + \frac{1}{4})}{\alpha^2 + \beta^2 (f/f_0 - f_0/f)^2} \right\} \text{ dB},\qquad(4.26a)$$

where α is the resonator resistance, β is the resonator reactance and f_0 is the resonant frequency.

Two conditions may be noted which simplify equation (4.26a):

(i) At $f = f_0$,

$$R = 20 \lg \left\{ \frac{(\alpha + \frac{1}{2})}{\alpha} \right\}\qquad(4.26b)$$

and

(ii) for $\alpha \ll \frac{1}{4}$, $f_0 \gg f \gg f_0$,

$$R = 10 \lg \left\{ 1 + \frac{1}{4\beta^2 (f/f_0 - f_0/f)^2} \right\}^2.\qquad(4.26c)$$

Beranek (1960) shows two sets of curves (see Fig. 4.45a and b) relating R and f/f_0 for the conditions $\alpha = \beta$ and $\alpha = \beta/2$.

A procedure for sizing a resonator system is proposed by Beranek as follows:

(a) for a known R at a given frequency select values of α, β and f_0 from Figs. 4.45a and b; alternatively, use equations (4.26a) and (4.26c);

(b) knowing the duct area A and f_0, determine the resonator volume

$$V = \frac{Ac}{2\pi f_0 \beta} \text{ (m}^3)\qquad(4.27)$$

$(c = 342 \text{ m/s});$

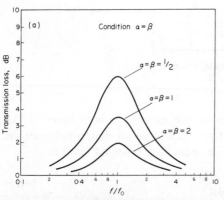
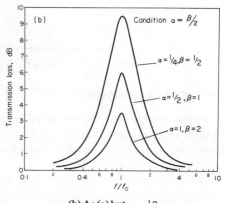

(a) Sound transmission loss of side branch attenuator, with resistance parameter α equal to reactance parameter β.

(b) As (a) but $\alpha = \frac{1}{2}\beta$.

FIG. 4.45. Side brand sound attenuator (Beranek, 1971).

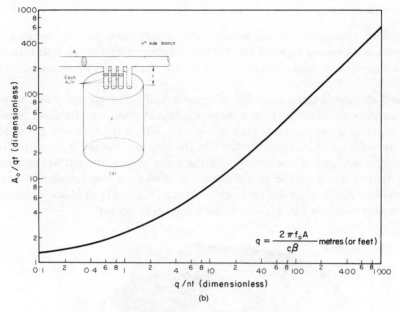

(a) Geometry of side branch resonator.
(b) Design chart for resonator aperture area A_0.

FIG. 4.46. Sound resonator design (Beranek, 1971).

(c) calculate a length parameter q, where

$$q = \frac{2\pi f_0 A}{c\beta} \; (m);$$

(4.28)

(d) choose a convenient aperture length t and number of apertures n (see Fig. 4.46a), calculate the parameter q/nt and find from Fig. 4.46b a value of the aperture area A_0;
(e) select a screen for the aperture having a flow resistance

$$R = \frac{A_0}{A} \alpha \rho c \; (rayls).$$

(4.29a)

Note: at a temperature of 20°C and normal pressure

$$R = \frac{408 A_0}{A} \alpha.$$

(4.29b)

(e) Relaxation Attenuators

This is a series arrangement of membrane absorbers (see Table 4.18) consisting of a number of cells with membranes having varying surface masses; absorbent material may be included in the cells. Such an arrangement offers great flexibility in achieving a narrow or wide bandwidth frequency attenuation spectrum. The word relaxation is misleading and only refers to the basic theory which uses the decay of reverberant sound energy within the cell to obtain the degree of sound absorption.

(f) General Comments

Absorption, resonance or relaxation type attenuators can be specified from manufacturers of sound control equipment, whereas methods of interference and reflection attenuation may be incorporated into the design of an airconditioning system.

Standard attenuators are available in circular and rectangular forms and may be installed at all or some of the following points in a system—the fresh-air intake; the exhaust outlet (in these positions they will need to filter any prominent external noise sources as well as those contributed by the airconditioning system); after the plant-room, at the end of the supply and the room end of the extract system. It is essential to pay heed to the manufacturer's advice on the manner of installing attenuators in order to gain the best performance (Fig. 4.47). It is not uncommon to find attenuators installed in a system the wrong way round!

FIG. 4.47. Cylindrical attenuators. These attenuators are for use in air-flow applications and especially for both "on fan" and "in duct" applications. Two types are available, a *straight through* design and one which is a *high-performance* version. Both are supplied in a wide range of sizes and flanged for direct attachment to most axial flow fans; they are equally suitable for centrifugal fan inlet connection (with acknowledgements to Sound Attenuators Ltd.).

It is worth bearing in mind the obvious point that sound travels with and against the direction of airflow and consequently in some situations attenuators may be required on the inlet as well as the discharge side of the fan.

Miscellaneous requirements such as resistance to airflow, fire, moisture, mould, bacteria, vermin and insects, non-erosion of fibrous material at high air velocities and the use of odour-free materials can easily be met by the range of attenuators available today.

Attenuation is rated in terms of measured sound power levels.

Dynamic insertion loss (*IL*) ratings of attenuators are established by taking sets of readings before and after insertion of the attenuator in a duct system with air flowing through it. This is the preferred method of rating.

Noise reduction (*NR*) ratings are measurements of the difference in sound levels inside the duct at points just before and just after the attenuator, made without air flowing. It has been found that when installed in an actual system with airflow the actual attenuation is about 3 dB less than the NR rating.

The attenuation required should be specified for each octave band and a particular note made of any discrete frequency peaks that are predicted to occur in any frequency band. The self-noise (due to airflow) characteristic for the attenuator should also be requested from the manufacturer.

The effectiveness of an attenuator can be limited if noise short-circuits the attenuator or the sound insulation downstream is insufficient to prevent the entry of external noise. Discontinuities between the connecting duct and the attenuator and use of damped attenuator casings are examples of the corrective treatment which may be necessary. For ducts passing through a noisy area the sound power level transmitted into the duct L_w may be estimated from

$$L_w = L_{w,0} - R + 10 \lg \left(\frac{A}{4} + \frac{AS_w}{F} \right),\qquad(4.30)$$

where $L_{w,0}$ is the sound pressure level outside the duct, R is the sound reduction index of duct wall (Table 4.19), A is the duct cross-sectional area (m^2), S_w is the area of duct wall exposed to noise (m^2), α is the absorption coefficient for internal duct surface and

$$F = \frac{(2A + \propto S_w)(2A + S_w)}{(2A + \propto S_w) + (2A + S_w)}.$$

The converse situation of noise-break-out through duct walls may be studied by considering the duct surface passing through the room S as a partition having a sound reduction index R. Hence the sound pressure level in the duct will be

$$L_p = L_w - 10 \lg A,\qquad(4.31a)$$

where L_w is sound power level in the duct of cross-sectional area A.

The insulation of the duct wall will be

$$L_p - L_{p_r} = R - 10 \lg \left(\frac{S_w}{S} \right)\qquad(4.31b)$$

where L_{p_r} is the sound pressure level outside the duct and S is the room absorption. Solving these equations for L_{p_r} a comparison can be made with the design sound pressure level.

Cross-talk has already been referred to in section 2.6.1. Figure 4.48 shows a cross-talk attenuator; notice that noise reductions of 30–60 dB are possible over the speech frequency range of 500–4000 Hz.

TABLE   4.19. SOUND   REDUCTION   INDICES   FOR   DUCT   WALLS
(IHVE GUIDE, 1970)

Mass of wall per unit area (kg/m²)	Sound reduction index R (dB)						
	Octave band centre frequency (Hz)						
	125	250	500	1000	2000	4000	5000
5	6	10	13	17	22	33	38
10	10	17	22	28	33	38	40
15	13	20	25	32	36	40	40
20	17	22	28	33	38	40	40

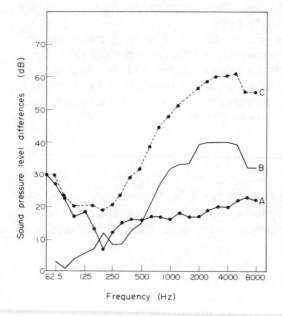

FIG. 4.48. *Cross-talk attenuator.* The graph shows a copy of actual laboratory results on the unit shown. Curve *A* shows the difference between the two reverberant chambers without the attenuator but with the grille alone in the orifice between the two; curve *B* gives the insertion loss of the unit, while curve *C* is indicative of the noise reduction including the grille end reflection (with acknowledgement to Sound Attenuators Ltd.).

Special care is required in controlling the noise at the mixing boxes of dual-duct airconditioning systems. A typical dual-duct unit is shown in Fig. 4.49.

4.7. Control of Structureborne Sound

This section should be read in conjunction with section 9.6.

All mechanical equipment used in airconditioning systems, particularly fans, pumps and compressors, vibrates to some degree. Vibration problems have

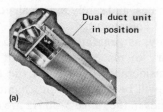

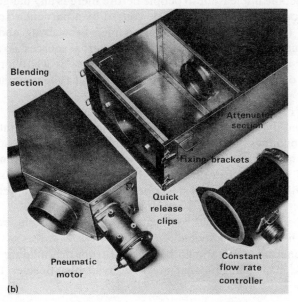

FIG. 4.49. Dual duct unit (with acknowledgements to Sound Attenuators Ltd.). (a) A dual duct unit installed above a false ceiling. (b) The blending section and constant flow-rate controller removed from the unit.

increased in recent years due to:

* increasing use of lightweight building structures;
* installation of mechanical equipment on intermediate and top floors, since lower floors and basements often have greater commercial value;
* adoption of high-pressure, high-velocity air distribution systems;
* construction of larger buildings and more centralization of plant requires use of larger equipment which usually generates greater vibratory forces;
* use of better soundproofing techniques for buildings reduces the effects of outside noise but makes the occupants more aware of internal sources.

Mechanical equipment to be isolated is generally of either the reciprocating or centrifugal type. It is impossible during manufacture to remove all the out-of-balance forces associated with reciprocating machines, and rarely practical, economically, to balance all those produced by centrifugal equipment.

To isolate a disturbing forced frequency, as generated by a piece of equipment, from a building structure, it is necessary to place the equipment on a resilient material having a natural frequency considerably lower than the disturbing frequency. The theory and practice of this will be discussed in Chapter 9.

The data given in Fig. 4.50 for frequencies in the range 5–50 Hz are taken from the *Building Research Station Digest*, **117** (1970) and is the most reliable information at present available for the estimation of structural damage from vibration. In general, vibration must become unpleasant to people before there is any danger of damage to the building, although minor cracks can be caused by a particular source operating for a short time (see section 9.7). Door slamming or walking about heavily are reckoned to produce greater vibration than most external sources, so attention to the selection of door-closing mechanisms and floor finishes is well worth while.

The reduction of structureborne and airborne sound follow similar patterns:

(a) Reduction of vibration at source: selection of equipment that has been statically and dynamically balanced, the use of high mass plinths or vibration isolation mountings, flexible joints on the inlet and discharge sides of fans, and other equipment are all measures that should be taken to minimise and contain the level of vibration. Consideration should also be given to using the floating floor or ceiling constructions described earlier in this chapter for the plant-room.

(b) Sources of vibration should be located as far as possible from people and equipment likely to be affected.

(c) Pipework and ductwork should be isolated from the building structure using the purpose-made hangers that are now available. Besides this, the blade-

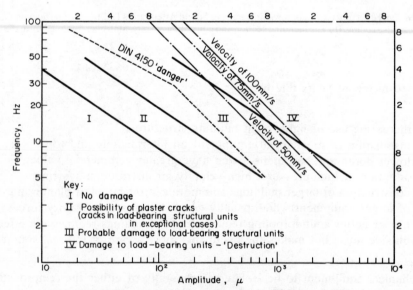

FIG. 4.50. Effect of low frequency vibration on building structure.

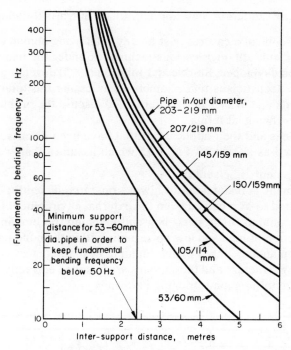

FIG. 4.51. Pipework inter-support distances to avoid bending frequencies (Müller, 1966).

passage frequency of the pump can excite the mains into their bending frequency modes, and Fig. 4.51 shows the fundamental bending frequency as a function of the support inter-distance for various pipe sizes. It is desirable that these support frequencies are chosen so that no fundamental exists over 50 Hz. Joints arising from pipes or ducts passing through structural elements need to be made soundtight by using suitable caulking materials such as fibreglass, neoprene or foam materials. Even if precautions have been made to minimise vibration transmission from the plant-room, in-line attenuators are often required at the end of the distribution system. Types suitable for airflow have already been discussed; Fig. 3.21 shows an attenuator suitable for use in water systems. Drumming of duct walls should be avoided by using a gauge of metal and bracing sufficient to stiffen the walls. It is not easy to predict ductwall resonances, but vibration damping materials are now available which may be sprayed on to the sheet metal to curb their effect (see section 9.6.2).

(d) Within the occupied space the use of floating floors with soft finishes to reduce impact noise has already been discussed; besides footsteps and slamming doors, water cisterns and lift motors are other common sources of airborne and structureborne sound encountered in building design.

4.8. Some Examples of Noise and Vibration Control in Building Design

Environmental control measures cost money. Balancing human and social values against cost is difficult although various techniques under the name of cost-benefit analysis have been developed. Starkie and Johnson (1975) develop noise cost models which will assist acousticians and planners in assessing the total impact of their decisions. Richards and Waller in Crocker (1972) discuss the problem of evaluating loss of amenity in financial terms.

Many of the ideas and their applications that have been discussed in this chapter can be summarised as shown on Fig. 4.52 which assumes that:

* good equipment is selected;
* the equipment is maintained to always operate efficiently;
* noise emitted to or received from the external environment to the plant-room will be accounted for in designing and planning the building;
* the interaction of high and low noise areas within the building is considered at the initial planning stages of the building design;
* the building structure and the services are installed correctly with no airborne or structureborne sound bridging pathways.

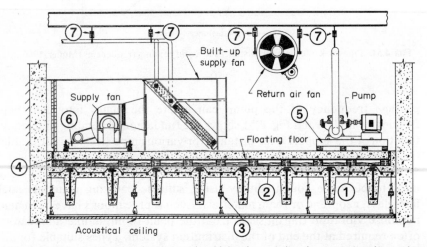

FIG. 4.52. A cut-away drawing showing the way in which noise and vibration control equipment is typically installed in a plant-room. For those critical applications requiring complete freedom from transmitted vibration and sound, a three-step approach is recommended. This consists of a floating floor, high deflection spring isolators on all equipment, and a resiliently suspended ceiling of a dense impervious material. The system components are: 1, isolation panel consisting of fibreglass isolation pads bonded to exterior grade plywood and supporting the floating floor; 2, low-density acoustic fibreglass blanket; 3, fibreglass ceiling isolation hangers; 4, fibreglass perimeter isolation board with mastic seal; 5, fibreglass-isolated inertia base for supporting pumps; 6, spring-isolated inertia base for supporting fans; 7, spring and fibreglass hangers for isolating fans and piping (with acknowledgements to Sound Attenuators Ltd.).

4.8.1. CASE STUDY 1: SNAPE CONCERT HALL, ALDEBURGH

Aldeburgh is world-famous for its annual Music Festival under the direction of the distinguished composer Benjamin Britten.[†] Snape Concert Hall (Fig. 4.53) was designed for an NR level of 25 when the ventilation system was delivering

†Britten died on 4 December 1976.

(a)

(b)

FIG. 4.53. (a) Snape Concert Hall, Aldeburgh, and (b) the interior of the Hall where a level of NC 25 was achieved using a Sound Attenuator Ltd. type PG 854 attenuator. The air distribution outlets are seen on the left of the picture.

17000 ft^3/min at low speed; under boost conditions this level rises to NR 35, the supply air volume flow rate being doubled. A double-inlet double-width centrifugal fan is used to distribute the air. Air distribution in the hall is via seven large feature grilles, each 7 ft high by 1 ft wide, acoustically a very demanding layout, for grilles of this size have a very high directivity factor and, in addition, the grille closest to the fan is only 3 ft from a seated listener.

A full acoustic analysis was carried out on both paths of airborne and structureborne sound transmission. Calculations revealed that the level in the concert hall if the fan were not attenuated would be as follows:

Frequency (Hz)	125	250	500	1000	2000	4000
Sound pressure level (dB)	64	60	59	53	44	35
Sound pressure level at NR 25 for comparison (dB)	45	38	31	27	25	22

An attenuator having nominal dimensions of 96 in. long by 64 in. wide by 84 in. high was recommended to provide the following static insertion loss:

Frequency (Hz)	125	250	500	1000	2000	4000
Loss (dB)	20	37	50	50	50	50

Tests after installation gave the following results:

(a) High speed (NR 35 design criterion)						
Measured level (dB)	43	40	39	36	30	21
Sound pressure level of NR 35 (dB)	53	46	40	37	35	33

(b) Low speed (NR 25 design criterion)						
Measured level (dB)	37	36	29	25	23	19
Sound pressure level of NR 25 (dB)	45	38	31	27	25	22

The data for this project was provided by Sound Attenuators Limited.

4.8.2. CASE STUDY 2: SOUNDPROOF BOOTHS

In a factory it would seem obvious that reasonably quiet working conditions are essential for such shopfloor employees as the works supervisory staff handling production and personnel administration. It is evident, however, that many factories, with their noisy machinery and thin timber works offices, afford very little aural protection for these people. It is perhaps sobering to wonder how many industrial disputes could be avoided if only the shopfloor negotiators could clearly hear each other speaking without having to shout to be heard; or even how production would benefit if the foreman's concentration was not disturbed by the noise of the factory equipment. An easy and economic answer to the problem of providing such acoustically controlled conditions is to use a soundproof office. The office shown in

FIG. 4.54. Soundproof office (with acknowledgements to Sound Attenuators Ltd.).

Fig. 4.54 can be erected on site from a standard range of prefabricated acoustic panels made of galvanised steel with a sound-absorbent filling. The panels snap together quickly and securely with a simple locking device, giving an unbroken internal surface. The whole structure can be quickly taken down, if necessary, and relocated. A fan provides the necessary ventilation. Lighting systems are designed in consultation with the client to allow for density of occupation, nature of work, siting of work stations, and so on.

Although intended primarily for works supervisory personnel, its flexibility of design makes the office suitable for several industrial applications. For example, if it should prove necessary to locate an area for delicate assembly work near to a machine area, whose noise could seriously threaten concentration, it would be a simple matter to erect a soundproof workroom to house and protect the assembly workers. Furthermore, the problem of providing rest rooms, medical rooms and control rooms can now be easily overcome.

Tests carried out to determine acoustic performance have shown that an average sound reduction index of 28–30 dB can be expected, depending on the amount of glazing involved. Steel doors are offered as standard, with timber as an alternative.

4.8.3. CASE STUDY 3: DESIGNING AGAINST SONIC BOOM

The sound insulation recommendations described in the Building Regulations and shown in Fig. 4.29 do not cover every eventuality. The problem of assessing the amount of roof insulation required in broadcasting studios, situated in areas where supersonic as well as subsonic aircraft fly overhead, has been studied by Burd *et al.* (1968). The problem arises because the background noise levels permitted in studios are very low, lying in the range 15–25 dB approximately at 1000 Hz (Fig. 4.55), whereas aircraft noise may measure 80–120 dB at 1000 Hz, hence building sound insulation of 55–105 dB would be necessary at this frequency, well in excess of the values shown for party-wall grade in Fig. 4.29a.

A sonic boom is the audible feature of a shock wave generated by an aircraft flying at supersonic speeds. The pressure distribution can be represented by an N-wave as shown in Fig. 4.56a, which resembles an actual boom shock-wave as can be seen by comparing Fig. 4.56a and b. The shape of this wave changes with altitude as the pressure jump at the leading edges of the wings is added to the overpressures following the low shock-wave, eventually catching up with the low shock-wave near the ground; the rarefaction at the tail lags behind this high pressure. At any instant the leading and trailing shock waves lie approximately along a pair of cones each intersecting the ground area in hyperbolae as shown in Fig. 4.56c. These hyperbolae sweep across the country marking out an area where buildings and people will feel the effects of the boom. It should be noted that a theoretical analysis of shock-waves would require an understanding of non-linear acoustics, an aspect referred to in Chapters 7 and 8 but not dealt with in this book. Overpressure amplitude spectra for

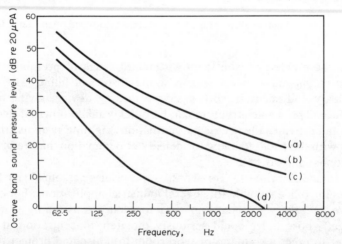

FIG. 4.55. Permissible background noise in studios from all sources for (a) sound studios for light entertainment, (b) sound studios (except drama) and all television studios, (c) sound drama studios, and (d) threshold of hearing for continuous spectrum noise (Burd *et al.*, 1968).

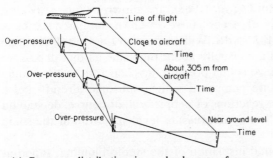

(a) Pressure distribution in a shock wave from a supersonic aircraft.

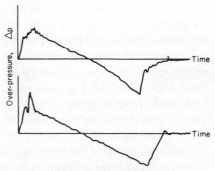

(b) Observed pressure waves of two sonic booms.

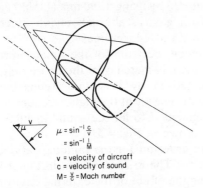

(c) Aircraft shock waves intersect ground in hyperbolae which sweep over the ground.

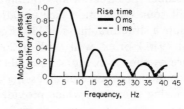

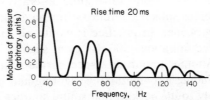

(d) Amplitude spectra of N-waves of duration 120 ms.

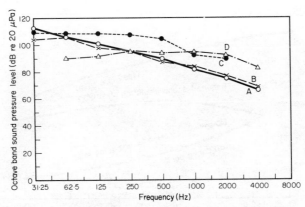

(e) Octave band sound pressure level spectra for A and B, two actual booms of 95.8 Pa (2 lb/ft²) peak overpressure; C, theoretical values for a 95.8 Pa (2 lb/ft²) N-wave-duration 120 ms—rise time 1 ms; D, Comet aircraft, 110 PNdB, measured.

FIG. 4.56. Characteristics of sonic booms (Burd et al., 1968).

typical *N*-waves, as measured by Burd *et al.* (1968), are shown in Fig. 4.56d. An overpressure of 95·8 Pa on the basis of American research in the early sixties is equivalent to a sound level of about 102 PNdB; Wilhemsen and Larsson (1973) state that a nominal overpressure of 100 Pa is usually about the same as that due to jet engine noise at a level of 100 dBA. Two sound pressure level spectra for two sonic booms are given in Fig. 4.56e and are compared with a subsonic aircraft and a simulated *N*-wave boom. Annoyance reactions of people will, of course, depend on other factors than the amplitude of the sound pressure level spectrum of the sonic boom (see Fig. 2.44).

In order to assess the required sound insulation of the studio building, superimpose a studio sound criterion, which is likely in this case to be chosen by the client, from Fig. 4.55, and a sonic boom spectrum from Fig. 4.56e. The difference between these curves gives the sound reduction index required as shown in Fig. 4.57.

Remember the effect of absorption in the studio can alter the effectiveness of the resultant sound insulation (see equation (11.24)). To achieve this degree of sound insulation, a structural mass of 1300 kg/m^2 is required by using a 290 mm (11½ in.) and 228 mm (9 in.) brick wall with a 200 mm (8 in.) cavity; alternatively, a kinetic structure as shown in Fig. 4.36 could be used with a reduced mass.

Notice the importance of assessing the frequency spectrum for the sound source and choosing structures which provide an acoustic mis-match over the entire range. The noise from supersonic aircraft is characterised by very low frequencies, whereas subsonic aircraft often have more acoustic energy at the higher frequencies (Fig. 4.56e). In practice it would not be recommended to let the sound insulation curve fall above 500 Hz, even slightly as it is shown in Fig. 4.57.

Another important aspect is the design noise criterion for the building services including airconditioning, lighting and lift systems. The criteria shown in Fig. 4.55 apply to *all* background sound sources including the effect of the external environment. It is wise to set the building services noise criteria at least 5 dBA lower than the criterion chosen from Fig. 4.55.

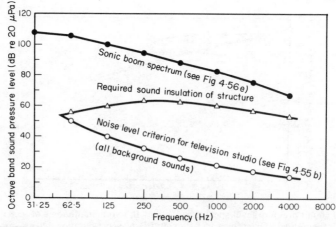

FIG. 4.57. Required sound insulation of television studio building subjected to sonic boom; notice that the building services systems should be designed to a noise level criterion at least 5 dB below the criterion shown here for all background sources (based on data of Burd *et al.*, 1968).

This case study shows that as new problems arise that affect the designing and planning of buildings—in this selected example sonic boom, it is necessary to consult authoritative research and to make sound surveys at the building site in question.

To conclude this case study some conclusions reached by Wilhelmsen and Larsson (1973) are quoted from their report which are applicable to buildings having a similar construction to those used in their tests, i.e. a prefabricated timber-frame building. Most of the tests were performed by means of supersonic overflights at altitudes varying between 100 and 13 000 m which generated booms with overpressures of up to 1740 Pa.

Damage to wallpaper may be expected at the angle between the external and internal wall at overpressures in excess of 400 Pa, over joints between sections at overpressures in excess of 1000 Pa and over joints between fibre boards nailed to a continuous backing only at overpressures higher than those recorded here.

Cracks can occur in the jointing compound between tiles on a timber backing at overpressures higher than 400 Pa and cracks in the paint film at the angle of the ceiling at overpressures higher than 300 Pa.

In order that cracks of the above types should occur as a result of the small movements which take place, the surface must be brittle. The cracks will therefore be hair-cracks and very difficult to detect with the naked eye.

The movements in building components due to some everyday stresses were also measured. When a door was shut the movements recorded 50 cm from the door were of the same order as those caused by a sonic boom with an overpressure of 500 Pa.

It was only when the overpressure was as much as 1680 Pa that visible damage was detected in the course of the visual inspections. This consisted of extension of existing cracks in a window-pane. A casement nailed to the external wall from the outside was also loosened.

A comparison between the design wind pressures applicable in Sweden and the effect of sonic booms on buildings shows that booms generated by the J 35 Draken flying at an altitude of 3000 m correspond to the lowest design wind pressure (500 Pa) and those due to flights at 1000 m to the highest design wind pressure (1500 Pa) according to Swedish Constructional Standards SRN 67.

It should be noted that the lowest permitted altitude for military supersonic flights over land in Sweden is 10 000 m, and over the sea 5000 m.

4.8.4. A PICTURE GALLERY OF SOUND CONTROL IN ACTION

(With acknowledgement to Industrial Acoustics Co. Ltd.).

FIG. 4.58. Duct attenuators on the intake of forced-draft boiler fans installed at St. Crispin's Hospital, Northamptonshire (with acknowledgements to Industrial Acoustics Co. Ltd.).

FIG. 4.59. A view of attenuator modules being installed on a twin-fan cooling tower (with acknowledgements to Industrial Acoustics Co. Ltd.).

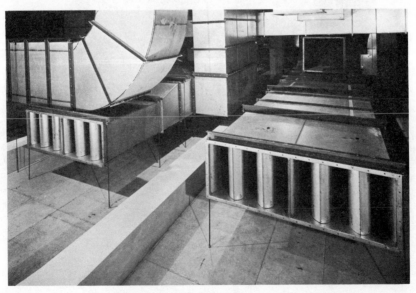

FIG. 4.60. Duct attenuators installed in a mechanical services system of the large Wills Hartcliffe factory by Industrial Acoustics Co. Ltd.

FIG. 4.61. Modula fan plenum formed from 100-mm-thick Noishield panels at the Halifax Building Society headquarters: these plenums operate at a 9 in. (220 mm) s.w.g. pressure and measure 56 ft (17 m) long by 29 ft (8·8 m) wide by 10 ft (3 m) high (with acknowledgements to Industrial Acoustics Co. Ltd.).

(a) Acoustic housing for standby diesel generator set showing the Noishield panel walls, double-leaf shield door and power flow cooling air attenuator. The equipment housed is a Dorman 8QTCA-V8 engine coupled to a McFarlane alternator; standby rating 44 kW. Sound control equipment by Industrial Acoustics Co. Ltd.

(b) Inside view of diesel generator housing.

FIG. 4.62. Diesel generator housing.

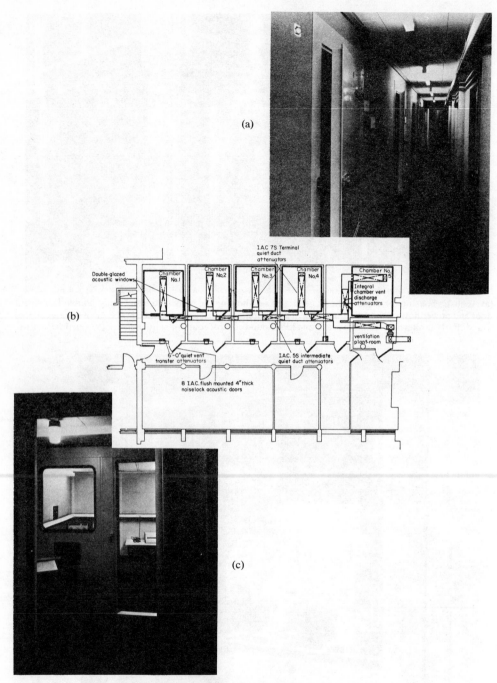

(a) View down the corridor between audiometric rooms.
(b) Plan of a modular acoustic audiometric facility designed, manufactured and installed for the Royal National Throat, Nose and Ear Hospital by the Industrial Acoustics Co. Ltd.
(c) View inside an audiometric room.

FIG. 4.63. Audiometric rooms (with acknowledgement to Industrial Acoustics Co. Ltd.).

CHAPTER 5

SOME ACOUSTICAL DESIGN
TECHNIQUES FOR BUILDINGS

5.1. Acoustical Design Briefs and Specifications for Buildings

The brief and specification for a building design must include a comprehensive summary of the acoustical criteria. In order to make sure that the criteria have been selected wisely and can be achieved in practice, an integrated design team working from the design inception to building commissioning is essential. This is obvious when one considers that the required internal noise climate of a building depends on what the work activity is, and this will be specified by the client; the external noise climate will depend upon the site chosen by the client; the choice of building structure will dictate the relationships between the inside and outside noise climate, and hence the architect and building environmental engineer, together with the structural engineer, must evolve a joint decision on this matter; the orientation of the building affects the noise exposure of occupied spaces; the internal planning of a building must also take into account the noise critical areas and place them far away, or shield them, from external and internal noise sources; various degrees of privacy within the building will necessitate selecting internal wall partitions, ceilings and floor with this acoustical requirement borne in mind; design of services systems have to meet noise level requirements; the choice of surface finishes will affect the sound distribution throughout the space; in the special case of a concert hall, a theatre or an opera house, the shape of the space must be taken into account when designing the sound distribution. Thus planning, layout, materials, shape, choice and design of systems, surface finishes are keywords in acoustic design. The problem is not made easier by other factors:

* cost constraints;
* successful acoustic design depends upon the quality of building materials and of building construction;
* the acoustical requirements interact with other needs—for instance the choice of building structure also depends upon the thermal insulation, the solar protection, the day lighting, the natural ventilation requirements and the structural requirements; the choice of surface finishes also depends upon aesthetics, cleaning costs and durability; the choice of the space shape depends upon the social environment and also governs the quality of the air movement distribution, besides acoustic considerations;
* the external noise climate changes with time, usually increasing from year to year.

363

These problems cannot be overcome if good communication between the client, the architect, the building environmental engineer, the structural engineer and, at a later stage, the builder, does not occur. Table 5.1 shows a list of the factors which must be accounted for in a building brief and specification in which the noise environment is a first-order priority factor.

5.2. Assessment of the External Noise Environment

A visit to the site where the building is to be erected is essential to measure the existing nature and level of noise and to see the relationship of the building being designed to others that are existing. Future plans for the area which may alter the noise climate need to be discussed with the local planning authority, e.g. new buildings, industrial development, new roadways, new airports, increased road, rail or air traffic. Measurements should be recorded in dBA; dBD scales are provided on some sound level meters for measuring aircraft noise in PNdB. *Building Research Establishment Digest*, **153**, May 1973, entitled *Motorway Noise and Dwellings*, makes recommendations on the measurement of L_{10} and L_{90} levels as described in the following two paragraphs:

To measure L_{10} accurately involves a statistical analysis of the fluctuating traffic noise levels and requires elaborate equipment. With the microphone one metre from the external façade of a dwelling the noise levels are recorded continuously for a period of several minutes at precisely one-hour intervals throughout the 18-hour period from 6 a.m. to midnight of a typical week-day, preferably Tuesday to Thursday. The hourly sampling period should be long enough to include the passage of at least 50 vehicles and preferably 100 or more. On busy traffic routes of 3000 or more vehicles per hour, this means recording for about two minutes in every hour; longer sampling periods are needed for roads with sparse traffic. A period should be chosen when the wind is adverse (from road to microphone) and the average wind component about 10 km/h. Having derived individual L_{10} values from the results for each hourly period, the arithmetic mean of the 18 values is found. The result may be designated L_{10} (18 h) unless otherwise stated. The procedure described will give an accurate measure of L_{10} under the conditions prevailing at the time of the measurements.

It is not always practicable to adopt this prolonged method of measurement, and in these circumstances a simpler measuring procedure using a sound level meter may be found useful. Basically, the procedure consists of noting sound level meter readings in rapid but regular succession over a reasonable short-term period, and converting the results to L_{10} either by a simple statistical analysis or, more simply, by crossing out the top 10 per cent of the readings and taking the highest remaining level. Readings can be facilitated by a simple attachment to the sound level meter to cue the readings by a light-flash at regular intervals. Some recent trials (see B. R. E. Publication EN/49/72) indicate that by this means readings of motorway noise at four-second intervals over a period of 15 minutes give short-term values of L_{10} that are accurate within about 1 dB(A) of the results obtained by more complex methods. Moreover, on roads with heavy traffic flows, the approximate L_{10} (18 h) level can be obtained by measuring the short-term value of L_{10} during the middle of

TABLE 5.1. MODEL ACOUSTICAL SPECIFICATION FOR A BUILDING SPACE

Building orientation
Range of sound levels throughout building
Internal location
Plant-room location

Space dimensions
  Volume m^3
  Roof surface area m^2
  Wall surface areas m^2
  Floor surface area m^2

Sound levels surrounding space
  External: present L_{10}, L_{90} dBA
            future (20 years ahead) L_{10}, L_{90} dBA
  Adjacent internal areas L_{10}, L_{90} dBA

Nature of sound sources
  External: traffic L_{NP}, L_{10}, L_{90} dBA
            aircraft L_{NP}, NNL   PNdB, L_{10}, L_{90} dBA
            industrial L_{NP},   CNL, L_{10}, L_{90} dBA
            people L_{NP}, L_{10}, L_{90} dBA
            construction L_{NP}, L_{10}, L_{90} dBA
            future levels L_{NP},   PNdB, L_{10}, L_{90} dBA
  Internal: conversations: with telephone L_{10} dBA
                            without telephone L_{10} dBA
            machines L_{10} dBA
            footsteps L_{10} dBA
            plant-rooms L_{10} dBA

General information from client
  Describe level of privacy required
  Work activity in space
  Work activity in surrounding spaces (including corridor, stairways, doorways, toilets, lifts)
  Machines in spaces
  Number of people in space
  Occupancy hours
  Any future changes envisaged
  Degree of visual contact needed between employees
  Degree of aural contact needed between employees
  Preferred surface finishes
  Comment on any other notable aural features that are important

Selected criteria
  Traffic noise index (external) L_{10} dBA (18 h)
  Noise and number index (external)   NNI   dB
  Noise pollution level (external) L_{NP} dB
  Background sound level (internal) L_{90} dBA
  Peak sound level (internal) L_{10} dBA
  Sound reduction index of external structure R   dB
  Sound reduction index of internal structures R   dB
  Articulation index   0·1   to   0·8
                                            (privacy)   (good communication)
  Design noise rating for services   NR $\doteq L_{10}$ dBA $- 12$
     (i.e. assume background sound level is about 6 dB below activity noise level)
  Sound absorption coefficients of each surface   metric Sabines
  Reverberation time at: 1000 Hz T_{1000} s
                          500 Hz T_{500}
                          250 Hz T_{250}
  Time delay between direct and first reflected sound t ms
  Vibration criteria   mm deflection or
                                              m/s^2 acceleration units

TABLE 5.1 *(cont.)*

Comments on specific problems
  Services plant-room
  Services distribution system (e.g. damper noise, service voids)
  Terminal noise
  Cross-talk
  Other noise source problems

the day, but avoiding peak traffic periods and subtracting 2 dBA. Clearly L_{90} readings can also be obtained from these analyses.

A typical noise recording sample is shown in Fig. 5.1f and shows how the choice of the L_{10} level takes into account the fluctuating nature of the noise climate.
Several factors need recording when measuring external sound levels:

(a) weather conditions with particular reference to wind speed and air temperature and whether it is wet or dry, clear or foggy;
(b) ground conditions, i.e. the surface material and whether it is dry or wet;
(c) variation of L_{10} and L_{90} sound levels with distance and where possible with height.

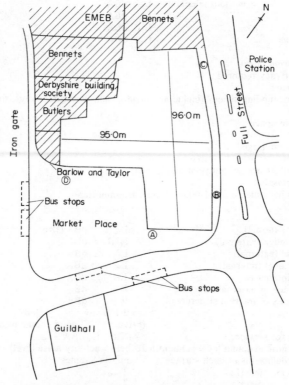

(a) Site plan showing points used for acoustic survey in Derby.

FIG. 5.1. A typical site investigation of the acoustic climate; for (b), (c), (d), and (e) writing speed was 160 mm/s and chart speed was 3 mm/s.

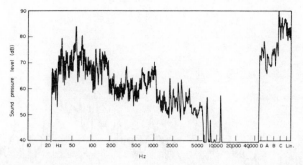

(b) Third octave band analysis of sound pressure level recording sample measured at point *A* on (a).

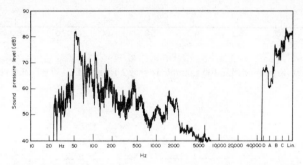

(c) Third octave band analysis of sound pressure level recording sample measured at point *B* on (a).

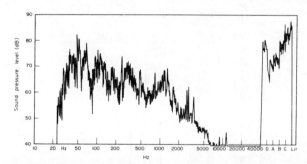

(d) Third octave band analysis of sound pressure level recording sample measured at point *C* on (a).

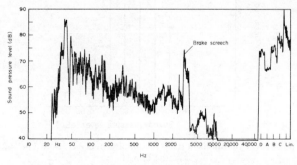

(e) Third octave band analysis of sound pressure level recording sample measured at point *D* on (a).

FIG. 5.1. (*Continued*).

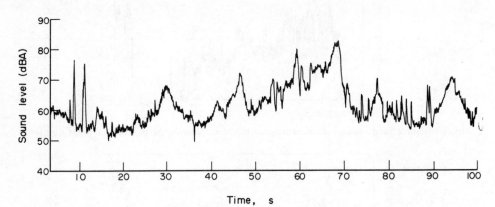

(f) Sound level variation within 100 s sampled every 2 s at site; $L_{10} = 71\cdot4$ dBA, see Fig. 5.1h.

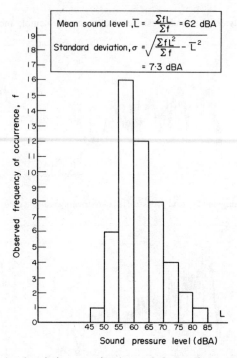

$$\text{Mean sound level}, \bar{L} = \frac{\Sigma fL}{\Sigma f} = 62 \text{ dBA}$$

$$\text{Standard deviation}, \sigma = \sqrt{\frac{\Sigma fL^2}{\Sigma f} - \bar{L}^2}$$
$$= 7\cdot3 \text{ dBA}$$

(g) Histogram of the noise level variations at point A sampled every 2 s over a period of 100 s; data taken from Fig. 5.1f.

FIG. 5.1. (*Continued*).

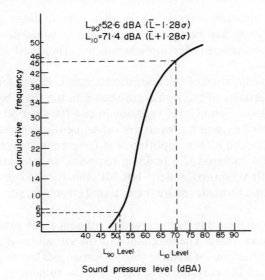

(h) Cumulative probability plot for the noise level variations occurring at point A for 100 s: data taken from Fig. 5.1g.

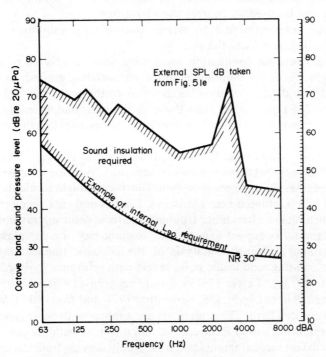

(i) External structure, siting and orientation of building must reduce the external level to below the internal background requirement.

FIG. 5.1. (*Continued*).

Remember that measurements taken in open surroundings need 3 dB or 3 dBA adding to them if they are being used to predict a sound level outside a future building to allow for surface reflection which is superimposed on the sound incident upon the surface.

A typical site investigation of the acoustic climate is illustrated in Fig. 5.1a–i. The frequency characteristics of the traffic are shown in the third-band octave analyses recorded in Fig. 5.1b–e. Sound level variation over a 100 s period sampled every 2 s is shown in Fig. 5.1f, the L_{10} and L_{90} levels are subsequently derived from such samples as shown in Fig. 5.1g and h. The importance of frequency analysis is stressed in Fig. 5.1i which shows an example of a building requiring an internal background sound level (L_{90}) of 35 dBA (approximately NR 30); the hatched area gives the sound reduction which must be achieved by the external structure, siting and orientation of the building.

Tholén (1975) considers the factors which influence the propagation of ground vibrations due to road traffic. The magnitude of the vibrations is governed mainly by the type of subsoil, the type of vehicle (heavy lorries and buses give rise to velocity amplitudes more than 10 times those for cars), and the roughness of the road surface; the velocity of the vehicle has a linear relationship with vibration amplitude up to 50 km/h acceleration and deceleration seldom affected the velocity amplitude; differences in pavement types and variations in load carried by the same vehicle were negligible. The distance from the road, the subsoil and the building foundations are factors which can be used to control the vibrations.

For distances between 5 and 50 metres the velocity amplitude was inversely proportional to distance from the source.

In a building where the foundation walls were built directly on rock the velocity amplitudes were 1% of the amplitudes in the surrounding ground. In two buildings with piles and foundation columns taken down to the rock, the highest measured velocities were 2–5 times lower than those in the surrounding ground. Sometimes vibrations in certain parts of a building may be more intense than those at the surface of the ground.

Vibrations in soft clay (range 0.3–1 mm/s) were about 10 times those in sandy till (range 0.01–0.03 mm/s) but these results varied greatly from one location to another.

It may be necessary to take measurements during the night particularly in the case of houses and hotels situated near roadways. Lorries and cars accelerating, braking, reversing or turning are often more troublesome than vehicles travelling at a steady speed. For planning occupied spaces near loading bays or car parks underneath hotels, for instance, a careful analysis of the acoustic implications is required. Methods of calculating road traffic noise levels with reference to dwellings are given by Grandjean (1973b), Lewis (1973a, b), Ljunggren (1973) and in the *Building Research Establishment Digest*, **135**, November 1971, and *Digest*, **153**, May 1973†; the BRE data is presented here. These methods are applicable to other types of buildings using different internal sound level criteria and different building constructions, but the data are based on measurements taken for freely moving traffic, whereas in towns and cities this is unlikely to be the case. The most important factors are now detailed.

---

†These Digests have been superseded by *BRE Digests*, **185** and **186** (January/February, 1976) and published as a booklet titled *Predicting Road Traffic Noise* by HMSO.

(a) Traffic Flow under Varying Conditions of Traffic Speed, Traffic Composition and Road Gradient

The noise exposure to be expected at 1 m from the façade of a building 30 m from the kerb line of the nearest traffic lane for various traffic densities comprising 20% heavy vehicles moving at a mean speed of 75 km/h on a level road assuming an adverse wind component from the road to the building of about 10 km/h is shown in Fig. 5.2a. For other speeds and distances and for combining the noise exposures from two roads, use Fig. 5.2b–d respectively.

An increase of 10% in the percentage of heavy vehicles causes the L_{10} level to rise by about 1 dBA. Thus if the proportion of heavy vehicles is 40%, increase the noise prediction from Fig. 5.2a by 2 dBA; conversely, if it is only 10%, reduce the values given in Fig. 5.2a by 1 dBA. Note that Grandjean (1973b) suggests an increase of 5 dB for 20% heavy traffic composition.

For road gradients make adjustments to the values obtained from Fig. 5.2a as shown in Table 5.2.

Blitz (1973) could not obtain any conclusive evidence of variations in the L_{10}, L_{50} or L_{90} levels with gradient on single carriageway urban main roads for traffic travelling at a steady speed. Grandjean (1973b) recommends increasing the L_{50} level by 4 dB for a 5% road gradient.

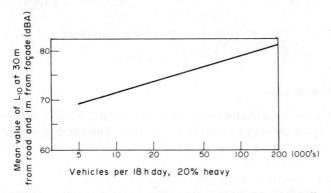

(a) Basic chart for prediction of noise exposure at 30 m from road carrying traffic at 75 km/h.

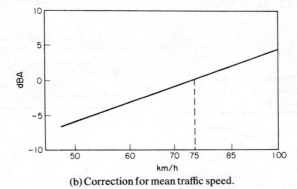

(b) Correction for mean traffic speed.

FIG. 5.2. Calculating sound levels of traffic on level roads; adverse wind condition of 10 km/h assumed (*Building Research Establishment Digest*, **153**, May 1973; refer to later *Building Research Establishment Digest*, **185** and **186**, January/February 1976).

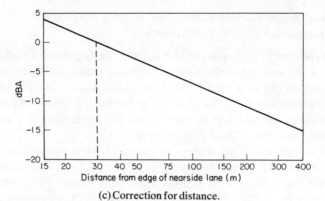

(c) Correction for distance.

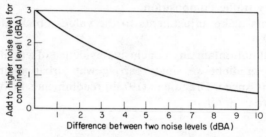

(d) Combination of exposures for two roads.

FIG. 5.2. (*Continued*).

(b) Ground Absorption

Sound propagated over distances greater than about 20 m at heights below 10 m is absorbed to varying degrees by the ground surface. The amount of absorption which occurs at particular frequencies depends on the ground surface material, i.e. grassland and earth are acoustically soft, whereas paving stones are classed as hard. If the surface is soft, use the corrections shown in Table 5.3.

No allowance should be made if the ground surface is paved or if there is to be screening by buildings or barriers.

TABLE 5.2. EFFECT OF ROAD GRADIENT ON ROAD TRAFFIC SOUND LEVELS

Road gradient	Addition
> 1 in 50 } < 1 in 25	1 dBA
up to 1 in 12	2 dBA

Note: 30 mph ≡ 48 km/h; 40 mph ≡ 64 km/h; 50 mph ≡ 80 km/h; 60 mph ≡ 96 km/h; 70 mph ≡ 112 km/h.

TABLE 5.3. GROUND ABSORPTION CORRECTIONS FOR ROAD TRAFFIC NOISE ASSUMING SOURCE IS 0·7 m ABOVE THE LEVEL OF THE MOTORWAY; ONLY USE FOR ACOUSTICALLY SOFT GROUND SURFACES (e.g. GRASSLAND)

Mean height of propagation (m)	Ground attenuation (dBA)
6	1
4·5	2
3	3
1·5	4
0·7	5

(c) Crossroads and Traffic Lights

Grandjean (1973b) advocates an increase of sound level (L_1, L_{10} or L_{50}) by 3–5 dB at road junctions and crossroads.

(d) Continental Practice

Projects undertaken in other countries may involve using L_1, L_{50} and L_{eq} levels. Traffic data using these levels are presented on Fig. 4.5 in Chapter 4.

Example

Estimate the sound level at 40 m from a road having a traffic density of 100000 vehicles per 18-hour day.

Other data:

Mean traffic speed	85 km/h
Heavy traffic	40%
Road gradient	1 in 25

Solution

Using Fig. 5.1a, mean L_{10} value is $\bar{L}_{10} = 79$ dBA.
Using Fig. 5.1b, speed correction is $+2$ dBA.
Using Fig. 5.1c, distance correction is -2 dBA.
Heavy traffic correction (see section 5.1) is $+2$ dBA.
Road gradient correction (see Table 5.2) is $+1$ dBA.
Resultant average L_{10} sound level

$$L_{10} = 79 + 2 - 2 + 2 + 1 = 82 \text{ dBA}.$$

For offices in towns or cities add 2 dBA to allow for the heavier traffic during office hours and for the fact that Fig. 5.1a really only applies to freely moving traffic. Residential development should not proceed in areas where the *maximum* L_{10} level exceeds 68 dBA. In this example, houses would have to be built at a further distance from the road or a noise barrier having a minimum noise reduction of 14 dB would have to be placed between the road and the houses. More recent design data is presented in the Building Research Establishment Report *Predicting Road Traffic Noise* published by HMSO and the 1975 Department of the Environment and Welsh Office Joint Publication *Calculation of Road Traffic Noise* (HMSO).

5.3. Control of Road Traffic Noise

Control of noise from road traffic is achieved by the following means:

* modification of any of the factors discussed under (a) in preceding section, in particular *increasing the distance* between the traffic and the building;
* *building orientation*: side roads give significant noise reductions (in the order of 10–15 dB); if it is possible to ensure that most people in the building are exposed to the side-road noise climate this is preferable to main-road

exposure; in cities the use of this factor is limited by land development and
the daylighting requirements because adjacent buildings in side roads will cut
off sunlight;
* *noise screens* positioned between the roadway and the building; roadways
landscaped as cuttings offer some protection to nearby buildings but insufficient reliable data is yet available to give quantitative guidelines here;
* *the building façade* will play an important part in insulating the occupants of the
building from road noise, but the choice of building structure is also important
in determining the variation and level of indoor temperature and ventilation.

The Department of the Environment have issued a Design Bulletin, No. 26,
entitled *New Housing and Road Traffic Noise*, which is very helpful and gives
examples of the work discussed so far.

5.3.1. NOISE SCREENS

The most reliable design data in this country is summarised again in *Digest*, **153**, **185**
and **186** issued by the Building Research Establishment. The screen should create as
large an angle of acoustic shadow as possible by placing the screen as close to the
source as possible and by making it as long and as high as is practicable.

Diffraction limits the attenuation that can be achieved to about 20 dBA. A solid
panel screen having a mass of about 20 kg/m² provides a noise reduction of 25–30 dB
through the screen itself; a 10 kg/m² screen would give a 20–25 dB reduction, and a
10 dB attenuation can be achieved with a screen mass of 5 kg/m².

Figure 5.3a enables the attenuation to be estimated for various path differences
assuming that the barrier is very long. The nomogram illustrated in Fig. 5.3b
overcomes the need to calculate the path difference; the example plotted on Fig. 5.3b
shows that for a source–receiver distance of 100 m and with a 2 m high barrier placed
at a relative position 15% between source and receiver, the sound attenuation will be
about 12 dB. If a larger reduction is required, then the nomogram can be used and the
relevant parameters manipulated until a satisfactory solution is obtained. Note that if
the sight-line (i.e. the straight line connecting the source point to the receiver point)
is very steep, errors occur which may be minimised by using sight-line distances, not
horizontal distances, and perpendicular barrier heights.

Traffic forms a longitudinal noise source, and it is essential that the noise screen is
long if flanking of noise around the ends of the barrier is not to occur. Figure 5.3c
enables a reassessment of the noise reduction to be made when the barrier is not as
long as is assumed for estimation of Fig. 5.3a and b. Referring to Fig. 5.3c, when the
angle is small the barrier length BB^1 is short, and the resultant noise reduction is very
small, but as α approaches 90° the potential noise reduction as estimated from Fig.
5.3a and b is achieved. If the sound reception point is not equidistant from each end
of the barrier, then a correction needs to be added to the lower of the two attenuation
values obtained for each angle subtended from Fig. 5.3a or b, and c; this is explained
in Fig. 5.3d. As a general guide the distance from the reception point to the end of the
screen should be at least ten times the shortest path to the screen (i.e. in Fig. 5.3c
$RB^1 = RB \geq 10\, RO$). As an example of a non-symmetrical partial screen the
Building Research Establishment Digest, **153** describes a 3 m high screen 200 m long

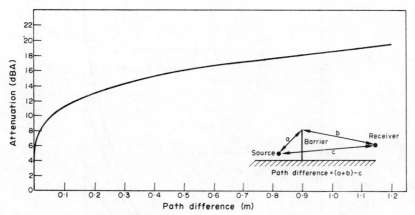

(a) Reduction of L_{10} by very long barriers; source 0.7 m above ground at centre of the motorway; no allowance has been made for ground absorption.

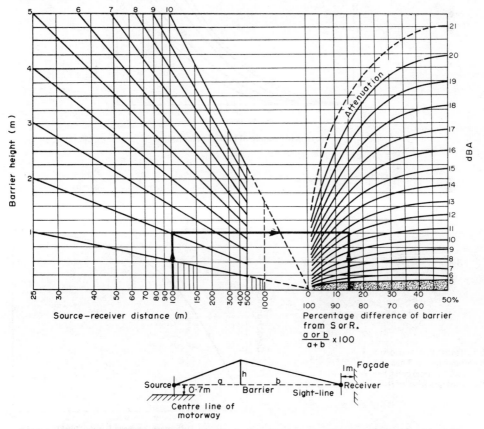

(b) Nomogram for attenuation of traffic noise L_{10} by long barriers; conditions described for (a) apply.

FIG. 5.3. Influence of noise screens on traffic noise (*Building Research Establishment Digest*, **153**, May 1973; also see *Building Research Establishment Digest*, **185** and **186**, January/February 1976).

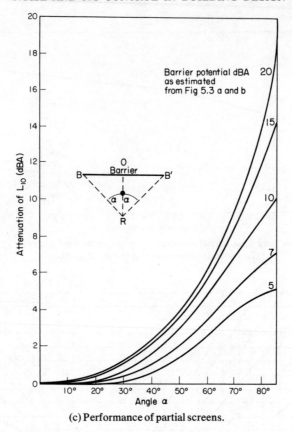

(c) Performance of partial screens.

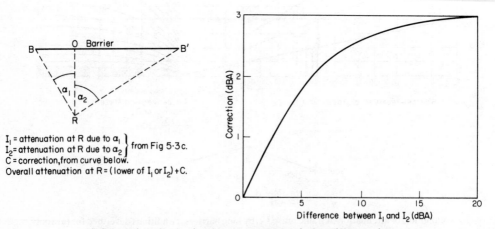

I_1 = attenuation at R due to α_1
I_2 = attenuation at R due to α_2 } from Fig 5·3c.
C = correction, from curve below.
Overall attenuation at R = (lower of I_1 or I_2) +C.

(d) Calculation of reduction due to non-symmetrical partial screening.

FIG. 5.3. (*Continued*).

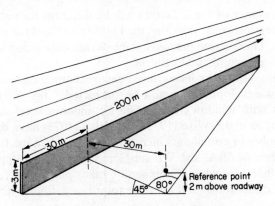

(e) Example of non-symmetrical partial screen.

FIG. 5.3. (*Continued*).

and 15 m from the middle of the motorway (Fig. 5.3e) with the reception point 2 m above road level at 30 m from the screen opposite a point 30 m from one end of it, thus subtending angles to the ends of the screen of 45° and 80°. From Fig. 5.3b it can be seen that if the noise screen had been very long in both directions it would have given a reduction of 12 dBA at the reception point. This is the screen potential to be used in Fig. 5.3c. Reading from Fig. 5.3c the attenuations associated with angles of 45° and 80° at the potential of 12 dBA, values of 2·5 and 10 dBA respectively are obtained, a difference of 7·5 dBA. From Fig. 5.2d the difference 7·5 dBA gives $C = 2$ dBA and the net reduction is $(2·5 + 2)$ dBA $= 4·5$ dBA, say 5 dBA, which is considerably less than the 12 dBA reduction of a long screen of the same design.

5.3.2. SELECTION OF BUILDING STRUCTURE TO SATISFY ENVIRONMENTAL CRITERIA

The sound insulation of various forms of building construction was discussed in Chapter 4. A building structure is not selected for its acoustic specification alone, although in cities and near airports, for example, this is a matter of primary concern. Factors that have to be taken into account when selecting materials for the external building structure are:

* solar protection;
* thermal insulation;
* condensation risk;
* sound insulation;
* daylighting requirements;
* ventilation rate;
* fire protection;
* structural considerations;
* cleaning, durability and weather protection;
* capital and operating costs;
* aesthetics.

The relative importance of these factors will be different for each building design depending on the location of the building, the building use and the money available to build and operate it. This section presents some design data to assist in arranging and meeting the chosen environmental criteria.

68 dBA on the L_{10} (18-hour) index has been recommended as the upper limit to which residential development should be subjected. This corresponds to about 58–63 dBA indoors with windows open. The L_{10} indoors level should not exceed 35 dBA in bedrooms at night. Wise (1973) and Utley (1973) have drawn attention to the fact that the number of complaints increases rapidly when the L_{10} sound level in a general office reaches 60 dBA; the Wilson Committee set an upper limit of $L_{10} = 55$ dBA in spaces where speech communication takes place. For concentrated work in private offices, classrooms and lecture rooms, an upper limit of 45 dBA is desirable.

Note that all these recommendations refer to *upper* levels, and in all cases lower levels are preferable; in Chapter 2 it was declared that registering complaints is a poor and unreliable indicator of desirable environmental standards. Waller (1971) has suggested the noise criteria listed in Table 5.4 which are expressed in terms of the noise pollution level as well as dBA (see Chapter 2).

Table 5.5 shows the sound insulation required for the structure for a given external noise climate and a specified internal noise level; buildings such as houses, hotels and factories employing night-shift workers require analysis L_{10} daytime (6 a.m. until

TABLE 5.4. GENERAL INTERNAL NOISE CRITERIA (WALLER, 1971)

	Minimum steady (dBA)	Optimum steady (dBA)	Maximum desirable adverse noise $(L_{NP}dBA)$[a]	
			Bedroom	Living-room
Houses: Rural	25	30	40	50
Suburban	25	35	45	55
Urban	25	40	50	60
Offices: General	35	45	55	
Private	30	N[b]	55	
Hospital: Wards	35	45	55	
Schools: Lecture rooms	25	30 or N[b]	50[c]	
Colleges, universities: Mixed activity rooms	35	45	55	
Factories: General			90	
Areas where communication is important			65	

EXTERNAL NOISE CRITERIA IN HOUSING AREAS

	Urban	Suburban
No advantage below these levels	50(L_{NP}dBA)	45(L_{NP}dBA)
Generally acceptable levels	70	65

[a]$L_{NP}dBA = L_{50} + (L_{10} - L_{90}) + \dfrac{(L_{10} - L_{90})^2}{60}$. L_{10}, L_{50} and L_{90} all expressed in dBA.

[b]$N = 70$—attenuation of partition.

[c]Depends on speech interference and size.

TABLE 5.5

External noise level (L_{10} dBA)	Sound insulation of structure (IdB)[a]		
	Internal noise level (L_{10} dBA)		
	35	45	55
60	25 dB	15 dB	5 dB
70	35	25	15
80	45	35	25

[a]Using equation (11.24) sound insulation is

$$I = R - 10 \lg\left(\frac{S}{A}\right)$$

for a structure having sound reduction index R, area S and absorption in the receiving space A.

midnight) *and* L_{10} night (midnight until 6 a.m.) conditions; the latter may be the determining factor in some situations.

The sound reduction index does not give the effective sound insulation because the sound absorption of the space behind the structure plays a part in the sound energy transmission process (see section 11.4.3). Other limitations to the theoretical estimation of sound insulation are flanking transmission and the quality of construction. Typical building constructions and types of glazing which will achieve the sound insulation values listed in Table 5.5 were described in Chapter 4.

Cracks permit sound to leak through the structure, but at the same time they allow infiltration of fresh air to occur and contribute towards, or fulfil, the ventilation rate of an occupied space (note: *ventilation rate* is the *fresh*-air requirements for a space, whereas the *air supply rate* is the *total air* supply rate to a space, and includes recirculated air as well as fresh air). If the external noise climate dictates that sealed windows should be used, mechanical ventilation is necessary to provide the air needed to dilute the carbon dioxide and remove the odours released into the space. Figure 5.4 shows that up to about three air changes per hour ventilation also assists significantly in reducing overheating in summer.

The thermal admittance procedure set out in the *IHVE Guide, 1970*, is used to calculate the level and variation of indoor temperature for various combinations of building location, building orientation, window size, solar protection, thermal response, ventilation rate and heat gain. The effect of some various forms of solar control on indoor temperature level and daily range are illustrated in Fig. 5.5. The meaning given to lightweight and heavyweight building types is interpreted in Table 5.6.

Reflective films applied to glazing are more effective than heat-absorbent glasses in minimising the effect of solar radiation on indoor temperature. Likewise mid-pane venetian blinds are preferable to internal blinds, but the minimum overheating will occur with external sunbreakers. The final choice of the solar control system is not easy. There is a high cost involved in cleaning and maintaining external solar protective devices. Reflective films reduce the daylight factor considerably (Fig. 5.6). Reflecting solar energy away from building surfaces is wasteful, but at present there

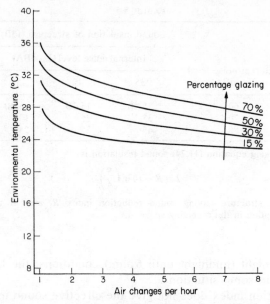

FIG. 5.4. Effect of air supply rate on indoor temperature (Ridpath, June 1973. Environmental Temperature Symposium, Institution of Heating and Ventilating Engineers).

is no economic way of using *all* the internal building fabric as a solar collector and a thermal store whilst providing sufficient ventilation and daylight with a minimum of noise transfer.

Wise (1973) has proposed the use of the diagrams shown in Fig. 5.7 for assessing the interactive effects of ventilation rate, glazing proportion and type of solar protection on the indoor thermal conditions in lightweight and heavyweight buildings. The assumptions that were made in producing these diagrams are given by Wise (1973). Briefly they are applicable in situations where the following conditions exist:

* building orientation south-west;
* lights not in use;
* room heights of 2·4–3·0 m;
* ventilation rates occurring in daytime (8.00–18.00 hours BST);
* night-time ventilation rate 0·5 ach;
* metabolic rate 60 W/m^2;
* local air velocity 0·1 m/s;
* comfort environmental temperature 25°C with a daily range of 8°C for occupants wearing shirt and trousers;
* comfort environmental temperature 23°C with a daily range of 4°C for occupants wearing lightweight suit.

(See Building Research Establishment Current Paper CP/4/71 for calculation of environmental temperature.)

For other window types, such as tinted and reflective glazing, curtains and roller blinds, it is practicable to interpolate between the limits given using relative values of solar gain factor from Book A of *IHVE Guide, 1970.*

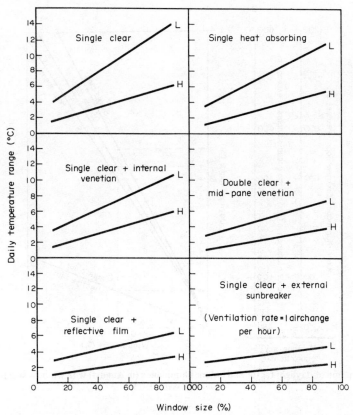

FIG. 5.5. Solar control schemes and their effect on indoor temperature in lightweight (L) and heavyweight (H) buildings (Wise, 1973).

The mean temperatures derived from the diagrams are likely to be exceeded on average for about 2 working days per year. In practice, a higher failure rate may be acceptable. For example, if the indicated mean temperatures are reduced by 2°C the diagrams then relate to conditions which are likely to be exceeded for about 6 working days per year.

The design steps for selecting type and amount of glazing compatible with the environmental aspects of building design are as follows:

(a) select indoor L_{10} sound level;
(b) select window with required external–internal sound insulation;
(c) select peak indoor environmental temperature and daily range;

TABLE 5.6. TYPICAL DESCRIPTIONS OF TWO WEIGHTS OF ROOM

	Ceiling	Internal walls	External wall	Floor
Lightweight	Suspended	Dry partitions	Lightweight	Carpeted or cork-tiled
Heavyweight	Exposed slab	100 mm brick	Cavity brick	Concrete, linoleum-covered

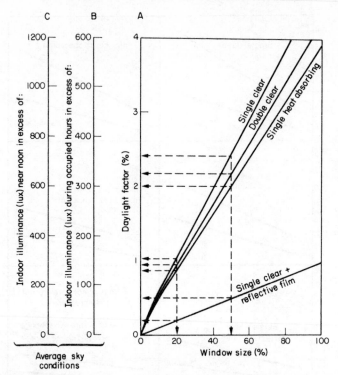

FIG. 5.6. Variation of daylight factor and indoor illuminance (in August) with window size for four types of glazing (Wise, 1973).

(d) select daylight factor;
(e) select solar control compatible with conditions in (c) and (d) (see Figs. 5.5 and 5.6);
(f) check ventilation rate and see if mechanical ventilation is necessary (see Fig. 5.7).

5.4. Acoustic Design Problems of Internal Spaces

Each building presents its own unique problems which can only be discovered as the design is formulated. But there are many problems which keep recurring, and it is these problems which are summarised here in the identification matrix shown later in Table 5.7; many of them having already been discussed in detail. It should be emphasised that the problems in Table 5.7 are the minimum number that will occur in practice. Some problems can be subjected to a more detailed analysis than others. This chapter will conclude by recording the design techniques being used at present to (a) achieve a specified level of privacy in an office, (b) check sound quality in auditoria and (c) assess the noise output of airconditioning systems.

Background noise criteria for various types of buildings are shown in Fig. 5.8. The American NC curves and the preferred NC curves, both due to Beranek, are shown in addition to the European NR standard curves. The NC and NR curves are similar. The PNC curves have values about 1 dB lower than the NC curves in the 125, 250,

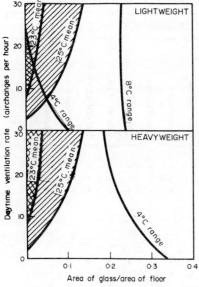

(a) Unprotected single clear glazing.

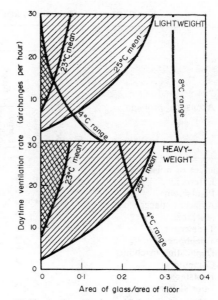

(b) Single clear glazing with internal venetian blind.

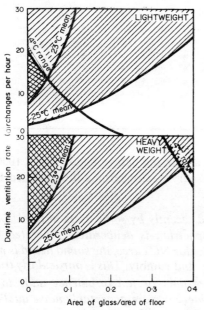

(c) Double clear glazing with mid-pane venetian blind.

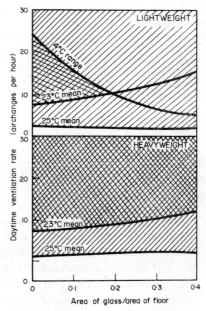

(d) Single clear glazing with external blind.

FIG. 5.7. Interaction of building mass, ventilation rate and solar protective schemes on indoor environmental temperature level and temperature variation (*Building Research Establishment Digest*, **162**, February 1974, and also BRE Paper CP 6/73). Note: *IHVE Guide, 1970*, considers a minimum openable window area to be 5% of the floor area to give sufficient ventilation for smoke relief in the case of fire.

TABLE 5.7. COMMON PROBLEMS IDENTIFICATION MATRIX

Acoustical problem	All	Dwellings	Hotels	Offices (Large)	Offices (Private)	Schools	Universities	Hospitals	Court-rooms	Theatres	Concert hall	Recording studios	Television studios	Factories	Others
Sources of noise:															
External	*	*	*	*	*	*	*	*	*	*	*	*	*	*	
Plant-room	*	*	*	*	*	*	*	*	*	*	*	*	*	*	
Adjacent spaces	*	*	*	*	*	*	*	*	*	*	*	*	*	*	
Machines				*									*		
Typewriter				*	*										
Telephones				*	*										
Duplicating machines				*											
Computers				*											
Footsteps		*	*	*	*	*	*	*	*	*	*	*	*		
Door slamming		*	*	*	*	*	*	*	*						
Heating ⎫ Ventilation ⎬ Airconditioning ⎭		*	*	*	*	*	*	*	*	*	*	*	*		
Toilets		*	*	*	*	*	*	*	*	*	*	*	*		
People talking		*	*	*	*	*	*	*	*						
Others	?														?
Other problems:															
Privacy		*	*	*	*	*	*	*	*						
Deafness														*	
Speech communication				*	*	*	*	*	*	*	*	*	*	*	
Speech reinforcement						*	*		*	*				*	
Echoes				*	*	*	*	*	*	*	*	*	*	*	
Delay time										*	*				
Reverberation				*	*	*	*	*	*	*	*	*	*	*	
Sound quality										*	*	*	*		

500 and 1000 Hz octave bands; in the 63 Hz and in the octave bands above 1000 Hz they are about 4 or 5 dB lower than the NC curves. Concerning the modified PNC curves Beranek (1971) states:

> The 1957 NC curves have been criticised recently by some users in the USA. It has been demonstrated that if a noise spectrum is deliberately generated with octave-band levels equal to those of a particular NC curve, the sound heard is not a pleasant, or neutral sound, but is both hissy and rumbly. This is particularly true of noise from airconditioning systems, where duct or fan turbulence adds to the rumble of the random noise sound. To achieve a more acceptable noise quality it sometimes has been necessary to lower the levels by about 5 dB in both the very low and the very high frequency bands as compared to those permitted by the 1957 NC curves.

Since acoustic design consultants in this country often encounter NR and NC specifications, both are included here to remove any idea that they are very different.

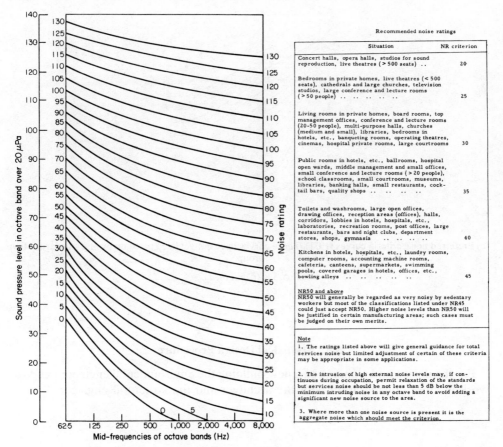

The table to the right of the NR curves:

Recommended noise ratings

Situation	NR criterion
Concert halls, opera halls, studios for sound reproduction, live theatres (> 500 seats) ..	20
Bedrooms in private homes, live theatres (< 500 seats), cathedrals and large churches, television studios, large conference and lecture rooms (> 50 people)	25
Living rooms in private homes, board rooms, top management offices, conference and lecture rooms (20–50 people), multi-purpose halls, churches (medium and small), libraries, bedrooms in hotels, etc., banqueting rooms, operating theatres, cinemas, hospital private rooms, large courtrooms	30
Public rooms in hotels, etc., ballrooms, hospital open wards, middle management and small offices, small conference and lecture rooms (> 20 people), school classrooms, small courtrooms, museums, libraries, banking halls, small restaurants, cock-tail bars, quality shops	35
Toilets and washrooms, large open offices, drawing offices, reception areas (offices), halls, corridors, lobbies in hotels, hospitals, etc., laboratories, recreation rooms, post offices, large restaurants, bars and night clubs, department stores, shops, gymnasia	40
Kitchens in hotels, hospitals, etc., laundry rooms, computer rooms, accounting machine rooms, cafeteria, canteens, supermarkets, swimming pools, covered garages in hotels, offices, etc., bowling alleys	45

NR50 and above

NR50 will generally be regarded as very noisy by sedentary workers but most of the classifications listed under NR45 could just accept NR50. Higher noise levels than NR50 will be justified in certain manufacturing areas; such cases must be judged on their own merits.

Note

1. The ratings listed above will give general guidance for total services noise but limited adjustment of certain of these criteria may be appropriate in some applications.

2. The intrusion of high external noise levels may, if continuous during occupation, permit relaxation of the standards but services noise should be not less than 5 dB below the minimum intruding noise in any octave band to avoid adding a significant new noise source to the area.

3. Where more than one noise source is present it is the aggregate noise which should meet the criterion.

(a) NR curves and recommendations.

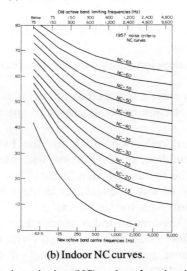

(b) Indoor NC curves.

FIG. 5.8. Noise rating (NR), noise criterion (NC) and preferred noise criterion curves with design recommendations for various building types (Beranek, 1971, *IHVE Guide* 1970, Book A).

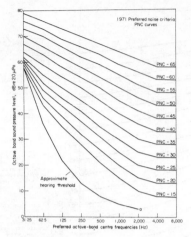

(c) Preferred noise criteria curves (1971). These curves have not yet been acted on by any standardisation group.

FIG. 5.8. (*Continued*).

5.5. Acoustical Design Procedure

For the design and planning of a new building the environmental engineer needs to work in close collaboration with the architect and the structural engineer.

5.5.1. PLANNING STAGE

(a) assess external noise climate;
　　(i) measure 10% and 90% sound levels and the frequency spectrum of particular sources;
　　(ii) ascertain future developments which may affect the nature or level of the noise climate;
(b) decide on the design criteria required within the building and plan the layout of the internal spaces attempting to separate the quieter areas from the noisier ones;
(c) decide on the sound reduction index for the external structure, and the orientation of the building, in conjunction with the client, the architect and other members of the building design team;
(d) decide on the shape of any auditoria, or ceilings in the case of landscaped offices, in conjunction with the architect;
(e) decide on the sound reduction index required for internal walls, partitions, doors, ceilings and floors (also finish required), with the architect; consult with the structural engineer concerning the effect of the mass of the building elements on the structure;
(f) decide on the amount of absorption material required especially in critical

areas such as auditoria, recording studios or landscaped offices in conjunction with the architect;

(g) assess the acoustical performance of all equipment to be used in the building (to cover all mechanical and electrical services besides office equipment) in terms of octave band (or narrow band) spectra with details about any discrete frequencies and any special installation or operating device.

5.5.2. ASSESSMENT OF NOISE CONTROL EQUIPMENT REQUIRED

(a) estimate dynamic insertion loss required on inlet and discharge side of the system to achieve the design acoustical conditions in the building (use method systematically laid out in the *IHVE Guide, 1970*, Book B);

(b) detail final proposals for plant-room structure and any floating floor constructions that may be required (see Chapter 4);

(c) check to see if either breakout noise from the system to a quiet area, or the converse case, will occur;

(d) check to see if cross-talk is likely to occur;

(e) calculate transmissibility required for vibration isolation of equipment.

5.5.3. SELECTION OF NOISE CONTROL EQUIPMENT

Having attempted to obtain the spectrum of attenuation required, equipment should be selected in conjunction with a manufacturer, and remembering that interference and reflection type sound filters can be easily incorporated as part of the airconditioning network design, to give the degree of airborne and structureborne sound control required.

5.6. Designing for Sound Privacy in Offices

5.6.1. SMALL OFFICES

Beranek (1971) describes the method devised by Young (1965). The basis of this method is the following equation:

$$\text{Sound excess} = \left(\begin{array}{c}\text{sound level} \\ \text{in source room}\end{array}\right) - \left(\begin{array}{c}\text{Effect of} \\ \text{partition}\end{array}\right) - \left(\begin{array}{c}\text{Background sound} \\ \text{level in receiving} \\ \text{room plus privacy} \\ \text{number}\end{array}\right)$$

$$= \left(\begin{array}{c}\text{Sound level of} \\ \text{voice}\end{array} + \begin{array}{c}\text{Room} \\ \text{effect}\end{array}\right) - \left(\begin{array}{c}\text{Insulation} \\ \text{of partition}\end{array}\right) - \left(L_{90} + P\right)$$

$$= \{68 \text{ dBA} + F \text{ (Fig. 5.9a)}\} - \{R + K \text{ (Fig. 5.9b)}\}$$

$$- \{NR \text{ (Fig. 5.8)} + 6 - P\}$$

For satisfaction the sound excess $\leqslant 0$; R is the sound reduction index of the partition being used and P is the privacy number. Beranek (1971) recommends the following values:

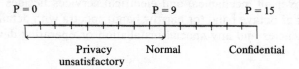

Various values of sound excess can be achieved by altering R or L_{90} (i.e. NR) for a given P-value. Clearly the sound excess should be negative; the probability of satisfaction increases as the magnitude of this negative quantity increases.

Example. Two furnished management offices each having a floor area of 18 m² are separated by a partition having a sound reduction index of 35 dB. The partition also has an area of 18 m². Confidential privacy is required. Can this be achieved?

Sound excess $= 68 + 5$ (F from Fig. 5.8a) $- \{35 + (0)$ (K from Fig. 5.8b)$\}$
$$- \{25 \text{ (NR from Fig. 5.9)} + 6 - 15 \ (P)\}$$

Sound excess $= +22$ dB.

This is not satisfactory. Try a partition with a sound reduction index of 45 dB.

Sound excess $= +12$ dB.

This remains unsatisfactory. Let the background sound level increase to NR = 30 ($L_{10} \simeq 36$ dBA) and decrease the partition area so that the rate of *receiving floor area/common partition area* is 5:1, giving $K = +7$. The sound excess using a 45 dB partition is now reduced to zero providing a satisfactory solution.

This example shows how important it is that the architect and building environmental engineer work closely together, not only in selecting partitions but also in establishing economical criteria and sizes of rooms.

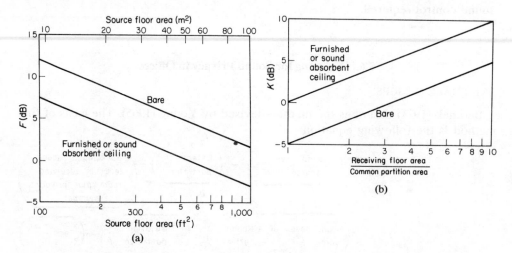

(a) Estimate of source room function F from room dimensions for stud-and-plaster construction and room heights of about 3 m.

(b) Estimate of K from room dimensions for stud-and-plaster construction and room height of about 3 m.

FIG. 5.9. Estimation of F and K factors used in designing acoustical privacy for offices (Beranek, 1971).

5.6.2. LANDSCAPED OFFICES

A detailed study of the acoustical problems that arise in landscaped offices was made in section 2.6.3. Beranek (1971) gives some design data for selecting noise screens and also for providing a controlled background noise.

In a large office people become aware of other people near them, and tend to speak more quietly than if they were in a small office on their own, but this is offset to some extent because when people use the telephone their voice level often unconsciously increases. Measurements taken by the author suggest that a voice level of 65 dBA is common, and also that a satisfactory background sound level is 50 dBA. Taking a privacy number of 9, Beranek (1971) defines the equivalent sound reduction index for the combination of source–receiver distance and a screen placed between two people requiring privacy from each other's conversations as

$$R_{eq} = 66^* + 9 - 50^* - G.$$

(*These values are chosen from the experience of the author; designers may have evidence which leads them to prefer other values.)

The factor G is chosen from Table 5.8.

TABLE 5.8

	Factor G in dB			
	Distance between speaker and nearest unintended listener D			
Office absorption	1 m	2 m	4 m	6 m
Office with absorbent ceiling only	0	5	10	15
Office with absorbent ceiling and carpet	0	6	12	18

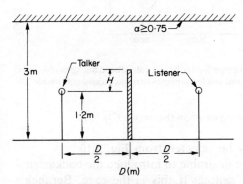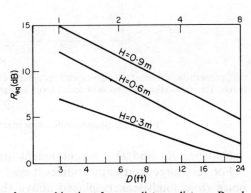

FIG. 5.10. Determination of the sound reduction index for a combination of source–listener distance D and excess screen height H. The quantity H is the excess in barrier height above 1.2 m; the room height is assumed to be 3 m and the ceiling is covered with a sound-absorbing material with $\alpha \geqslant 0.75$ (Beranek, 1971).

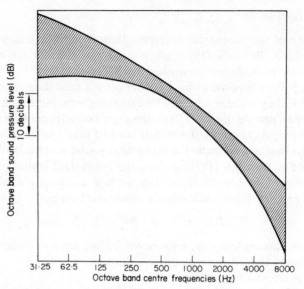

(a) Suggested shape of background noise spectrum for landscaped offices. If the noise spectrum is exactly that of the upper edge of the shaded area, add 7 dB to the level for the 1000 Hz band to obtain L_A, the A-weighted level for the noise, in dBA; for a noise with the shape of the lower edge of the shaded area,
$$L_A = L_{1000\,Hz} + 6\,dBA.$$

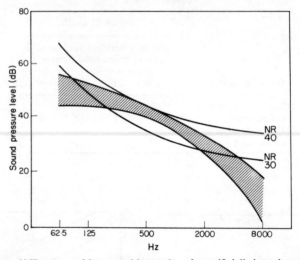

(b) Comparison of normal NR curves with acceptable envelope for artificially introduced background sound levels. This shows the depression of sound levels at high and low frequencies to reduce apparent rumble and hiss.

FIG. 5.11. Sound conditioning in landscaped offices (Beranek, 1971).

Knowing R_{eq} and D, a screen height may be selected from Fig. 5.10.

For very large landscaped offices it may be desirable to introduce the background noise from loudspeakers placed above the ceiling. If this is the case, Beranek's experience suggests that the background noise spectrum shown in Fig. 5.11 should be used; a comparison with the shape of the NR curves is also shown.

5.7. Sound Quality in Auditoria

Chapter 10 discusses in depth the various parameters that contribute towards sound quality. Criteria were given for reverberation times; the formula of Marshall (1967a), stated as equation (10.44), enables a height–width ratio to be ascertained which allows good spatial responsiveness. Simple ray tracing techniques can be very helpful in assessing whether obvious acoustic faults will occur, and can also be helpful in designing ceiling reflectors.

Figure 5.12a shows ray tracing in an auditorium that exhibits echo, sound shadow and non-uniform sound distribution.

Uniform sound quality can only result if:

(i) there are unobstructed sight–sound lines;
(ii) balconies are not too large;
(iii) rooms are of correct size and proportion;
(iv) concave surfaces are to be avoided because they are likely to produce sound concentration.

Sound diffusion can be achieved by the use of (a) surface irregularities and (b) absorptive and reflective treatments applied to various surfaces.

The defects shown in Fig. 5.12a can be corrected by applying an absorption material on the back vertical surface; by serrating this surface or by tilting this surface; by altering the curvature of the roof to prevent sound concentration; by

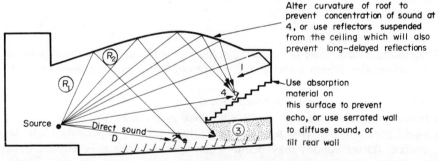

Alter curvature of roof to prevent concentration of sound at 4, or use reflectors suspended from the ceiling which will also prevent long–delayed reflections

Use absorption material on this surface to prevent echo, or use serrated wall to diffuse sound, or tilt rear wall

(a) Auditorium acoustical defects: 1, echo; 2, long-delayed reflections: time delay $= (R_1 + R_2 - D)/0.34$ ms, use maximum of 30 ms; 3, sound shadow (see (b)); 4, sound concentration.

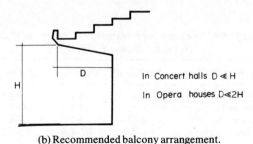

In Concert halls $D \ll H$

In Opera houses $D \ll 2H$

(b) Recommended balcony arrangement.

FIG. 5.12. Problems in auditoria acoustics.

using reflectors suspended from the ceiling to prevent overlong delayed reflection; by using a balcony with the proportions shown in Fig. 5.12b the sound shadow can be eliminated.

Doelle (1972) recommends that:

(a) shape auditorium so that the audience is as close to the sound source as possible;

(b) raise the sound source as much as is practicable so that the direct sound to the listeners is unimpeded;

(c) rake the floor (up to a maximum gradient of 1 in 8 but consult local building bylaws) because sound is absorbed when it travels over the audience at grazing incidence; a method of establishing the floor slope for good vision as well as good sound is shown in Fig. 5.13;

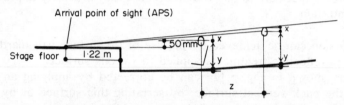

FIG. 5.13. Method of establishing good sight lines and shaping floor based on one-row vision (Doelle, 1972); $x = 125$ mm maximum; $y = 1 \cdot 12$ m; $z =$ row spacing. If resulting slope is too steep, raise APS, reduce x, or arrange for two-row vision.

(d) surround source with large sound reflective surfaces to give sufficient reverberation; the dimensions of the reflectors need to be comparable to the wavelength of the sound for reflection to occur; curved ceilings give more useful reflections than horizontal ones;

(e) recommended volume per auditorium size is given in Table 5.9;

(f) avoid parallelism between sound reflective surfaces especially close to the source; flutter echoes may occur when a found source is located between parallel surfaces, or even between non-parallel surfaces if they are highly reflective.

TABLE 5.9. VOLUME PER AUDIENCE SEAT (m³) AS
GIVEN BY DOELLE (1972)

Type of auditorium	Minimum	Optimum	Maximum
Speech rooms	2·3	3·1	4·3
Concert halls	6·2	7·8	10·8
Opera houses	4·5	5·7	7·4
Churches	5·1–5·7	7·2–8·5	9·1–12
Multi-purpose halls	5·1	7·1	8·5
Cinemas	2·8	3·5	5·1

A procedure for designing ceiling reflectors is illustrated in Fig. 5.14. The steps taken are as follows:

(a) assume point A is fixed and consider a point source at X;
(b) select point P in audience where first ceiling reflection is required;
(c) join PA and AX and bisect angle PAX; the bisector is the normal to the plane of the first ceiling reflector;
(d) find the image of X in this plane; the image is X_A^1;
(e) determine the cut-off point B by selecting the last audience position for this reflector T;
(f) repeat the process from point B, drawing QB and BX and bisecting QBX, hence derive second reflector BC;
(g) continue this successive development of the ceiling reflectors to find DC and DE.

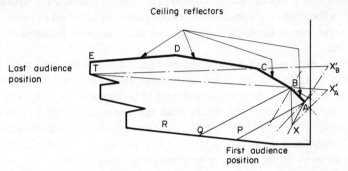

FIG. 5.14. Design method for ceiling reflectors to ensure adequate and uniform distribution of reflected sound.

5.8. Noise from Airconditioning Systems

The various sources of noise that are prevalent in airconditioning systems were reviewed in Chapter 3. Airconditioning noise is, of course, only one part of the background noise spectrum for a space, but has it become significant over the last decade as more high-velocity systems are being installed and as people have become more sensitive to internal noise sources in buildings that are well insulated from the external environment. Research by Hay (1963) has been noted in section 2.6.3 which indicates the varying degrees of subjective acceptability in landscaped offices for various differences in sound level between the airconditioning system and the office with people working in it.

5.8.1. CASE STUDY 1: NOISE CONDITIONS IN UNIVERSITY LECTURE ROOMS

In order to ascertain the importance of the problem a questionnaire was circulated to forty-nine universities and colleges throughout Great Britain. The questionnaire consisted of ten questions requesting details of any ventilation or airconditioning services, the acoustic acceptability of the lecture rooms provided with ventilation, steps taken at the design stage to ensure satisfactory noise levels, specific cases worthy of investigation and details of any amplification systems used. Replies from sixteen universities were received from architects and building and planning officers concerning some 120 lecture rooms. In about a third of these the acoustics were considered unsatisfactory from some point of view by lecturers, building officers or students. Although built for the sole purpose of allowing communication between lecturer and student, the inclusion of background noise level limits in lecture room specifications appears to be rare. In only one instance was a noise level specified; NC 25 was chosen. Nearly half the cases made no mention in a vague manner that the noise level should be kept to a minimum.

It is difficult to describe and locate noise sources, but people thought that the fans and motors were responsible for noise in over half the cases, and that the airflow in the ducts in about a third of the cases. Duct wall resonance, grille noise, structural transmission and traffic noise were contributory factors.

Often the ventilation systems were designed to provide a greater air-change rate when the lecture rooms were fully occupied. This meant that the fan speed increased in proportion to the extra air supplied. In every case this resulted in noise complaints and the fans were run at the slower speed; the outcome was inadequate ventilation and overheating.

Acoustic treatment was not usually recommended at the design stage. Frequently the system was installed, noise complaints were then received and further money had to be spent to provide acoustic treatment. This usually took the form of sound absorption material placed in the ducts near the room concerned or at bends.

Several instances were quoted where the fans were positioned adjacent to the lecture room. Sometimes insufficient thought had been given to plant-room construction resulting in high noise levels in neighbouring work rooms.

As a result of the questionnaire survey it was decided to carry out a site appraisal of the acoustic conditions in seventy-four of the lecture rooms. The results of the measurements superimposed on a set of NR curves are shown in Fig. 5.15 and in Table 5.10.

In 13% of the cases for single-speed systems the acoustic conditions were intolerable and in another 21% unsatisfactory. It is difficult to say which of the rooms classified as moderate would be unsatisfactory, but according to the standard recommendations those which exceed NR 30 by more than 2 dB should be classed as unsatisfactory. This means that about 50% of the rooms are acoustically inadequate. The field studies exposed two broad categories of noise sources—those originating from the basic nature of the airflow system and those arising from bad planning or installation.

Magnetic tape recordings were made in the lecture rooms, unoccupied, with the ventilation system on and off. These were analysed using 10 Hz constant bandwidth and 6% bandwidth audiospectrometers. In order to identify the noise sources, measurements were taken in some of the plant-rooms. The spectra for the room and the plant-room were then compared and the spectral components transmitted from the plant identified. Figure 5.16 shows a typical plant-room and measurement positions.

The noise spectra for the plant-room are shown in Fig. 5.17a–d for two fan speeds. It can be seen that at fast speed fewer peaks are present than from the lower speed. This suggests that as the speed increases the broadband aerodynamic noise in the fan

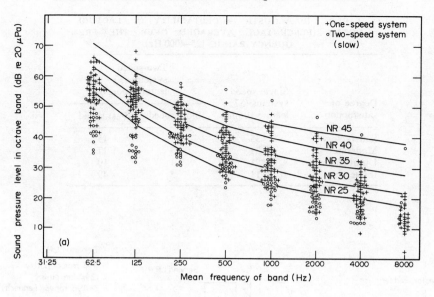

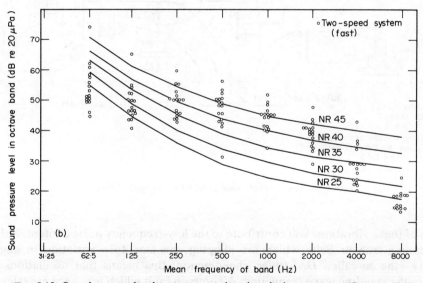

FIG. 5.15. Sound pressure level measurements in university lecture rooms (Croome, 1969).

gradually masks the discrete noise from mechanical sources and even the fundamental fan blade passage frequency and its harmonics. In this ventilation plant the major
mechanical noise sources were the V-belt and the fan bearings.

Many factors contribute to the noise which can be emitted from V-belt drive
systems. Firstly, the belt travels over pulleys and can vibrate along its passage due to
the formation of standing waves. The fundamental frequencies for this installation
are 10 Hz and 3 Hz on low speed and twice these values on high speed; the higher

TABLE 5.10. ACOUSTIC ACCEPTABILITY OF LECTURE ROOMS (PERCENTAGE AVERAGED OVER THE FREQUENCY RANGE 125–4000 Hz)

Degree of satisfaction	Single speed systems (57 lecture rooms)	Two-speed systems (17 lecture rooms)	
		Slow speed	Fast speed
Satisfactory	35	81	15
Moderate	31	13	17
Unsatisfactory	21	1	19
Intolerable	13	5	49

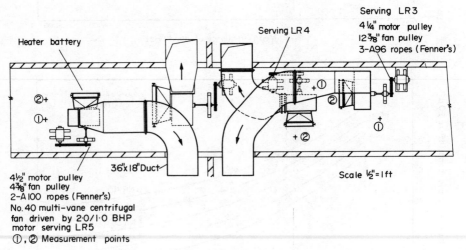

FIG. 5.16. Typical ventilation plant-room (Croome, 1969).

modes of these vibrations will contribute to the low-frequency noise content. Pulleys not running true or belts which are slipping can lead to a variation in angular velocity—the so-called *Drehfehler* phenomenon; this means that oscillations will occur in the standing wave frequency. A common fault which was very prevalent on this system was due to belt joints which gave a characteristic belt squeak each time contact was made with the pulley. Referring to Fig. 5.17a and b it can be seen that at 420–510 Hz discrete frequencies are emitted which correspond to the belt-joint noise. The reason for the wide spread is that three belts are used which will not necessarily all sound at the same instant, and, further, each joint will not necessarily cause the same note to be emitted on contact with the small or large pulley. This sound has a high harmonic content which can be seen at around 900, 1350, 1800, 2250 and 2700 Hz.

This plant delivered air to the lecture room whose ventilation system noise spectrum is shown in Fig. 5.17e. It can be seen that at the lower fan speed some discrete frequency components are still present, otherwise the spectrum has a broadband character. The geometrical form of the air distribution duct influences the

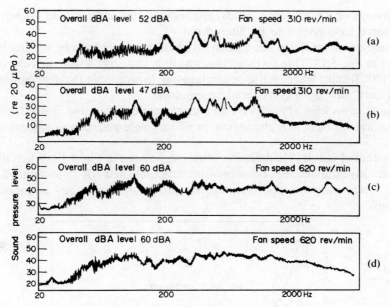

Sound spectra measured in plant-room shown in Fig. 5.16; 6% bandwidth resolving power. (See Fig. 3.7a for a narrow band analysis (Croome, 1969)).

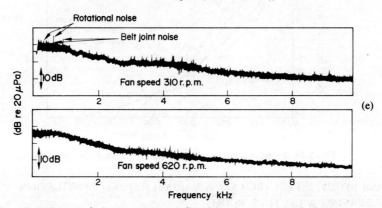

Soundspectra in lecture room ventilated by system **e** (a)–(d) above.

FIG. 5.17.

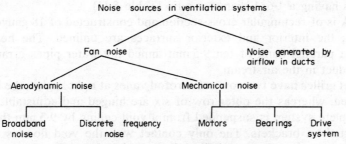

FIG. 5.18. Noise sources in ventilation systems.

transmission of sound since bends, expanders, contractions and branch off-takes cause some frequencies to be filtered out.

The noise sources in the ventilation installations studied can be summarised by the diagram in Fig. 5.18. This investigation was only concerned with air velocities below 5 m/s (1000 ft/min) where noise generation due to airflow in the distribution network has up until now been considered unimportant.

In some cases high noise levels are due to careless planning. Some lecture rooms were positioned next to a plant-room or had a single glazed wall adjacent to a main road.

The effect of careless installation work can be seen in Fig. 5.19. When the damper was correctly adjusted the sound pressure level satisfied NR 25 to within 2 dB above 500 Hz. Commissioning and maintenance procedures must pay attention to noise problems.

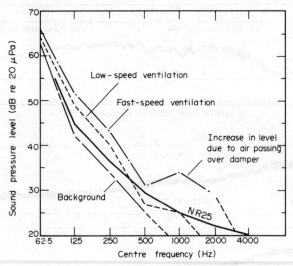

FIG. 5.19. Effect of incorrect damper adjustment on noise spectrum.

5.8.2. CASE STUDY 2: RESEARCH ON A VARIABLE-SPEED VENTILATION SYSTEM SERVING A LECTURE ROOM

The system shown in Fig. 5.20 serves a seventy-four seat lecture theatre in a university. Air is distributed from a unit comprising a 444 mm diameter centrifugal fan with ten backward-curved blades. The drive shaft is mounted on ball-bearings and is driven by a variable-speed motor with four fixed speeds via a double V-belt and pulleys having a 1·6 : 1 diameter ratio (Table 5.11).

Ductwork is of rectangular cross-section and constructed of 18 gauge galvanised sheet steel; the interior and exterior surfaces are unlined. The heater battery comprises a parallel bank of ten 9·5 mm finned hot-water pipes arranged evenly across the duct in the airstream.

The outlet grilles have two rows of aerofoil vanes at right angles; the inner row of five are fixed, whereas the outer row of six are hinged and adjustable.

The complete system is suspended from a void ceiling by 9·5 mm threaded rods and supported on brackets. The only contact with the void floor is at the grille outlets.

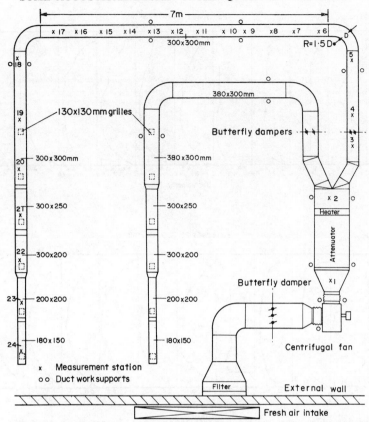

FIG. 5.20. Plan of plenum ventilation system in ceiling void over lecture room in the Department of Civil Engineering, University of Technology, Loughborough.

Measurements of the air velocity across the duct cross-section revealed the airflow profiles shown in Fig. 5.21 for the two lower fan speeds only. Air turbulence is evident at the bend between test stations 5 and 6. Later measurements, not recorded on Fig. 5.21, verified turbulence at test station 18 after the bend following test station 17. The profiles are consistent for the two fan speeds.

Now look at the in-duct sound profiles shown in Fig. 5.22. Attenuation of sound occurs at all frequencies, but this is offset almost entirely by generation of sound which is evident at each bend and at each fan speed except at the first bend for fan speed 2. The generation happens for all frequencies, but is most marked in the

TABLE 5.11. PERFORMANCE SPECIFICATION OF CENTRIFUGAL FAN UNIT

	Speed 1	Speed 2	Speed 3	Speed 4
Motor:				
No. of poles	4	6	8	12
Horse power	1·2	1·3	2·5	5·5
Speed (rev/min)	480	720	960	1440
Fan:				
Speed (rev/min)	760	1145	1530	2290
Volume flow rate (m³/s)	0·73	1·17	1·55	2·33

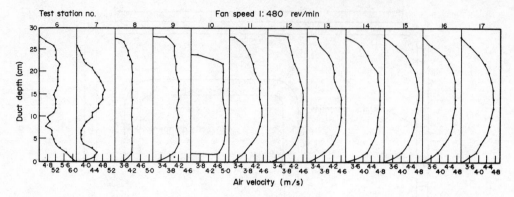

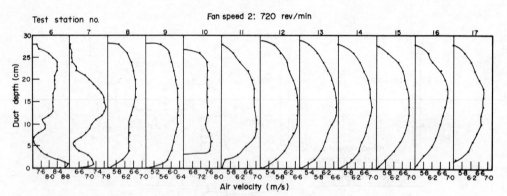

FIG. 5.21. Airflow profiles across duct between test stations 6 and 17; at test station 10 there is a constriction
due to a duct joint (Croome, 1976).

500–1000 Hz frequency range. Sound is generated as the air travels through the
attenuator at the three higher fan speeds. One significant conclusion is that *sound
generation is just as likely to occur in a low speed ventilation system (average air
velocity in this example of about 4 m/s) as well as in a high-speed system (average
velocity in this example of about 13 m/s).*

It is well established that ventilation systems noise has a large low-frequency
content. This work reaches a similar conclusion. At the lowest fan speed, sound in
the 31·5–250 Hz range has sound levels along the system of 63–88 dB (linear),
whereas sound in the frequency range 500–1000 Hz has levels in the range 39–75 dB
(linear). In either case the peak levels are due to sound generation. This pattern of
behaviour is similar at the higher fan speed although the frequencies in the range
500–1000 Hz gradually increases their share of the sound energy as the fan speed
increases.

The effect of sound generation commences at some distance before the bend; at
low speeds this distance is about 1 m but extends to 2 or 3 m as the air velocity
increases. The generation at the first bend (test stations 5 and 6) shows some
interaction with that which has preceded it at the dampers (test stations 3 and 4), the
breeches piece (test stations 2 and 3) and through the attenuator (test stations 1 and

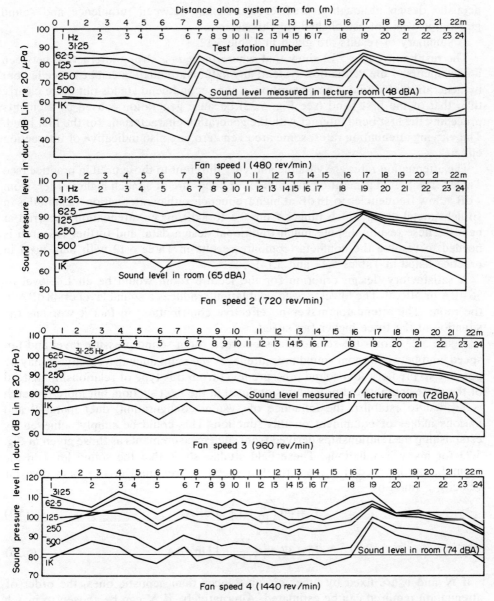

FIG. 5.22. Sound level profiles measured inside the duct in dB linear (Croome, 1976).

2); such an interaction also happens to some extent between the generation effects at the bends themselves. Duct resonances do not appear to be significant in this system as no fundamental of harmonic frequencies are filtered.

Notice that a given rise in sound level due to generation occurs over a much smaller distance than a similar decrease in sound level.

This work confirms the point that airflow systems should be designed to have the most simple configurations wherever possible. Good aerodynamic design will assist acoustic design although a precise relationship between turbulence and sound generation was not established for this system.

A summary of results are given in Table 5.12.

The results in Table 5.12 show that where there is a strong interaction between any elements of the ventilation system; any natural attenuation can be cancelled out by generation of sound. The generation at the second bend stands out more clearly than that of the first bend (see Fig. 5.22) because generation caused by elements preceding the first bend interact with the generation characteristic for the first bend. The varying attenuation rates, some are even zero, are also indicative of interaction effects.

The attenuation rates are variable and very different to those used in practice (see Table 5.13). Sound generation magnitudes at bends are of a much higher order, from 4 dB at low frequencies to 18 dB at high frequencies, than attenuation magnitudes in straight metal ducts which ranges from 0·5 to 3 dB/m; no attenuation occurred at bends. These results dispute much traditional design data, and further research is needed to resolve the conflicts (compare results in Table 5.12 with data used in practice listed in Tables 5.13 and 5.14).

A satisfactory design criterion for the lecture room would be an L_{10} level of 36 dBA or NR 30. The lowest speed ventilation produces a sound level of 48 dBA in the room. The attenuator makes no effective contribution; in fact it worsens the situation at the three higher fan speed.

Figure 5.23 shows the relationships between room sound pressure level and fan speed, and between room sound pressure level and main duct air velocity. It may be possible to predict room sound pressure level from this type of relationship instead of the more complicated procedures outlined in the next section, but more research is needed to establish the influence of system configuration, duct material and various makes of equipment on this function. This could be simply achieved by establishing the relationships derived from such measurements as those given in Fig. 5.23 for many installations. These field studies show that the sound level in the room L, can be predicted from the fan speed N (rev/min) or the main duct air velocity v (m/s) by

$$L = 40 \left(\frac{v}{17} + 1 \right) \text{ in dBA} \tag{5.1}$$

or

$$L = 37 \cdot 4 \left(\frac{N}{2110} + 1 \right) \text{ in dBA.} \tag{5.2a}$$

If N and v are fixed by other circumstances than acoustic ones, the order of attenuation required can be estimated. Alternatively, if N can be chosen to match the selected criterion for L, then this value of L can also be chosen to determine v, and hence the size of the main distribution duct can be calculated. Using Darcy's formula (see chapter 5 of Croome and Roberts, 1975) the room sound level may be related to pressure drop per unit length of main duct Δp by

$$L = 40 \left\{ \frac{1}{17} \left(\frac{2gm \Delta p}{f} \right)^{1/2} + 1 \right\} \text{ in dBA} \tag{5.3}$$

for a duct having a mean hydraulic depth m, friction factor f and where $g = 9 \cdot 81 \text{ m/s}^2$.

TABLE 5.12. SOUND ATTENUATION AND GENERATION IN VENTILATION SYSTEM OPERATING AT AIR SPEEDS OF 4-14 m/s

Fan speed (rev/min)	Average air velocity in 300×300 mm duct between test stations 6 and 17 (m/s)	Sound attenuation (−) in dB/m, and sound generation (+) in dB per bend[a]						Sound pressure level					
								Test station 1 (dB lin)		Test station 24 (dB lin)		Room	
		31·5 Hz	63 Hz	125 Hz	250 Hz	500 Hz	1000 Hz	<500 Hz	≥500 Hz	<500 Hz	≥500 Hz	dB lin	dBA
System off	—	—	—	—	—	—	—	—	—	—	—	—	25
480	4	−1·4 0+10 −2·0	−1·1 0+10 −2·0	−1·1 0+12 −2·0	−2·0 0+14 −2·5	−2·0 −1·2+16 −3·0	−3·0 −1·2+16 −3·0	76–88	62	72	54	62	48
720	7	0 0+0 −0·5	0 0+4 −0·5	0 0+8 −1	0 0+10 −1	0 0+14 −2·3	0 0+14 −2·3	86–100	66–74	76–84	54–64	67	65
960	8	0 −0·4+8 −2·0	0 −0·8+8 −2	0 −0·8+10 −2	0 −0·8+10 −2	−0·5 −1·6+16 −3·5	−0·5 −1·6+16 −3·5	92–104	72–82	82–90	64–70	78	72
1440	14	0 −0·8+4 −1·5	0 −0·8+4 −1·5	0 −0·8+10 −1·5	0 −0·8+10 −2	0 −1·2+18 −2·5	0 −2·0+18 −2·5	94–110	80–85	92–96	76–84	81	74

[a]These are average values; the sound generation figures are based on values measured at each 90° bends, and have neglected the sound generated in the attenuator, the breeches piece and the dampers. At each frequency there are three attenuation values given—one for the first 7 m of the system, the second for test stations 9–17 and the third for test stations 20–24.

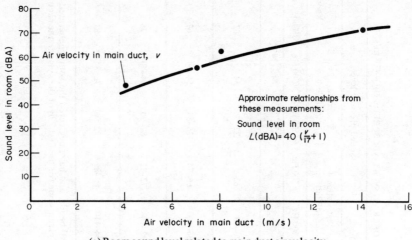

(a) Room sound level related to main duct air velocity.

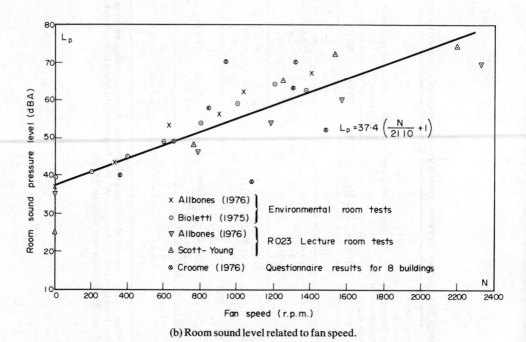

(b) Room sound level related to fan speed.

FIG. 5.23. Prediction of sound level in mechanically ventilated or airconditioned rooms (Croome, 1976).

Further work has established the following conclusions
(a) Formula (5.2) is more reliable than formula (5.1) and only applies to centrifugal fans.
(b) Natural attenuation and airflow generation effects are included in the empirical derivation of formula (5.2a), i.e. the system attenuation and generation which occurs between the fan and the measuring point in the room; additional attenuation, α, may be required and formula (5.2a) will be modified to

$$L = \left\{ 37 \cdot 4 \left(\frac{N}{2110} + 1 \right) \right\} - \alpha. \tag{5.2b}$$

Further analysis will be necessary to determine the frequency characteristics of α.
(c) An overall natural attenuation rate of 0·5 dB per metre is a conservative estimate which may be used in planning the distance between airconditioning plant and occupied spaces but any straight sections in the network which measure less than 7 m in length between fittings such as bends and branches, should be ignored when making this estimate. If L_p is the sound level in the plant-room and D is the length of the airflow ducts between the plant-room and the occupied space, less any straight runs which are less than 7 m, then the sound level in the room is

$$L = L_p - 0 \cdot 5D - \alpha. \tag{5.2c}$$

Thus plant selection, choice of fan speed, distance planning can be decided for a specified value of L. These factors may be combined in such a way that $\alpha = 0$ and no cost for sound control equipment is necessary, but in practice it is more likely that $\alpha > 0$. Good maintenance is also necessary to maintain the design operating conditions.
(d) The results give reliable predictions (± 5 dB) for centrifugal fan speeds below 900 rpm. Higher fan speeds give less reliable predictions.
(e) German research reported by Eck (1973) and illustrated in Chapter 3, Fig. 3.35b, shows that the loudness level of an axial fan is dependent on speed according to

$$L = 54 \left(\frac{N}{2160} + 1 \right) \text{ phons.}$$

(f) Adjustment of louvres in ceiling grilles can have a significant effect on the sound level measured in a room.

5.8.3. ESTIMATION OF BACKGROUND SOUND PRESSURE LEVEL IN A
ROOM SERVED BY AIRCONDITIONING OF VENTILATION SYSTEMS

The latter case study has shown that at this time the existing method of estimating the noise emission from an airconditioning system is not ideal; other studies by the author, but not related here, have also confirmed this. The method is described in the *Institution of Heating and Ventilation Engineers Guide, 1970*, Book B, chapter B 12, and also in the *American Society of Heating, Refrigeration and Airconditioning*

Engineers Guide 1973, Systems Volume, chapter 35; Reese and Gupta (1974) have modified the American method.

Until more experimental evidence is available, preferably in the nature of field studies, this method will continue to be used. The method is now outlined using data from the *IHVE Guide*.

Step 1. Estimate fan sound power level (L_W)

Use one of equations (3.9a), (3.9b) or (3.9c) repeated here:

$$L_w = 67 + 10 \lg P + 10 \lg p_s$$

or

$$L_w = 40 + 10 \lg Q + 20 \lg p_s$$

or

$$L_w = 105 + 20 \lg P - 10 \lg Q,$$

where Q is the delivered volume (m³/s), p_s is the fan static pressure (N/m²) and P is the rated fan motor power (kW).

L_w is power level from fan discharge *or* inlet; hence add 3 dB to allow for noise emitted from other end of fan.

Correct this value of L_w by using the typical fan noise spectra data shown in Fig. 5.24.

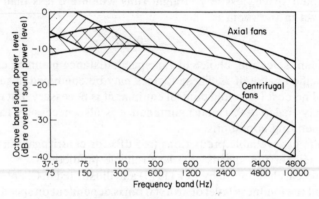

FIG. 5.24. Typical fan noise spectra (*IHVE Guide, 1970*).

This method of estimating the sound power levels of a fan assumes a broadband spectrum. There may be pure tones arising from the blade passage frequency f_b and therefore it is recommended that 5 dB should be added to the computed level of the octave band containing the blade passage frequency and 3 dB to its first harmonic h_1. These frequencies are given by

$$f_b = n \cdot N, \qquad h_1 = 2f_b,$$

where n is number of blades and N is fan speed in rev/s.

Step 2. Deduct arithmetically the attenuation of ducts, bends, branches, etc.

The attenuation of sound through the components of a ducted system is as follows:

Straight sheet metal ducts. Data on the attenuation of conventional plain rectangular and circular sheet metal ducts are given in Table 5.13.

TABLE 5.13. ATTENUATION IN STRAIGHT SHEET METAL
DUCTS (*Ihve Guide*, 1970)

Section	Mean duct dimension or diameter (mm)	Attenuation (dB/m) Octave band centre frequency (Hz)		
		125	250	500 and above
Rectangular	Up to 300	0·6	0·5	0·3
	300–450	0·6	0·3	0·3
	450–900	0·3	0·3	0·2
	Over 900	0·3	0·2	0·1
Circular	Up to 900	0·1	0·1	0·1
	Over 900	0·03	0·03	0·06

Bends. Radiused bends, with or without splitters, provide negligible attenuation. Square-back bends reflect some of the sound energy back towards the noise source, and this can significantly reduce the noise being propagated away from the source; this attenuation is shown in Table 5.14.

TABLE 5.14. ATTENUATION PROVIDED BY 90° MITRED
BENDS WITHOUT TURNING VANES (*Ihve Guide*, 1970)

Minimum duct side (mm)	Attenuation (dB) Octave band centre frequency (Hz)						
	125	250	500	1000	2000	4000	8000
150	0	0	1	6	7	4	4
300	0	1	6	7	4	4	4
450	0	3	7	5	4	4	4
600	1	6	7	4	4	4	4
750	3	7	5	4	4	4	4

Branches. For most typical duct branches it can be assumed that the acoustic energy divides in direct proportion to the areas of the duct branches (Fig. 5.25). That is the total acoustic energy remains constant but it is divided amongst the branches so that, in any one branch, the acoustic energy is less than that in the approach duct. The effective attenuation for a branch duct is given by

$$A = 10 \lg \left(\frac{a_1 + a_2}{a_1} \right),$$

where a_1 is the area of branch (m²) and a_2 is the area of main duct after branch (m²).

When the volumes divide and velocities are approximately equal, the calculation may be simplified by using air volume flow ratios instead of area ratios. Where two or more such branches ultimately serve the same room, the acoustic energy which is divided amongst the branches becomes additive again within the room.

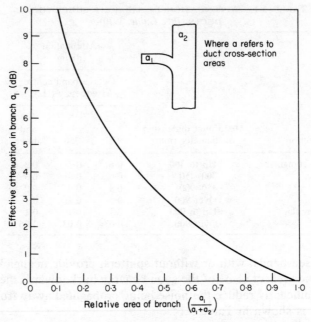

FIG. 5.25. Effective attenuation of a duct branch (*IHVE Guide, 1970*).

End Reflection. Sounds of wavelengths that are long with respect to the dimensions of the duct outlet tend to be reflected back within the duct rather than pass into the room. The theoretical value of the attenuation provided by end reflection at an opening is shown graphically in Fig. 5.26. It is recommended that this information be applied as follows:

(a) *Grilles.* Use Fig. 5.26 with duct dimension determined from core area of grille. For high aspect ratio grilles (linear grilles) this is not strictly accurate but may be used in absence of specific information.

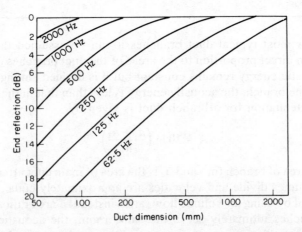

FIG. 5.26. Theoretical value of end reflection attenuation (*IHVE Guide, 1970*).

(b) *Ceiling diffusers.* Use Fig. 5.26 with duct dimension taken as $1 \cdot 25 \times$ neck diameter.

(c) *Induction units.* Figure 5.26 should not be applied to induction units. These have substantial insertion loss and this includes some end reflection. This insertion loss can be at least 10 dB in all octave bands; accurate data should be obtained from the manufacturer.

If a bend immediately precedes a grille or diffuser, the end reflection may be reduced. Where such an arrangement occurs the attenuation provided by end reflection should be taken as 50% of that recommended above.

Step 3. Add, logarithmically (decibel sum) the sound power level generated by fittings, etc.

Each item must be added at the correct stage of the path from fan to room so that it will receive the attenuation of all remaining "room side" components. The overall sound power level generated by a duct fitting can be approximately expressed by the equation:

$$L_w = C + 10 \lg A + 60 \lg v,$$

where C is the constant that depends on fitting and the flow turbulence, A is the minimum flow area of fitting (m²) and v is the maximum velocity in fitting (m/s).

Typical values of C for a range of duct fittings in low turbulence flow conditions and the corrections to obtain the octave band power levels are given in Table 5.15. These data provide some guidance on airflow generated noise for velocities between 10 and 30 m/s.

The octave band sound power levels for airflow generated noise at grilles and diffusers should be obtained direct from the manufacturer. This information will refer to uniform flow conditions at entry to the terminal, for a bad entry higher generated noise levels can be expected.

In the absence of manufacturer's test data some idea of the generated noise level can be obtained from the following empirical relationships. However, these should not be used for an accurate assessment:

grilles:

$$L_w = 17 + 10 \lg A + 60 \lg v;$$

diffusers:

$$L_w = 32 + 13 \lg A + 60 \lg v,$$

where A is the minimum open area of fitting (m²) and v is the air velocity at opening (m/s).

If the vanes of the grille are set for a wide spread, the constant in the first of these equations should be increased to 21. To determine the spectrum from these equations use Table 5.16.

Step 4. Add, logarithmically, the sound power level entering the room from all outlets served by the fan

Step 5. Calculate sound pressure level at a receiving position

Note that the combination of steps 1, 2 and 3 equals the sound power level

TABLE 5.15. CORRECTION FOR LOW TURBULENCE DUCT FITTINGS (*Ihve Guide*, 1970)

Duct fitting	Value of C (dB)	Remarks	Octave band power level corrections (dB) Octave band centre frequencies (Hz)						
			125	250	500	1000	2000	4000	8000
Straight duct	−10	No internal projection	−2	−7	−8	−10	−12	−15	−19
90° radiused bend	0	Aspect ratio ≤2:1 Throat radius ≥w/2	−2	−7	−8	−10	−12	−15	−19
90° square bend with turning vanes	+10	Close spaced short radius single skin turning vanes	−2	−7	−8	−10	−12	−15	−19
Gradual contraction	+1	Area ratio 3:1A and v taken for smaller duct	0	−10	−16	−20	−22	−25	−30
Sudden contraction	+4	Area ratio 3:1A and v taken for smaller duct	0	−10	−16	−20	−22	−25	−30
Butterfly damper	−5	A and v apply to minimum free area at damper	−3	−9	−9	−10	−17	−20	−24

TABLE 5.16. CORRECTIONS TO OVERALL SOUND POWER LEVEL FOR DIFFUSERS AND GRILLES (*Ihve Guide*, 1970)

$\dfrac{\sqrt{A}}{v}$	Corrections (dB) Octave band centre frequency (Hz)						
	125	250	500	1000	2000	4000	8000
0·01	− 10	− 8	− 7	− 6	− 6	− 7	− 12
0·02	− 8	− 7	− 6	− 6	− 7	− 12	− 21
0·04	− 7	− 6	− 6	− 7	− 12	− 21	− 30
0·06	− 6	− 5	− 6	− 10	− 17	− 26	− 35
0·08	− 6	− 6	− 7	− 12	− 21	− 30	− 40
0·10	− 5	− 6	− 8	− 15	− 24	− 33	− 43

radiated from one terminal. This data will be required for the calculation of sound pressure level at a particular receiving position.

The sound pressure level at a receiver in a room due to sound being radiated from a terminal device (or unity equipment) is given by

$$L_p = L_w + 10 \lg \left(\frac{4}{R} + \frac{Q}{4\pi r^2} \right)$$

where R is the room constant (m²), Q is the directivity factor for terminal and r is the distance of receiver from terminal (m).

L_p can be computed as follows:

Let $L_p = L_w + C_1 + C_2 + C_3 + C_4$,

where C_1 is the correction for room acoustic characteristics given in Table 5.17; C_2 is the correction for total room surface area given in Table 5.18; C_3 is the correction for distance between terminal and receiver given in Table 5.19, and C_4 is the correction for directivity given in Table 5.20.

The correction for directivity depends on both the mounting position and the area of the outlet. For surface-mounted terminals three possible arrangements are

TABLE 5.17. CORRECTIONS FOR ROOM ACOUSTIC CHARACTERISTICS (*Ihve Guide*, 1970)

Description of room acoustic characteristics	Correction C_1 (dB) Octave band centre frequency (Hz)						
	125	250	500	1000	2000	4000	8000
Live	+ 16	+ 15	+ 14	+ 12	+ 13	+ 15	+ 16
Medium live	+ 11	+ 11	+ 9	+ 7	+ 6	+ 6	+ 6
Average	+ 11	+ 9	+ 7	+ 5	+ 4	+ 3	+ 3
Medium dead	+ 9	+ 6	+ 5	+ 3	+ 2	+ 1	+ 1
Dead	+ 6	+ 4	+ 2	+ 0	+ 1	+ 1	+ 1

specified as shown in Fig. 5.27. These are:

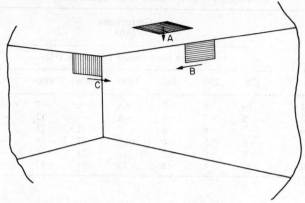

FIG. 5.27. Outlet locations (*IHVE Guide*, 1970).

Position *A*—more than 1 m from any other major room surface.
Position *B*—closer than 1 m to one other major room surface.
Position *C*—closer than 1 m to two other major room surfaces

TABLE 5.18. CORRECTIONS FOR TOTAL ROOM SURFACE AREA (*Ihve Guide*, 1970)

Total room surface area (m²)	Correction C_2 (dB)
25	− 8
50	− 11
100	− 14
250	− 18
500	− 21
1000	− 24
2500	− 28
5000	− 31
10000	− 34

TABLE 5.19. CORRECTIONS FOR THE DISTANCE BETWEEN THE TERMINAL AND RECEIVER (*Ihve Guide*, 1970)

Distance (m)	Correction C_3 (dB)
1	− 11
1·5	− 14
2	− 17
2·5	− 19
3	− 21
4	− 23
5	− 25
7	− 28

Live. Rooms with hard and heavy surfaces, with no soft furnishings and without any acoustical treatment or fittings of absorbing material.

Medium live. Rooms with hard surfaces (e.g. panel construction with no special acoustical treatment) but with some absorbent content (e.g. people, covered chairs, or limited soft furnishings).

Average. Rooms which have acoustical ceilings or appreciable soft furnishings (e.g. carpeted or upholstered and soft drapes).

Medium dead. Rooms which have both acoustical ceilings and appreciable soft furnishings.

Dead. Rooms which have been specially treated to absorb sound.

The sound pressure level at a position in a room is computed as the decibel sum of the reverberant and direct sound pressure levels. In practice for a particular position

TABLE 5.20. CORRECTIONS FOR DIRECTIVITY (*Ihve Guide*, 1970)

Mounting position	Outlet area (m²)	Correction C_4 (dB) Octave band centre frequency (Hz)						
		125	250	500	1000	2000	4000	8000
A	0·01	+3	+4	+5	+6	+7	+8	+8
	0·05	+4	+5	+6	+7	+8	+8	+9
	0·1	+4	+6	+7	+8	+9	+9	+9
	0·25	+5	+7	+7	+8	+9	+9	+9
	0·1	+7	+8	+8	+9	+9	+9	+9
	10·0	+9	+9	+9	+9	+9	+9	+9
B	0·01	+6	+6	+7	+7	+8	+8	+9
	0·05	+7	+7	+8	+8	+9	+9	+9
	0·1	+7	+8	+8	+9	+9	+9	+9
	0·25	+8	+8	+8	+9	+9	+9	+9
	1.0	+8	+9	+9	+9	+9	+9	+9
	10·0	+9	+9	+9	+9	+9	+9	+9
C	All areas	+9	+9	+9	+9	+9	+9	+9

the nearest single outlet only is normally considered. However, if another outlet is no more than a further 1 m distance this should be included in the determination of the direct sound pressure level for that position. If there are a number of similar terminals in the room this gives the total sound power level from all the terminals. The total permissible sound power output is equal to

$$L_{wm} = L_w + 10 \lg N,$$

where L_{wm} is the total sound power level, L_w is the sound power level for one outlet and N is the number of terminals in room.

Estimate contribution of external noise to internal sound pressure level.

The sound pressure level in a room due to outside noise is given by

$$L_{p2} = L_{p1} - R + 10 \lg \frac{S}{A},$$

where L_{p1} is the sound pressure level outside, L_{p2} is the sound pressure level inside, R is the sound reduction index of wall or roof, A is the total absorption in room and S is the area of, say, wall exposed to external noise (m²).

Step 6. Compare background sound pressure level obtained in step 5 with design criterion

At frequencies where the background sound pressure level exceeds the design, criterion attenuation or redesign is necessary.

A worked example is given in the *IHVE Guide, 1970*, Book B, chapter 12.

PART III

Some Fundamentals of Acoustics

CHAPTER 6

CAUSES AND PATHWAYS OF
MECHANICAL VIBRATIONS

6.1. Sources and Pathways of Sound

Referring to Fig. 6.1a, during source operation some energy, from that imported into the system by the periodic driving force, will be transferred to the source structure. From here the energy of vibration in the structure will then disturb the air surrounding it and hence generate *airborne sound*. Vibrations will also be conducted into any solids with which the source has contact, thus originating *structureborne sound* (referred to in the building services trade simply as "vibration"—a misleading term since it could equally apply to airborne sound). The apportionment of the vibrational energy depends on the mass, damping capacity and stiffness of the source structure, of the air surrounding it, and of the solid with which the source has contact, besides the operating conditions (e.g. speed of operation, the presence of out-of-balance forces). The sound propagated through the air or the structure may be sensed either by a measuring instrument or the human sensory organs; struc-tureborne sound may be sensed as surface displacements or as airborne sound resulting from these surface displacements (Fig. 6.1b).

The action of vibration force $F_0 \sin \omega t$, on a body of mass m, is opposed by the stiffness and the damping forces, $r\dot{x}$ and kx respectively, of the medium, the mass and its surroundings, besides the inertia force $(m\ddot{x})$ of the mass itself. Thus balancing the force causing motion with those forces opposing it gives the fundamental equation of motion

$$m\ddot{x} + r\dot{x} + kx = F_0 \sin \omega t \qquad (6.1)$$

for a displacement x, velocity $\dot{x}$, acceleration $\ddot{x}$ and angular frequency ω over a time t. This is usually represented by the mass, spring, damper models which will be referred to again later in this chapter shown in Fig. 6.2. Any vibration problem needs a consideration of this equation in one form or another; sometimes it is a simpler version but often it is more complex.

The vibrating source may take several forms. For building environmental engineers centrifugal fans, axial-flow fans, centrifugal and reciprocating refrigerating machines are the chief sources of concern, but in everyday life there are many more—the human voice speaking or singing, an object falling on a surface, a violin playing, a church bell ringing. They all share a common pattern—*excitation of the source then partial energy transfer via its structure to the surroundings as struc-tureborne and airborne sound*, the remaining energy is dissipated as heat within the structure of the source. Other types of sound source exist; these have an *aerodynamic*

417

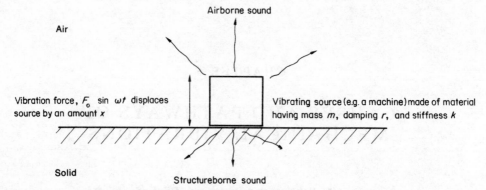

(a) For source $m\ddot{x} + r\dot{x} + kx = F_0 \sin \omega t$. Amount of sound transferred to air and solid depends on the degree of mismatch between the mass, damping and stiffness properties of the source and the air for airborne sound, and the source and the solid for structureborne sound. For good sound control:

$$(m\ddot{x} + r\dot{x} + kx)_{air} \ll (m\ddot{x} + r\dot{x} + kx)_{source},$$

$$(m\ddot{x} + r\dot{x} + kx)_{solid} \ll (m\ddot{x} + r\dot{x} + kx)_{source}.$$

Sound control equipment is needed usually to achieve this.

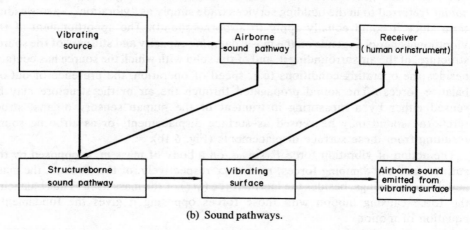

(b) Sound pathways.

FIG. 6.1. Source sound emission and pathways.

or an *hydrodynamic* nature, and are referred to by Lighthill (1952) as *pseudosound* sources in order to distinguish them from sound radiated from vibrating structures already described. Pseudosound originates from eddies, vortices and general turbulence within the fluid which cause fluctuations in the static pressure field (section 3.3). Specific instances found in airconditioning systems are the impact of flow on sound-absorbent linings, turbulent flow occurring in bends or other changes of duct geometry, boundary layer turbulence and wakes produced behind fan blades.

Sound can disturb the human operator by interfering with speech, communication, sleep or other everyday tasks; it can cause annoyance and it may also be physiologically damaging, producing temporary or permanent deafness. In these situations sound is harmful and is then referred to as *noise*. Just as there are optimum patterns of temperature, relative humidity, air movement and lighting, preferable

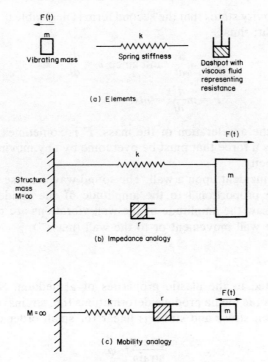

FIG. 6.2. Mass, spring, damper models and analogies.

auditory environmental patterns also exist. Sound is an important parameter in the environmental building design process.

6.2. Mass, Stiffness and Damping

The essence of this chapter can be appreciated by studying the diagram shown in Fig. 6.1a illustrating a source of vibration which might be a fan or some other machine. The machine, the building structure with which it is in contact, and the air around it, all possess the inherent properties of *mass, damping capacity* and *stiffness*. These three properties will each be considered in turn.

6.2.1. MASS

The influence of *mass* in any dynamic system is evident from Newton's Second Law of Motion which postulates that the force acting on a body is equal to its rate of change of momentum.

For a mass m travelling with a velocity v the force acting is

$$F = \frac{d(mv)}{dt}$$

$$= m\frac{dv}{dt} + v\frac{dm}{dt}.$$

The theory of relativity states that the second term is negligible if $v \ll v_L$, where v_L is the velocity of light; thus

$$F = m\frac{dv}{dt} \text{ and since } v = \frac{dx}{dt} = \dot{x},$$

$$F = m\frac{d^2x}{dt^2} = m\ddot{x},$$

where d^2x/dt^2 is the acceleration of the mass. F is sometimes referred to as the *inertia* force and is a force that must be overcome by any moving body in order to attain a given velocity.

Consider sound incident upon a wall. The sound waves cause the wall to vibrate with an amplitude proportional to the amplitude of the incident sound. But by Newtons Second Law the amplitude of the wall vibrations are proportional to the acceleration of the wall movement or to the wall (mass^{-1}).

6.2.2. STIFFNESS

Stiffness is related to the elastic properties of a medium. Stresses applied to homogeneous isotropic media produce deformations (i.e. strains) in the medium; the relationship between stress and strain is linear for small order changes, and

$$\frac{\text{stress}}{\text{strain}} = \phi,$$

where ϕ is an elastic modulus. If for a given stress the strain is small, then ϕ is high, and consequently the stiffness is also high. When a load is removed the medium will revert to its original dimensions for a fatigue and creep-free specimen acting in the elastic range. If the stress is of a tensile or compressive nature and applied unidirectionally, ϕ is called *Young's modulus*, but if applied omnidirectionally producing volumetric strains, ϕ is referred to as a *bulk modulus*; unidirectional shear stresses produce shear strains related by a shear (or rigidity) modulus. In acoustics the bulk modulus is of prime concern because the acoustic pressure fields are usually omnidirectional.

The bulk modulus of elasticity K for a gas or a solid is given by

$$K = -\frac{p}{\Delta V/V}.$$

This expression states that a pressure rise p acting on an element having a volume V produces a decrease in volume or *condensation* ΔV; likewise, a reduced pressure within the medium will cause an increase in volume or dilatation. The reciprocal of K is termed the compressibility. If ΔV is small, then the compressibility is low (i.e. the stiffness is high).

A force acting on a material in the direction i deforms it by an amount dx_i, the stress is defined as

$$\text{stress} = F_i/x_i x_j$$

and the strain by

$$\text{strain} = dx_i/x_i.$$

If the material has an elastic modulus Y, then

$$F_i = \frac{Y x_i x_j \, dx_i}{x_i}$$

$$= k \, dx_i.$$

The force F_i is related to the strain energy stored in the material, and the *stiffness factor* k is defined as the force per unit deformation (i.e. units of k will be N/m). This system is usually modelled by a spring of stiffness k and deflection or displacement x. Notice that the force and displacement are in phase and that the stiffness factor of a component depends on its shape and the material from which it is made. As an example the shear stiffness of a rubber block thickness h and cross-sectional area A is $k = GA/h$, where G is the shear modulus for a rubber having a given hardness. Often it is assumed that there is a linear relationship between the force and the displacement for a given material, but real systems often exhibit non-linear characteristics.

Forces originating from mechanical vibration sources are usually cyclical in nature and materials used to control the propagation of these vibrations are dynamically tested and performance specified in terms of their *dynamic* as well as their *static* properties. Baker (1974) relates the dynamic stiffness k_d and static stiffness k_s by

$$k_d = \alpha k_s$$

and gives α as 1·2–1·5, and even as high as 2 in some cases.

6.2.3. DAMPING

Damping capacity refers to the rate of loss of vibrational energy in a substance. Everyone is familiar with the "ring" of a wine glass; customers clink pottery to hear if it rings, and if it does this indicates that the sound is travelling through a relatively unstressed structure. The secret of preparing "pure" structures is endowed in the art of bell casting for example.

The following extract is a translation taken from the Latin work *De campanis fundendis* (*How to Cast a Bell*) by the eleventh-century monk, Theophilus:

> Take great care in mixing the tin and copper in the proper proportions before casting the bell. Only in this way will it ring properly. If you don't take care, the chime will be out of tune. The metal should be one fifth tin, and four fifths copper to sound right. Make sure the metal is well refined.

And later:

> When you have cast your bell in the way I have described, if it is out of tune, you can correct it. If you want the note higher, file the bell on its underside; if you want it lower, file around the outside of the rim.

H. B. Walters in his book *Church Bells of England* published in 1912 by Oxford

University Press, refers to the treatise on bell casting by Walter de Odyngton, a
monk of Evesham, in the reign of Henry III and concludes:

> *The best bells have what is called a silvery tone, and stories are constantly told of*
> *silver tankards and the like being cast into the furnace to improve the sound. When*
> *Tancho, a monk of St. Gall, was casting a fine bell for Charlemagne's church at*
> *Aachen, he asked for 100 lbs. of silver. But this may only have been to purchase*
> *copper and tin, or to defray the expenses of casting. If he had thrown the silver into*
> *the furnace, it would certainly not have produced a fine bell, in fact the result would*
> *have been much the same if he had thrown in a hundred pounds of lead. It is the tin*
> *in the alloy which gives brilliancy of tone, and the addition of silver would only*
> *impair it.*

These familiar cases indicate the nature of internal damping capacity; when it is
low the vibrations travel easily through the material. The motor-car engine is usually
enclosed by a steel or cast-iron casing; when the casing material is a manganese–
copper alloy the sound emission is considerably reduced due to the increased
damping capacity of the material. Another example is the vibratory parts feeder
which is a common noise source in many manufacturing plants. Noise is radiated
from the bowl alone, from the parts fed in to it impacting the bowl and by the parts
impacting each other. Figure 6.3, from the work of Warnaka *et al.* (1972), shows how
the noise radiated from the bowl alone is reduced by 15 dBA by constructing the
bowl from cast aluminium which has a higher damping capacity than steel. The
impact noise was reduced by applying a damping material to the inside surface of the
bowl.

Physicists often refer to internal damping as *internal friction*; certainly it can be
considered as an internal resistance to the motion of the vibrations. The main causes
of this phenomenon in the case of solids arise from the ordering of the atoms within a
material, lattice imperfections, slip along grain boundaries, thermoelasticity, scatter-

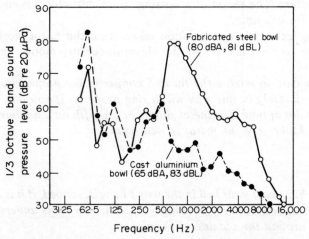

FIG. 6.3. Noise radiated from 18 in. (46 cm) metal bowl of a vibratory feeder with no parts in the bowl
showing the difference due to material of construction of the bowl (Warnaka *et al.*, 1972).

ing of acoustic waves by the grains of a material and energy losses due to interactions with internal magnetic fields. Damping in polymers like rubber and plastics is mainly of a viscous nature, the long-chained molecules acting with some similarity in behaviour to a fluid. Gases attenuate sound energy due to internal absorption caused by viscous effects, heat transfer by conduction and radiation between compression and rarefraction regions, diffusion of gas molecules from compression to rarefraction areas and energy exchanges occurring between the internal degrees of freedom of the gas.

The effects of damping in structures excited by vibrating forces are to reduce the vibration amplitudes at resonance and consequently reduce the stresses, the structural fatigue and the airborne sound radiated from the structure; to increase the rate of decay of the free vibrations; to reduce structureborne vibration (see Chapter 9); to reduce the effects of coincidence in the sound transmission characteristic of the structure (see Chapter 11).

When a material is deformed, part of the energy is stored as elastic strain energy. On removal of the deforming forces some of the energy is recoverable, some is lost and dissipated as heat. It is this energy loss which is referred to as *internal* or *material damping*.

A vibrating structure has kinetic energy and potential or elastic strain energy. For a structure mass m, stiffness k, displaced by x, the total energy is

$$E = \tfrac{1}{2} m \dot{x}^2 \text{ (kinetic)} + \tfrac{1}{2} k x^2 \text{ (potential)}. \tag{6.2}$$

In practice the damping energy ED is lost, thus

$$E = \tfrac{1}{2} m \dot{x}^2 + \tfrac{1}{2} k x^2 + E_D, \tag{6.3}$$

$$E = E_k + E_p + E_D. \tag{6.4}$$

In the case of lead, most of the energy used to deform it is lost, and consequently the internal damping capacity is high. The nature of the damping force is likened to the viscous damping mechanism more commonly spoken of with respect to liquids, but is present, of course, in gases and solids too. In this case, the force in a direction i is defined as

$$F_i = \mu x_i x_j \frac{dv_i}{dx_j} \tag{6.5}$$

or
$$F_i = r \, dv_i, \tag{6.6}$$

where r is the damping constant related to the material's viscous properties designated by the factor μ; dv_i/dx_j is the velocity gradient and $x_i x_i$ is an elemental cross-sectional area. In contrast to the stiffness force ($F \propto x$), the damping force ($F \propto \dot{x}$) is $\pi/2$ out of phase with the displacement. This force is usually modelled by a dashpot having a plunger operating against a fluid with given viscous properties.

Under the action of a sinusoidal force $F_0 \sin \omega t$, a material has an in-phase modulus Y_T in tension, and an out-of-phase modulus Y_C in compression. For a material of cross-sectional area A and thickness h, the spring stiffness is

$$k = \frac{Y_T A}{h} \tag{6.7}$$

and for an angular frequency ω, the damping constant is given by

$$r = \frac{Y_C A}{\omega h}.$$ (6.8)

It should be remembered that k and r may vary with amplitude, frequency and temperature.

The resultant modulus Y_R is

$$Y_R = (Y_T^2 + Y_C^2)^{\frac{1}{2}}$$ (6.9)

and the phase difference between the stress and strain is called the *loss angle* ϕ, the *loss factor* being defined as

$$\eta = \tan \phi = \frac{Y_C}{Y_T}$$ (6.10)

where $Y_T = Y_R \cos \phi$ and $Y_C = Y_R \sin \phi$. For shear tension and compression corresponding equations apply using shear moduli.

The damping properties of a material may also be stated in terms of energy lost per cycle of vibration, closed loop stress–strain patterns occurring when a material is vibrated; internal friction can be treated as a mechanical hysteresis phenomenon. The *specific damping capacity* of a uniformly stressed material is defined as

$$\psi = \frac{\text{specific damping energy}}{2 \text{ maximum potential energy per unit volume}}$$

$$= \frac{E_D}{2E_p}$$ (6.11)

which for a stress amplitude δ_0 and strain ϵ_0 can be expressed as

$$\psi = \frac{\pi \delta_0 \epsilon_0 \sin \phi}{2(\frac{1}{2}\delta_0 \cos \phi \, \epsilon_0)}$$

$$\psi = \pi \tan \phi.$$ (6.12)

The *loss factor* η is the energy dissipated per radian

$$\eta = \frac{E_D/2\pi}{E_p}.$$ (6.13)

If the displacement is expressed in the sinusoidal form $x = x_0 \sin (\omega t + \phi)$, then the kinetic energies and potential energies take the form

$$E_k = \tfrac{1}{2}m\dot{x}^2 = \tfrac{1}{2}mx_0^2\omega^2 \cos^2 (\omega t + \phi),$$

$$E_p = \tfrac{1}{2}kx^2 = \tfrac{1}{2}kx_0^2 \sin^2 (\omega t + \phi).$$

It has been seen already that the equation of motion for a forced-damped vibrating system is $m\ddot{x} + r\dot{x} + kx = F(t)$, where $r\dot{x}$ is the damping force and r is the viscous damping coefficient (see equation (6.1)) thus the specific damping energy E_D is given by

$$E_D = (r\dot{x})(dx) = r\dot{x}^2 \, dt.$$

Integrating over a whole cycle,

$$E_D = rx_0^2\omega \int_0^{2\pi} \cos^2(\omega t + \phi)\, d(\omega t),$$

$$E_D = \pi rx_0^2\omega \quad \text{but} \quad \psi = \frac{E_D}{\frac{1}{2}kx_0^2}.$$

Therefore $$\psi = \frac{2\pi\omega}{k}.$$ (6.14)

This can be expressed in terms of the damping ratio D, where $D = r/r_c$ and r_c is the *critical damping coefficient*, and since $r_c = 2\sqrt{km} = 2\,m\omega_n$ then remembering that the natural angular frequency $\omega_n = \sqrt{k/m}$, the specific damping capacity can be expressed as

$$\psi = 4\pi D\left(\frac{\omega}{\omega_n}\right).$$ (6.15)

The loss factor can also be expressed in terms of the damping ratio

$$\eta = \frac{E_D/2\pi}{\frac{1}{2}kx_0^2} = \frac{\pi rx_0^2\omega/2\pi}{\frac{1}{2}kx_0^2}$$

$$= \frac{\omega r}{k}$$

$$= 2D\left(\frac{\omega}{\omega_n}\right).$$ (6.16)

At resonance $\omega = \omega_n$ and $\eta = 2D$.

The decay rate Δ_t is defined as the rate of decrease of sound pressure level L_p with time.

$$\Delta_t = -\frac{dL_p}{dt}$$

and L_p is defined by $20\lg(x/x_0)$ in decibels, hence

$$\Delta_t = -\frac{20}{2\cdot3}\frac{d}{dt}\left(\ln\frac{x}{x_0}\right),$$

$$\Delta_t = -8\cdot69\frac{\dot{x}}{x},$$

$$\Delta_t = -8\cdot69\left(\frac{r}{r_c}\right)\omega_n \quad \text{in dB/s},$$

$$\Delta_t = -8\cdot69D\omega_n.$$ (6.17)

This assumes that the displacement variation in time can be expressed in the form

$$x = x_0\exp(-D\omega_n t)\cos(\omega t + \phi),$$

where $\omega = \omega_n\sqrt{1 - D^2}$. If at times t_0 and t_N, $\cos(\omega t + \phi) = 1$, then for $N = 1, 2, 3, \ldots$,

$$\frac{x(t_0)}{x(t_N)} = \exp(2\pi ND)$$

because $t_N = t_0 + (2\pi N/\omega_n)$. The *logarithmic decrement* is defined as

$$\delta = \frac{1}{N} \ln \frac{x(t_0)}{x(t_N)} = 2\pi D. \tag{6.18}$$

In Chapter 10 the *reverberation time* T will be defined as $\Delta_t = 60/T$ dB/s and thus may be related to the damping ratio by using equation (6.17).

Summarising, the loss factor of a material can be expressed in several ways which do not just apply to viscous damping, although in practice an equivalent viscous damping coefficient is commonly used. For small damping at resonance, and assuming a uniform stress distribution,

$$\eta = \frac{\psi}{2\pi} = 2D = 2\frac{r}{r_c} = \frac{2 \cdot 2}{f_n T} = \frac{\Delta_t}{27 \cdot 3 f_n} = \frac{\delta}{\pi} = b = \frac{1}{Q}. \tag{6.19}$$

All the symbols have been defined in the text and also on Fig. 6.4.

Many metals have low damping properties but some alloys (e.g. manganese copper) exhibit high damping. Lead is an exceptional metal in that it has high damping properties; plastics have high values also. Applications of damping will be considered in Chapter 9.

At this point the distinction between *material* and *Coulomb* damping should be made. Material damping depends on the material composition and internal molecular structure, on the stress amplitude and the modulus properties; it acts throughout the solid, whereas Coulomb damping, or slip damping, occurs when two bodies held together by pressure move relative to one another. As such it is an interface process and depends on the coefficient of friction, pressure, shear stress–strain distribution

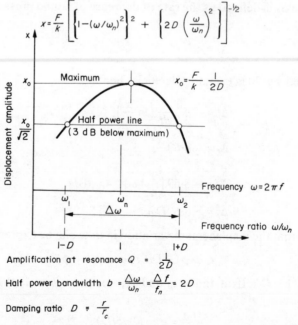

FIG. 6.4. Relationship between displacement and vibration properties at given frequencies.

and the interface geometry force. Thus for an interface area A and pressure p the Coulomb damping force is

$$F_c = \beta p A,$$

where β is the coefficient of friction which at the commencement of a movement is higher (the static coefficient of friction) than a short time later when a lower steady-state value defines the force of the motion (the dynamic coefficient of friction). It has already been seen that material damping is modelled by an equivalent viscous force expressed as (see equation (6.5)),

$$F_v = \mu A \frac{dv_i}{dx_j}.$$

The equation of motion for a mass-spring dashpot isolator system has been given as

$$m\ddot{x} + r\dot{x} + kx = F_0 \sin \omega t,$$

whereas for a Coulomb damper system the equation of motion is

$$m\ddot{x} + k\dot{x} \pm F_c = F_0 \sin \omega t.$$

6.3. Impedance Model of Sound Sources and Pathways

Sound pathways can be illustrated by means of an impedance model. The impedance concept will first be explained using electrical and mechanical analogies.

Any energy transfer depends on the potential levels and the resistance offered by the medium. In electrical theory the classical law due to Ohm states that the current flowing between any two points in a purely resistive circuit is directly proportional to the voltage difference between the points and indirectly proportional to the resistance:

$$I = \frac{V}{R};$$

similarly, in the case of steady-state heat flow,

$$q = \frac{\Delta \theta}{R},$$

where $\Delta \theta$ is the temperature differential causing a heat flow rate q through the thermal resistance R. A similar law holds true in acoustics:

$$u = \frac{p}{r},$$

where p is the *acoustic pressure* (Pa), r is the *acoustic resistance* in acoustical ohms and u is the *volume velocity* (m³/s) describing the movement of a volume element under the action of p. Only rarely, however, is an electrical, thermal or mechanical circuit purely resistive. In the acoustical case the action of mass and stiffness have also to be taken into account; this is analogous to the behaviour of an L–C–R series electric circuit with an applied sinusoidal electromotive force (Fig. 6.5). At any instant a current i (or flow rate of charge $\dot{q}$) flows in the circuit under the action of a periodic electromotive force $E_0 \sin \omega t$; potential drops are established across L, R

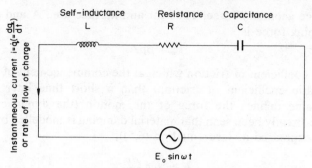

FIG. 6.5. L–R–C series circuit.

and C which when added will give

$$L\ddot{q} + R\dot{q} + \frac{q}{C} = E_0 \sin \omega t. \qquad (6.20)$$

This equation can be solved to give the electrical impedance Z defined as

$$Z = \frac{E_0 \sin \omega t}{\dot{q}},$$

$$Z = R + j\left(\omega L - \frac{1}{\omega C}\right), \qquad (6.21)$$

where $j = \sqrt{-1}$ and ω is the angular frequency (for frequency f the angular frequency is $2\pi f$).

Readers with a background in mechanical engineering will prefer to recall that the second-order linear differential equation (6.1) describes the motion of a one degree of freedom system written as

$$m\ddot{x} + r\dot{x} + kx = F_0 \sin \omega t,$$

where m, r and k are the effective mass, damping coefficient and stiffness respectively; x is the linear displacement and $F_0 \sin \omega t$ the forcing function. For readers unfamiliar with equation (6.1) it is derived in sections 9.2 and 9.3. The mechanical impedance is given as

$$Z = r + j\left(\omega m - \frac{k}{\omega}\right). \qquad (6.22)$$

The acoustic resistance r is analogous to R in equation (6.20). Self-inductance (L) in an electrical circuit opposes the change in current; the inertia force ($m\ddot{x}$) in a mechanical system opposes any change in linear velocity. If an acoustic pressure p disturbs a volume element V, mass M, over a surface area S, and the particles of the medium undergo a volume displacement $\Delta V = Sx$, then the kinetic energy acquired is

$$\frac{1}{2}\frac{M}{S^2}\left\{\frac{d(Sx)}{dt}\right\}^2;$$

the quantity M/S^2 is called the *acoustical mass* (or sometimes the *inertance*) and opposes any change in volume velocity ($u = S\dot{x}$).

Stiffness in mechanical systems has been discussed in section 6.2.2 and was seen to vary inversely with compressibility. Electrical capacitance C_E is defined by the charge on the conductor q required to raise its potential V, hence $C_E = q/V$. In acoustics the equivalent quantity C_A is represented by the volume displacement in the medium required to withstand an applied pressure p, hence $C_A = \Delta V/p$. In section 6.2.2 the bulk modulus was defined as $K = -p/(\Delta V/V)$, hence the analogous quantity to electrical capacitance is V/K or simply $1/K$ if K is expressed for unit volume of the medium (i.e. compressibility is analogous to the electrical capacitance).

If ψ represents a volume displacement, $\dot{\psi}$ a volume velocity, and $\ddot{\psi}$ a volume acceleration in the medium the equation of motion for a periodic acoustic pressure $p_0 \sin \omega t$ is

$$m\ddot{\psi} + r\dot{\psi} + k\psi = p_0 \sin \omega t ; \qquad (6.23)$$

solving the acoustic impedance is

$$Z = r + j\left(\omega m - \frac{k}{\omega}\right). \qquad (6.24)$$

Often this is written as

$$Z = r + jX, \qquad (6.25)$$

where X is called the acoustic reactance. Equation (6.24) indicates that the mass and stiffness elements in an acoustic circuit are effective in different frequency ranges, mass being predominant at high frequencies and stiffness at low frequencies. When the mass and stiffness reactances are equal, $\omega m = k/\omega$, the acoustic impedance is a minimum and a condition of *resonance* is said to have occurred. In general, the acoustic impedance function can be expressed

$$Z = Z(r, m, k). \qquad (6.26)$$

This function shows the interrelation of r, m and k which were first described qualitatively in connection with Fig. 6.1a and can be represented by the model shown in Fig. 6.6, which also includes the aerodynamic noise sources. The model strips our present study of all its decorations and exposes the key which may be used to unlock the environmental problems arising from structureborne and airborne sound. The key is that *each vibrating source and its surroundings have calculable acoustic impedances ; the optimum sound environment can be attained by arranging the appropriate mis-match between these impedances by sound control impedances* (shown as Z_α and Z_β in Fig. 6.6) *or by designing the vibrating source to have a suitable acoustic impedance*; the designer can take the former pathway, but it is the manufacturer designing his equipment who must effect the latter.

The symbolic notation used for acoustic mass, stiffness, resistance was shown in Fig. 6.2. Elastic action is described by a spring of stiffness k and viscous resistance r by a dashpot. It should be noted that the symbol for r also includes the resistance due to the acoustic radiation pressure and the internal damping of the medium; in the initial stages of the acoustic flow, the Coulomb damping force will operate, but its effect is transitory.

In parallel arrangement of these elements (see Fig. 6.2b) the volume velocity at the

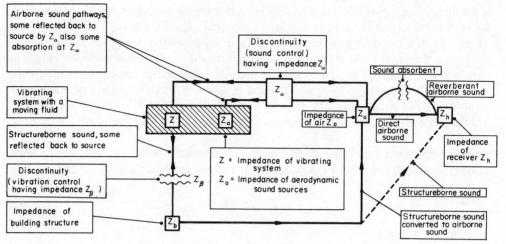

FIG. 6.6. Impedance model of sound pathways.

ends of k and r at any instant is the same, and the total force acting on the system is the sum of that acting on k, r and m separately (for ease, the mass M has been taken to be infinite and therefore, immovable); this is often termed the *classical impedance analogy*.

A series arrangement of the elements means that the same acoustic pressure acts upon them all, and the resultant volume velocity is the sum of those for k, r and m (see Fig. 6.2c); this is usually referred to as a *mobility analogy*.

This work forms the basis for the design of acoustic filters. By writing down the equations for the various combinations of k, r and m the frequency transmission characteristics may be deduced, or given requirements for the latter, suitable values of k, r and m may be found.

6.4. Analogues

We have seen that the behaviour of mechanical systems may be studied using electrical circuits. In building environmental engineering, analogue computers have advanced our understanding of heat, moisture, air and sound transfer between spaces and across building structural elements. It should be mentioned that computer studies nowadays apply the benefits of analogue and digital systems using hybrid computers to the investigation of problems, but here we wish to confine our attention to the relevance of analogues because they demonstrate the value of using a coherent unified approach to energy problems rather than treating sound, heat, fluid and electricity flows as completely independent subjects.

Kirchhoff's laws state that:

(a) the algebraic sum of all the voltages around any closed circuit is equal to zero;
(b) the algebraic sum of all the currents flowing into and from any point in a circuit is zero.

These laws are used to evolve two kinds of electrical analogies for mechanical systems—*voltage–force* (*mass–inductance*) and *current–force* (*mass–capacitance*) analogies. The voltage–force comparison is more commonly used. Table 6.1 shows the elements which form the analogue of one another in the various systems.

TABLE 6.1

Mechanical system	Electrical system	
	Voltage–force analogy	Current–force analogy
d'Alembert's principle	Kirchhoff's voltage law	Kirchhoff's current law
Degree of freedom	Loop	Node
F-Force (N)	V-voltage (volts)	i-Current (amps or Coulombs per sec)
m-Mass (kg or N s^2/m)	L-inductance (Henrys)	C-Capacitance (Farads)
x-Displacement (m or mm)	q-charge (Coulombs)	$\int V\,dt$
$\dot{x}$ Velocity (m/s)	i (or $\dot{q}$)-loop current (amps)	V-node voltage (volts)
r-Damping (N s/m)	R-resistance (Ohms)	$1/R$-conductance (mho)
k-Stiffness (N/m)	$1/C$-capacitance^{-1} (Farads^{-1})	$1/L$-inductance^{-1} (Henrys^{-1})

In general, if the forces act in series in the mechanical system, then the electrical elements representing these forces are put in parallel; the converse of this is true.

The classical impedance analogy is frequently used to model vibrating systems encountered in building environmental engineering. Figure 6.7 shows the mechanical model and the corresponding electrical analogies. Thus the equation of motion is

$$m\ddot{x} + r\dot{x} + kx = f(t), \tag{6.27}$$

or

$$m\frac{dx}{dt} + r\dot{x} + k\int \dot{x}\,dt = f(t). \tag{6.28}$$

For a current–force analogy (Fig. 6.7b) the circuit equation is

$$C\ddot{V} + \frac{1}{R}\dot{V} + \frac{1}{L}V = \frac{di(t)}{dt}, \tag{6.29}$$

where $i = CV$, $i = V/R$ and $i = (1/L)\,V\,dt$.

Integrating with respect to time,

$$C\dot{V} + \frac{V}{R} + \frac{1}{L}\int V\,dt = i(t). \tag{6.30}$$

Using Kirchhoff's voltage law the following equations are represented by the

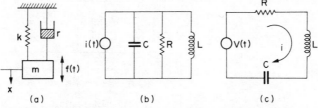

FIG. 6.7. Mechanical and electrical analogues of a single degree of freedom system. (a) Spring, damper mass vibrating system. (b) Current–force analogy. (c) Voltage–force analogy.

circuit diagram shown in Fig. 6.7c:

$$L\frac{di}{dt} + Ri + \frac{1}{C}\int i\,dt = V(t) \qquad (6.31)$$

or

$$L\ddot{q} + R\dot{q} + \frac{1}{C}\int \dot{q}\,dt = V(t). \qquad (6.32)$$

Equations (6.28), (6.30) and (6.31) are analogous, as are equations (6.27), (6.29) and (6.32).

Using the same approach, the analogies shown in Table 6.2 can be derived.

TABLE 6.2. ANALOGIES ENCOUNTERED IN PRACTICE

Mechanical system	Electrical analogy
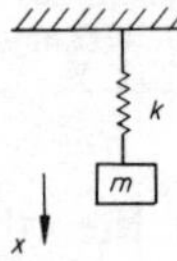 Single degree of freedom mass–spring system $m\ddot{x} + kx = 0$	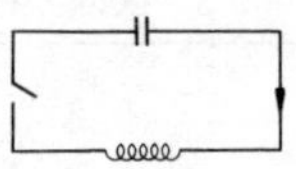 Voltage – force equation $L\ddot{q} + q/C = 0$
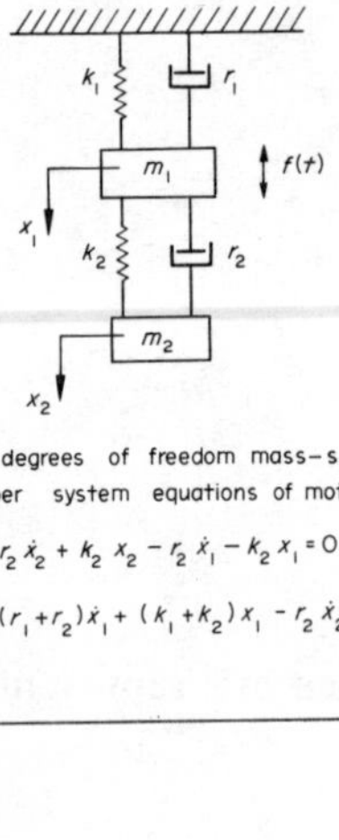 Two degrees of freedom mass–spring damper system equations of motion are $m_2\ddot{x}_2 + r_2\dot{x}_2 + k_2 x_2 - r_2\dot{x}_1 - k_2 x_1 = 0$ $m_1\ddot{x}_1 + (r_1+r_2)\dot{x}_1 + (k_1+k_2)x_1 - r_2\dot{x}_2 - k_2 x_2 = f(t)$	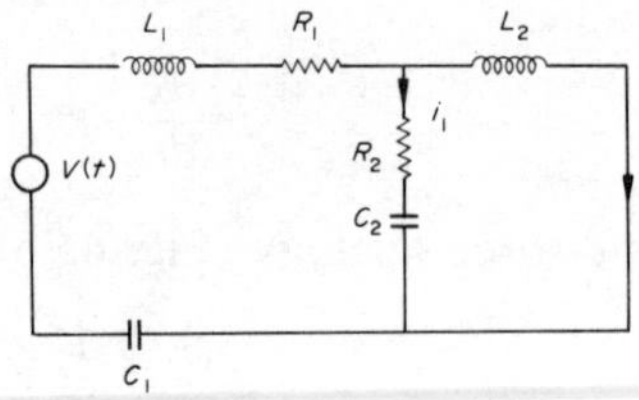 Voltage – force circuit equations are $L_2\ddot{q}_2 + R_2\dot{q}_2 + \frac{1}{C_2}\int \dot{q}_2\,dt - R_2\dot{q}_1 - \frac{1}{C_2}\int \dot{q}_1\,dt = 0$ $L_1\ddot{q}_1 + (R_1+R_2)\dot{q}_1 + (\frac{1}{C_1}+\frac{1}{C_2})\int \dot{q}_1\,dt - R_2 i_2 - \frac{1}{C_2}\int \dot{q}_2\,dt = V(t)$
	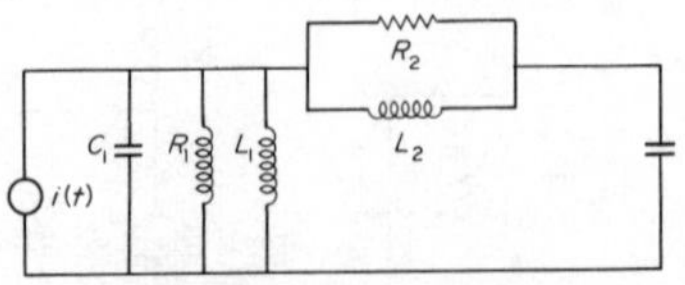 Current – force circuit equations are: $C_2\dot{V}_2 + \frac{V_2}{R_2} + \frac{1}{L}\int V_2\,dt - \frac{V_1}{R_2} - \frac{1}{L_2}\int V_1\,dt = 0$ $C_1\dot{V}_1 + (\frac{1}{R_1}+\frac{1}{R_2})V_1 + (\frac{1}{L_1}+\frac{1}{L_2})\int V_1 dt - \frac{V_2}{R_2} - \frac{1}{L_2}\int V_2 dt = i(t)$

TABLE 6.2 (*cont.*)

Mechanical system	Electrical analogy

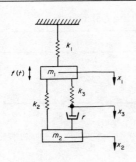

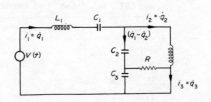

Three degrees of freedom mass-spring damper system

Equations of motion:

$$m_1\,\ddot{x}_1 + k_1\,x_1 + k_2\,(x_1 - x_2) + k_3\,(x_1 - x_3) = f(t)$$

$$m_2\,\ddot{x}_2 + k_2\,(x_2 - x_1) + r\,(\dot{x}_2 - \dot{x}_3) = 0$$

$$k_3\,(x_3 - x_1) = r\,(\dot{x}_2 - \dot{x}_3)$$

Voltage–force analogy cicuit equations:

$$L_1\,\ddot{q}_1 + \frac{1}{C_1}\int \dot{q}_1\,dt + \frac{1}{C_2}\int (\dot{q}_1 - \dot{q}_2)\,dt + \frac{1}{C_3}\int (\dot{q}_1 - \dot{q}_3)\,dt = V(t)$$

$$L_2\,\ddot{q}_2 + (\dot{q}_1 - \dot{q}_3)R + \frac{1}{C_2}\int (\dot{q}_2 - \dot{q}_1)\,dt = 0$$

$$R\,(\dot{q}_3 - \dot{q}_2) + \frac{1}{C}\int (\dot{q}_3 - \dot{q}_1)\,dt = 0$$

Viscoelastic behaviour in some materials
 may be represented by

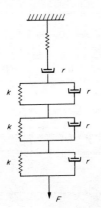

Current – force analogy

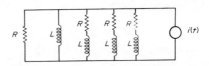

Energy transfer in a three-leaf
structure may be represented by

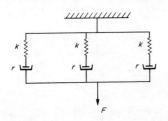

Voltage – force analogy

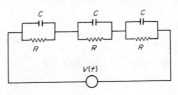

The basic elements in an analogue computer are the resistor, the capacitor and the d.c. amplifier, the two variables being voltage and current. These elements are combined to operate in the ways shown in Table 6.3.

TABLE 6.3. BASIC OPERATIONS OF AN ANALOGUE COMPUTER

Operation	Representation
Sign inversion $$e_o = - e_i$$	e_i — R — amplifier — R — e_o Equal values for input and output resistances of amplifier
Summation $$e_o = -\left(\frac{R_o}{R_1} e_1 + \frac{R_o}{R_2} e_2 + \frac{R_o}{R_3} e_3\right)$$	e_1 — R_1; e_2 — R_2; e_3 — R_3 — R_o — e_o
Integration $$e_o = -\frac{1}{RC} \int e_i \, dt + E_o$$ Time constant is RC in seconds; E_o is initial condition	E_o, C, e_i — R — e_o Output resistance replaced by capacitor
Multiplication $$e_o = - (R_o / R_i)\, e_i$$ A potentiometer may be used so that $$e_o = - k\, (R_o / R_i)\, e_i$$	R_o, e_i — R — e_o — potentiometer k

Magnitude scaling is achieved by multiplying or dividing the equation by a constant so that the output voltage remains within that range required for accuracy.

Time scaling may be needed so that some physical behaviour may be observed or recorded. To slow a process down

$$\frac{d^n}{dt^n} = a^n \frac{d^n}{dT^n},$$

whereas to speed a process up

$$\frac{d^n}{dt^n} = \frac{1}{a^n} \frac{d^n}{dT^n},$$

TABLE 6.4. SIMPLE ANALOGUE COMPUTER CIRCUITS

Process	Circuit
$x = -\int Ax\, dt + x_o$	$e_o = -\dfrac{1}{RC}\int e_i\, dt + e_o$ 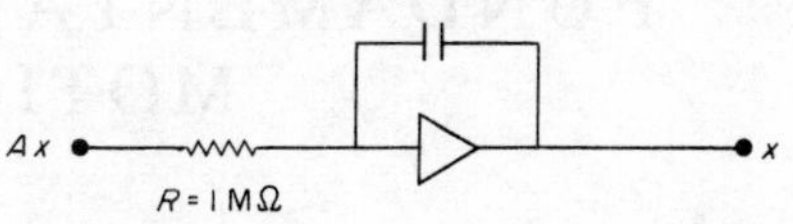

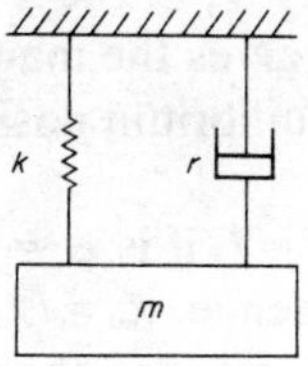

$m\ddot{x} + r\dot{x} + kx = 0$

Therefore $\ddot{x} = -\dfrac{r}{m}\dot{x} - \dfrac{k}{m}x$

Integrating once with respect to time,

$\dot{x} = -\int \left(\dfrac{r}{m}\dot{x} + \dfrac{k}{m}x\right)\, dt + \dot{x}_o.$

If the initial conditions are
$\dot{x}\ (0) = \dot{x}_o$ and $x\ (0) = x_o$, then these
can be set by two amplifiers; the inputs
$(r/_m)\dot{x}$ and $(k/_m)x$ can be made
available from potentiometers with $1/RC$
ratios; the third amplifier acts as a
sign inverter

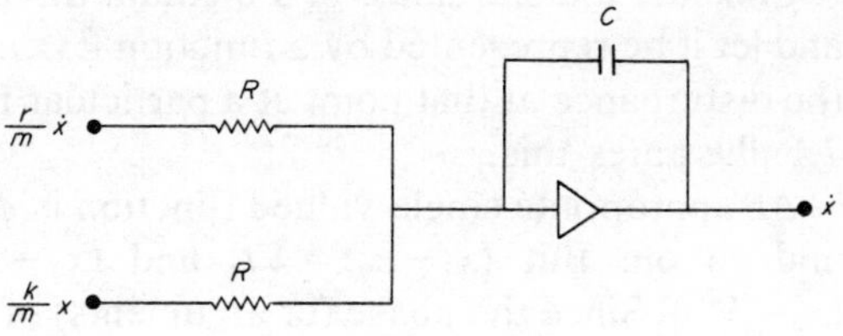

Basic circuit diagram

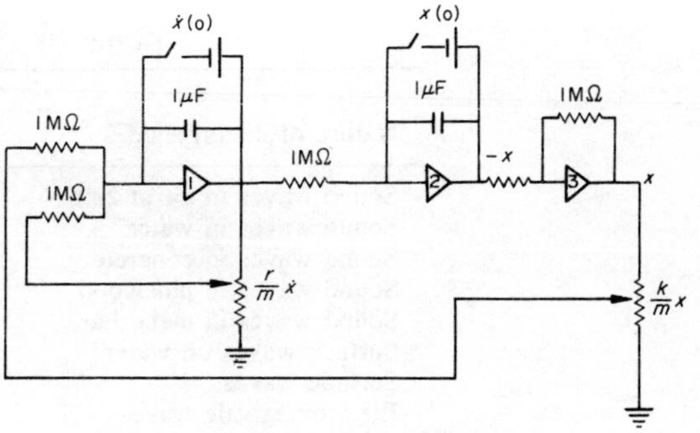

Circuit in practice

where t is the real time of process, T is the machine time and a is the magnification factor. Any change must be applied throughout the entire problem.

Some simple examples of computer use are given in Table 6.4.

CHAPTER 7

FUNDAMENTALS OF WAVE MOTION

7.1. Mathematical Representation of a Wave

Consider a disturbance in a medium travelling with velocity V, in the x-direction and let it be represented by a function $\phi(x)$. Each value of x gives the magnitude of the disturbance at that point at a particular time t from an equilibrium position; Fig. 7.1 illustrates this.

An appropriate single-valued function is $\phi(x)$, thus when $t = t_1$ it is $\phi = (x_1 - Vt_1)$ and so on. But $(x_1 - x_0) = Vt_1$ and $(x_2 - x_1) = V(t_2 - t_1)$, hence $x_0 = (x_1 - Vt_1) = (x_2 - Vt_2)$. Since the phase (or arguments) of these functions are equal, the functions are equal, therefore

$$\phi(x_0) = \phi(x_1 - Vt_1) = \phi(x_2 - Vt_2).$$

We can conclude that $\phi(x - Vt)$ represents a disturbance travelling in the $+x$-direction with a velocity V; a wave travelling in the $-x$-direction would be represented by the function $\phi(x + Vt)$. In practice, disturbances may change their magnitude or even their form due to *attenuation* in the medium. Typical values of V are given in Table 7.1. V usually refers to the propagation velocity in a medium which is stationary relative to the observer. Occasionally moving media are encountered like sound transmission in windy conditions or in the oceans.

TABLE 7.1

Nature of disturbance	V (m/s)
Sound waves in air at 20°C	330
Sound waves in water	1500
Sound waves in concrete	1700
Sound waves in pinewood	4179
Sound waves in metal bar	~ 5000
Surface waves on water	~ 0.5
Seismic waves	> 5000
Electromagnetic waves	3×10^8

Three-dimensional waves spread out from their source, and at any instant the disturbance will have a wavefront in the form of a sphere. Compared with our initial unidimensional disturbance the three-dimensional one does not propagate itself unchanged and the wave function has to be modified to account for this. A simple

436

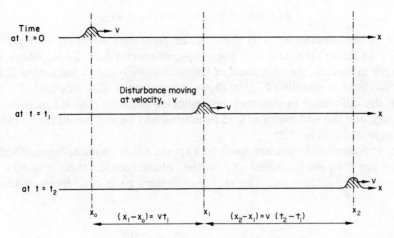

FIG. 7.1. Propagation of a sound pulse.

case is $(1/r)\phi(r - Vt)$ for a point in the medium at a distance r from the source, but other conditions govern the precise form of the function.

The *amplitude* of a wave is its maximum value taken from an equilibrium position. For plane and spherical waves the general progressive wave function for three-dimensional propagation is

$$\psi\{g(x, y, z)\}\phi\{g(x, y, z) - Vt\}, \tag{7.1}$$

where ψ is the amplitude function and ϕ is the phase function. For a spherical wave $g(x, y, z) = (x^2 + y^2 + z^2)^{\frac{1}{2}}$ and for a plane wave $g(x, y, z) = \alpha x + \beta y + \gamma z$.

A disturbance may be represented by a sum of periodic waves. The simplest type of periodic function is the sinusoidal type where for one dimension:

$$\psi(x)\phi(x - Vt) \equiv A \cos(kx - \omega t) \tag{7.2}$$

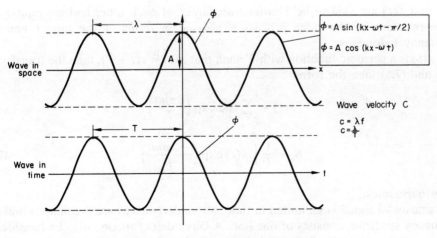

FIG. 7.2. Periodic wave function in space and time.

or
$$\psi(x)\phi(x - Vt) \equiv A \sin(kx - \omega t), \qquad (7.3)$$

where A is the amplitude, ω is the *angular frequency* $\omega = 2\pi f$, where f denotes *frequency* in Hertz (Hz) and k is the *propagation constant* $2\pi/\lambda$, where λ is the *wavelength* in metres; the reciprocal of the frequency is called the *period T*. In order for the functions in equation (7.2) or those in equation (7.3) to be equal, $V = \omega/k$ (or $V = f\lambda$); the difference in sign may be neglected because cosine functions possess even parity, and the odd parity of sine functions may be accounted for by including a phase angle of $\frac{1}{2}\pi$ (Fig. 7.2).

Often, exponential forms are used to express wave motion. Remembering that $\exp(j\theta) = \cos\theta + j\sin\theta$, where θ is the phase angle, then $\exp j(kx - \omega t) = \cos(kx - \omega t) + j\sin(kx - \omega t)$. The real or imaginary parts can represent a harmonic wave.

7.2. Complex Signals

Any complex signal can be considered as a superposition of sinusoidal signals. Fourier's theorem states that if $s(t)$ is a real time function (e.g. variation of sound pressure with time) and is sufficiently steady, or continuous, the integral

$$\int_{-\infty}^{+\infty} \{s(t)\}^2 \, dt$$

having a finite value, then

$$s(t) = \int_{-\infty}^{+\infty} S(f) \exp(2\pi j f t) \, df, \qquad (7.4)$$

and the spectral function, or complex amplitude spectrum $S(f)$, is

$$S(f) = \int_{-\infty}^{+\infty} s(t) \exp(-2\pi j f t) \, dt. \qquad (7.5)$$

$S(f)$ and $s(t)$ are said to be *Fourier transforms* of each other and are equivalent representations of the same process expressed in terms of the time t and the frequency f domains.

If $s(t)$ is a periodic function with T such that $s(t) = s(t + T)$, then the transforms (7.4) and (7.5) take the form

$$s(t) = \sum_{n=-\infty}^{n=+\infty} S_n \exp\left(\frac{2\pi j n t}{T}\right) \qquad (7.6)$$

and

$$S_n = \frac{1}{T} \int_0^T s(t) \exp\left(\frac{-2\pi j n t}{T}\right) \qquad (7.7)$$

for n harmonics.

A sinusoidal signal is unlimited in time; all its derivatives are continuous and the frequency spectrum consists of one line. A Dirac delta function may be considered as a counterpart to this because it has a single spectral line in the time domain whereas the frequency domain is one of constant amplitude, i.e. $S(f) = 1$, thus

$$s(t) = \lim_{t_0 \to \infty} \int_{-t_0}^{+t_0} \exp(2\pi jft)\, df. \tag{7.8}$$

It can be shown that

$$s(t) = \int_{-\infty}^{+\infty} s(\tau)\delta(t - \tau)\, d\tau. \tag{7.9}$$

When $s(t) = 1$, then

$$\int_{-\infty}^{+\infty} \delta(t)\, dt = 1 \quad \text{and} \quad \delta(t) = 0 \quad \text{for} \quad t \neq 0.$$

Any signal can be thought of as a succession of short pulses each represented by a delta function. If a room is excited by one of these pulses, the corresponding output signal is denoted by the impulse response function $g(t)$. If $g(t)$ is known (e.g. by measuring the resultant sound pressure level in a room) the output signal $s^1(t)$ resulting from a series of input signal responses $s(\tau)$ is

$$s^1(t) = \int_{-\infty}^{+\infty} s(\tau)g(t - \tau)\, d\tau. \tag{7.10}$$

The equivalent output signal in the frequency domain is

$$S^1(f) = G(f)S(f), \tag{7.11}$$

where $G(f)$ is the transmission function related to the impulse response by the Fourier transforms

$$G(f) = \int_{-\infty}^{+\infty} g(t)\exp(-2\pi jft)\, dt, \tag{7.12}$$

$$g(t) = \int_{-\infty}^{+\infty} G(f)\exp(2\pi jft)\, df. \tag{7.13}$$

7.3. The Behaviour of Waves in a Bounded Medium

If a wave strikes a fixed boundary it is reflected. Imagine the action of sound in an airconditioning duct or a room; the incident and reflected waves striking the various surfaces set up multi-reflections. A series of *standing* (or *stationary*) waves are said to exist as distinct from progressive waves. A simple example is shown in Fig. 7.3. At any point between the two boundaries the resultant disturbance can be found by *superposition* (i.e. add the effects of one wave on the other). This theory is only permissible for so-called *linear* waves; the meaning of this will be explained later.

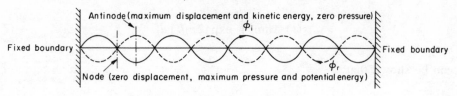

Antinode(maximum displacement and kinetic energy, zero pressure)
ϕ_i

Fixed boundary

ϕ_r

Node (zero displacement, maximum pressure and potential energy)

Fixed boundary

Reflected wave ϕ_r = B cos (kx + ωt)

FIG. 7.3. Standing wave formation.

Thus the standing wave disturbance at any point will be given by

$$\phi = \phi_{\text{incident}} + \phi_{\text{reflected}},$$

$$\phi = A \cos(kx - \omega t) + B \cos(kx + \omega t).$$

The frequency and velocity of the waves are the same but the amplitudes might be different due to absorption and transmission of energy at the surface. Expanding this function,

$$\phi = A(\cos kx \cos \omega t + \sin kx \sin \omega t) + B(\cos kx \cos \omega t - \sin kx \sin \omega t). \qquad (7.14)$$

The boundaries at $x = 0$ and $x = l$ are fixed so that the disturbance will be zero at these points and this will be true for all values of t (e.g. $t = 0$ and $t = n/4f$ for $n = 1, 3, 5, \ldots$). Inserting these conditions in equation (7.14) gives $A + B = 0$ and $(A - B) \sin kl = 0$, hence $\sin kl = 0$ and in general $\sin kx = 0$ which means $kl = n\pi$, where $n = 0, 1, 2, 3 \ldots$. The principal conclusion is that *a linear disturbance travelling in a finite medium sets up a series of standing or (stationary) waves having wavelengths $2l/n$*. These waves are called the *normal modes*. If the velocity of a sound wave being propagated in a bounded medium is c, then the frequency of the normal modes will be $f = nc/2l$ giving the *fundamental frequency* $f = c/2l$ for $n = 1$, sometimes referred to as the *first harmonic*, $n = 2$ gives the second harmonic and so on.

Standing waves, like progressive ones, are functions of space and time but remain distinct by their characteristic *nodes* (places of zero particle displacement or maximum particle pressure) and *antinodes* (places of maximum particle displacement or minimum particle pressure). In a three-dimensional medium the standing wave condition ($\sin kl = 0$) is applied in the x-, y-, z-directions (Fig. 7.4). The space is bounded by surfaces giving linear dimensions a, b, c; hence the standing wave condition is

$$\sin(k_x x) \sin(k_y y) \sin(k_z z) = 0$$

and
$$k_x a = n_x \pi, \qquad k_y b = n_y \pi, \qquad k_z c = n_z \pi.$$

Therefore
$$\sin\left(\frac{n_x \pi x}{a}\right) \sin\left(\frac{n_y \pi y}{b}\right) \sin\left(\frac{n_z \pi z}{c}\right) = 0.$$

For a wave propagated in direction OR (Fig. 7.4) with a velocity c, allowance has to be made for the fact that the wavefront is at angles α, β, γ to the x-, y-, z-axes.

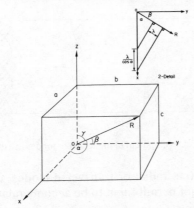

FIG. 7.4. Standing wave formation in a room.

The inset on Fig. 7.4 shows that for the x-axis the standing wave condition will be

$$\frac{a}{\lambda/2\cos\alpha} = n_x$$

and, similarly, will be

$$n_y = \frac{b}{\lambda/2\cos\beta}, \qquad n_z = \frac{c}{\lambda/2\cos\gamma}$$

for the y- and z-axes. By geometry $\cos^2\alpha + \cos^2\beta + \cos^2\gamma = 1$, hence

$$\left\{\left(\frac{n_x}{a}\right)^2 + \left(\frac{n_y}{b}\right)^2 + \left(\frac{n_z}{c}\right)^2\right\}^{1/2} = \frac{2}{\lambda} = \frac{2f}{c}. \tag{7.15}$$

The system of standing waves given by equation (7.15) is characterised by:

(a) *oblique waves* if $n_x \neq 0$, $n_y \neq 0$, $n_z \neq 0$ and are formed from eight travelling waves undergoing reflection at six surfaces;
(b) *tangential waves* if, for example, $n_x = 0$ and $n_y \neq 0$, i.e. are formed from travelling waves undergoing reflections at four surfaces;
(c) *axial waves* if, for example, $n_x = 0 = n_y$, then a z-axial wave is created consisting of two waves undergoing reflections at two surfaces.

7.4. Wave Propagation in Pipes and Airconditioning Ducts

The wave modes which are propagated depend on the nature of the vibration source (e.g. fan, pumps), the properties of the pipe or duct wall (i.e. sheet metal ducts conduct sound easily, whereas builders' work ducts do not), the fluid flow velocity and the cross-sectional dimensions of the pipe or duct. The acoustic flow in a duct can be considered as being two-dimensional due to sound radiated in a longitudinal mode n_x, and that in a radial mode n_y; thus for a rectangular duct having dimensions a, b the frequencies propagated, using equation (7.15) will be

$$f_{n_x, n_y} = \frac{c}{2}\left\{\left(\frac{n_x}{a}\right)^2 + \left(\frac{n_y}{b}\right)^2\right\}^{1/2}. \tag{7.16}$$

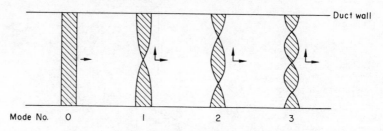

FIG. 7.5. Development of higher order modes in a duct.

The highest order modes (i.e. radial or transverse modes) can only exist when the cross dimensions of the duct permit them to be accommodated as shown in Fig. 7.5. For a circular duct of radius r,

$$f_{n_x,n_y} = \frac{c}{2} r\alpha_{n_x,n_y},$$

(7.17)

where α_{n_x,n_y} is the solution of the Bessel equation

$$\frac{\partial J_{n_x}(\pi\alpha)}{\partial \alpha} = 0$$

(7.18)

(solutions to this equation are given in Table V, p. 444 of *Vibration and Sound* by Morse in the second edition published by McGraw-Hill in 1948).

CHAPTER 8

PROPAGATION OF SOUND

THIS chapter is applicable to sound travelling through any medium and there is no need at this stage to confine the discussion to propagation through air. So far we have discussed wave motion in terms of some undefined disturbance, but our concern now is to consider a specific kind of disturbance—sound waves.

8.1. What is Sound?

Let us try and picture what is happening when a sound wave passes through a medium. In Fig. 8.1 a vibrating surface is shown reacting on the particles of the medium. Initially the surface moves against the particles, displaces them and crowds them together to form a dense high-pressure region called a *compression* (Fig. 8.1b). As the surface returns to its initial position the compression moves on but leaves a low-density and low-pressure zone called a *rarefaction* (Fig. 8.1c). These events continue so that the energy from the vibrating surface is transmitted through the medium as a pressure wave. Notice that the particle pressure and particle's displacement waves are $\frac{1}{2}\pi$ out of phase and that the motion imposed on the particles by the vibrating surface is superimposed on the natural random movement, called Brownian motion, of the particles.

If the frequency of the pressure wave is in the audio range it is sensed by the ear and called sound. Table 8.1 gives the frequency ranges generally accepted for mechanical vibrations and shows that it is the human sensory system which distinguishes between the vibrations that are classified as infrasonic, audio or ultrasonic waves. Below 3 Hz the human body responds as a unit mass; in the range 3–100 Hz various body resonances occur. These ranges vary from person to person and there is also some overlap between the ranges.

In section 7.1 we defined a wave function as

$$\phi = \psi\{g(x, y, z)\}\phi\{g(x, y, z) - Vt\} \quad \text{(see equation (7.1)).}$$

In acoustics ϕ can be a particle pressure p; a particle displacement η or a particle velocity $\dot{\eta}$; these fundamental quantities will be related to one another in the next section. Simple harmonic functions of these quantities would be

$$p = p_0 \cos(kx - \omega t) \tag{8.1a}$$

or
$$p = p_0 \sin(kx - \omega t), \tag{8.1b}$$

where p_0 is the particle pressure amplitude and $\omega/k = c$ is the velocity of sound.

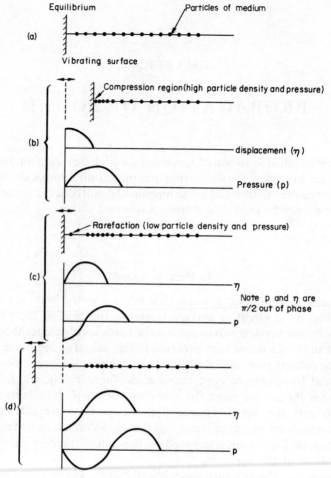

FIG. 8.1. Sound propagation.

Since the pressure and displacement are $\frac{1}{2}\pi$ out of phase, the corresponding displacement to $p = p_0 \cos(kx - \omega t)$

is $$\eta = \eta_0 \sin(kx - \omega t), \qquad (8.2)$$

and the particle velocity will be

$$\dot{\eta} = -\omega \eta_0 \cos(kx - \omega t), \qquad (8.3)$$

the strain will be

$$\frac{d\eta}{dx} = k\eta_0 \cos(kx - \omega t). \qquad (8.4)$$

Notice that $\dot{\eta}$ and p are in phase with one another.

The changes of pressure that occur when sound travels through air are usually taken to follow an adiabatic law $pV^\gamma = $ constant because there is not enough time during a compression or rarefraction for heat to flow away. If this law is

TABLE 8.1

Classification	Frequency range (Hz)	Human sensor
Infrasonic	0–15	Tactile receptors in skin; proprioreceptors in muscles; vestibular system (i.e. otolith bones in the ear and the semicircular canals)
Low-frequency vibration	15–100	
Audio	15–20 000	Ear; vestibular system at very high sound levels; overlaps with low-frequency vibration range from 15 to 100 Hz
Ultrasonic	> 20 000	Absorbed to various degrees by body tissues—can produce chemical, thermal and cavitation effects in the body

differentiated it gives

$$dp\,V^{\gamma} + p\gamma V^{\gamma-1}\,dV = 0;$$

hence

$$\gamma p = \frac{-dp}{dV/V},$$

where dp is the acoustic pressure change. In section 6.2.2 $-dp/(dV/V)$ was defined as the bulk modulus K, hence

$$K = \gamma p. \tag{8.5}$$

This is a first-order solution and is only valid for small linear changes in the acoustic pressure; this point is discussed further in the following section.

8.2. Sound Propagation in Gases

We now look at the relationship between the cause (i.e. the acoustic pressure) and the effect (i.e. the expansions and contractions which take place) in gaseous mediums (e.g. air). Consider the effect of a sound wave on a volume of gas at two instants in time (Fig. 8.2).

The sound pressure wave causes the volume elements of gas V_0 ($= ABCDEFGH$) to expand acquiring a new volume V ($= A'B'C'D'E'F'G'H'$), i.e. a dilation D occurs. A little later a condensation s will take place. Hence we can write

$$V = V_0(1 + D).$$

If ρ_0 is the density corresponding to V_0, then assuming conservation of mass ($\rho_0 V_0 = \rho V$),

$$\rho_0 = \rho(1 + D).$$

The condensation is defined by $\rho = \rho_0(1 + s)$, thus neglecting small order terms

$$(D + s) = 0. \tag{8.6}$$

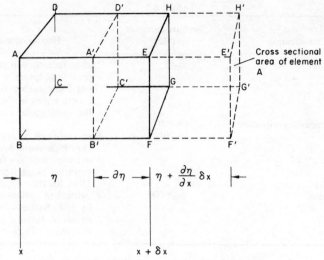

FIG. 8.2. Dilatation in medium due to sound.

Since we are considering dilation $V > V_0$ and $\rho < \rho_0$, hence $\rho = \rho_0 - \delta\rho$.
Equating the mass of gas in the two states

$$\rho_0 \delta(xA) = (\rho_0 - \delta\rho)\left\{\partial(xA) + \frac{\partial(\eta A)}{\partial x}\,\delta x\right\}$$

and neglecting the order of the small order term $\delta\rho(\partial(\eta A)/\partial x)\,\delta x$

$$\frac{\delta\rho}{\rho_0} = \frac{1}{A}\frac{\partial(\eta A)}{\partial x}. \tag{8.7}$$

Now
$$s = -\frac{\delta\rho}{\rho_0} \quad \text{but} \quad D = +\frac{\delta\rho}{\rho_0}.$$

Therefore
$$D = \frac{1}{A}\frac{\partial(\eta A)}{\partial x}. \tag{8.8}$$

If the cross-section of the wavefront is not a variable,

$$D = \frac{\partial\eta}{\partial x} \tag{8.9}$$

and
$$s = -\frac{\partial\eta}{\partial x}. \tag{8.10}$$

It can be concluded that the dilation (or condensation) represents the strain imposed on the gas by the passage of a sound wave through it. Note also the wave velocity is $\dot{x}$ and the particle velocity is $\dot{\eta}$; hence

$$\dot{\eta}/\dot{x} = \frac{\partial\eta}{\partial x} = D.$$

At the end of the previous section it was noted that adiabatic pressure volume changes are more likely to occur in sound transmission and it was established that the bulk modulus is

$$K = \gamma p = -\frac{dp}{dV/V} \quad \text{(see equation (8.5))}.$$

This relationship will now be derived more rigorously for the volume change of the gas element shown in Fig. 8.2. For an adiabatic process

$$\frac{p}{\rho^\gamma} = \frac{p_0}{\rho_0^\gamma}$$

and for an increase in volume (decrease in density), the acoustic pressure will decrease by an amount δp; hence

$$(p_0 - \delta p) = (\rho_0 - \delta\rho)^\gamma \frac{p_0}{\rho_0^\gamma}$$

$$= p_0 \left(1 - \frac{\delta\rho}{\rho_0}\right)^\gamma,$$

and since $s = -\delta\rho/\rho_0$,

therefore $\qquad (p_0 - \delta p) = p_0(1 + s)^\gamma$;

expanding by the Binominal theorem gives

$$(p_0 - \delta p) = p_0 \left\{1 + \gamma s + \frac{\gamma(\gamma - 1)s^2}{2!} + \cdots\right\}.$$

Neglecting s^2 and higher powers of s

$$-\delta p = \gamma p_0 s \text{ (or } \gamma p_0 D = +\delta p) \qquad (8.11)$$

comparing this with equation (8.5)

$$-\delta p = Ks \text{ (or } KD = +\delta p). \qquad (8.12)$$

8.3. The Wave Equation (Including the Effect of Damping)

The propagation of sound can be quantified by means of a general wave equation. In section 7.1 a unidimensional wave function was described by

$$\phi = \phi(x - ct),$$

where V has been replaced by the velocity of sound c. Differentiating ϕ with respect to x and t,

$$\frac{\partial \phi}{\partial x} = \phi' \quad \text{and} \quad \frac{\partial \phi}{\partial t} = -c\phi',$$

where ϕ' denotes the first derivative $\phi^1 (x - ct)$.

Therefore $\qquad\qquad\qquad c = -\frac{\partial \phi/\partial t}{\partial \phi/\partial x}.$

This gives c for a wave travelling in the $+x$-direction. In other words the wave velocity is not defined independently of direction because a wave travelling in the

$-x$-direction would have a velocity $+(\partial\phi/\partial t)/(\partial\phi/\partial x)$. Equations of motion invariably include an acceleration term $\partial^2\phi/\partial t^2$. These are good reasons for differentiating a second time with respect to x and t giving

$$c^2 = \frac{\partial^2\phi/\partial t^2}{\partial^2\phi/\partial x^2} \tag{8.13}$$

which when written in the form

$$\frac{\partial^2\phi}{\partial x^2} - \frac{1}{c^2}\frac{\partial^2\phi}{\partial t^2} = 0 \tag{8.14}$$

is referred to as the *general wave equation*.

This result is independent of the wave direction since it satisfies both wave functions $\phi(x - ct)$ and $\phi(x + ct)$. If ϕ is a particle displacement in the medium, then $\partial^2\phi/\partial t^2$ is an acceleration (i.e. force per unit mass) and $\partial^2\phi/\partial x^2 = \partial/\partial x(\partial\phi/\partial x)$ is a spatial rate of change of strain (units will be length^{-1}). The nature of the velocity of sound c already becomes evident. From equation (8.13) it can be seen that c is a function of the force per unit mass to bring about a unit spatial rate of strain in the medium. The case for a gas and solid will now be described:

(a) Gas

Referring to Fig. 8.3 the net force acting on the element due to the acoustic pressure is $-(\partial(\delta p)/\partial x)\,\delta x$.

By Newton's Second Law

$$\rho_0\delta x\frac{\partial^2\eta}{\partial t^2} = -\frac{\partial(\delta p)}{\partial x}\,\delta x,$$

where ρ_0 is the density before the acoustic pressure takes effect and η is the resulting particle displacement. Using equation (8.12)

$$\rho_0\delta x\frac{\partial^2\eta}{\partial t^2} = -\frac{\partial(Ks)}{\partial x}\,\delta x.$$

Since

$$s = -\frac{\partial\eta}{\partial x},$$

then

$$\frac{\partial^2\eta}{\partial x^2} - \frac{\rho_0}{K}\frac{\partial^2\eta}{\partial t^2} = 0. \tag{8.15}$$

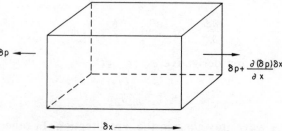

FIG. 8.3. Forces acting on acoustic element, no damping.

Comparing this with equation (8.14), the wave function ϕ in this case being the particle displacement, it can be seen that

$$c = \sqrt{\frac{K}{\rho_0}} \qquad (8.16)$$

or, since $K = \gamma p$,

$$c = \sqrt{\frac{\gamma p}{\rho_0}} \qquad (8.17)$$

and the general gas law $p/\rho_0 = mRT$ so that

$$c = \sqrt{m\gamma RT}, \qquad (8.18)$$

which shows that c is temperature dependent (note: m is mass of gas, R is Universal Gas Constant, T is absolute temperature).

(b) Solids

Consider a long, thin elastic rod made from a homogeneous material of density ρ and having unit cross-sectional area. A tensile stress F acts at one end. For an element of thickness δx the net force acting will be $(F + (\partial F/\partial x)\,\delta x) - F$ and since the rod is elastic, $Y = F/\partial\phi/\partial x$, where Y is Young's modulus. Applying Newton's Second Law to the element,

$$\rho \delta x \frac{\partial^2 \phi}{\partial t^2} = \frac{\partial F}{\partial x}\,\delta x$$

$$= \frac{\partial}{\partial x}\left(Y\frac{\partial \phi}{\partial x}\right)\delta x.$$

Therefore
$$\frac{Y}{\rho} = \frac{\partial^2 \phi / \partial t^2}{\partial^2 \phi / \partial x^2}.$$

Comparing this with the general wave equation (8.14) we see that

$$c = \sqrt{\frac{Y}{\rho}}.$$

The general wave equation (equation (8.14)) can be written in the three-dimensional form as

$$\frac{\partial^2 \phi}{\partial x^2} + \frac{\partial^2 \phi}{\partial y^2} + \frac{\partial^2 \phi}{\partial z^2} = \frac{1}{c^2}\frac{\partial^2 \phi}{\partial t^2}.$$

Now
$$\left(\frac{\partial^2}{\partial x^2} + \frac{\partial^2}{\partial y^2} + \frac{\partial^2}{\partial z^2}\right)$$

is known as the Laplacian (or Del) operator written ∇^2, thus

$$\nabla^2 \phi - \frac{1}{c^2}\frac{\partial^2 \phi}{\partial t^2} = 0 \qquad (8.19)$$

or
$$\nabla^2 \phi - \frac{1}{c^2}\ddot{\phi} = 0. \qquad (8.20)$$

This is a linear second-order differential equation. Sound waves that create large amplitude disturbances in the medium (e.g. sonic booms) cannot be represented by this type of equation on account of non-linear effects (e.g. change of wavefront shape, large volume changes).

The properties of the general wave equation can be summarised as follows:

* it is linear and therefore all solutions obey the principle of superposition, thus

$$\phi = \phi(x - ct) + \phi(x + ct);$$

* it is assumed that the wave does not change its shape or size;
* it has been derived using a plane wave function;
* it is satisfied by the solution

$$\phi = \phi_1(x - ct) + \phi_2(x + ct);$$

* the wave function ϕ may be a *vector* wave function (e.g. particle displacement, particle velocity or particle velocity or particle pressure) possessing a magnitude and direction in space at a specified instant in time, or a *scalar* wave function (e.g. density) which is specified by magnitude only at any instant in time;
* in the case of sound waves, changes of position of the particles in the medium take place in a direction parallel to the direction of propagation and therefore are said to be of a *longitudinal* type in contrast to *transverse* waves (e.g. electromagnetic) which have wave vectors at right angles to the propagation direction;
* it assumes that no other sound sources are encountered during propagation; if airborne generation occurs in the system then

$$\nabla^2 \phi - \frac{1}{c^2} \ddot{\phi} \neq 0;$$

  this will be discussed in a later section;
* if damping occurs in the medium it is assumed that the damping force R is proportional to the particle velocity

$$R \propto \frac{\partial \eta}{\partial t} \quad \text{or} \quad R = r \frac{\partial \eta}{\partial t}.$$

Thus the equation of motion for the element shown in Fig. 8.3 is

$$\rho_0 \frac{\partial^2 \eta}{\partial t^2} = -\frac{\partial}{\partial x}\left(-K \frac{\partial \eta}{\partial x}\right) - r \frac{\partial \eta}{\partial t}. \tag{8.21}$$

Therefore

$$r \frac{\partial \eta}{\partial t} = K \frac{\partial^2 \eta}{\partial x^2} - \rho_0 \frac{\partial^2 \eta}{\partial t^2},$$

$$\frac{r}{\rho_0 c^2} \dot{\eta} = \nabla^2 \eta - \frac{1}{c^2} \ddot{\eta}.$$

If a solution of this is $\eta = \eta_0 \sin(\chi x - \omega t)$ where χ is the propagation constant, then its substitution in equation (8.21) will give

$$\chi = \pm\left(k - j \frac{r}{2\rho c}\right). \tag{8.22}$$

It is easier to show this by using the exponential form of the wave function,

$$\eta = \eta_0 \exp\{j(\omega t + \chi x)\}.$$

The general solution is

$$\eta = A \exp(-\alpha x) \exp j(\omega t - kx) + B \exp(\alpha x) \exp j(\omega t + kx), \qquad (8.23)$$

where α is the attenuation coefficient $r/(2\rho c)$.

The particle velocity $\dot{\eta}$ and the particle pressure p ($p = -Ks = -\rho c^2 (\partial \eta / \partial x)$) may be found by differentiating equation (8.23) appropriately.

8.4. Energy in a Progressive Sound Wave

At any instant the energy (i.e. rate of doing work) at a point in the wave will be the product of the particle pressure and the particle velocity at that point. When the acoustic pressure is compressing, the medium energy flows into the compression; energy flows away from the rarefaction regions. We can write the energy flow at any instant as

$$\epsilon = p\dot{\eta}.$$

The energy flow through unit area in unit time is called the *sound intensity* I (in W/m²).

The average intensity over a time period T will be the time average of the product hence

$$\bar{I} = \frac{1}{T} \int_0^T \overline{p\dot{\eta}} \, dt. \qquad (8.24)$$

The instantaneous sound pressure is

$$p = -K \frac{\partial \eta}{\partial x}.$$

This can be written as

$$= -K \frac{\partial \eta}{\partial t} \frac{\partial t}{\partial x}$$

since combining equations (8.3) and (8.4) gives

$$\frac{\partial \eta}{\partial x} = \frac{-k}{\omega} \dot{\eta} = \frac{-1}{c} \dot{\eta}$$

and the wave velocity c is $\partial x / \partial t$.

Therefore

$$p = \frac{K}{c} \dot{\eta}.$$

Substituting this expression for $\dot{\eta}$ in equation (8.24),

$$\bar{I} = \frac{c}{KT} \int_0^T p^2 \, dt.$$

Now the mean square pressure

$$\overline{p^2} = \frac{1}{T} \int_0^T p^2 \, dt.$$

Hence

$$\bar{I} = \frac{c}{K} \overline{p^2} \quad \text{and} \quad K = \rho c^2.$$

Therefore

$$\bar{I} = \overline{p^2}/\rho c, \tag{8.25}$$

but using equation (8.1)

$$\bar{I} = \frac{1}{\rho c} \left\{ \frac{1}{T} \int_0^T p_0^2 \cos^2 (kx - \omega t) \, dt \right\}, \tag{8.26}$$

$$\bar{I} = p_0^2 / 2\rho c.$$

Now the peak particle pressure p_0 is related to the root mean square pressure, $p_{r.m.s.}$, by $p_0/\sqrt{2} = p_{r.m.s.}$; hence

$$\bar{I} = \frac{p_{r.m.s.}^2}{\rho c} \tag{8.27}$$

where $p_{r.m.s.}$ is the effective value of the sound pressure (i.e. the value sensed by the human ear or a measuring instrument). Note that since on average over a period of time there will be as many negative pressures as positive ones, then the mean square pressure must be used in order to find the effective sound pressure.

If $p_1, p_2, \ldots, p_i$ are acoustic pressures at various instants in time, then

$$p_{r.m.s.} = \left(\frac{1}{n} \sum_{i=1}^{i=n} p_i^2 \right)^{1/2} = (\overline{p^2})^{1/2}.$$

Now electrical energy $E = V^2/R$, hence ρc corresponds to the electrical resistance R. This result was mentioned by analogy with thermal and electrical quantities in section 6.3.

For air at normal pressure and temperature (20°C) the acoustic resistance or impedance is

$$\rho c = 1 \cdot 2 \, (\text{kg/m}^3) \times 344 \, (\text{m/s}) = 408 \text{ acoustic ohms or rayls.}$$

How can acoustic energy be measured? Ideally, the acoustic energy requires an instrument which can simultaneously measure the particle pressure and particle velocity. Such an instrument, an acoustic wattmeter, has been developed by the Indian researcher Kiyoshi Awaya but is not available on the commercial market. In practice a microphone is used which responds either to the particle pressure or the particle velocity. The microphone measures the root mean square value of the pressure (or the velocity) in the sound field, the averaging taking place over a much longer time period than the period of the acoustic waves.

8.5. Energy in a Standing Wave

In section 7.3 the standing wave disturbance at any point was expressed as

$$\phi = A \cos (kx - \omega t) + B \cos (kx + \omega t).$$

Inserting the standing wave conditions for a sinusoidal displacement $\eta = \eta_0 \sin(k_n x - \omega_n t)$ in the nth mode,

$$\eta_n = A_n \sin\left(\frac{n\pi x}{l} - \omega_n t\right) + B_n \sin\left(\frac{n\pi x}{l} + \omega_n t\right).$$

This can be rewritten as

$$\eta_n = C_n \sin\frac{n\pi x}{l} \cos \omega_n t.$$

For all modes

$$\eta_{\Sigma n} = \sum_{n=i}^{n=\infty} C_n \sin\frac{n\pi x}{l} \cos \omega_n t.$$

Energy density is expressed as energy per unit volume so that the kinetic energy density is

$$\epsilon_k = \frac{1}{2}\rho \sum \dot{\eta}_n^2.$$

Hence the average kinetic energy density over the time T will be

$$\bar{\epsilon}_k = \frac{1}{2}\rho \frac{1}{T} \int_0^T \sum \omega_n^2 C_n^2 \sin^2\left(\frac{n\pi x}{l}\right) \sin^2(\omega_n t)\, dt$$

$$= \frac{1}{8}\rho \sum \omega_n^2 C_n^2.$$

At the nodal positions in a standing wave the particle displacement, and therefore the kinetic energy, is zero; for conservation of energy the potential energy will be a maximum. The converse condition is true at the antinode. Hence we can describe the potential energy by an equation exactly similar to that for the kinetic energy with the sine and cosine terms interchanged to preserve the $\frac{1}{2}\pi$ phase difference. Thus

$$\bar{\epsilon}_p = \frac{1}{2}\rho \frac{1}{T} \int_0^T \sum \omega_n^2 C_n^2 \cos^2\frac{n\pi x}{l} \cos^2 \omega_n t\, dt$$

$$= \frac{1}{8}\rho \sum \omega_n^2 C_n^2.$$

Hence the total average energy density is the sum of the average potential and kinetic values, i.e.

$$\bar{\epsilon} = \bar{\epsilon}_k + \bar{\epsilon}_p,$$

$$\bar{\epsilon} = \frac{1}{4}\rho \sum \omega_n^2 C_n^2. \tag{8.28}$$

8.6. The Fundamental Nature of a Simple Pulsating Spherical Sound Source

In deriving the general wave equation a general function ϕ was used as the dependent variable (see equation (8.14)). ϕ is termed the velocity potential (analogous to electrical potential) and is defined by

$$v_i = -\frac{\partial \phi}{\partial x_i}.$$

It can be seen that this satisfies the general wave equation now iterated

$$\nabla^2 \phi - \frac{1}{c^2} \frac{\partial^2 \phi}{\partial t^2} = 0.$$

The solution of this equation for spherical waves is

$$\phi = \frac{1}{r} \phi(r \pm ct).$$

Using a sinusoidal function for ϕ,

$$\phi = \frac{A}{r} \cos\{(\omega t - kr) + \alpha\},$$

where A and α are constants of integration to be determined from the known behaviour of the sound source.

$$p = \rho \frac{\partial \phi}{\partial t} \left(= \rho \frac{\partial \phi}{\partial r} \frac{\partial r}{\partial t} = -\rho c \dot{\eta} = p \right),$$

$$p = -\rho \frac{\omega A}{r} \sin\{(\omega t - kr) + \alpha\},$$

$$\frac{\partial p}{\partial r} = +\rho \frac{\omega A k}{r} \cos\{(\omega t - kr) + \alpha\} + \frac{\rho \omega A}{r^2} \sin\{(\omega t - kr) + \alpha\},$$

where r is large, the second term is negligible and the rate of change of pressure with distance $\partial p/\partial r$, is in phase with ϕ, but near the source the $1/r^2$ term must be taken into account, and as a consequence $\partial p/\partial r$ is no longer in phase with ϕ. The particle velocity is given by

$$\dot{\eta} = -\frac{\partial \phi}{\partial r} = -k \frac{A}{r} \sin\{(\omega t - kr) + \alpha\} + \frac{A}{r^2} \cos\{(\omega t - kr) + \alpha\}.$$

Again, it can be seen that when r is large p and $\dot{\eta}$ are in phase, but in the near field of the source they become out of phase. For low values of r the particle velocity has two components $90°$ out of phase with one another which means that the particle motion is not fully directed along the longitudinal direction of the sound wave; on reaching the far field all the energy is propagated radially outwards.

A pulsating sound source exerts a force on the fluid medium in which it is immersed. From Newton's Third Law an equal and opposite force is exerted on the source by the medium due to the acoustic radiation pressure. It is the dynamic balance between these forces which determines the flow of energy out from the source at any instant in time or at any position on the surface of the source.

Using complex notation the velocity potential

$$\phi = \frac{A}{r} \exp\left[j\{(\omega t - kr) + \alpha\}\right],$$

hence the particle pressure is

$$p = \rho \frac{\partial \phi}{\partial t} = j \frac{A\omega}{r} \rho \exp\left[j\{(\omega t - kr) + \alpha\}\right]$$

and the particle velocity is

$$\dot{\eta} = -\frac{\partial \phi}{\partial r} = \left(\frac{1}{r} + jk\right) \frac{A}{r} \exp\left[j\{(\omega t - kr)\} + \alpha\right],$$

hence the specific acoustical impedance is

$$Z = \frac{p}{\dot{\eta}} = \rho c \left(\frac{k^2 r^2}{1 + k^2 r^2} + j \frac{kr}{1 + k^2 r^2}\right). \tag{8.29}$$

As r increases the imaginary term, which represents the radiation impedance, can be neglected and $Z \to \rho c$, hence the sound intensity $I = p^2/\rho c$. In the near field, the wavefront has greater curvature and the radiation term is important; notice that it is 90° out of phase with the real term. The sound intensity in the near field has a resistance and reactance component

$$I = I_R + jI_I = \left\{\frac{\rho c}{p^2}\left(\frac{k^2 r^2}{1 + k^2 r^2} + j \frac{kr}{1 + k^2 r^2}\right)\right\}^{-1}. \tag{8.30}$$

Certainly the relationship between sound pressure level and distance will depend on the source power, spectrum and geometry besides the medium in contact with its surface.

8.7. Propagation of Sound Through the Atmosphere
(see Rayleigh, volume II, page 129)

The structure of a building acts as a climatic moderator and as such will be selected with its thermal (including solar), light, moisture and acoustic transmission characteristics borne in mind. Environmental engineers consider the effects of noise incident upon and emanating from a building. Airconditioning plants are often situated on roof tops, and sound from the fresh-air intakes or exhaust outlets must not offend neighbouring premises; buildings situated near roadways or airports must be constructed of materials which have a sufficiently high sound reduction indeed. For these reasons it is useful to have an acquaintance with the propagation of sound through the atmosphere.

Equations (8.17) and (8.18) gave the velocity of sound as

$$c = \sqrt{\frac{p\gamma}{\rho}} \quad \text{and} \quad c = \sqrt{\gamma R T m},$$

so that meteorological changes in the temperature, or the ratio of p/ρ, will alter c; changes in moisture content of the atmosphere alter the values of R and consequently γ. The effects of dust, smoke, turbulence and fogs have not been extensively investigated. Delany (1971) quotes typical attenuation values of 1 dB/km for reasonably dense fogs and a similar figure for turbulence effects.

In a similar manner to light, sound undergoes refraction when passing through a non-uniform medium. Snell's law states that

$$\frac{\sin i}{\sin r} = \frac{c_i}{c_r},$$

where i is the angle of incidence in the air in which the velocity of sound is c_i and r is the angle of refraction in the air in which the velocity of sound is c_r. This law is valid if angles i and r lie in the same plane as the normal. Rewriting this law in terms of absolute temperature and density gives

$$\frac{\sin i}{\sin r} = \left(\frac{T_i}{T_r}\right)^{1/2} = \left(\frac{\rho_r}{\rho_i}\right)^{1/2}.$$

Temperature gradients in the atmosphere cause the sound wavefront to deviate from its original direction. Two cases are shown in Fig. 8.4a and b. Condition (a) is more usual, but that in (b) occurs when a temperature inversion is present in the atmosphere which is quite common on frosty nights and over water on summer nights. In the latter case a sound channel is formed by refraction at the inversion layer and reflection from the ground, which can transmit sound over long distances; the refracted and reflected waves form a *reverberant sound field* (i.e. a bounded field) and the sound intensity I varies inversely with distance r; thus

$$I \propto \frac{1}{r}. \tag{8.31}$$

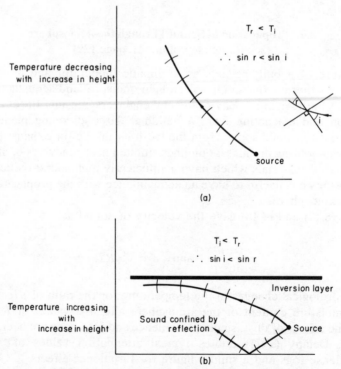

FIG. 8.4. Refraction of sound due to temperature gradient.

In an unbounded field (usually referred to as a *free* or *direct sound field*) the inverse square law is valid and

$$I \propto \frac{1}{r^2}. \tag{8.32}$$

The ideal free field occurs out of doors on a calm day free from atmosphere perturbations; in the laboratory an attempt is made to simulate this by using *anechoic rooms*, the highly absorbent boundaries minimising reflection of sound.

Apart from temperature effects, wind gradients and selective absorption of sound by particular ground surfaces (e.g. grass surfaces have been shown to absorb sound in the 250–350 Hz band) can affect the transmission of sound through the atmosphere. Of particular interest is the emanation of sound from the airconditioning plant in a building which can reverberate in the spaces in between adjacent buildings and can also interact with the turbulent air movement patterns that may occur. Similarly, an environmental services plant sited at roof level can be a source of annoyance to neighbouring premises if correct remedial measures are not employed.

8.8. Examples

1. Show that the maximum stress and the maximum strain in a medium due to a plane harmonic sound pressure wave occur at the same instant in time. Establish the phase difference between the stress imposed and the particle displacement.

Solution

Let the displacement in the medium due to the sound wave be

$$\eta = \eta_0 \sin(\omega t - kx),$$

then the strain is

$$\frac{d\eta}{dx} = -k\eta_0 \cos(\omega t - kx),$$

$$\left(\frac{d\eta}{dx}\right)_{max} = -k\eta_0.$$

The maximum strain occurs at time, $\quad t = \dfrac{kx}{\omega}$.

The stress, or pressure, can be represented as

$$p = p_0 \cos(\omega t - kx),$$

$$p_{max} = p_0.$$

The maximum stress also occurs at time, $\quad t = \dfrac{kx}{\omega}$.

Stress is related to the strain by the modulus property of the medium

$$p = Y \frac{d\eta}{dx}$$

$$\int_0^\eta d\eta = \frac{p_0}{Y} \int_x^{\omega t/k} \cos(\omega t - kx)\, dx$$

$$\eta = \frac{p_0}{Yk} \sin(\omega t - kx).$$

Thus the stress and displacement are $\frac{1}{2}\pi$ out of phase.

---

2. A plane harmonic sound wave progresses along a solid rod for which Young's modulus is Y, the density is ρ and the wave velocity c. The excess stress in the rod is given by

$$X = A \cos(\omega t - kx).$$

What is the instantaneous displacement η from equilibrium at any point? Establish an expression for the average energy transmission.

Solution

The stress is $\qquad\qquad X = A \cos(\omega t - kx)$

or, since $Y(d\eta/dx)$, then the displacement is

$$\eta = \frac{A}{-kY} \sin(\omega t - kx).$$

Now the particle velocity is

$$\dot\eta = \frac{\omega A}{-kY} \cos(\omega t - kx).$$

The average energy over a time period T is

$$\bar\epsilon = \frac{1}{T} \int_0^T X\dot\eta\, dt$$

$$= \frac{1}{T} \frac{\omega A^2}{-kY} \int_0^T \cos^2(\omega t - kx)\, dt$$

$$= \frac{1}{T} \frac{\omega A^2}{-2kY} \int_0^T \{1 + \cos 2(\omega t - kx)\}\, dt$$

$$\bar\epsilon = -\frac{\omega A^2}{2kY} \left(1 + \frac{\sin 2\omega T}{2\omega T}\right)$$

---

3. Superposition the two harmonic waves

$$f_1(x, t) = \cos \omega_1 \left(t - \frac{x}{c}\right),$$

$$f_2(x, t) = \cos \omega_2 \left(t - \frac{x}{c}\right),$$

having angular frequencies ω_1, ω_2 and wave velocity c. What is the effect of letting ω_1 and ω_2 lie close together so that

$$\omega_1 = \omega_2 + \epsilon,$$

where $\epsilon \ll \omega_2$?

Solution

At every point in space the motion is valid, hence at $x = 0$

$$f(0, t) = f_1(0, t) + f_2(0, t),$$
$$f(0, t) = \cos \omega_1 t + \cos \omega_2 t$$
$$= 2 \cos \{\tfrac{1}{2}(\omega_1 + \omega_2)t\} \cos \{\tfrac{1}{2}(\omega_1 - \omega_2)t\},$$

$$f(0, t) = 2 \cos \{\tfrac{1}{2}(2\omega_2 + \epsilon)t\} \cos \left(\frac{\epsilon}{2} t\right),$$

$$f(0, t) = \left(2 \cos \frac{\epsilon t}{2}\right)\{\cos \tfrac{1}{2}(2\omega_2 + \epsilon)t\}.$$

Beat frequency $= \dfrac{(\epsilon/2)}{2\pi}$

$T_1 = \dfrac{2\pi}{(\epsilon/2)}$

$f = \left(\dfrac{\omega_2 + \epsilon/2}{2\pi}\right)$

$\therefore T_2 = \left(\dfrac{2\pi}{\omega_2 + \epsilon/2}\right)$

4. The general wave equation is

$$\nabla^2 \phi - \frac{1}{c^2} \frac{\partial^2 \phi}{\partial t^2} = 0.$$

Show that for a spherical wave of radius $r^2 = x^2 + y^2 + z^2$ the wave equation takes the form

$$\frac{\partial^2 (r\phi)}{\partial r^2} - \frac{1}{c^2} \frac{\partial^2 (r\phi)}{\partial t^2} = 0.$$

Solution

$$\frac{\partial \phi}{\partial x} = \frac{\partial \phi}{\partial r} \frac{\partial r}{\partial x} \quad \text{since } r^2 = x^2 + y^2 + z^2$$

$$= \frac{x}{r} \frac{\partial \phi}{\partial r},$$

$$\frac{\partial^2 \phi}{\partial x^2} = \frac{\partial}{\partial x}\left(\frac{\partial \phi}{\partial x}\right)$$

$$= \frac{1}{r}\frac{\partial \phi}{\partial r} + x\frac{\partial}{\partial r}\left(\frac{1}{r}\frac{\partial \phi}{\partial r}\right)\frac{\partial r}{\partial x}$$

$$= \frac{1}{r}\frac{\partial \phi}{\partial r} + \frac{x^2}{r}\left(\frac{1}{r}\frac{\partial^2 \phi}{\partial r^2} - \frac{1}{r^2}\frac{\partial \phi}{\partial r}\right).$$

Therefore
$$\nabla^2 \phi = \frac{3}{r}\frac{\partial \phi}{\partial r} + \frac{x^2 + y^2 + z^2}{r}\left(\frac{1}{r}\frac{\partial^2 \phi}{\partial r^2} - \frac{1}{r^2}\frac{\partial \phi}{\partial r}\right)$$

$$= \frac{2}{r}\frac{\partial \phi}{\partial r} + \frac{\partial^2 \phi}{\partial r^2}$$

$$= \frac{1}{r}\frac{\partial^2}{\partial r^2}(r\phi).$$

Hence the general wave equation is

$$\frac{1}{r}\frac{\partial^2(r\phi)}{\partial r^2} - \frac{1}{c^2}\frac{\partial^2(r\phi)}{\partial t^2} = 0,$$

$$\phi = \frac{1}{r}f_1(r - ct) + \frac{1}{r}f_2(r + ct).$$

Considering the diverging wave

$$\phi = \frac{1}{r}\cos k(r - ct)$$

$$-\frac{\partial \phi}{\partial r} = \frac{k}{r}\sin k(r - ct) + \frac{1}{r^2}\cos k(r - ct).$$

If ϕ is the velocity potential then the particle velocity is $v = -\partial \phi/\partial r$. Notice the cosine term is important near the source in the direct sound field, whereas the sine term is important in the far field.

---

5. Show that the root mean square acoustic pressure is given by

$$p_{\text{r.m.s.}} = p_{\text{peak}}/\sqrt{2}.$$

Solution

$$p_{\text{r.m.s.}} = \sqrt{\frac{1}{T}\int_0^T \{p_0 \cos(\omega t - kx)\}^2\, dt}$$

$$= \sqrt{p_0^2 \frac{1}{T}\frac{1}{2}\left\{t + \frac{\sin 2(\omega t - kx)}{2\omega}\right\}_0^T},$$

$$p_{\text{r.m.s.}} = \frac{p_0}{\sqrt{2}}.$$

6. Calculate the speed of sound in:

(a) air at 20°C, $1·013 \times 10^5$ Pa and having $\gamma = 1·4$;
(b) water having a bulk modulus of $2·1 \times 10^9$ Pa and a density of 998 kg/m²;
(c) copper having a Young's modulus of $12·2 \times 10^{10}$ Pa and a density of 8900 kg/m³.

(a) $c_{air} = \sqrt{\dfrac{1·013 \times 10^5 \times 1·4}{1·2}} = 344$ m/s

assuming the density of air at 20°C is $1·2$ kg/m³;

(b) $c_{water} = \sqrt{\dfrac{2·1 \times 10^9}{998}} = 1460$ m/s;

(c) $c_{copper} = \sqrt{\dfrac{12·2 \times 10^{10}}{8900}} = 3700$ m/s.

---

7. The intensity of a plane sinusoidal sound wave in air of frequency 1 kHz is $2·41 \times 10^{-5}$ W/m². Calculate the r.m.s. value of excess pressure in Pa and also in terms of decibels above the reference pressure of $0·00002$ Pa. Also calculate the r.m.s. particle velocity and maximum value of displacement. (Noise Conference, UWIST, 1973.)

Pressure $= \sqrt{\rho c \times \text{intensity}}$

$\qquad\qquad = \sqrt{1·2 \times 343 \times 2·41 \times 10^{-5}}$

$\qquad\qquad = 0·1$ Pa.

Sound pressure level $= 20 \lg \left(\dfrac{0·1}{0·00002}\right)$

$\qquad\qquad = 74$ dB.

R.M.S. velocity $= \dfrac{\text{Pressure}}{\rho c}$

$\qquad\qquad = \dfrac{0·1}{1·2 \times 343}$

$\qquad\qquad = 2·41 \times 10^{-4}$ m/s.

Maximum displacement $= \dfrac{\sqrt{2} \times \text{r.m.s. velocity}}{\omega}$, where $\omega = 2\pi f$

$\qquad\qquad = \dfrac{\sqrt{2} \times 2·41 \times 10^{-4}}{2\pi 10^3}$

$\qquad\qquad = 5·43 \times 10^{-8}$ m.

This problem requires manipulation of the relationships between intensity, pressure, sound pressure level, velocity and displacement. It also illustrates the smallness of the physical quantities involved.

---

8. A factory manager wishes to install a group of machines in a room. Each machine raises the noise level from a background sound power level of 60 dB when measured at the centre of the room. How many machines can be installed if the sound power level shall not exceed 80 dB, assuming diffused conditions with an otherwise quiet background. Calculate the percentage drop in production as the level attributed to each machine rises to 65 dB. Assume each machine has exactly the same acoustic performance. (Noise Conference, UWIST, 1973.)

Sound power level for one machine $= 10 \lg \left(\dfrac{W_1}{W_0}\right)$ dB.

Sound power level for two identical machines $= 10 \lg \left(\dfrac{2W_1}{W_0}\right)$ dB.

Hence each time the number of machines is doubled the sound level rises 3 dB and a tenfold change produces a 10 dB rise.

TABLE 8.2

Number of machines	Sound power level generated	
1	60	65
2	63	68
4	66	71
8	69	74
10	70	75
16	72	77
20	73	78
32	75	80
40	76	
64	78	
80	79	
100	80	

Table 8.2 shows that 100 machines would give a sound level of 80 dB if one such machine gives 60 dB. Further, if one machine gives rise to 65 dB, the 80 dB level is reached with 32 machines, i.e. a percentage production drop of 68%.

CHAPTER 9

STRUCTUREBORNE SOUND

THIS chapter is concerned with the basic concepts which underly the motion of vibrating bodies. In environmental engineering these concepts are important in applying control measures to limit transmission of vibrations from equipment to the building structure.

We have already seen in section 6.1 that the equation of motion for a periodic exciting force $F_0 \sin \omega t$, acting on a system with one degree of freedom, is

$$m\ddot{x} + r\dot{x} + kx = F_0 \sin \omega t.$$

The symbols m, r and k represent mass, resistance offered by a viscous medium and the stiffness of a spring respectively (Fig. 6.2). Every vibrating system can be resolved into these basic components although in systems having multi-degrees of freedom a complexus of parallel and series elements may be present. In this chapter we consider the solutions to this equation which suggest suitable values of r and k to be used in any vibration control design aimed to limit the transmission of the motion of the mass to the building structure (note: sometimes the converse may be required, e.g. an undersea structure can undergo buffeting from circulation currents, hence systems *within* the structure need protection from any structural vibrations). The solutions given only apply to linear vibrations; the effects of the mass of the spring have been neglected.

9.1. Undamped, Free Vibrations

In this case $r\dot{x}$ and $F_0 \sin \omega t$ are zero, hence

$$m\ddot{x} + kx = 0. \tag{9.1}$$

This equation represents the motion of a mass on a spring, the inertia force $m\ddot{x}$ and the restoring force oppose the motion of the mass.

The general solution of equation (9.1) can be written as

$$x = A \sin \omega_n t + B \cos \omega_n t, \tag{9.2}$$

where $\omega_n = \sqrt{k/m}$. A and B are constants which can be found from the initial static conditions, thus at $t = 0$, $\dot{x} = 0$ and $x = x_0$,

$$x = x_0 \cos \omega_n t. \tag{9.3}$$

The natural frequency of the motion is

$$f_n = \frac{1}{2\pi} \sqrt{\frac{k}{m}}, \tag{9.4}$$

and the period T is $1/f_n$; the natural angular frequency $\omega_n = 2\pi f_n$. A common case in practice occurs when a refrigeration plant is to be installed; the static deflection δ of the floor due to the equipment is given by

$$\delta = \frac{mg}{k} \quad \text{(i.e. using equation (9.1)).}$$

The natural frequency of motion imparted to the building structure before the plant is switched on is

$$f_n = \frac{1}{2\pi} \sqrt{\frac{g}{\delta}}. \tag{9.5}$$

Since $g = 9\cdot81$ m/s² and δ = static deflection in millimetres,

$$f_n = 15\cdot76 \sqrt{\frac{1}{\delta}} \text{ Hz.} \tag{9.6}$$

Figure 9.1 compares the action of a single spring system with one containing several springs acting in series or parallel.

(a) Parallel systems

Each spring experiences the same displacement, hence the restoring force F is

$$F = (k_1 + k_2 + \cdots k_n)x,$$

i.e. this system is equivalent to a single spring of stiffness

$$k = k_1 + k_2 + \cdots k_n.$$

(b) Series systems

Restoring force $\qquad\qquad F = k(x_1 + x_2).$
Each spring experiences the same force, hence

$$F = k \left(\frac{F}{k_1} + \frac{F}{k_2} \right),$$

thus a series system is equivalent to a single spring with stiffness

$$k = 1 \Big/ \left(\frac{1}{k_1} + \frac{1}{k_2} + \cdots \frac{1}{k_n} \right).$$

Petrusewicz and Longmore (1974) give stiffness formulae for common components.

The *Helmholtz resonator* is a well-known type of acoustic filter. The natural frequency is given by

$$f_n = \frac{1}{2\pi} \sqrt{\frac{k}{(al\rho)}}, \tag{9.7}$$

where a and l are the cross-sectional area and the length of the neck (Fig. 9.2) and ρ is the gas density; the effects of friction in the neck of the resonator are neglected here. (See Alfredon in Crocker, 1972).

(a) Single spring system

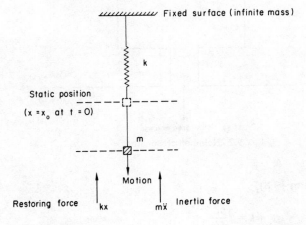

(b) Parallel systems

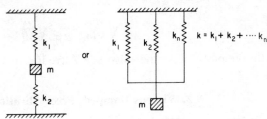

(c) Series systems

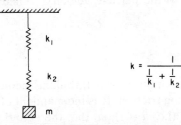

FIG. 9.1. Series and parallel mass-spring systems.

Consider the motion of an acoustic pressure wave in the neck of the resonator. For a pressure change dp in a distance dx, the restoring force is given by

$$F = a\, dp = k\, dx. \tag{9.8}$$

Assuming adiabatic compressions and rarefactions in the acoustic pressure wave

$$pV^{\gamma} = c,$$
$$dp\, V^{\gamma} + \gamma p\, V^{\gamma-1}\, dV = 0.$$

Therefore
$$\frac{dp}{dV} = -\frac{\gamma p}{V} \quad \text{and} \quad dV = a\, dx,$$

hence
$$\frac{dp}{dx} = -\frac{\gamma p a}{V}.$$

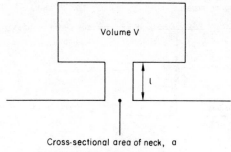

Cross-sectional area of neck, a
FIG. 9.2. Helmholtz resonator.

Substituting equation (9.8),

$$k = \frac{\gamma p a^2}{V}.$$

Therefore

$$f_n = \frac{1}{2\pi} \sqrt{\frac{\gamma p a}{V l \rho}} = \frac{c}{2\pi} \sqrt{\frac{a}{V l}}, \qquad (9.9)$$

where c is the velocity of sound (also see Alfredon in Crocker, 1972).

9.2. Viscous Damped, Free Vibrations

In section 8.3 the general wave equation was derived using a damping force R taken to be proportional to the particle velocity, hence

$$R = r\dot{x}.$$

In general

$$R \propto \dot{x}^n,$$

where n approaches 2 for motion in a turbulent medium; for systems with high values of Coulomb or sliding friction, R is independent of $\dot{x}$. We shall make the same assumption as previously (i.e. $n = 1$), so that the equation of motion for a damped but freely vibrating system will be

$$m\ddot{x} + r\dot{x} + kx = 0. \qquad (9.10)$$

This can be solved by substituting $x = \exp(st)$ as a trial solution, thus

$$ms^2 + rs + k = 0,$$

which gives

$$s = \frac{-r \pm (r^2 - 4mk)^{1/2}}{2m}$$

$$= -\frac{r}{2m} \pm \left\{ \left(\frac{r}{2m}\right)^2 - \left(\frac{k}{m}\right) \right\}^{1/2},$$

inserting the natural angular frequency

$$s = \frac{-r}{2m} \pm \left\{ \left(\frac{r}{2m}\right)^2 - \omega_n^2 \right\}^{1/2}.$$

When $r = 0$, $s = \omega_n$ then the simple harmonic function derived in section 9.1 is obtained. The three other cases to be considered are:

(i) Light damping when $\omega_n > r/2m$

$$s = -\frac{r}{2m} \pm j \left\{ \omega_n^2 - \left(\frac{r}{2m} \right)^2 \right\}^{1/2},$$

and the general solution is

$$x = \exp\left(-\frac{r}{2m} t\right) \left[C \cos \left\{ \omega_n^2 - \left(\frac{r}{2m} \right)^2 \right\}^{1/2} t + D \sin \left\{ \omega_n^2 - \left(\frac{r}{2m} \right)^2 \right\}^{1/2} t \right];$$

since $(r/2m)$ must have the dimensions of angular frequency, let $\omega = r/2m$, then, expressing the solution more simply,

$$x = E \exp(-\omega t)\{\cos(\omega_n^2 - \omega^2)^{1/2} t + \epsilon\}, \tag{9.11}$$

where E and ϵ are constants.

(ii) Critical damping when $\omega_n = r/2m$

$$s = -\frac{r}{2m} = -\omega,$$

and the general solution is

$$x = (G + Ht) \exp(-\omega t). \tag{9.12}$$

The motion is said to be *aperiodic* (non-oscillatory) and comes to rest in the minimum time.

(iii) Heavy damping when $\omega_n < r/2m$

$$s = -\omega \pm (\omega^2 - \omega_n^2)^{1/2}$$

and the general solution is

$$x = \exp(-\omega t)\{J \exp(\omega^2 - \omega_n^2)^{1/2} t + K \exp - (\omega^2 - \omega_n^2)^{1/2} t\}.$$

This can be expressed in the form

$$x = L \exp(-\omega t)\{\cosh(\omega^2 - \omega_n^2)^{1/2} t + \phi\}, \tag{9.13}$$

where L and ϕ are constants. The motion is again aperiodic.

The three possible responses are shown in Fig. 9.3. Referring to Fig. 9.3 the decrement δ is defined as

$$\delta = \frac{x_n}{x_n + 1} = \frac{\exp(-\omega t)}{\exp\{-\omega(t + T)\}} = \exp(\omega T)$$

and the *logarithmic decrement* is the parameter often used to denote the degree of damping

$$\ln \delta = \omega T = \frac{r}{2m} T. \tag{9.14}$$

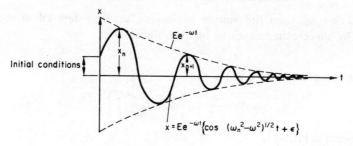

(a) Light damping

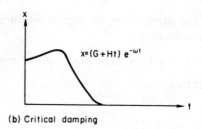

(b) Critical damping

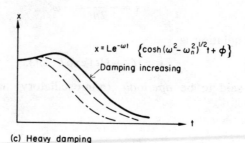

(c) Heavy damping

FIG. 9.3. Damped free vibrations.

The characteristic time constant τ for the motion is

$$\tau = \frac{1}{\omega} = \frac{2m}{r}.$$
(9.15)

9.3. Periodic Excitation in Linear Systems

The equation of motion for such a system is

$$m\ddot{x} + r\dot{x} + kx = \psi(t),$$
(9.16)

where $\psi(t)$ is the exciting force function.

We shall only consider periodic excitation because step and impulse functions are not normally encountered in environmental engineering. Thus for a driving angular frequency ω and force amplitude F_0 the equation of motion will be

$$m\ddot{x} + r\dot{x} + kx = F_0 \sin \omega t.$$
(9.17)

The general solution to this type of differential equation comprises a *complementary function*, which is the solution for the transient state vibrations taken when $F_0 \sin \omega t = 0$ (i.e. the solutions of $m\ddot{x} + r\dot{x} + kx = 0$ which have already been found), and a *particular integral* which is the solution for the steady-state vibrations. Two cases will be considered.

(a) In the absence of damping

$$m\ddot{x} + kx = F_0 \sin \omega t. \tag{9.18}$$

The particular integral may be found by letting $x = A \sin(\omega t - \phi)$ and substituting for $\ddot{x}$ and x in equation (9.18). The resulting equation will be valid for all values of ωt and yields the solution

$$F_0 = x_0(k - m\omega^2)$$

on taking $x = x_0$ when $\omega t = \frac{1}{2}\pi$.

Therefore
$$x_0 \sin \omega t = \frac{F_0 \sin \omega t}{(k - m\omega^2)}. \tag{9.19}$$

The complete general solution will be the algebraic sum of equations (9.3) and (9.19):

$$x = x_0 \cos \omega_n t + x_0 \sin \omega t,$$

$$x = x_0 \cos \omega_n t + \frac{F_0 \sin \omega t}{(k - m\omega^2)},$$

$$x = x_0 \cos \omega_n t + \frac{F_0}{m(\omega_n^2 - \omega^2)} \sin \omega t. \tag{9.20}$$

Plotting the displacement x against ω (Fig. 9.4) it can be seen that:

(i) a resonance condition occurs when $\omega = \omega_n$;
(ii) the amplitude resonance is infinite because there is no damping;
(iii) the amplitudes at any driving frequency decrease with increasing stiffness factor k.

(b) With damping the equation of motion is

$$m\ddot{x} + r\dot{x} + kx = F_0 \sin \omega t. \tag{9.21}$$

Again, the general solution will consist of the complementary function (see equations (9.11), (9.12) and (9.13)) and the particular integral. The latter is obtained in exactly the same manner as in (a) above. Confining our interest to the steady-state solution, the particular integral is amplitude,

$$x = \frac{(F_0/m) \sin(\omega t - \phi)}{\left\{ \left(\dfrac{r\omega}{m}\right)^2 + (\omega_n^2 - \omega^2)^2 \right\}^{1/2}}.$$

This solution can very simply be obtained by putting $x = A \sin(\omega t - \phi)$, $\dot{x} = \omega A \cos(\omega t - \phi)$ and $\ddot{x} = -\omega^2 A \sin(\omega t - \phi)$ in equation (9.21). This expression can

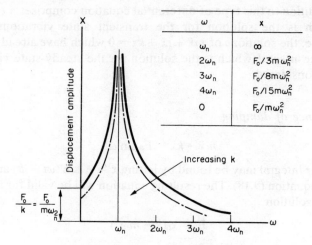

ω	x
ω_n	∞
$2\omega_n$	$F_0/3m\omega_n^2$
$3\omega_n$	$F_0/8m\omega_n^2$
$4\omega_n$	$F_0/15m\omega_n^2$
0	$F_0/m\omega_n^2$

(a) Displacement–frequency function.

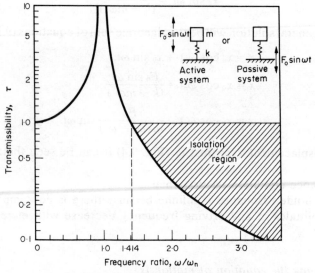

(b) Transmissibility frequency function.

FIG. 9.4. Response of forced undamped system.

be written in the form

$$x = \frac{(F_0/\omega_n^2 m)[\sin(\omega t - \phi)]}{\left\{\left(\dfrac{r}{m\omega_n}\right)^2\left(\dfrac{\omega}{\omega_n}\right)^2 + \left(1 - \dfrac{\omega^2}{\omega_n^2}\right)^2\right\}^{1/2}}. \qquad (9.22)$$

Phase angle ϕ is given by

$$\tan\phi = \frac{r\omega/m}{(\omega_n^2 - \omega^2)}$$

or
$$\tan \phi = \frac{(r/m\omega_n)(\omega/\omega_n)}{\{1 - (\omega^2/\omega_n^2)\}}.$$ (9.23)

The amplitude and phase functions are shown in Figs. 9.5a and b for various degrees of damping.

The following conclusions can be made:

(i) the resonance amplitude decreases with increased damping;

(ii) when $r/m\omega_n > 1.0$ no significant maximum occurs in the displacement amplitude;

(iii) when $\omega/\omega_n > 2.0$ the effect of various degrees of damping (e.g. $0.5 < (r/m\omega_n) < 4$) on the displacement amplitude is small, so that it is uneconomical to use materials with large damping coefficients with systems having high excitation frequencies and low natural frequencies;

(iv) the displacement lags behind the exciting force, three distinct areas of phase lag occur:

when $\dfrac{\omega}{\omega_n} < 1$, ϕ lies between 0 and $\frac{1}{2}\pi$,

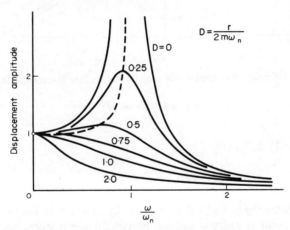

(a) Displacement–frequency function for various degrees of damping.

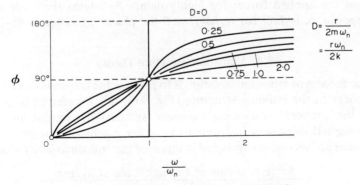

(b) Phase function for various degrees of damping.

FIG. 9.5. Response of forced damped system.

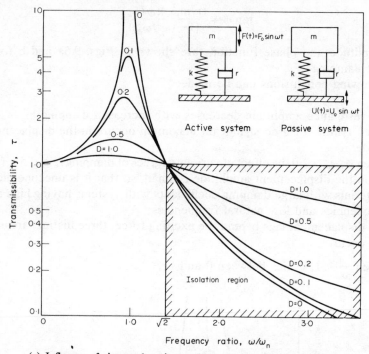

(c) Influence of viscous damping on the value of the transmissibility.

FIG. 9.5. (*Continued*).

$$\frac{\omega}{\omega_n} = 1, \quad \phi = \tfrac{1}{2}\pi,$$

$$\frac{\omega}{\omega_n} > 1, \quad \phi \text{ lies between } \tfrac{1}{2}\pi \text{ and } \pi;$$

(v) when the damping is small ϕ is practically zero for $(\omega/\omega_n) < 1$, but when this frequency ratio is unity a sudden phase change occurs, and for $(\omega/\omega_n) > 1$, $\phi \sim \pi$, i.e. at high frequencies the system moves in the opposite direction to that of the applied force; for highly damped systems there are no sudden changes in ϕ, in fact for $(\omega/\omega_n) > 0.5$ the phase lag range is 85–110°.

9.4. Vibration Isolation Theory

In practice the aim of vibration isolation is to prevent the transmission of vibrations from machinery to the building structure (Fig. 9.6); this is referred to as an *active* system. In the converse situation—a passive system—the motion of an excited structure is limited to protect instruments or other machinery mounted on it. The effectiveness of an isolator is expressed in terms of the *transmissibility* τ defined by

$$\tau = \frac{\text{force transmitted through isolator system}}{\text{exciting force applied to mass on isolator system}}.$$

Isolation efficiency is defined as $(1 - \tau)$.

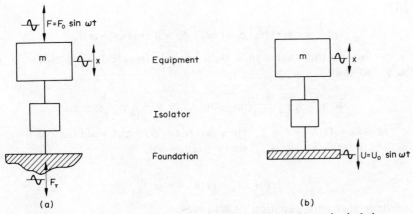

FIG. 9.6. Concept of vibration isolation: (a) active isolation, (b) passive isolation.

9.4.1. FORCED, UNDAMPED SYSTEM

In this case the ratio of the force amplitudes will be

$$\tau = \frac{|kx|}{|F_0 \sin \omega t|}.$$

Using equation (9.20) (but without the complementary function) and rearranging the terms,

$$\tau = \frac{1}{1 - (\omega/\omega_n)^2}. \tag{9.24}$$

The transmissibility function is plotted in Fig. 9.4b. It can be seen that:

(i) isolation is only achieved when $(\omega/\omega_n) > \sqrt{2}$;
(ii) infinite resonance occurs when $(\omega/\omega_n) = 1$;
(iii) $\tau = 1$ at $(\omega/\omega_n) = 0$ and when $(\omega/\omega_n) = \sqrt{2}$;
(iv) for $0 < (\omega/\omega_n) < 1$, the natural frequency $\omega_n > \omega$ and the transmissibility reaches very high values; the system is very stiff.

In design attempt to keep the natural frequency low so that ω/ω_n is high; in practice this means using a low stiffness factor (i.e. soft springs) and a large static deflection.

9.4.2. DAMPED ACTIVE SYSTEM

The ratio of the force amplitudes is

$$\tau = \frac{|r\dot{x} + kx|}{|F_0 \sin \omega t|}.$$

$|F_0 \sin \omega t|$ is given in equation (9.22); for

$$|r\dot{x} + kx| \quad \text{let} \quad x = G \sin(\omega t - \phi)$$

and then

$$r\dot{x} + kx = G\{r\omega \cos(\omega t - \phi) + k \sin(\omega t - \phi)\}.$$

Let γ be one of the angles in a right-angled triangle having sides $r\omega$, k and $(k^2 + r^2\omega^2)^{1/2}$ so that

$$\sin \gamma = \frac{r\omega}{(k^2 + r^2\omega^2)^{1/2}} \quad \text{and} \quad \cos \gamma = \frac{k}{(k^2 + r^2\omega^2)^{1/2}},$$

$$r\dot{x} + kx = G(k^2 + r^2\omega^2)^{1/2}\{\sin \gamma \cos(\omega t - \phi) + \cos \gamma \sin(\omega t - \phi)\}$$
$$= G(k^2 + r^2\omega^2)^{1/2} \sin\{\gamma + (\omega t - \phi)\},$$

hence

$$|r\dot{x} + kx| = G(k^2 + r^2\omega^2)^{1/2}$$

and combining this with equation (9.22) gives

$$\tau = \left\{ \frac{(k^2 + r^2\omega^2)}{\left(\dfrac{r}{m\omega_n}\right)^2 \left(\dfrac{\omega}{\omega_n}\right)^2 + \left(1 - \dfrac{\omega^2}{\omega_n^2}\right)^2} \right\}^{1/2} \frac{1}{k}$$

Rearranging the terms,

$$\tau = \left\{ \frac{1 + \left(\dfrac{r}{m\omega_n}\right)^2 \left(\dfrac{\omega}{\omega_n}\right)^2}{\left(\dfrac{r}{m\omega_n}\right)^2 \left(\dfrac{\omega}{\omega_n}\right)^2 + \left(1 - \dfrac{\omega^2}{\omega_n^2}\right)^2} \right\}^{1/2}. \tag{9.25}$$

A plot of τ against ω/ω_n is shown in Fig. 9.5c.
Conclusions that can be drawn from this relationship are:

(i) no effective isolation (i.e. $\tau < 1$) take place until $(\omega/\omega_n) > \sqrt{2}$;
(ii) when $(\omega/\omega_n) > \sqrt{2}$ increased viscous damping adversely affects the isolation; for example, at $\omega/\omega_n = 2$ the following values can be read from Fig. 9.8c:

Damping factor D	Transmissibility τ	Isolation efficiency $(1 - \tau)$ (%)
1	0·82	18
0·5	0·62	38
0·2	0·40	60
0·1	0·36	64
0	0·33	67

With a little damping, say $0 < D < 0.2$, the isolation efficiency varies from 67 to 60% but increase the damping factor to 0·5 and the isolation efficiency falls to 38%; it is possible to design an isolation system with variable characteristics such that D decreases as ω/ω_n increases; in practice attempt to make $\omega/\omega_n \geqslant 3$;
(iii) when $(\omega/\omega_n) < \sqrt{2}$ damping is necessary to decrease the amplitude resonance which occurs at $(\omega/\omega_n) = 1$; this condition will occur in all systems operating at $(\omega/\omega_n) \simeq 1$ when starting up and running down;

(iv) when $(\omega/\omega_n) > \sqrt{2}$, damping in the range $0 < D < 0\cdot1$ has very little effect;
(v) when $(\omega/\omega_n) < 0\cdot25$ damping has very little influence on the system although τ remains above unity.

Table 9.1, based on Harris and Crede (1961), summarises the principal properties of vibration isolators.

9.5. Limitations of Theory

Two principal factors limit the application of the theory discussed so far. They are (a) isolators may display non-linear characteristics and (b) a one-degree-of-freedom system has been represented whereas in practice a system will often execute many degrees of freedom.

9.5.1. NON-LINEAR SYSTEMS

Springs and dampers may have non-linear characteristics as, for example, those shown in Fig. 9.7 on page 480. Rubber in compression and conical springs have the hardening property (see curve OA on Fig. 9.7a) in which the rate of change of force with displacement $\partial F/\partial x$, increases as x increases, i.e. the spring becomes stiffer. Curve OB on the same diagram shows the converse case. The natural frequency will under these circumstances vary with the vibration amplitude (Fig. 9.7b). The response curve of the non-linear system with a hard spring and with a linear viscous damper is shown in Fig. 9.7c. Notice that the response curve leans to the right, whereas for a system with a soft spring it would lean to the left; the normal linear response curve is upright as discussed in section 9.4. The behaviour of this non-linear system is unstable in the region between BC (increase of frequency) and D to E (decrease of frequency). To the left of ED and to the right of BC the amplitude function is single-valued, but in between, a region of instability exists in which the amplitude has three possible values. In practice the curve DB is unattainable.

Other problems can arise in non-linear systems from (a) subharmonics especially in lowly damped complex systems and (b) interactive effects in complex systems.

The equation of motion for a forced, damped non-linear system can be represented by

$$m\ddot{x} + r\dot{x} + k(x) = F_0 \sin \omega t,$$

where the stiffness function can be expressed as a polynomial in x, say

$$k(x) = a + bx + cx^2 + dx^3.$$

Solutions may be obtained using a perturbation method. In this way it can be shown that the displacement amplitude $|x|$ is

$$\omega^2 = 1 - \frac{1}{|x|} + \frac{3}{4}\,\epsilon|x|^2, \tag{9.26}$$

where ϵ is the spring characteristic. When $\epsilon < 0$ the spring is soft; $\epsilon > 0$ signifies the spring is hard and the linear case has $\epsilon = 0$. A plot of equation (9.26) displays the

(continued on page 479)

TABLE 9.1. TYPES OF IDEALIZED ACTIVE VIBRATION
ISOLATOR (BASED ON HARRIS AND CREDE, 1961)

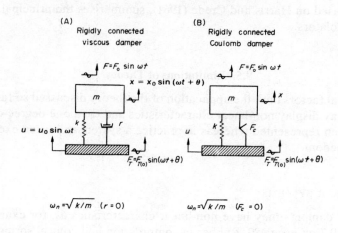

(A)
Rigidly connected
viscous damper

$$\omega_n = \sqrt{k/m} \quad (r = 0)$$

$$r_c = 2\sqrt{km}$$

$$D = r/r_c$$

(B)
Rigidly connected
Coulomb damper

$$\omega_n = \sqrt{k/m} \quad (F_c = 0)$$

$$D_c = \frac{F_c}{ku_0}$$

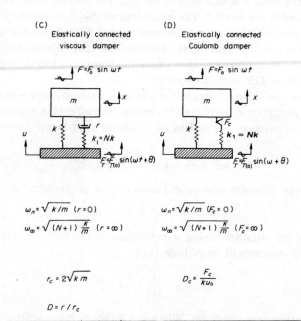

(C)
Elastically connected
viscous damper

$$\omega_n = \sqrt{k/m} \quad (r = 0)$$

$$\omega_\infty = \sqrt{(N+1)\frac{k}{m}} \quad (r = \infty)$$

$$r_c = 2\sqrt{km}$$

$$D = r/r_c$$

(D)
Elastically connected
Coulomb damper

$$\omega_n = \sqrt{k/m} \quad (F_c = 0)$$

$$\omega_\infty = \sqrt{(N+1)\frac{k}{m}} \quad (F_c = \infty)$$

$$D_c = \frac{F_c}{ku_0}$$

TABLE 9.1 (*cont.*). ABSOLUTE TRANSMISSIBILITY CURVES

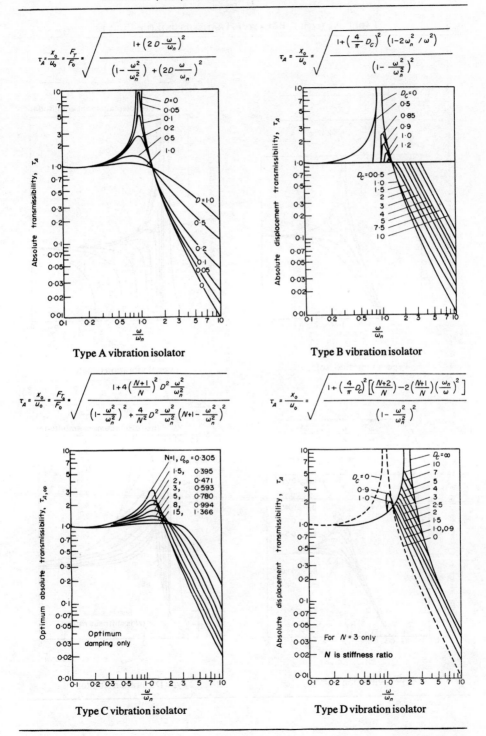

$$\tau_A = \frac{x_0}{u_0} = \frac{F_T}{F_0} = \sqrt{\frac{1+\left(2D\frac{\omega}{\omega_n}\right)^2}{\left(1-\frac{\omega^2}{\omega_n^2}\right)+\left(2D\frac{\omega}{\omega_n}\right)^2}}$$

Type A vibration isolator

$$\tau_A = \frac{x_0}{u_0} = \sqrt{\frac{1+\left(\frac{4}{\pi}D_C\right)^2\left(1-2\omega_n^2/\omega^2\right)}{\left(1-\frac{\omega^2}{\omega_n^2}\right)^2}}$$

Type B vibration isolator

$$\tau_A = \frac{x_0}{u_0} = \frac{F_{T_n}}{F_0} = \sqrt{\frac{1+4\left(\frac{N+1}{N}\right)^2 D^2\frac{\omega^2}{\omega_n^2}}{\left(1-\frac{\omega^2}{\omega_n^2}\right)^2+\frac{4}{N^2}D^2\frac{\omega^2}{\omega_n^2}\left(N+1-\frac{\omega^2}{\omega_n^2}\right)^2}}$$

Type C vibration isolator

$$\tau_A = \frac{x_0}{u_0} = \sqrt{\frac{1+\left(\frac{4}{\pi}D_C\right)^2\left[\left(\frac{N+2}{N}\right)-2\left(\frac{N+1}{N}\right)\left(\frac{\omega_n}{\omega}\right)^2\right]}{\left(1-\frac{\omega^2}{\omega_n^2}\right)^2}}$$

Type D vibration isolator

TABLE 9.1 *(cont.)*. RELATIVE TRANSMISSIBILITY CURVES

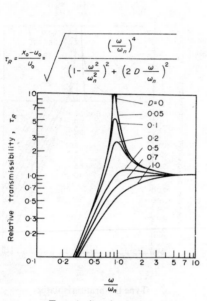

$$\tau_R = \frac{x_0 - u_0}{u_0} = \sqrt{\frac{\left(\frac{\omega}{\omega_n}\right)^4}{\left(1 - \frac{\omega^2}{\omega_n^2}\right)^2 + \left(2 D \frac{\omega}{\omega_n}\right)^2}}$$

Type A vibration isolator

$$\tau_R = \frac{x_0 - u_0}{u_0} = \sqrt{\frac{\left(\frac{\omega}{\omega_n}\right)^4 - \left(\frac{4}{\pi} D_C\right)^2}{\left(1 - \frac{\omega^2}{\omega_n^2}\right)^2}}$$

(use displacement amplitude)

Type B vibration isolator

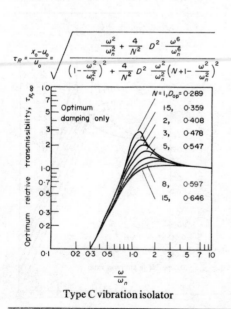

$$\tau_R = \frac{x_0 - u_0}{u_0} = \sqrt{\frac{\frac{\omega^2}{\omega_n^2} + \frac{4}{N^2} D^2 \frac{\omega^6}{\omega_n^6}}{\left(1 - \frac{\omega^2}{\omega_n^2}\right)^2 + \frac{4}{N^2} D^2 \frac{\omega^2}{\omega_n^2}\left(N + 1 - \frac{\omega^2}{\omega_n^2}\right)^2}}$$

Type C vibration isolator

$$\tau_R = \frac{x_0 - u_0}{u_0} = \sqrt{\frac{\left(\frac{\omega}{\omega_n}\right)^4 + \left(\frac{4}{\pi} D_C\right)^2 \left[\frac{2}{N} \frac{\omega^2}{\omega_n^2} - \left(\frac{N+2}{N}\right)\right]}{\left(1 - \frac{\omega^2}{\omega_n^2}\right)^2}}$$

(use displacement amplitude)

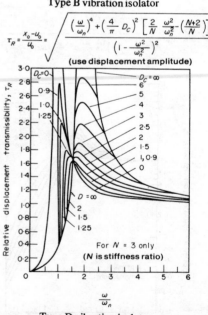

For $N = 3$ only
(N is stiffness ratio)

Type D vibration isolator

TABLE 9.1 (*cont.*). MOTION RESPONSE

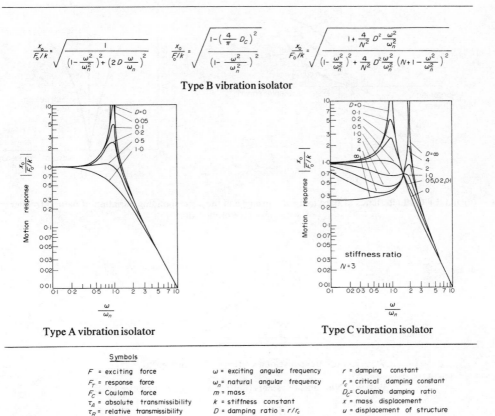

$$\frac{x_0}{F_0/k} = \sqrt{\frac{1}{\left(1 - \frac{\omega^2}{\omega_n^2}\right)^2 + \left(2D\frac{\omega}{\omega_n}\right)^2}}$$

$$\frac{x_0}{F/k} = \sqrt{\frac{1 - \left(\frac{4}{\pi}D_C\right)^2}{\left(1 - \frac{\omega^2}{\omega_n^2}\right)^2}}$$

$$\frac{x_0}{F_0/k} = \sqrt{\frac{1 + \frac{4}{N^2}D^2\frac{\omega^2}{\omega_n^2}}{\left(1 - \frac{\omega^2}{\omega_n^2}\right)^2 + \frac{4}{N^2}D^2\frac{\omega^2}{\omega_n^2}\left(N + 1 - \frac{\omega^2}{\omega_n^2}\right)^2}}$$

Type B vibration isolator

Type A vibration isolator

Type C vibration isolator

Symbols		
F = exciting force	ω = exciting angular frequency	r = damping constant
F_T = response force	ω_n = natural angular frequency	r_c = critical damping constant
F_C = Coulomb force	m = mass	D_C = Coulomb damping ratio
τ_A = absolute transmissibility	k = stiffness constant	x = mass displacement
τ_R = relative transmissibility	D = damping ratio = r/r_c	u = displacement of structure

(continued from page 475)

familiar jumping and hysteresis physical phenomena introduced by the non-linearity (Fig. 9.8). Subharmonic solutions may also be obtained using perturbation methods.

Taylor (1973) gives the natural frequency of non-linear deflection isolators (e.g. cork and rubber in compression, polymers in shear) as

$$f_n = 0.5\sqrt{\frac{W}{k}} \quad \text{in Hz,} \tag{9.27}$$

where W is the isolation load (kg) and k is the isolator dynamic stiffness (kg/m). This formula may be compared to the one used for linear isolators (e.g. some springs, polymers in compression)

$$f_n = 0.5\sqrt{\frac{1}{\delta}} \quad \text{in Hz,} \tag{9.28}$$

where δ is the static deflection in metres or

$$f_n = 15.76\sqrt{\frac{1}{\delta}} \tag{9.29}$$

if δ is given in millimetres.

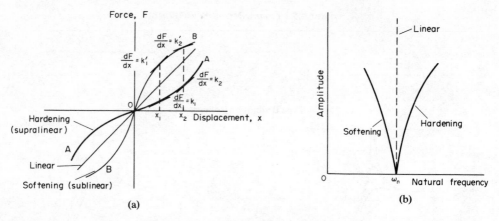

(a) and (b) Characteristics of hard and soft springs and the corresponding variation of natural frequency with amplitude of vibration.

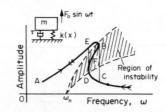

(c) Response of a system with a hard spring.

FIG. 9.7. Non-linear systems (Petrusewicz and Longmore, 1974).

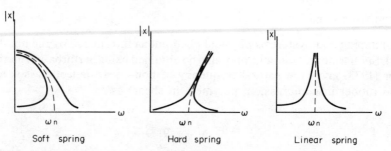

FIG. 9.8. Plots of the equation: $\omega^2 = 1 - (1/|x|) + \frac{3}{4}\epsilon|x|^2$, where ω is the angular frequency, ω_n the natural angular frequency and $|x|$ the displacement amplitude.

9.5.2. MULTI-DEGREES OF FREEDOM SYSTEMS

All the vibrating systems that have been considered in this chapter have been simplified by confining the motion to one direction (i.e. a single-degree-of-freedom system) but in practice there is freedom of movement in all directions. For instance airconditioning or heating equipment in operation can move in vertical or horizontal translational modes, and also in the rotational and rocking modes. The number of degrees of freedom n corresponds to the number of independent coordinates measured from a convenient datum; if q coordinates are chosen arbitrarily and there exist p equations of constraint, then

$$n = q - p.$$

Figure 9.9 shows some examples. Such systems possess n natural frequencies and n *normal modes* of vibration. Note that for a normal mode to exist one of the normal coordinates will be stationary whilst one other, or others, oscillate; any other vibration is an arbitrary combination of other normal modes. Often combinations of translational and rotational motions occur giving rise to *coupled modes* of vibration.

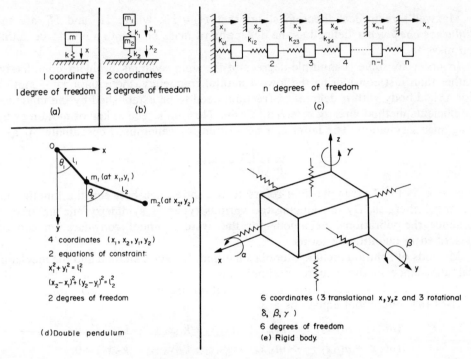

Fig. 9.9. Multi-degrees of freedom systems.

In multi-degrees-of-freedom systems the lowest natural frequency is usually the most important. Approximate methods exist for calculating the lowest natural frequency of the system which bypass the necessity for solving n simultaneous equations. Two methods commonly used are those due to Dunkerley and Rayleigh. The former method involves treating a complex system as a series of one-degree-of-freedom systems and calculating the natural frequency of each. Rayleigh's method is more accurate but more difficult to apply.

The basis of Rayleigh's method is founded upon the Rayleigh principle which states that in the fundamental mode of vibration the mean values over a time period of the kinetic and potential energies are equal. The lowest natural frequency for a system of masses m_i having deflections x_i is given by

$$\omega_n = \left(\frac{g \sum m_i x_i}{\sum m_i x_i^2}\right)^{1/2}.$$

The deflections are found by using the equations:

$$x_1 = g(H_{11}m_1 + H_{12}m_2 + H_{13}m_3 + \cdots H_{1i}m_i),$$
$$x_2 = g(H_{21}m_1 + H_{22}m_2 + H_{23}m_3 + \cdots H_{2i}m_i),$$
$$x_3 = g(H_{31}m_1 + H_{32}m_2 + H_{33}m_3 + \cdots H_{3i}m_i),$$
$$\cdots \cdots \cdots \cdots \cdots \cdots \cdots \cdots \cdots \cdots \cdots \cdots,$$
$$x_i = g(H_{i1}m_1 + H_{i2}m_2 + H_{i3}m_3 + \cdots H_{ii}m_i).$$

Maxwell's reciprocal theorem states that $H_{ij} = H_{ji}$, where H_{ij} and H_{ji} are the *influence coefficients* defined as the deflection of mass m_i due to a unit force acting on mass m_j.

In order to solve a multiple-degrees-of-freedom systems problem completely, rather than just considering the lowest natural frequency, the equations of motion for a rigid body with n degrees of freedom need to be established by the influence coefficients method already referred to using Newton's second law of motion or by Lagrange's equation. The latter can be written for generalised coordinates x_i as

$$\frac{d}{dt}\left(\frac{\partial L}{\partial x_i}\right) - \frac{\partial L}{\partial q_i} = 0,$$

where L is the Lagrangian and refers to the difference between the kinetic and potential energies. By introducing the various types of symmetry encountered in practice the vibrational behaviour for the systems under consideration can be described in quantitative terms.

Methods of solution include Stodola's method, Holzer's method and the mechanical impedance method; more attention, however, has been given to the matrix approach (Smollen, 1966). Equations of motion are put into matrix form. For n masses the differential equations are:

$$(m_{11}\ddot{x}_1 + m_{12}\ddot{x}_2 + \cdots m_{1n}\ddot{x}_n) + (k_{11}x_1 + k_{12}x_2 + \cdots k_{1n}x_n) = 0,$$
$$(m_{21}\ddot{x}_1 + m_{22}\ddot{x}_2 + \cdots m_{2n}\ddot{x}_n) + (k_{21}x_1 + k_{22}x_2 + \cdots k_{2n}x_n) = 0,$$
$$\cdots \cdots \cdots \cdots \cdots \cdots \cdots \cdots \cdots \cdots \cdots \cdots \cdots \cdots \cdots,$$
$$(m_{n1}\ddot{x}_1 + m_{n2}\ddot{x}_2 + \cdots m_{nn}\ddot{x}_n) + (k_{n1}x_1 + k_{n2}x_2 + \cdots k_{nn}x_n) = 0,$$

which in matrix notation become

$$\begin{bmatrix} m_{11} & m_{12} \ldots m_{1n} \\ m_{21} & m_{22} \ldots m_{2n} \\ \cdots \cdots \cdots \cdots \\ m_{n1} & m_{n2} \ldots m_{nn} \end{bmatrix} \begin{bmatrix} \ddot{x}_1 \\ \ddot{x}_2 \\ \cdot \cdot \\ \ddot{x}_n \end{bmatrix} + \begin{bmatrix} k_{11} & k_{12} \ldots k_{1n} \\ k_{21} & k_{22} \ldots k_{2n} \\ \cdots \cdots \cdots \cdots \\ k_{n1} & k_{n2} \ldots k_{nn} \end{bmatrix} \begin{bmatrix} x_1 \\ x_2 \\ \cdot \cdot \\ x_n \end{bmatrix} = \begin{bmatrix} 0 \\ 0 \\ \cdot \cdot \\ 0 \end{bmatrix}$$

and may be simply expressed as

$$[M](\ddot{x}) + [K](x) = 0,$$

where $[M]$ is the *inertia matrix* and $[K]$ the *stiffness matrix*.

$$(\ddot{x}) + [M]^{-1}[K](x) = 0,$$
$$(\ddot{x}) + [C](x) = 0,$$

where $[C]$ is called the *dynamic matrix*. The natural frequencies can be obtained from the characteristic equation

$$I - C = 0$$

for the unit diagonal matrix I.

Matrix iterative techniques require the influence coefficients to be substituted in the matrix equation. Expansion and normalisation procedures are carried out until the first mode repeats itself. For the next higher modes the *orthogonality principle* is used to obtain a new matrix which is free from the lower order modes. For n-degrees of freedom the orthogonality principle states that

$$\sum_{n=1}^{i=n} m_i A_i^r A_i^s = 0,$$

where A_i^r, A_i^s represent the amplitudes of vibration in the r, s modes; $r \neq s$ are the principal modes for the system.

Seto (1964) works out examples using several of the aforementioned methods.

A simple two-degrees system is shown in Fig. 9.10a in which the effect of coupling can be clearly seen. The case for a rigid body with one, two or three planes of symmetry is shown in Fig. 9.10b. Notice that in the case with two planes of symmetry the natural frequencies of the coupled horizontal and rocking modes can

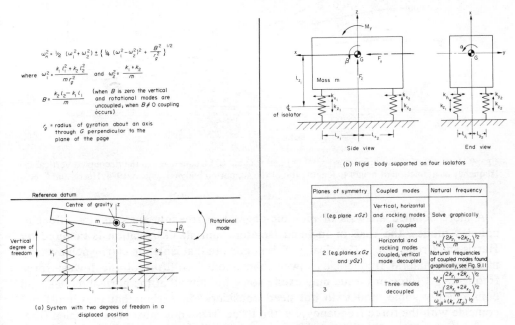

FIG. 9.10. Isolation of multi-degrees of freedom systems (Grosjean, 1974).

be estimated approximately by reference to the equation and the corresponding curves shown in Fig. 9.11. The curves are based on the assumption that the isolators have identical properties ($k_{x_1} = k_{x_2}$, $k_{z_1} = k_{z_2}$), that they are positioned at the extreme corners of the equipment ($l_{x_1} = l_{x_2}$), and that the mass of the equipment is uniformly distributed. These curves show that low values of stiffness in the horizontal or vertical directions can lead to instability in that particular direction, and also that when $\omega_{nx\beta}/\omega_{nz}$ is high there is a strong dependence on the stiffness ratio.

$$\frac{\omega_{nx\beta}}{\omega_{nz}} = \frac{1}{\sqrt{2}} \sqrt{\frac{4\eta\lambda^2 + \eta + 3}{\lambda^2 + 1} \pm \sqrt{\left(\frac{4\eta\lambda^2 + \eta + 3}{\lambda^2 + 1}\right)^2 - \frac{12\eta}{\lambda^2 + 1}}}$$

where $\eta = k_x/k_z$ and $\lambda = 2l_x/2l_z$

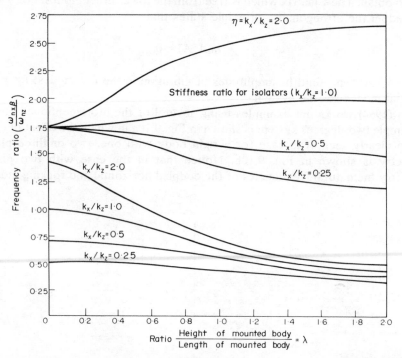

FIG. 9.11. Ratio of the coupled horizontal (x) and rocking (β) frequencies to the decoupled vertical (z) frequency as a function of height to length ratio of the supported body (Grosjean, 1974; Harris and Crede, 1961).

A more detailed analysis of multiple-degrees-of-freedom systems, the effect of coupled modes, the case of inclined isolators and shock isolation is covered by Harris and Crede (1961). Grosjean (1974) points out that isolation can be achieved by making all natural frequencies lower than about two-fifths of the lowest force frequency. A force or torque may excite some but not all of the normal modes, and consequently some modes do not need consideration except that they must not coincide with the force frequency, i.e. the force frequency must be uncoupled from the mode being considered.

9.6. Vibration Isolation Practice

9.6.1. VIBRATION ISOLATORS

In selecting suitable forms of isolation many factors are involved besides isolation performance. Cost, environmental conditions, ageing, corrosion, access maintenance, fire risk, durability, serviceability are some of the aspects to which attention needs to be given. The degree of resilience necessary is, however, a principal factor. For instance if a 100 mm deflection is necessary a rubber isolator would need to be 1 m thick if the rubber is not to be excessively strained; this would be too costly and would take up too much space, so that a metal spring is a more realistic choice. Ranges of deflection and natural frequency obtained by various types of isolators are shown in Fig. 9.12a and b; notice there is some overlap between the two sets of data because some manufacturers will always be able to provide some isolators with a performance outside a given range. In general, metal-coil or leaf-springs and air isolators are used for high deflection, and hence low, natural frequency applications, whereas cork and composite pads are utilised where low deflections are required. Rubbers, covering a range of natural rubber, neoprene, silicone, butyl and nitrile elastomers, are suitable for intermediate ranges of application.

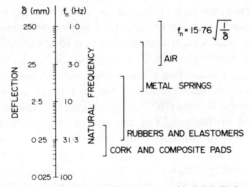

(a) Ranges of application of different types of isolator (Baker, 1974).

Isolator type	Approximate range of deflection (mm)		Approximate range of natural frequency (Hz)	
	Minimum	Maximum	Minimum	Maximum
Metal springs (load capacity up to 18,000 kg)	10	114	2	100
Polymers (load capacity up to 9000 kg)	0·25	23	3	31
Slab (cork, rubber, neoprene)	2	24	11	44

(b) Approximate range of isolator performance based on a survey of British manufacturers made by Taylor (1973).

FIG. 9.12. Isolator performance data.

The following practical points are for general guidance in designing vibration isolation systems.

* Aim to keep the frequency ratio $(\omega/\omega_n) \geqslant 3$. For simple systems ω can be estimated from the speed of rotation (N rev/s) of the machinery, remembering that $\omega = 2\pi f = 2\pi N$; harmonics can be neglected. Systems comprising two units (e.g. fan and motor) should estimate ω from the lowest speed unit. More complex systems need detailed analysis as given in Harris and Crede (1961).

* Use a minimum of three support points; in practice at least four supports are used. For optimum efficiency each isolator should carry an equal load.

* The dynamic performance data for isolators should be used in design calculations, because the dynamic stiffness can achieve values from 20 to 200% more than the static stiffness values. This means that the natural frequency is higher in operation since $f_n = 1/2\pi \sqrt{k/m}$.

* Remember that isolators suffer static deflection due to the mass of the equipment. Woods (1972) advocates the use of a "rule of thumb" for isolation mounting selection which is related to the static deflection of the supporting structure. In critical areas the static deflection of the mounting should be about twenty times that of the supporting structure under the dead load of the equipment, whereas for non-critical areas this factor may be reduced to three.

* Inertia blocks can be used to reduce the displacement amplitude, to minimise the effect of unequal weight distribution and to stabilise resiliently mounted plant having unbalancing forces. The use of inertia blocks to lower the centre of gravity is of great value when the mounted natural frequency is below 3 Hz; ideally the centre of gravity should lie in a plane lower than that of the isolation mountings. Woods (1972) suggests that the weight of the block should be at least equal to the weight of the equipment, and should be two to five times the weight if reciprocating machines are being used. Do not support machines and their ancillary drive equipment on separate bases; it is probably better to use an inertia block about twice the weight of the total assembly. In practice, inertia blocks have a minimum thickness of 150 mm.

* It has been found (see Woods, 1972) that high-pressure fans can exert high back pressures which exert a high turning moment on the fan base. This may be reduced by increasing the base frame length or by using an inertia block in conjunction with stiffer mountings.

* Use flexible connections on the inlet *and* the discharge side of fans and pumps.

* Some isolators are non-linear and this results in high natural frequencies.

* Isolate pipework and ductwork from the building structure as well as from heating and airconditioning plant. Observance of this simple rule, often neglected in practice, can save the environmental systems designer embarrassment and headaches at a later date. Beware of any possibilities of isolation bridging.

* Check with the manufacturer of the isolating equipment that the environmental conditions of the installation will not harm the isolators in any way. Certain

temperature and moisture levels, oil or chemical-laden atmospheres can be detrimental to isolators. Also check the fire-precaution requirements.
* Ensure that there is sufficient access to the isolators for maintenance inspections.
* Use of floating floors was discussed in Chapter 4.

A brief summary of the principal features of the isolators commonly used in building environmental engineering is given in Table 9.2. For detailed accounts of isolator performance the reader should refer to the following references.

In Harris and Crede (1961):
    Chapter 32: Application and design of isolators by Crede.
    Chapter 33: Air suspension and servo-controlled isolation systems by Cavanaugh.
    Chapter 34: Mechanical springs by Wahl.
    Chapter 35: Rubber springs by Frye.
    Chapter 36: Material and interface damping by Lazan and Goodman.
    Chapter 37: Vibration control by applied damping treatments by Hamme.
Also:
    Dart, S. L., and Guth, E., 1946, Elastic properties of cork, *J. Appl. Phys.* **17** (5) 314–18.
    Dart, S. L., and Guth, E., 1947, *J. Appl. Phys.* **18** (5) 470–8.
    Tyzzer, F. G., and Hardy, H. C., 1947, Properties of felt in the reduction of noise and vibration, *J. Acoust. Soc. Am.* **19**, 872–8.

It is worth noting that considerable progress is taking place in the development of resilient materials. Newland and Liquorish (1974) have classified resilient materials into three principal categories:

(a) *cork modified elastomeric pads* for use with static loading pressures of 6×10^4 to 2×10^5 Pa and for vertical natural frequencies of 16–30 Hz;
(b) *moulded synthetic or natural rubber pads or mountings* which can cover the same static loading pressure range as (a) but for lower natural frequency applications of 8–18 Hz;
(c) *composite rubber pads with inter-laminar reinforcement* which can operate at much higher static loading pressures of $1–7 \times 10^6$ Pa over a natural frequency range of about 18–30 Hz.

The natural frequencies quoted here apply to a 50 mm thick pad; doubling the thickness will decrease the natural frequency to a value $1/\sqrt{2}$ of those quoted above.

Development of resilient materials is not simply a case of finding new materials. In Chapter 6 we saw that stiffness is a function of the material and the shape factor. The shape factor can be altered on a macroscopic scale by geometrical arrangement, such as ribs or laminated forms, or on a microscopic scale by the addition of microcellular particles (e.g. cork particles can be added to rubber). The result is a composite pad with a different performance under varying stress.

TABLE 9.2. VIBRATION ISOLATOR CHARACTERISTICS

Metal springs	Rubbers and elastomers	Cork, felt, composite pads

Handle large deflections
High load capacity
Corrosion resistant
Temperature effects small
Low internal damping
 	
Transmit high frequencies; hence, normally insert rubber of felt pads between the spring and the building structure
Coned disc springs (known as *Belleville* springs) have many advantages:
  (i) high energy storage for small space;
  (ii) variable stiffness, by stacking;
  (iii) controlled Coulomb damping, by parallel stacking;
  (iv) controlled non-linear stiffness characteristic.
But springs fatigue easily, and consideration must be given to the possibility of sideways buckling
	Used for medium deflections (0·23–25 mm)
	Medium load capacity
	Can vary stiffness by altering hardness
	Unsuitable where ozone and oil present; neoprene can withstand oil but more expensive
	
	Low fatigue but service life less in compression than in shear
	Small energy-storage capacity, higher in compression than in shear-strain natural rubber by 10% maximum
	Can easily be bonded to steel
	Dynamic stiffness is greater than in static condition and is temperature, frequency, strain amplitude dependent
	Silicones have even higher internal damping and can be resistant to oil, but have low tear strength and high cost
		Cork and felt only applicable for low deflections (cork 150–200 mm thick for 5 mm, 75–115 mm thick for 2–4 mm)
		Cork: used in compression or in compression-shear; becomes less stiff at high load; also frequency dependent
		Recommended loading 5000–1500 kg per m^2
		
		Compresses with age under load
		Can be made impervious to oil and water
		Normal temperatures (<30°C) have little effect
		Felt; use smallest area (5% of flat base) of thickness (10–25 mm)
		High damping
		Only use for natural frequencies over 40 Hz
		Reinforced neoprene pads: These have been used for isolating buildings (cinemas, offices) from underground railway vibrations (see Tico, 1967)

9.6.2. VISCOELASTIC MATERIALS AND DAMPING APPLICATIONS

In Chapter 6 the meaning of the term damping was discussed, together with the concepts of stiffness and mass. Here the applications of damping will be discussed. Air and water are transmitted through buildings in ducts and pipes which are more likely to be made out of steel—a low damping material, although fibreglass, plastics, asbestos or builders' work concrete—brick ducts are sometimes used. With the use of higher fluid velocities, aerodynamic and hydrodynamic sources of sound have become prevalent in environmental services systems, and these can cause duct boom as the air or water pulses along the conduit. This means that even if the transmission system is well isolated from the plant, vibration can remain a problem.

Before any damping treatment can be applied to a vibrating surface to reduce the amplitude of vibration it is necessary to know how much a surface made of a given material will vibrate.

Ungar (1971) gives the maximum stress amplitude δ_{max} in a uniform homogeneous panel as

$$\delta_{max} \simeq \frac{\sqrt{3}\,vY}{c_L},$$

where v is the amplitude of the panel velocity and the material is specified by its modulus of elasticity Y and density ρ_p, the speed of sound in the material being $c_L = \sqrt{Y/\rho_p}$.

The velocity distribution over the panel is obtained by superposition of the modal velocity contributions $v_{m,n}$

$$v_{(x,\,y)} = \sum_{m,\,n=0}^{\infty} v_{m,n} \psi_{m,n}(x,\,y),$$

where $\psi_{m,n}(x,\,y)$ represents the deflection shape corresponding to a sound wave in the m, n mode. By considering the flexural motion of a uniform panel the modal velocity $v_{m,n}$ of a resonant mode can be shown to be dependent on the modal pressure $p_{m,n}$, the loss factor η, the surface mass of the panel m, and the resonance angular frequency of the m,n mode

$$v_{m,n} = \frac{p_{m,n}}{\eta m \omega_{m,n}}$$

or in terms of the panel density ρ_p and thickness h

$$v_{m,n} = \frac{p_{m,n}}{\eta \rho_p h \omega_{m,n}}.$$

$p_{m,n}$ are constants obtained from a knowledge of the pressure distribution over the panel $p(x, y)$, hence for an area A

$$P_{m,n} = \frac{\int_A p(x,\,y)\psi(x,\,y)\,dA}{\int_A \psi_{m,n}^2(x,\,y)\,dA}.$$

Broadband (random) excitation of the panel is likely to excite several modes, and the total response is approximated by superposition of the resonant modal responses. Ungar (1971) shows that in the case of panels subject to random excitation the root means square displacement varies inversely as $\sqrt{\eta}$.

Typical loss factors for material damping in metals and common building materials are given in Table 9.3. Manganese–copper, cobalt (in the form of Nivco 10 consists

TABLE 9.3. VALUES OF LOSS FACTORS FOR BUILDING MATERIALS (CREMER AND HECKL, 1967: VER AND HOLMER, 1971)

Material	Loss factor η		
	$\times 10^{-4}$	$\times 10^{-3}$	$\times 10^{-2}$
Aluminium	1		
Brass	<10		
Brick			1–2
Concrete:			
light; porous			1·5
dense			1–5
Copper		2	
Cork			13–17
Glass		0·6–2	
Gypsum board			0·6–3
Lead		0·5–2	
Magnesium	1		
Masonry blocks		5–7	
Oak, fir			0·8–1
Plaster		5	
Plexiglass, Lucite			2–4
Plywood			1–1·3
Sand, dry			12–60
Steel, iron	1–6		
Tin		2	
Wood fibreboard			1–3
Zinc	3		

of 72% Co, 23% Ni), titanium and other alloys exhibit a high degree of damping. The loss factor covers the many known inelastic mechanisms and hysteresis phenomena which may be classified into two main groups of material damping—*dynamic hysteresis* and *static hysteresis*. Dynamic hysteresis (also known as viscoelastic, rheological or state hysteresis) is caused by molecular thermal behaviour and slip at grain boundaries, is a linear and time-dependent effect; it includes the effect of anelasticity also known as internal friction; the phenomenon is modelled by a spring–dashpot damper system (see Chapter 6). Static hysteresis is due to magneto elasticity and plastic strain, is essentially a non-linear effect and excludes the time derivatives of stress and strain; the phenomenon is modelled by spring and Coulomb friction elements. Damping in *soft* materials such as mastics, rubbers and plastics is strongly dependent on frequency and temperature.

Vibrating surfaces may be treated by adding a layer or more of a high damping material. *Mastics*—asphalt-based, water-soluble or high-polymer types—are commonly used; asphalted felt is also used and damping tapes are also available. When applying mastics it is a good rule to always concentrate the greatness thickness of the mastic on regions of the surface where the largest vibrations occur; a typical asphalt base mastic may have a vibration decay rate of 25 dB/s with a surface weight of 4 kg/m² reducing to 3 dB/s when 1 kg/m². The performance of viscoelastic layers is strongly temperature dependent, but materials are available which yield optimum performance over a wide range of temperatures.

Asphalted felts are manufactured by impregnating paper, rag or asbestos felts with asphalt. The common range of weight is 1–5 kg/m² with decay rates of 1–12 dB/s for one-ply treatments; this range may be increased to 40 dB/s by using four-ply layers or even to 400 dB/s by covering the viscoelastic layer with a sheet metal or fibreglass loading septum. The manner of cementing the felts to the vibrating surface is very important. Spot cementing rather than "all over" cementing is better to ensure maximum energy dissipation by friction between the panel and the layer. The felts are often indented. Figure 9.13 shows some of the mechanisms that operate when

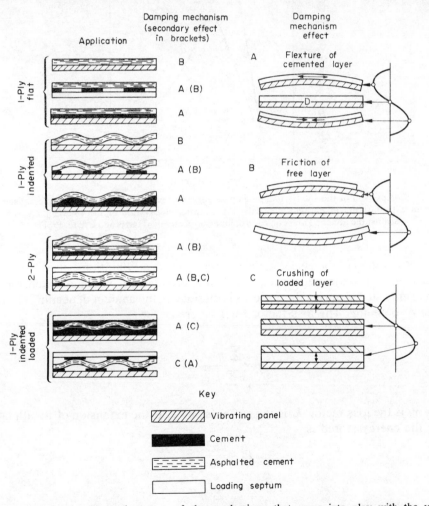

FIG. 9.13. Schematics illustrating some of the mechanisms that come into play with the various applications of asphalted felt; for each of the treatment sections at the left there is indicated the damping mechanism at the right which probably contributes most to the treatment's effectiveness (Harris and Crede, vol. 3, 1961).

using asphalted felt. *Damping tapes* are also available. Harris and Crede (1961) describe a special damping configuration (Fig. 9.14) which is frequency selective. A chart showing the relative effectiveness of the vibration damping materials and structures discussed is given in Fig. 9.15.

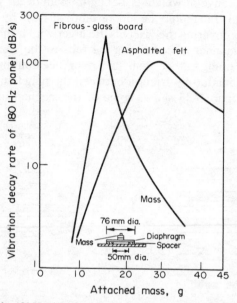

FIG. 9.14. Comparison of the 180 Hz damping effectiveness of two 76 mm diameter frequency-selective deadeners with different diaphragm materials; the vibration decay rate of the 180 Hz plate is plotted against the attached mass on each damper respectively (Harris and Crede, 1961).

The basic action of the viscoelastic layer is simple. The vibration causes bending of the panel and its associated viscoelastic layers producing flexure and extension; the strain energy stored during this process is dissipated, the amount depending upon the degree of damping. The loss factor for i layers is

$$\eta = \frac{\sum \eta_i k_i x_i^2}{\sum k_i x_i^2},$$

where η_i is the loss factor, k_i is the stiffness and x_i is the extension of the ith layer. Since the energy stored is

$$E_i = \frac{k_i x_i^2}{2}$$

$$\eta = \frac{\sum \eta_i E_i}{\sum E_i}.$$

For an uncovered viscoelastic layer attached to a vibrating panel, the bending

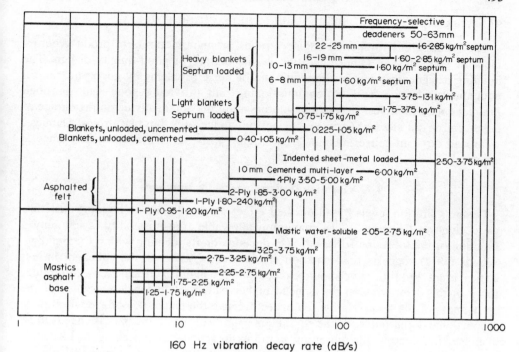

FIG. 9.15. Effectiveness of various vibration damping materials and structures (Harris and Crede, 1961).

produces flexure and extension of the two layers. Ungar (1971) gives the loss factor
as

$$\eta = \beta_2 \Big/ \Big\{ 1 + \frac{k^2(1+\beta_2^2) + (r_1/H_{12})^2\{(1+k^2) + (\beta_2 k)^2\}}{k[1 + (r_{2/H_{12}})^2\{(1+k)^2 + (\beta_2 k)^2\}]} \Big\},$$

where β_2 is the loss factor of the viscoelastic layer, $k = Y_2 H_2 / Y_1 H_1$ is the ratio of
extensional stiffnesses, $r_1 = H_1/\sqrt{12}$ is the radius of gyration of elastic layer (m),
$r_2 = H_2/\sqrt{12}$ is the radius of gyration of viscoelastic layer (m), $H_{12} = (H_1 + H_2)/2$ is the
distance between neutral planes of the two layers, Y_i is the elastic modulus of ith layer
and H_i is the thickness of ith layer.

The extensional strain energy storage can be increased by using a spacing layer.

Shear energy in addition to flexural and extensional energies is involved in the
damping process if a septum or constraining layer is added on top of the viscoelastic
layer. Practical design procedures for panel damping applications is considered in
detail by Ungar (1971).

The effect of increasing the damping of building constructions by means of
viscoelastic layers should not be overlooked. Sylwan (1973) shows that the impact
noise isolation and the sound reduction index above the coincidence frequency
(defined in Chapter 11) will increase by 10 dB for an increase by a factor of 10 in the
loss factor. The theory and some empirical evidence is discussed by Sylwan (1973).

Finally, it should be noted that Coulomb damping occurs at joints as surfaces slip
over one another. Bolted, riveted joints reduce resonance peaks more than one-piece
welded joints.

9.6.3. VIBRATION CONTROL BY REDUCING THE EXCITING FORCES

The building environmental engineer relies on manufacturers to produce equipment of good design. Quiet operation is one feature of good quality design. The building designer can lessen the likelihood of noise problems by good and careful equipment *selection* and by commissioning and providing good and thorough *maintenance* schedules. In selecting motors and fans obtain advice on the choice of either ball or oil sleeve bearings, and on the choice of V-fan-belt drives. Damage, wear and dirt can cause imbalance of moving parts, so that good maintenance is essential.

9.7. Vibration Design Criteria for Buildings

Design vibration levels based on their effect on people and on building structure were given in Chapter 2. The maximum allowable transmissibility is not only a function of the vibrating force but also depends on its location within the building. Woods (1972) suggests that isolation efficiencies of 75% may be sufficient in less critical areas, but this value should exceed 90–95% in noise critical areas. Mountings to be installed on wooden floors need higher isolation efficiencies because of the springiness of the floor relative to concrete. Since the magnitude of the vibration is proportional to the square of the equipment rotational speed, special care needs to be exercised in selecting fan, motor and compressor speeds and matching the force this creates with the appropriate degree of isolation. Table 9.4. lists values of isolation efficiencies as advocated by Woods (1972).

Three standards are currently applicable.

* *BS 4675: 1971. Recommendations for a basis for comparative evaluation of vibration in machinery.* This standard attempts to classify the degree of severity of vibration arising from various types of machine.
* *MIL-STD 167. Mechanical vibrations of shipboard equipment.* Formulae are quoted for calculating the limits of imbalance in rotating machinery. The maximum allowable imbalance U depends on the weight of the rotating part W and the maximum rotational speed N such that

$$U \propto \frac{W}{N} \text{ for speeds} > 1000 \text{ rev/min,}$$

$$U \propto \frac{W}{N^2} \text{ for speeds between 150 and 1000 rev/min,}$$

$$U \propto W \text{ for speeds} < 150 \text{ rev/min.}$$

* *National Electrical Manufacturers Association (NEMA) Standard MG1-12·05 June 1972, Dynamic Balance of Motors.* This indicates the maximum vibration allowed for a given motor size and speed.

TABLE 9.4. RECOMMENDED ISOLATION EFFICIENCIES (CONCRETE FLOOR SLAB) (WOODS, 1972)

		Transmissibility (%)	Isolation efficiency
Critical areas			
Centrifugal compressors		0·5	99·5%
Centrifugal fans	>25 hp	2	98%
Reciprocating compressors	>50 hp		
Pumps	>5 hp		
Axial flow fans	>50 hp	4	96%
Centrifugal fans	5–25 hp		
Reciprocating compressors	10–50 hp		
Pumps	3–5 hp		
Unit airconditioners	Supported		
Fan-coil units	Supported		
Axial flow fans	10–50 hp	6	94%
Centrifugal fans	up to 5 hp		
Reciprocating compressors	up to 10 hp		
Pumps	up to 3 hp		
Air-handling units			
Axial flow fans	up to 10 hp	10	90%
Unit Airconditioners	Hung		
Fan-coil units	Hung		
Pipes	Hung		
Gas-fired boilers	>25 kw		7–12 Hz
Oil-fired boilers	>15 kw		4–7 Hz
Less critical areas			
Centrifugal compressors		6	94%
Centrifugal fans	>25 hp		
Reciprocating compressors	>50 hp	10	90%
Pumps	>5 hp		
Unit airconditioners	Supported		
Fan-coil units	Supported		
Axial flow fans	>50 hp	20	80%
Centrifugal fans	5–25 hp		
Reciprocating compressors	10–50 hp		
Pumps	3–5 hp		
Air-handling units			
Unit airconditioners	Hung		
Fan-coil units	Hung		
Axial flow fans	10–50 hp	25	75%
Axial flow fans	up to 10 hp	30	70%
Centrifugal fans	up to 10 hp		
Reciprocating compressors	up to 10 hp		
Pumps	up to 3 hp		
Pipes	Hung		
Gas-fired boilers	>25 kw		12–20 Hz
Oil-fired boilers	>15 kw		12–20 Hz

CHAPTER 10

THE BEHAVIOUR OF SOUND
IN ROOMS

10.1. The Nature of Sound Fields

If a source of sound is prolonged in time a steady sound wave pattern will be set up in the space. When the sound is stopped a finite time is required for the sound wave pattern to decay. The prolongation of the sound after the source has ceased to emit energy is the *reverberant* sound in contrast to the *direct* sound. Direct sound energy transmitted directly from a source to a receiver decreases with the square of the distance in accordance with the classical spherical wavefront energy distribution law; other energy attenuation factors in addition to this are people or objects in the direct sound pathway absorbing the sound, any unusual atmospheric effects (e.g. temperature gradients of velocity gradients) and interference by other sound or image sources. The reverberant or indirect component depends on the interaction of the sound with the surfaces of the room and any obstructions that cause scattering and diffraction. The reverberant sound is thus a function of the sound absorption of the surface materials besides the size of the space. The ears receive a superposition of the direct and reverberant sound waves. In the case of speech the time delay between the direct and reverberant sound is in the range of 10–30 ms, whereas for music this is a little higher, say 20–40 ms. If this time delay is extended further, *echoes* are perceived. Short, sharp sounds can also cause echoes because the energy is transferred around the room as a discrete group of waves and this discreteness sets up clear auditory impressions spaced out in time.

Equation (7.15) gave the characteristic frequencies of a space having a volume abc as

$$f_{x,y,z} = \frac{c}{2} \sqrt{\left(\frac{n_x}{a}\right)^2 + \left(\frac{n_y}{b}\right)^2 + \left(\frac{n_z}{c}\right)^2}, \tag{10.1}$$

the angles between the respective vibrations being α, β, γ where

$$\frac{a}{\lambda/2 \cos \alpha} = n_x \quad \text{and} \quad \frac{b}{\lambda/2 \cos \beta} = n_y \quad \text{and} \quad \frac{c}{\lambda/2 \cos \gamma} = n_z.$$

It will be recalled that if either $n_x = 0$ or $n_y = 0$ then propagation takes place in two planes perpendicular to the axis for which the mode number is zero and a *tangential* wave pattern exists, whereas if two mode numbers are zero then an *axial* wave pattern occurs, and if $n_x = n_y = n_z \neq 0$ then an *oblique* wave pattern is present.

Kuttruff (1973) shows the eigenfrequencies for a room with dimensions $3 \cdot 1$ m $\times$ $4 \cdot 1$ m $\times 4 \cdot 7$ m using equation (10.1); these are shown in Table 10.1.

TABLE 10.1. EIGENFREQUENCIES OF A RECTANGULAR ROOM WITH DIMENSIONS
3·1 m × 4·1 m × 4·7 m AS CALCULATED BY KUTTRUFF (1973) USING EQUATION (10.1)

Mode type	f_n	n_x	n_y	n_z	Mode type	f_n	n_x	n_y	n_z
Axial	35·169	1	0	0	Tangential	90·471	1	2	0
Axial	41·463	0	1	0	Tangential	90·776	2	0	1
Axial	54·840	0	0	1	Tangential	99·419	0	2	1
Tangential	55·022	1	1	0	Oblique	99·799	2	1	1
Tangential	65·694	1	0	1	Oblique	105·793	1	2	1
Tangential	68·751	0	1	1	Axial	108·511	3	0	0
Axial	72·341	2	0	0	Axial	109·676	0	0	2
Oblique	77·685	1	1	1	Tangential	110·047	2	2	0
Axial	82·925	0	2	0	Tangential	115·488	1	0	2
Tangential	83·381	2	1	0	Tangential	116·164	3	1	0

Equation (10.1) can be expressed in terms of wave propagation numbers thus since $f_{x,y,z} = k_{x,y,z} c /2\pi$,

$$k_{x,y,z} = \pi \sqrt{\left(\frac{n_x}{a}\right)^2 + \left(\frac{n_y}{b}\right)^2 + \left(\frac{n_z}{c}\right)^2}$$

or

$$k_{x,y,z}^2 = k_x^2 + k_y^2 + k_z^2, \tag{10.2}$$

which represents a sphere in k-space having radius $k_{x,y,z}$ and volume $\frac{4}{3}\pi k_{x,y,z}^3$ of which the first octant volume only, $\frac{\pi}{6}k_{x,y,z}^3$, is of interest. Since

$$k_x = \frac{n_x \pi}{a}, \qquad k_y = \frac{n_y \pi}{b} \quad \text{and} \quad k_z = \frac{n_z \pi}{c},$$

then the distances between adjacent lattice points (i.e. when $n_x = n_y = n_z = 1$) in k-space are

$$k_x = \frac{\pi}{a}, \qquad k_y = \frac{\pi}{b} \quad \text{and} \quad k_z = \frac{\pi}{c}.$$

Hence the k-volume per lattice point is π^3/abc and the number of lattice points in the first octant volume having a radius $k_{x,y,z}$ is, therefore,

$$N_f = \frac{\pi k_{x,y,z}^3}{6} \Big/ \frac{\pi^3}{abc}$$

$$= \frac{abc\, k_{x,y,z}^3}{6\pi^2}$$

$$= \frac{4\pi}{3}(V)\left(\frac{f}{c}\right)^3, \tag{10.3}$$

where abc is the room volume V and N_f is equivalent to the number of eigenfrequencies from 0 to $f = k_{x,y,z} c /2\pi$. The average density of the eigenfrequencies is

$$\frac{dN_f}{df} = 4\pi V \frac{f^2}{c^3} \quad \text{eigenfrequencies per Hertz.}$$

Errors have been made in the above account because each spherical lattice point in a coordinate plane has only been regarded as a half of a sphere, and each lattice point lying on an axis has only been counted as a quarter of a sphere. Kuttruff (1973) gives a corrected form for N_f as

$$N_f = \frac{4\pi}{3} V \left(\frac{f}{c}\right)^3 + \frac{\pi}{4} S \left(\frac{f}{c}\right)^2 + \frac{L}{8} \left(\frac{f}{c}\right),$$

(10.4)

where S is the total surface area, i.e. $S = 2(ab + bc + ca)$ and L is the total length of all the edges, i.e. $L = 4(a + b + c)$.

Notice that if equation (10.3) is used to estimate the number of eigenfrequencies for the room detailed in Table 10.1, then $N_f = 10$, but using equation (10.4) there are 20 eigenvalues as given. In the case of the Berlin Philharmonie Concert Hall, which has a volume of $25\,000$ m³, there will be in the order of 10^9 eigenfrequencies with a density of about 10^4 eigenfrequencies per Hertz at 1000 Hz.

With so many possible modes, interference between them is inevitable. If the sound pressure level of a slowly varying frequency sound source is fed through a loudspeaker in a room and is recorded, then a series of maxima and minima will be observed which correspond to the interference of vibrational modes—those in-phase displaying a maximum and those out of phase a minimum (Fig. 10.1). The spatial variation of sound pressure shows a similar pattern. Schroeder *et al.* (1962) gives a formula for estimating the average frequency spacing of adjacent maxima as

$$(\Delta f)_{max} = \frac{\bar{\delta}}{\sqrt{3}},$$

(10.5)

where $\bar{\delta}$ is the mean value of many damping constants associated with neighbouring eigenfrequencies and may be related to the reverberation time T by the form given by Kuttruff (1973) as

$$\bar{\delta} = \frac{6 \cdot 91}{T}.$$

(10.6)

Reverberation time is derived and discussed further later in this chapter. These

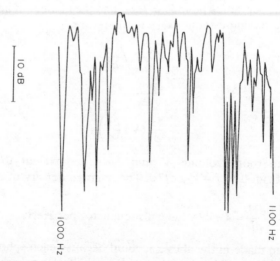

FIG. 10.1. Logarithmic record of the sound pressure level at steady state conditions from 1000 to 1100 Hz measured in a lecture room (Kuttruff, 1973).

formulae are valid for frequencies higher than

$$f \simeq \frac{5000}{\sqrt{V\delta}} \tag{10.7}$$

for a room volume of Vm^3 and where f denotes the lower limit for which the statistical treatment of the interference of modes is valid. Kuttruff (1973) also gives another interesting result in showing that the level difference between the most probable maximum value and the mean value of the frequency curve is

$$\Delta L_{max} \simeq 4 \cdot 3 \ln (\ln BT) \, dB$$

which for a concert hall that needs to respond to a frequency bandwidth B of 10 000 Hz and has a reverberation time of $T = 2s$, means that the absolute maximum sound level of the frequency curve is about 10 dB above the average level.

In practice, room acoustics are treated in a geometrical fashion. A sound wave becomes restricted to a ray. Diffraction, refraction and interference phenomena are neglected and reflection becomes all important. As with light-rays, incident sound rays reflected from a surface remain in the same plane, and the normal to the surface bisects the angle formed by the ingoing and the outgoing rays.

Image sources can be used, as in optics and electrostatics, to illustrate the reflection of sound rays. The construction of an image source is shown in Fig. 10.2a, where the effect of the surface is replaced by the image source. Not all the sound energy striking the wall is reflected from it; some is transmitted through the surface and some is absorbed by the surface. If the surface is irregular the sound will be scattered if the sizes of the irregularities are comparable to or are smaller than the wavelength of the incident sound. The reflections are said to be *diffuse*, but in practice reflections depend to some extent on the direction of the incident sound and partially diffuse reflections occur.

The energy of reflected rays is dissipated by absorption at the walls of a room, other surfaces and the air itself in the room. The total path length of successive reflections which may occur at the walls, floor and ceiling in an order of precedence that depends on the shape of the room, may be found by constructing first and higher order image sources as shown in Fig. 10.2b. Of course, as we have already discussed, several simultaneous reflection patterns may be set up by a sound source or by an array of sound sources such as an orchestra.

In practice only first-order images are considered because of their limited field of

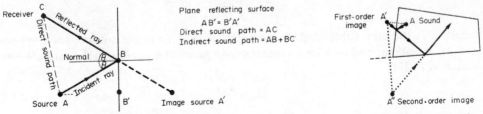

(a) Construction of an image sound source.   (b) Mirror sources of first and second order.

FIG. 10.2. Sound image sources.

view, but in rectangular rooms several image sources of higher order may coincide (Cremer, 1948). This replica effect is shown in two dimensions for a rectangular room in Fig. 10.3.

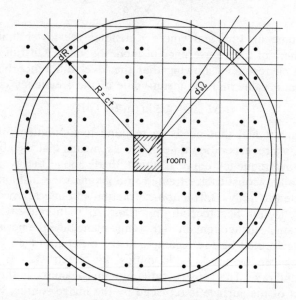

FIG. 10.3. Mirror sound source for a rectangular room; the pattern continues in an analogous manner in the direction perpendicular to the drawing plane (Cremer, 1948).

Consider a simple acoustic source of power W, emitting progressive spherical diverging waves in a free field (i.e. no reflecting boundaries). At a distance r from the source the sound intensity will be

$$I_r = \frac{W}{4\pi r^2} \tag{10.8}$$

or, more generally,

$$I_r = \frac{W}{\int_0^s dS}, \tag{10.9a}$$

$$I_r = \frac{W}{\int_0^\omega r^2 \, d\omega}, \tag{10.9b}$$

where S is the area subtended by the solid angle ω from the source; for a sphere $\omega = 4\pi$ steradians and $\omega = 2\pi$ steradians for a hemisphere. Now in the far field (see equation (8.27))

$$I_r = \frac{p_r^2}{\rho c}. \tag{10.10}$$

In the *near field* the nature of the source (i.e. geometry, mass, stiffness and damping) determines the relationship between I and p; phase variations usually occur over the surface of the source (see Section 8.6). The sound intensity may be expressed in the functional form

$$I = I(p, z_s, l),$$

where z_s is the impedance and l the characteristic dimension of the source.

Typical room sizes, furniture, even cracks around walls, doors or windows have dimensions corresponding to wavelengths of sound in the range 10 Hz ($\lambda \simeq 31$ m) to 10000 Hz ($\lambda \simeq 0.031$ m), hence they interfere with the direct passage of sound causing a proportion of it to be reflected from surfaces, the remaining portions being absorbed by and transmitted through them. The sound pressure at a point in a *reverberant field* is the resultant of that in the pressure wave direct from the source and that of all the reflected waves passing through that point; the sound intensity at a point in a standing wave is

$$I = \frac{p^2}{\rho c^2}. \tag{10.11}$$

Note the units of sound intensity which can be derived from the following:

The capacity for work, or energy	N m or W s or J
The rate of working, or power	N m/s or W or J/s
Force	N or kg m/s^2
Pressure	N/m^2 or Pa or kg/m s^2

Thus the sound intensity for a progressive wave is $I = p^2/\rho c$ and expressed in SI units

$$I = \left(\frac{N^2}{m^4}\right)\left(\frac{m^3}{kg}\right)\left(\frac{s}{m}\right)$$

$$= \left(\frac{Nm}{s}\right)\left(\frac{1}{m^2}\right)$$

$$= W/m^2.$$

Hence sound intensity is sound power per unit area.

In standing-wave conditions, such as in rooms, it is usual to speak of *sound energy density* ϵ or the energy stored in a small volume of the room air, where

$$\epsilon = \frac{\text{energy}}{\text{volume}} = \frac{W\,s}{m^3}.$$

In terms of the sound intensity,

$$\epsilon = \frac{I}{c},$$

$$\epsilon = \frac{p^2}{\rho c^2}.$$

Using equation (8.17),

$$\epsilon = \frac{p^2}{\gamma p_0},$$

where p_0 is the atmospheric pressure and γ is the ratio of the specific heats at constant pressure and volume respectively; for normal room air pressure and temperature $p_0 = 1.013 \times 10^5$ Pa and $\gamma = 1.4$. If the number of reflections is large and

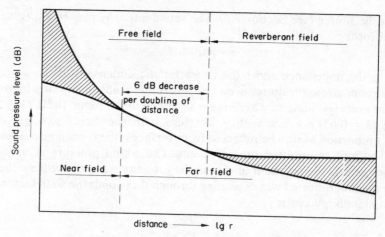

FIG. 10.4. Free (or direct) and reverberant sound fields.

combine in such a way that the average sound density is uniform, then an ergodic state exists and the reverberant field is said to be *diffuse*. Figure 10.4 shows the relationship between sound pressure level and distance in free and reverberant fields.

Combining equations (10.8) and (10.10) gives the sound pressure at a point r as

$$p_r^2 = \frac{\rho c W}{4\pi r^2} \qquad (10.12)$$

(note the direct field energy density $\epsilon_r = p_r^2/\rho c^2 = W/(4\pi r^2 c)$).

Hence the sound pressure level will be

$$L_p = L_\omega - 10 \lg (4\pi r^2) + 10 \lg \left(\frac{W_0 \rho c}{p_0^2}\right). \qquad (10.13)$$

Notice that the last term can be neglected because $W_0 = 10^{-12}$ W, $\rho c = 408$ Rayls and $p_0^2 = 4 \times 10^{-10}$ N²/m⁴. L_p decreases by 6 dB as r is doubled, but in the near field the influence of the source impedance invalidates this inverse square law, and special precautions are necessary if measurements are made in this region to deal with the large variations in sound pressure.

Most sources do not radiate omnidirectionally and exhibit a *directivity pattern*; at high frequencies directionality is most pronounced. Some directivity patterns for speech and musical instruments are shown in Fig. 10.5.

The *directivity factor Q*, is defined in a direction θ is defined as

$$Q_\theta = \frac{I_{\theta,r}}{\bar{I}_r}, \qquad (10.14)$$

where $I_{\theta,r}$ is the sound intensity measured in a direction θ and at a distance r from a directional source of sound power W; $\bar{I}_r$ is the average sound intensity at the same distance r around a non-directional source having the same sound power W. Hence, in terms of sound pressure level,

$$Q_\theta = \frac{\text{antilg}\,(L_{p,\theta,r}/10)}{\text{antilg}\,(\bar{L}_{p,r}/10)}$$

$$= \text{antilg}\,\frac{L_{p,\theta,r} - \bar{L}_{p,r}}{10}$$

and $\qquad\qquad 10\lg Q_\theta = (L_{p,\theta,r} - \bar{L}_{p,r}).$ $\qquad\qquad$ (10.15)

In general the *directivity index*

$$\text{DI} = 10\lg Q.\qquad\qquad (10.16)$$

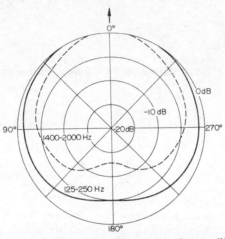

(a) Directional distribution of speech sources in a horizontal plane for two different frequency bands; the arrow points in the viewing direction (Kuttruff, 1973).

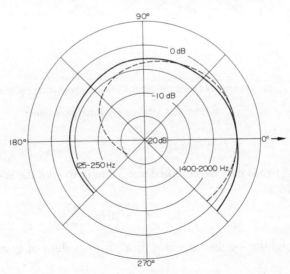

(b) Directional distribution of speech sounds in a vertical plane for two different frequency bands; the arrow points in the viewing direction (Kuttruff, 1973).

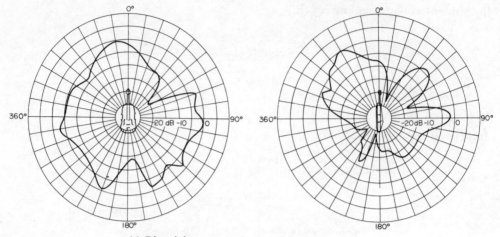

(c) Directivity response for a violinist (Cremer, 1965).

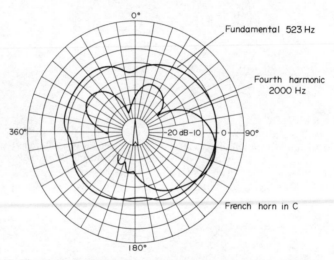

(d) Directivity characteristic for a French horn player (Cremer, 1973).

FIG. 10.5. Directivity patterns of sound.

Equation (10.13) may be modified accordingly to give the sound pressure level in a far free field as

$$L_p = L_\omega + 10 \lg \left(\frac{Q}{4\pi r^2} \right). \qquad (10.17)$$

For an airconditioning duct opening into a room values of Q may be taken from Fig. 10.6a and b.

It now remains to discover how the effect of the reverberant field may be accounted for and included in equation (10.17).

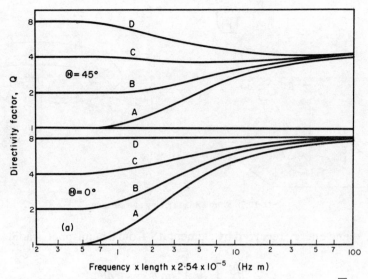

(a) Directivity factor at the four positions of the duct opening shown in (b); length is $\sqrt{A}$ where A is core area of guille.

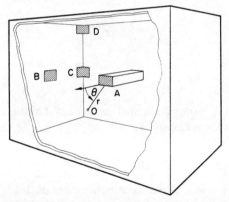

(b) Sketch showing duct opening positions A, B, C, D relevant to (a); O is the position of the listener.

FIG. 10.6. Directivity factors for airconditioning and ventilation terminals (Beranek, 1960).

10.2. Growth and Decay of Sound Fields

10.2.1. SOUND INTENSITY ARRIVING AT A SURFACE

Referring to Fig. 10.7 the energy density ϵ arriving at surface element dS from a torus element of volume dV, is

$$= \epsilon \, dV \frac{d\omega}{4\pi},$$

where the elemental solid angle $d\omega = (dS/r^2) \cos \theta$.

The total volume of the torus radius $r \sin \theta$, is $(2\pi r \sin \theta)(r \, d\theta \, dr)$, hence the total

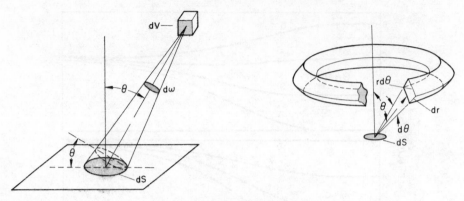

FIG. 10.7. Sound flux arriving at a surface.

amount of sound energy received by element dS from a hemisphere radius $c\,dt$ is

$$= \frac{\epsilon dS}{2} \int\limits_0^{c\,dt} dr \int\limits_0^{\frac{1}{2}\pi} \sin\theta \cos\theta\,d\theta$$

or, in terms of the sound intensity I (watts/m^2),

$$I = \frac{\epsilon c}{4}. \tag{10.18}$$

10.2.2. SOUND DECAY

If a room of volume V contains a surface area S having an average absorption coefficient $\bar\alpha$, the rate of energy decay will be

$$-\frac{d(\epsilon V)}{dt} = \left(\frac{\epsilon c}{4}\right) S\bar\alpha,$$

and if at some time $t = t$, $\epsilon \to \epsilon_t$ and at commencement of the decay $t = 0$, $\epsilon = \epsilon_m$

$$-\int\limits_{\epsilon_m}^{\epsilon_t} \frac{d\epsilon}{\epsilon} = \frac{cS\bar\alpha}{4V} \int\limits_0^t dt,$$

$$\ln\left(\frac{\epsilon_m}{\epsilon_t}\right) = \frac{cS\bar\alpha}{4V} t$$

$$\epsilon_t = \epsilon_m \exp\left(-\frac{cS\bar\alpha}{4V} t\right). \tag{10.19}$$

The reverberation time T is defined as the time taken for the energy to decay to 10^{-6} of its value when the source is switched off, i.e. $\epsilon_t = 10^{-6}\,\epsilon_m$,

$$T = \left(\frac{24 \ln 10}{c}\right)\frac{V}{S\bar\alpha} = 0.16\frac{V}{S\bar\alpha}, \tag{10.20}$$

where T is measured in seconds, V in m^3, S in m^2 units, and $S\bar\alpha$ is the total absorption in metre Sabines. This is known as Sabine's formula. Note that when all these surfaces are perfectly reflecting ($S\bar{\alpha} = 0$), then $T \to \infty$, but when complete absorption occurs ($S\bar{\alpha} = S$), then T should be zero, but calculation shows that a finite value of $0.16V/S$ is obtained.

10.2.3. GROWTH OF SOUND

For a source of sound power W the growth of sound may be estimated from the energy balance

$$\frac{d(\epsilon V)}{dt} = W - \frac{\epsilon c S\bar{\alpha}}{4}. \tag{10.21}$$

On integration,

$$\epsilon_t = \epsilon_m \left\{ 1 - \exp\left(-\frac{cS\bar{\alpha}}{4V} t \right) \right\}. \tag{10.21b}$$

10.2.4. SABINE AND EYRING FORMULAE

The assumptions that have been made in these derivations are that:

* the sound energy is distributed uniformly throughout the room;
* the energy is emitted from the source equally in all directions and at a constant rate;
* attenuation of sound only occurs at the boundaries, viscous damping of the air is neglected;
* no frequency dependence has been considered either for the energy emission or the absorption of energy;
* superposition has been neglected.

The remaining postulate has been that the energy decays in a continuous manner. Eyring (1930) took the view that decay of sound took place each time a reflection occurred, i.e. the energy was dissipated in a stepwise manner. Thus after n reflections per unit time the intensity would have decreased from ϵ_m when the sound source was switched off at $t = 0$, to $(1 - \alpha)^{nt} \epsilon_m$ at $t = t$. By equating the total energy density to that reflected and that absorbed,

$$\epsilon V = \epsilon V (1 - \bar{\alpha})^{nt} + \frac{\epsilon c S\bar{\alpha}}{4}, \tag{10.22}$$

hence in 1 second

$$n = \frac{cS}{4V} \text{ reflection/second.}$$

This corresponds to the velocity of sound divided by the mean free path $4V/S$. Now

$$\epsilon_t = \epsilon_m (1 - \bar{\alpha})^{nt}$$

and for $t = T$, $\epsilon_t = 10^{-6} \epsilon_m$, hence

$$10^{-6} = (1 - \bar{\alpha})^{nT}.$$

$$T = \frac{0 \cdot 16V}{S\{-\ln(1-\bar{\alpha})\}}. \tag{10.23}$$

When $\bar{\alpha} \ll 1$, $\{-\ln(1-\bar{\alpha})\} \to \bar{\alpha}$ † errors incurred being 13·7% when $\bar{\alpha} = 0\cdot259$, 9·1% when $\bar{\alpha} = 0\cdot2$, 4·8% when $\bar{\alpha} = 0\cdot1$ and zero when $\bar{\alpha} = 0\cdot02$. The Eyring equation (10.23) is recommended for use when $\bar{\alpha} > 0\cdot1$. Equations (10.19) and (10.21) may be expressed in terms of sound pressure level decay rate since $\epsilon_t \alpha(p_t^2)$ and $\epsilon_m \alpha(p_m^2)$, then the Sabine formula becomes

$$\frac{10cS\bar{\alpha}}{2\cdot3 \times 4V} = \frac{(L_{p,m} - L_{p,t})}{t} \, \text{dB/s} \tag{10.24}$$

when $t = T$, then $(L_{p,m} - L_{p,t}) = 60\,\text{dB}$. Eyring's formula will modify to become

$$\frac{10\,cS\,\{-\ln(1-\bar{\alpha})\}}{2\cdot3 \times 4\,V} = \frac{(L_{p,m} - L_{p,t})}{t} \, \text{dB/s} \tag{10.25}$$

At high frequencies air absorption due to heat conduction between the compression and rarefraction regions in the second waves, viscosity and moisture content becomes more significant particularly in large, reverberant spaces. Typical values are given in Table 10.2.

TABLE 10.2

Frequency (Hz)	Absorption/m³
1000	0·003
2000	0·007
4000	0·020

This absorption value should be added to the $S\bar{\alpha}$ or the $S\{-\ln(1-\bar{\alpha})\}$ term in the Sabine and Eyring formulae respectively.

In recent years the classical formulas of Sabine and Eyring have been increasingly criticised and computer simulation studies of sound fields even suggest that there may be fundamental errors in them. Calculations using these formulas ignore the location of the sound absorber; they assume completely diffuse sound fields; they assume that the probability of a ray hitting a surface depends on its size irrespective of the surface from which the ray originated. By treating the process of sound propagation within a room in a statistical way then the history of the sound rays, the room shape and the location of the sound absorber can be taken into account and the averaging of absorption coefficients for individual surfaces is no longer necessary.

Gerlach in Mackenzie (1975) derives an expression for the sound energy in a room having n surfaces having absorption coefficients $\alpha_1, \ldots, \alpha_n$ and where p_{ik} is the transitional probability for a sound ray coming from surface i having its reflection at the surface k.

---

†Remember

$$-\ln(1-\bar{\alpha}) = -\left(-\bar{\alpha} - \frac{\bar{\alpha}^2}{2} - \frac{\bar{\alpha}^3}{3} \cdots\right).$$

$$\epsilon_N = \sum_{k=1}^{n} \sum_{i=1}^{n} e_i^{(o)} e_{ik}^{(N)}$$

where ϵ_N is the total energy after N transitions or sound distributions have occurred; $e_{ik}^{(N)}$ are the elements of the matrix $(AP)^N = E^N$ for a diagonal matrix of reflection coefficients

$$A = \begin{bmatrix} (1-\alpha_k)_{k=1}^n & 0 & 0 & 0 \\ 0 & \cdot & 0 & 0 \\ 0 & 0 & \cdot & 0 \\ 0 & 0 & 0 & \cdot \end{bmatrix} \text{ and a matrix}$$

of transition probabilities

$$P = \begin{bmatrix} (p_{ik})_{i,k=1}^n & \cdot & \cdot & \cdot \\ \cdot & & & \\ \cdot & & & \\ \cdot & & & \end{bmatrix}$$

$e_i^{(o)}$ is the partial energy at surface i. This treatment of the reverberation process proceeds in discrete time intervals of length $\bar{t}$ after the source has been switched off at $t = 0$. Thus at times $\bar{t}, 2\bar{t}, 3\bar{t}, \ldots$ the sound particles change their distribution in accordance with the transition probability p_{ik}.

For the first transition at $\bar{t}$

$$e^{(1)} = e^{(o)}AP$$

where $e^{(o)}$ represents the energy distribution at $t = 0$. At $2\bar{t}$ the next transition occurs

$$e^{(2)} = e^{(1)}AP$$
$$= e^{(o)}(AP)^2$$

After N transitions $e^{(N)} = e^{(o)}(AP)^N$.

The transition probabilities are proportional to the solid angle spanned on average by wall k as viewed from the wall i. Computer programmes have been used (see Gerlach in Mackenzie, 1975) to calculate the P and A matrices, and also the reverberation times, $\bar{t}$.

10.2.5. OTHER ASPECTS OF SOUND GROWTH AND DECAY CURVES

The growth and decay of sound equations (10.19) and (10.21b) can be developed further. Expressing them in the form

$$(\epsilon_t)_g = \epsilon_m \{1 - \exp(-kt)\}$$

and
$$(\epsilon_t)_d = \epsilon_m \exp(-kt),$$

where
$$k = \frac{cS\bar{\alpha}}{4V} \quad \text{or} \quad k = \frac{0 \cdot 04c}{T} \approx \frac{13 \cdot 6}{T}$$

for a reverberation time T; the suffixes g and d denote sound growth and decay respectively. Table 10.3 summarises these equations with their corresponding growth and decay curves.

TABLE 10.3

Sound field	Equation

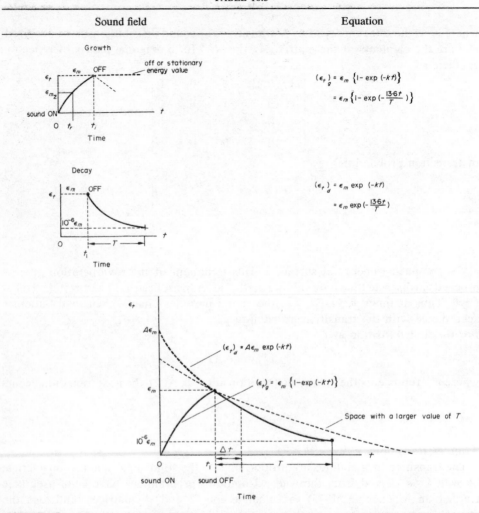

The *rise time* is the time it takes a sound pulse to reach a level 3 dB below the stationary sound level (Jordan, 1969). In terms of the sound energy curves shown in Table 10.3 the rise time is t_r, i.e. the time it takes the sound to reach half of its stationary or off energy level.

Jordan (1969) also defines the *steepness* or *slope* of the sound growth curve as being proportional to the reciprocal of the rise time and considers it as a measure of the average reflection intensity about 40–50 ms after excitation related to but not identical with clarity.

Delay time defines that period of time after the direct pulse in which 90% of all the sound reflections have a strength within 10 dB of that of the direct pulse. *Early decay time* is that time the sound level takes to fall by 10 to 15 dB from the off or stationary level (Jordan, 1975). The slope of the sound level–time curve for this time period defines the *initial reverberation time* T_{15}, whereas the slope defining the next 15–30 dB drop in level gives the *average reverberation time*, T_{15-30}. The usual overall reverberation time often given in T_{60} defining the time for the sound level to decay by 60 dB from its off or stationary level, but work by Jordan (1969, 1973), Atal *et al.* (1965) and Kuttruff (1973) has confirmed that it is the initial portion of a sound decay process which is responsible for our subjective impression of reverberation; the later reflections are usually masked by new sounds. Atal *et al.* (1965) shows that subjective reverberation time (T_s) correlates well with an initial reverberation time computed on the basis of the decay over the first 160 msec. If the impulse response does not have a strong direct sound component, T_s correlates better with a reverberation time based on the decay over the first 15 dB.

More attention has been given in recent years to differentiating between the varying decay rates rather than assuming an overall average.

For sound decay

$$(\epsilon_t)_d = A\epsilon_m \exp(-kt).$$

At time t_1, $\epsilon_t = \epsilon_m$, hence

$$1 = A \exp(-kt_1)$$

or

$$A = \exp(kt_1).$$

The time 0 to t_1 is related to the *rise time* t_r since at t_r, $\epsilon_t = \epsilon_m/2$ (Jordan, 1969).
At time $(t_1 + T)$, $\epsilon_t = 10^{-6} \epsilon_m$, hence

$$10^{-6} = A \exp\{-k(t_1 + T)\},$$
$$A = 10^{-6} \exp\{k(t_1 + T)\}.$$

The sound decay can be expressed in the form

$$(\epsilon_t)_d = \epsilon_m \exp(kt_1) \exp(-kt)$$

or

$$(\epsilon_t)_d = 10^{-6} \epsilon_m \exp\{k(t_1 + T)\} \exp(-kt).$$

Hence the decay of sound also depends upon the growth time t_1 as well as the time to decay. Jordan (1969) states that a complementarity exists between the slopes of the sound growth and decay processes.

The slope of the sound decay curve, or *rate* of sound decay, is given by

$$\left(\frac{d\epsilon_t}{dt}\right)_d = -k\epsilon_m \exp(kt_1) \exp(-kt)$$

or

$$\left(\frac{d\epsilon_t}{dt}\right)_d = -k\, 10^{-6}\epsilon_m \exp\{k(t_1 + T)\} \exp(-kt).$$

Similarly, the slope of the sound growth curve or *rate* of sound energy growth in a room is

$$\left(\frac{d\epsilon_t}{dt}\right)_g = k\epsilon_m \exp(-kt).$$

The difference in slopes is governed by the term $\{-\exp(kt_1)\}$ or $-10^{-6} \exp\{k(t_1 + T)\}$.

The ratio of the slopes is given by

$$R = \frac{(d\epsilon_t/dt)_d}{(d\epsilon_t/dt)_g} = -\exp(kt_1)$$

or

$$R = -10^{-6} \exp\{k(t_1+T)\}.$$

When $R = 1$,

$$\exp\{k(t_1+T)\} = -10^6$$

or

$$k(t_1+T) = -13\cdot6.$$

Direct sound is more important than the reverberant sound for speech intelligibility, whereas the balance between the direct and reverberant sound is very important in the case of spaces for musical performance and listening. In a concert hall the slope of the decay curve is less than that for a lecture room say, the reasons being due to the different sound sources used (voice/orchestra), the different reverberation times adopted for the designs and the different k-values. We shall see that for speech intelligibility it is essential that the voice sound energy/time curve shows distinct peaks in its pattern. This means that $R \leqslant 1$, whereas in the case of music $R > 1$. Notice that the sound source, for example the manner in which a person speaks as well as the room properties, govern the rate of sound energy growth.

The implications of studying the significance of the areas defined by the sound energy curves shown in Table 10.3 is also useful. Under the growth curve the area is

$$A_g = \int_0^{t_1} \epsilon_m\{1 - \exp(-kt)\}\, dt$$

$$= \epsilon_m\left[t_1 + \frac{1}{k}\{\exp(-kt_1) - 1\}\right].$$

The decay curve is defined by an area

$$A_d = \int_{t_1}^{t_1+\Delta t} A\epsilon_m \exp(-kt)\, dt$$

for a time delay Δt after the sound has ceased. Inserting values for the constant A already derived,

$$A_d = \int_{t_1}^{t_1+\Delta t} \epsilon_m \exp(kt_1)\exp(-kt)\, dt$$

or

$$A_d = \int_{t_1}^{t_1+\Delta t} 10^{-6}\epsilon_m \exp\{k(t_1+T)\}\exp(-kt)\, dt.$$

The significance of these expressions is that they interrelate the rise time, the reverberation time and the acoustical properties of the room. It may be that future criteria define values of (A_d/A_g) for different kinds of spaces (see work of Niese discussed on page 540).

Sound is perceived on a very small time scale. The integration time of the ear is in the order of 50 ms. Sound patterns vary from one place to another in a room, and from one instant to another at the same place. Each frequency responds differently to the space containing it. The pattern of each sound impulse depends on the mode of excitation, not only whether it is a voice speaking or an instrument playing, but also *how* the voice is spoken or the instrument played. Each sound impulse is unique. The ear can distinguish a lot of the fine spatial–temporal structure of sound. For these reasons more complex measurements are now usually made of a room's response to sound. Measurements are repeated many times under identical physical conditions.

Consider the standard sound growth and decay curves

$$(\epsilon_t)_g = \epsilon_m (1 - e^{-kt}),$$
$$(\epsilon_t)_d = \epsilon_m e^{-kt},$$

which are true for one pulse of sound. When repeated measurements are made it does not matter whether averages are evaluated by averaging over many free paths traversed by one particle (i.e. a *time average*) or by averaging for one instant a great number of different particles called an *ensemble average*. The difference between the growth and decay sound energy equations above for an ensemble average is

$$\langle (\epsilon_t)_g \rangle = \langle \epsilon_m \rangle - \langle (\epsilon_t)_d \rangle.$$

The brackets $\langle \rangle$ signify repeated measurements resulting in ensemble averages.

Jordan (1969) and Schroeder *et al.* (1965, 1966) define the slope or steepness of sound growth curves by the expression

$$\frac{d(\epsilon_t)_g}{dt} = \frac{d}{dt} (10 \lg \langle S^2(t) \rangle)_{t=t_0},$$

where $S(t)$ is the response of the space to white noise switched on at $t = 0$; t_0 is the time for the sound to reach 5 dB below its stationary level (t_0 is quoted as being typically one-fortieth of T_{60}).

Schroeder *et al.* (1965, 1966) describes a way of obtaining an ensemble average of the decay curves that would be obtained with bandpass filtered noise as a signal, using a simple integral over a tone burst response. If N is the noise power per unit bandwidth

$$\langle S^2(t) \rangle = N \int_t^\infty r^2(x) \, dx.$$

For sound growth

$$\langle S_g^2(t) \rangle = N \int_0^t r^2(x) \, dx$$

and for sound decay

$$\langle S_d^2(t) \rangle = N \int_t^\infty r^2(x) \, dx,$$

where $x = (t - T)$ and $r^2(t)$ is the squared impulse response of filter and amplifier connected in series with the room. A system diagram of the process is

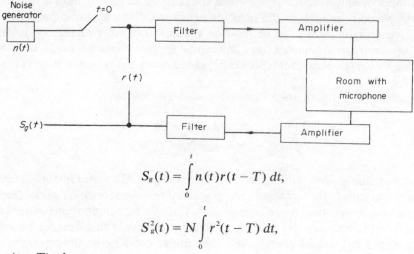

$$S_g(t) = \int_0^t n(t)r(t - T)\, dt,$$

$$S_g^2(t) = N \int_0^t r^2(t - T)\, dt,$$

let $x = (t - T)$, then

$$\langle S_g^2(t) \rangle = N \int_0^t r^2(x)\, dx,$$

is the ensemble average of the sound growth curves and

$$\langle S_d^2(t) \rangle = N \int_t^\infty r^2(x)\, dx$$

is the ensemble average of sound decay curves.

The time t will vary but can be ascertained by recording on tape then reversing.

10.3. Reverberant Field Energy

By definition the reverberant field energy is that energy remaining in the room after one reflection. For a source having a power W the total amount of energy absorbed for n reflections per second in time t is

$$(1 - \bar{\alpha})^{nt} W = (nt) \left(\frac{\epsilon_R c}{4} \right) S\bar{\alpha}$$

for $nt = 1$

$$(1 - \bar{\alpha}) W = \frac{\epsilon_R c}{4} S\bar{\alpha}.$$

Hence the reverberant field energy is

$$\epsilon_R = \frac{4W(1 - \bar{\alpha})}{c} \frac{}{S\bar{\alpha}}. \qquad (10.26)$$

The term $S\bar{\alpha}/(1 - \bar{\alpha})$ is known as the *room constant R*, which may be expressed in terms of the reverberation time thus:

$$R = \frac{0 \cdot 16 V}{(1 - \bar{\alpha})T} \tag{10.27}$$

or

$$R = \frac{S\bar{\alpha}}{(1 - \bar{\alpha})}; \tag{10.28}$$

hence

$$\epsilon_R = \frac{4W}{cR} \tag{10.29}$$

since

$$\epsilon_R = \frac{|p_R|^2}{\rho c^2},$$

then the mean square pressure contributions from the reverberant field will be

$$|p_R|^2 = \frac{4\rho c W}{R}. \tag{10.30}$$

From the preceding work it can be seen that the reverberant field increases the sound level. Transient sounds (i.e. those having time histories lasting less than the reverberation time) will be modified by reverberation. In the case of speech and music the reverberant energy from one transient sound must fall to a low enough level so as not to mask succeeding transients and thus preserve clarity and timbre; syllables or consonants usually follow one another at a rate of 10–15 per second, whereas musical notes can be delivered at a higher rate of 20 per second. The problem of designing rooms for speech communication was discussed in Section 2.5.

Within any room certain preferred sound wave patterns exist; the derivation of these modes was discussed in Section 7.2. Existence of such modes can lead to sound *coloration*, i.e. enhancement of specific frequencies which may continue to radiate after the source has ceased. This can also occur by the superposition of strong reflections onto direct sound.

Echoes occur when reflected sound is delayed and heard about 50 ms or later after the direct sound by the listener, i.e. if the path difference between the direct and first reflected sound is greater than about 17 m. Parallel reflecting surfaces can create a multiple echo of even rate called a *flutter echo*. Curved surfaces will focus sound having wavelengths equal or smaller than the radius of curvature.

The likelihood of coloration and echoes occurring in spaces is discussed further in Section 10.5.

The reverberant field theory outlined above has not taken modal behaviour or echo formation into account. Irregularity of room shape and use of absorption areas are means of controlling echoes, but decreasing the effects of modes is not straightforward (Doak, 1959).

10.4. The Resultant Sound Pressure Level in a Room

The contribution of the reverberant energy field to the direct energy field of the source may now be considered. The sound energy level at a point in the room will be the sum of the sound pressure due to the direct field p_r^2 (equation (10.12)) but including directivity, and that due to the reverberant field p_r^2 (equation (10.30)), hence

$$|p|^2 = |p_r|^2 + |p_R|^2,$$

$$|p|^2 = \frac{\rho c W Q}{4\pi r^2} + \frac{4\rho c W}{R}.$$

Therefore in terms of sound pressure level

$$L_p = L_\omega + 10 \lg \left(\frac{Q}{4\pi r^2} + \frac{4}{R}\right) + 10 \lg \left(\frac{W_0 \rho c}{p_0^2}\right).$$

At 20°C and 760 mmHg $\rho = 1 \cdot 2 \text{ kg/m}^3$ and $c = 342 \text{ m/s}$; hence with $W_0 = 10^{-12}$ (W) and $p_0 = 2 \times 10^{-5}$ (Pa)

$$L_p = L_\omega + 10 \lg \left(\frac{Q}{4\pi r^2} + \frac{4}{R}\right) + 10 \lg 1 \cdot 03.$$

The last term is negligible, hence

$$L_p = L_\omega + 10 \lg \left(\frac{Q}{4\pi r^2} + \frac{4}{R}\right). \tag{10.31}$$

The equation enables the sound pressure level in a room to be calculated for a source with a known power level and directivity factor positioned at distance r from an occupant in a room having a reverberant field characterised by the room constant R. In airconditioning systems L_ω will be the power level at the terminal resulting from the plant and airborne noise sources. A comparison of L_p at each octave band centre frequency with the design sound pressure levels derived from the noise rating curves (see Fig. 2.69) will enable the required attenuation to be estimated.

10.5. Multiple Noise Sources and Sound Distribution in Low-height Rooms

From equation (10.13) the direct sound pressure level from a non-directional source is

$$P_{p,D} = L_w - 20 \lg r - 10 \lg 4\pi,$$

$$L_{p,D} = L_w - 20 \lg r - 11. \tag{10.32}$$

The reverberant sound pressure level is derived from equation (10.30) giving

$$L_{p,R} = L_w - 10 \lg R + 10 \lg 4,$$

$$L_{p,R} = L_w - 10 \lg R + 6. \tag{10.33}$$

This expression suggests that the reverberant sound pressure level is independent of distance r, i.e. is uniform throughout the room. When $L_{p,D} = L_{p,R}$, then the *free-field radius* in metres is

$$r = 0 \cdot 14 \sqrt{R}. \tag{10.34}$$

At a distance r from a source the direct field is much less than the reflected sound field. The absorption of the room becomes effective in reducing the sound pressure level at distances from the source which are greater than r (also see Chapter 4, Section 4.6). Note that if the directivity factor $Q \neq 1$, then the free-field radius is modified accordingly by

$$r = 0 \cdot 14 \sqrt{RQ}. \tag{10.35}$$

If a number of noise sources are present, the direct and reverberant sound pressure levels are altered; the power output of each source and the distances between each source and the receiver must be known. For n sources of equal power all equidistant from the sound observation point,

$$L_{p,D} = L_w - 20 \lg r + 10 \lg n - 11, \tag{10.36}$$

$$L_{p,R} = L_w - 10 \lg R + 10 \lg n + 6. \tag{10.37}$$

Sabine in Harris (1957b) points out that this basic treatment of room acoustics assumes that the reverberant sound pressure level is uniform throughout the room, whereas when the floor linear dimensions are large compared to the ceiling height (i.e. more than five to ten times the ceiling height) the wall reflections become insignificant and multiple reflection takes place between the ceiling and floor only. Figure 10.8a shows the direct and reverberant sound pressure levels for a steady

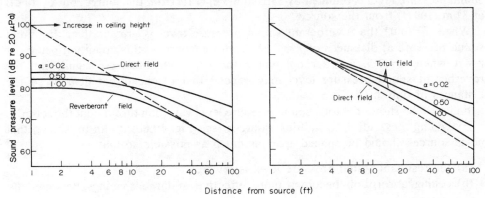

(a) Relation of pressure levels of direct and reflected sound to distance from source and absorption coefficient α of 10 ft (3 m) ceiling, assuming negligible wall reflection, due to single source of 0·0115 W acoustic power; coefficient of floor taken as 0·02.
(b) Relation of pressure levels of direct and total sound to absorption coefficient α of 10 ft (3 m) ceiling and distance from source, assuming same conditions as for (a).

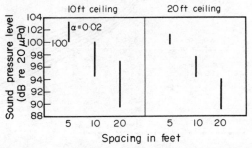

(c) Pressure level of combined direct and reflected sound in centre of bank of 100 sources as related to spacing of sources, ceiling height and ceiling coefficient α. Each source has acoustic power of 0·0115 W; floor coefficient taken as 0·02; wall reflections assumed negligible.

FIG. 10.8. Effect of absorption on sound fields radiated from single and multiple sources in low height rooms (Sabine in Harris, 1957b).

continuous source having a sound pressure level of 100 dB at 1 ft (0·3 m) and absorption coefficients of 0·02, 0·50 and 1. When the ceiling absorption coefficient equals 1 there are no ceiling reflections but just a single reflection from the floor. The assumptions made in deriving the curves shown in Fig. 10.8a are:

(a) the absorption coefficient of the floor is taken as constant at 0·02;
(b) the noise source and receiver are situated midway between the ceiling and floor;
(c) the absorption coefficients of the floor and ceiling are independent of the angle of incidence of the sound; the errors in this assumption are minimised by summating over many angles in carrying out the calculations;
(d) the wall reflections are negligible.

Notice that the reverberant sound pressure level is not constant with distance as inferred by equation (10.33), and also that changes in absorption alter the reverberant sound field by varying amounts according to the source distance, i.e. if a reflective ceiling ($\alpha = 0·02$) is altered to an absorbent one ($\alpha = 1·00$), then the reverberant sound pressure level is reduced by 5 dB at 0·3 m (1 ft) from the source, but by 14 dB at 31 m (100 ft) from the source.

When $\alpha = 0·02$ the reverberant sound pressure level is greater than the direct sound pressure at distances of $r = 0·6h$ from the sources; this condition occurs at $r = h$ when $\alpha = 0·5$. For a room where the wall reflections are significant, the reverberant sound pressure level may exceed that of the direct field at source distances of only $0·1h$.

Figure 10.8b shows the total sound pressure level variation throughout the room.

Extending these ideas to multiple sources (see Fig. 10.8c) it can be shown that noise sources should be spaced apart as much as possible so that:

(a) the resultant sound pressure level is lower;
(b) ceiling absorption becomes more effective at large spacings between the sources.

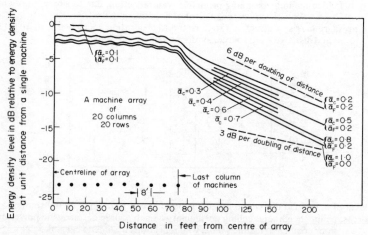

FIG. 10.9. Theoretical relation between energy density and distance for a square 20 by 20 array of incoherent point sources 8 ft (2·4 m) on centre in a room 20 ft (6 m) high. The curves show the effects of varying combinations of absorption coefficients $\bar{\alpha}_C$ and $\bar{\alpha}_F$, on ceiling and floor respectively. Where no value is specified, $\bar{\alpha}_F$ has the value of 0·2(Embleton and Dagg, 1962).

Embleton and Dagg (1962) measured the sound level in an array 20 columns by 20 rows of small machines 8 ft (2·4 m) apart in each direction (Fig. 10.9). The microphone was moved outwards from the centre of the array along a line half-way between the machines. In the array the sound levels can be seen to fluctuate within about 1·5 dB. Within the first 6 ft (1·8 m) of the array the sound level decreases by more than 6 dB per doubling of distance. Gradually the contributions to the sound level from each of the machines becomes more equal and the attenuation slopes more gradual. It can also be seen that the effect of the sound absorbent surface is greater at further distances from the array; also it appears to be more beneficial to treat one surface with absorption material rather than distribute it between the floor and the ceiling.

10.6. Impulse Response Analysis of Room Acoustics

If the impulse response function of a room, $g_1(t)$, consists of the direct sound plus one reflection which is weaker by q and occurs at a time delay t_0, after the direct sound

$$g_1(t) = \delta(t) + q\delta(t - t_0).$$ (10.38)

The corresponding squared transmission function given by the Fourier transformation (see equations (7.12) and (7.13)) is

$$|G_1(f)|^2 = |1 + q \exp 2\pi jft_0|^2$$
$$= 1 + q^2 + 2q \cos (2\pi ft_0).$$ (10.39)

A system with such an impulse resonance is known as a comb-filter with a ratio of maximum to minimum of $\{(1 + q)/(1 - q)\}^2$, the separation of adjacent maxima being $1/t_0$ (Fig. 10.10a). For a regular succession of n reflections

$$g_2(t) = \sum_{n=0}^{\infty} q^n\delta(t - nt_0)$$ (10.40)

which gives, in likewise manner, a transmission function

$$|G_2(f)|^2 = \{1 + q^2 - 2q \cos (2\pi ft_0)\}^{-1}.$$ (10.41)

Whether the colorations are audible or not depends on the time delay and on the relative heights of the maxima (Atal et al., 1962). Critical values of q are shown in Fig. 10.10b. When t_0 exceeds about 25 ms the regularity of the impulse function does not appear subjectively as coloration but rather as a regular repetition of the signal in the form of an echo or flutter echo (Kuttruff, 1973).

The likelihood of coloration occurring can be predicted empirically from reflectograms (or echograms) and autocorrelation analysis. For an experimental autocorrelation function ϕ_{gg}, weighted by a function $b(\tau)$,

$$\phi'_{gg} = b(\tau)\phi_{gg}(\tau),$$ (10.42)

where τ is the distance of the side maximum from the central maximum occurring at $t = 0$ and is measured from a reflectogram. Bilsen (1967/8) gives values of $b(\tau)$.

On the basis of research by Seraphim (1958) and Plenge (1965/6), Kuttruff (1973) concludes that it is not very meaningful to quote the reverberation time with a

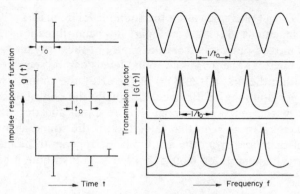

(a) Impulse responses and absolute values of transmission functions of various comb filters for $q = 0.7$ and time delay t_0 (Kuttruff, 1973).

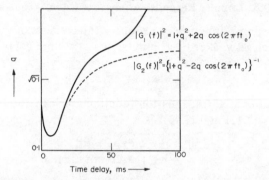

$$|G_1(f)|^2 = 1 + q^2 + 2q \cos(2\pi f t_0)$$

$$|G_2(f)|^2 = \{1 + q^2 - 2q \cos(2\pi f t_0)\}^{-1}$$

(b) Critical values of q resulting in just audible colouration of white noise passed through comb filters with transmission functions $G_1(f)$ and $G_2(f)$ (Kuttruff, 1973).

FIG. 10.10.

greater accuracy than 0.1 s for rooms which are not too heavily damped. The accuracy of reverberation time calculations is limited by the fact that it is difficult to be precise in calculating the total sound absorption of a space. For example the absorption effect of an audience is a major source of uncertainty in the prediction of reverberation times. Figure 10.11 shows values of audience absorption which have been calculated from the measurements of reverberation time made in the mentioned halls. The scattering of sound around the heads of the audience plays an important rôle in the audience sound attenuation process and no doubt has an influence on the ultimate effectiveness of side-wall reflections. Meyer *et al.* (1965), Sessler and West (1964) and Schultz and Walters (1964) have carried out experiments to establish data on selective audience attenuation effects. Their work suggests that it is advantageous to have rows of seats gradually rising from the stage, but Kuttruff (1973) adds the qualification that the sound absorption will be high.

10.7. Designing Spaces for Music

Sound shines like the dawn and the sun rises in the form of sound, sound seeks to rise in music, and colour is light.

(JEAN PAUL, quoted in *Schumann and the Romantic Age* by MARCEL BRION.)

Open-air Greek Herodes Atticus Theatre in Athens.

urfurstliche Oper in Munich named after its architect the Cu- lliés Theatre.

Philharmonie Hall, Berlin.

PLATE 10.1. (the author is grateful to Professor Lothar Cremer for providing these photographs; the theatres are discussed by Professor Cremer in *Auditorium Acoustics* (1975) edited by Mackenzie).

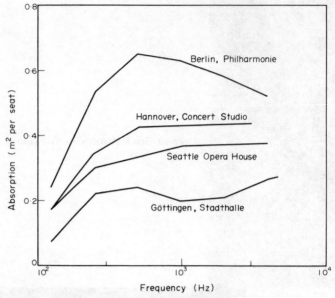

FIG. 10.11. Audience absorption in several halls as estimated from observed reverberation times (Kutruff, 1973).

All this being arranged, we must see with even greater care that a position has been taken where the voice falls softly and is not so reflected as to produce a confused effect on the ear. There are some positions offering natural obstructions to the projection of the voice, as for instance the dissonant, which in Greek are termed, κατηχοῦντες; the circumsonant, which with them are named περιηχοῦντες; and again the resonant, which are termed ἀρτηχοῦντες. The consonant positions are called by them συνηχοῦντες.

The dissonant are those places in which the sound first uttered is carried up, strikes against solid bodies above, and, reflected, checks as it falls the rise of the succeeding sound.

The circumsonant are those in which the voice spreading in all directions is reflected into the middle, where it dissolves, confusing the case endings, and dies away in sounds of indistinct meaning.

The resonant are those in which the voice comes in contact with some solid substance and is reflected, producing an echo and making the case terminations double.

The consonant are those in which the voice is supported and strengthened, and reaches the ear in words which are clear and distinct.

(VITRUVIUS, (*De Architectura*, see chapter 7 of *Collected Papers on Acoustics* by W. C. SABINE, Dover Publications Inc., 1964).)

In order that hearing may be good in any auditorium it is necessary that the sound should be sufficiently loud, that the simultaneous components of a complex sound should maintain their proper relative intensities, and that the successive sounds in rapidly moving articulation, either of speech or of music, should be clear and distinct, free from each other and from extraneous noises. These three are the necessary, as they are the entirely sufficient, conditions for good hearing. Scientific- ally the problem involves three factors: reverberation, interference, and resonance. As an engineering problem it involves the shape of the auditorium, its dimensions, and the materials of which it is composed.

Sound, being energy, once produced in a confined space, will continue until it is either transmitted by the boundary walls or is transformed into some other kind of energy, generally heat. This process of decay is called absorption.

(W. C. SABINE (Architectural acoustics, *The Journal of the Franklin Institute*, January 1915).)

History suggests that buildings have been unconsciously designed through the ages in styles that have been compatible with the music being performed in them. The purity of plainsong and Gregorian chant floating in the lofty and noble spaces of a Romanesque church; the emergence of opera in plush, aristocratic surroundings; the intimacy of chamber music in family-like salons; the awesome symphony orchestra with its great variety of tonal colours and intensity of sound used by such composers as Wagner, Mahler and Richard Strauss in concert halls which may have an audience of up to 10000 people, all seem to suggest some connection between building purpose and structure. Desmond Shawe-Taylor, music critic of the *Sunday Times*, wrote on 27 November 1966, with reference to the Royal Festival Hall:

I don't know whether the proposed extension of the Festival Hall's roof resonators to include the higher frequencies has yet begun; but it is clear that Bruckner of all composers requires a more resonant acoustic than the hall yet offers. Apart from the sound in fortissimo tuttis, there are those famous and much-debated bars of silence and "breaking off". The composer's ear evidently conceived the effect of such passages in the highly resonant ecclesiastical architec- ture to which he was accustomed; during the pause all the sounds of the previous climax were to be given time to die away before the new idea began—an effect very different from the dead blank we now experience. There is nothing that the conductor can do about this: the hall is here his master.

No doubt Bruckner's sense for sound in spaces was developed from his experi- ence of playing the organ in the Baroque churches of Vienna. In addition, history stamps its seal of tradition, so that many of the so-called new designs of today are attempting to recreate the success of the past in more precise technological terms; it is the methods used and the aesthetic vogues which have changed—there may also be less wealthy patrons of the arts, making economy a vital factor.

Hennenberg (1962) describes the reaction of musicians and listeners on hearing music in the new Gewandhaus Concert Hall at Leipzig completed in 1780:

Again and again its good acoustic properties are praised. For a long time one spoke in riddles about the acoustic secret of this hall. It was attributed to the fact that the relation between length, width and height corresponded to the "golden section"; that walls, ceiling and floor were of wood; that above, below and all round lay propitiously resonant hollow spaces; that supporting columns had been abandoned and the ceiling was allowed to be sustained by a strut-frame; that nothing impeded the diffusion of sound.

Our knowledge about concert hall acoustics today based on the scientific method suggests that these were astute observations. Winckel (1967) makes the point that it is no coincidence that the greatest orchestras have emerged from the places with the greatest concert halls such as Berlin, Vienna, Leipzig, Amsterdam and Boston.

The unconscious influence of environmental quality was enforced recently when during some acoustical tests at King's College Chapel in Cambridge the author met not only some members of the world-famed choir for that chapel but some of the less-often-heard choir belonging to the neighbouring chapel of St. John's College. Although the choirs are considered to be comparable, the excellent acoustics and perhaps also the organ of Kings, add some vital quality to the innate talent and sound of the choir.

An interesting observation from a music critic:

> *... just as the revealing acoustics of our Royal Festival Hall tidied up London orchestral string-playing overnight, so de Doelan (the concert hall home of the Rotterdam Philharmonic Orchestra) has played its part in improving its resident orchestra*

> (DESMOND SHAWE-TAYLOR writing in the
> *Sunday Times* on 9 November, 1969.)

Designing a concert hall is a great challenge because this type of building requires a much greater range of subjective responses to be satisfied than for spaces where prevention of speech interference is the prime concern.

It is worthwhile at this point recounting the differences between the acoustical requirements for rooms designed for music and those for speech. The acoustics of a space designed for speech must primarily ensure clarity and intelligibility (see Section 2.6). Understanding in the speech communication process depends also on gesture and lip reading besides voice projection. Our expectancy of the speech voice quality is not too critical except that it becomes important in telephone communication and sometimes as a form of attraction in the early stages of a human relationship, whereas listeners to and performers of music *expect*, assuming an adequate technique, a particular sound quality for the various styles and eras of music. This expectancy is inherited from musical experiences over the ages. There is an acutely high aesthetic sensitivity level in the case of music, whereas we are not quite so particular about a speaker's voice or accent as long as it is understandable. The environment must allow concentration to be unhindered in both circumstances. Clarity is a prerequisite for speech communication, but in music excessive clarity gives a subjective impression of brittleness and dryness in sound quality; besides, the musician has to work hard to make the music sound alive and he has little acoustical feedback from the reverberant field to identify any modification necessary in performance to match the acoustical conditions. Excessive clarity in music also allows unwanted bowing noises from the strings to be more easily heard. So reverberation in music is very important to a much greater extent than for speech, although it can make one room easier to speak in than another one, not only for providing feedback but because it helps to blend the musical sounds which may emanate from a much wider area (i.e. an orchestra and choir compared with a single speaking voice) at sometimes a much faster rate than for speech. As we will see, reverberation is related to *spatial responsiveness*, a factor which appears to be important in forming subjective impressions about musical sounds. All frequencies are important for music; the more limited range important for speech intelligibility was discussed in Chapter 2. Even in early Baroque music the composer intuitively knew that the addition of a bass, even if only doubling at an octave below the cello line, gave warmth of tone, whereas a high-frequency response is vital if the upper harmonics of instruments, especially the strings, are to be perceived and if feelings for timbre or sound quality and brilliance are to be fully realised.

What is known about the perception of sound arriving at the ears via direct and reflected pathways? Whether a reflection becomes an echo (i.e. a repetition of the signal) depends on several factors:

* the time delay between the direct and reflected sound; in general reflections arriving at the ears after 50 ms are perceived as echoes;
* the intensity level of the reflection relative to the sound signal and other reflections;
* the nature of the sound signal.

The importance of time delay is elucidated in *the law of the first wavefront* enunciated by Cremer (1948). This law states that *it is the sound signal to reach the listener first which subjectively determines the direction from which the sound comes*, even when a reflection is stronger than the direct sound by up to 10 dB provided that the time delay is less than 50 ms. Erroneous localisation of sounds is only likely to occur when curved surfaces focus sound thereby producing large spatial variations in sound energy, and in a space which is heavily lined with absorbents, thus producing a free field with few or no reflections. Haas (1951) and Meyer and Schodder (1952) carried out pioneer work in this field which supports these conclusions.

The absolute perception threshold of reflections is defined as the level, relative to the direct sound level, at which 50% of the test subjects perceive a difference (Burgtorf, 1961; Seraphim, 1961). Figure 10.12a shows the perception threshold for speech at a sound pressure level of 70 dB. As the delay time of the reflections is increased, the likelihood of hearing them increases. With a delay time of 20 ms, reflections at a level of 20 dB below the direct speech signal will be heard; whereas with a delay time of about 56 ms, reflections at even 40 dB below the direct sound will be heard. Reflections with higher levels than those illustrated will be audible to most people, whereas those at lower levels will not be. The effect of reverberant sound being received at different angles of incidence (θ) in the horizontal plane is shown in Fig. 10.12b, the direct sound being a short speech syllable at a sound pressure level of 55 dB received at normal incidence to the face (i.e. $\theta = 0°$). When the reflections arrive from the same directions as the direct sound ($\theta < 20°$), they are masked by the direct sound. Our hearing is more sensitive to lateral reflections ($30° < \theta < 90°$) than to those arriving from the front or the back of the head; Burgtorf

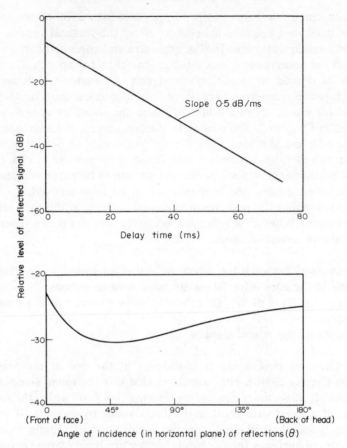

(a) Absolute perception threshold of reflections added to a direct sound signal (speech at 70 dB),
    as a function of delay time; both signals arrive from the front.
(b) Absolute perception threshold of a reflection delayed by 50 ms with respect to direct sound signal
    (speech at 55 dB), to which it is added.

FIG. 10.12. Perception threshold of speech signals (Kuttruff, 1973).

and Oehlschlägel (1964) have found that reflections arriving from above are also masked more easily by the direct sounds than are those incident from lateral directions. A multitudinal array of reflections does not create an entirely new situation; the threshold levels are based on the predominant unmasked reflections which occur at a given time delay; the threshold pattern falling then jumping back again as each new reflection is experienced.

Schubert (1969) has established the perception thresholds for reverberant sounds using music as a source. The results of this work are shown in Fig. 10.13. The slopes of the thresholds are much less, about 0·1 dB per ms of time delay, than those for speech which are about 0·5 dB per ms (Fig. 10.12a). They also display much more non-linearity than in the case of speech because they depend on the style of the music and on a greater variety of subjective factors. Notice again that lateral reflections ($\theta \simeq 90°$) are perceived more easily than those at frontal incidence.

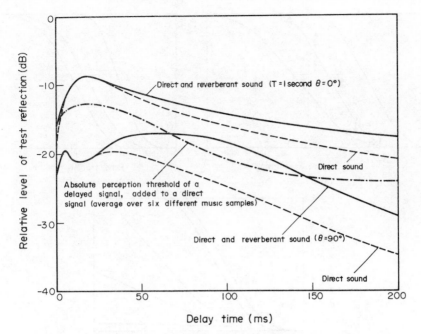

FIG. 10.13. Perception thresholds of music signals (Schubert, 1969).

Let us pursue the subjective aspects of perceiving reverberant sound further. Reflections can affect the sensation of loudness or judgements concerning sound quality (i.e. timbre). When will reflections annoy or distract a listener by being sensed as echoes? The famed work of Haas (1951) using speech has advanced our understanding of this problem; the principal results are shown in Fig. 10.14a–c. Speaking rate has a distinct effect on a listener's response, 50% of the subjects in Haas' experiment found a delay time of 40 ms disturbing for a speaking rate of 7·4 syllables per second, whereas the same proportion of subjects gave this response at 70 ms and 90 ms when the speech rates were 5·3 and 3·5 syllables per second respectively (Fig. 10.14a). Similarly, low reverberation times give more satisfaction than higher values. It must be remembered that long sound delays are not advocated for speech intelligibility. An echo will be more pronounced if the reverberation time is shorter; if the time delay is 20 ms then the reverberant sound is unlikely to cause echoes (Fig. 10.14b). The effect of raising and lowering the reflection sound level above and below that of the direct sound is shown in Fig. 10.14c. Raising the reflection level 10 dB above the direct sound level causes a comparatively small change in the echo disturbance response, and the so-called *Haas effect* refers to the fact that almost no one is bothered when the time delay is under 20 ms. When the reflection level is 10 dB below that of the direct sound, very few people will be disturbed even with high delay times.

Meyer and Schodder (1952) corroborated the findings of Haas. Muncey *et al.* (1953) carried out similar investigations for music and, as expected, showed that people are less sensitive to echoes when listening to music than to speech, although echoes are less acceptable in opera than in organ music. The suggestion is again that

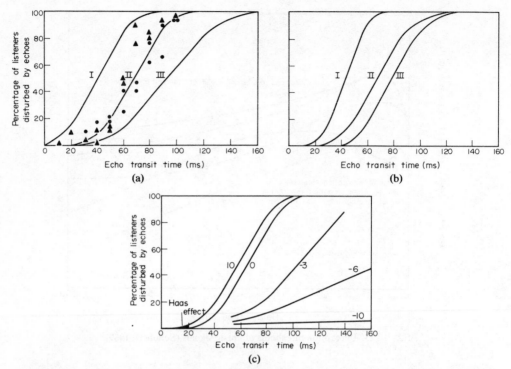

(a) Effect of speaking rate; reverberation time of listening room 0·8 s: I, 7·4; II, 5·3; III,3·5 syllables per second.

(b) Effect of reverberation time of listening room; speech rate 5·3 syllables per second; reverberation time (I) 0 s; (II) 0·8 s; (III) 1·6 s.

(c) Effect of reflection sound pressure level relative to direct speech level taken as 0 dB; speech rate 5·3 syllables per second; reverberation time 0·8 s.

FIG. 10.14. Disturbance due to reflections being sensed as echoes (Haas, 1951).

tradition has conditioned our responses; organ music has been often heard in churches with long reverberation times and echoes have been prevalent, whereas the plush surroundings of opera houses have made the presence of an echo taboo.

It is very difficult to completely define the acoustical qualities of a concert hall, a recital room or an opera house because they depend upon subjective impressions which many people find difficult to put into words. Poets and writers skilled in using the riches of language give clues as to what listeners of music intuitively seek. Frequently in literature effects in one sensory mode are evoked by a description which refers to another sense. For example Baudelaire in his essays on painting writes of "bright tones", "chords of tones" or describes a painter as "a harmonist"; we speak of "warm and cool colours", "golden tone", "warmth of tone", "dry sound", "tone like spun-silk", a "touch of velvet", reverberant sound evokes spaciousness and so on. A space has to respond to sound in a way which allows a variety of subjective impressions to be evoked in the listener.

Tables 10.4 and 10.5 show subjective values to be satisfied and their dependence on some physical factors such as reverberation time, initial time delay between direct and first reflected sound, and the pattern of reflected sound in the hall.

TABLE 10.4. VOCABULARY OF SUBJECTIVE ATTRIBUTES OF MUSICAL ACOUSTICAL
QUALITY (BERANEK, 1962)

Quality		Antithesis	
Noun form	Adjectival form	Noun form	Adjectival form
intimacy, presence	intimate	lack of intimacy lack of presence	non-intimate
liveness, fullness of tone	live	dryness deadness	dry dead
reverberation	reverberant	lack of reverberation	unreverberant
resonance	resonant	dryness	dry
warmth	warm	lack of bass	brittle
loudness of the direct sound	loud direct sound	faintness... weakness...	faint... weak...
loudness of the reverberant sound	loud...	faintness... weakness...	faint... weak...
definition, clarity	clear	poor definition	muddy
brilliance	brilliant	dullness	dull
diffusion	diffuse	poor diffusion	non-diffuse
balance	balanced	imbalance	unbalanced
blend	blended	poor blend	unblended
ensemble	—	poor ensemble	—
response, attack	responsive	poor attack	unresponsive
texture	—	poor texture	—
no echo	echo-free anechoic	echo	with echo echoic
quiet	quiet	noise	noisy
dynamic range	—	narrow dynamic range	—
no distortion	undistorted	distortion	distorted
uniformity	uniform	non-uniformity	non-uniform

Well-established relationships between the physical and subjective factors do not exist, but the work of Beranek (1962) in the United States, Cremer and Kuttruff (1965) in Germany, Parkin and Morgan (1971) and Hawkes (1970) in England, and Marshall (1967b, 1968) in Australia, is helping designers to achieve a better match between the acoustical environment and subjective experience.

The parameters used for concert hall design will now be discussed.

(a) Reverberation time

This factor can:

(a) increase the fullness of tone;
(b) enhance the bass;
(c) contribute towards the blending of instruments;
(d) increase the range of diminuendo and crescendo;
(e) help to diffuse the sound field (think of "clouds of sound").

TABLE 10.5. THE INTERRELATIONS BETWEEN THE AUDIBLE FAC-
TORS OF MUSIC AND THE ACOUSTICAL FACTORS OF THE HALLS IN
WHICH THE MUSIC IS PERFORMED (BERANEK, 1962)

Musical factors	Acoustical factors
Fullness of tone	Reverberation time
Clarity	Ratio of loudness of direct sound to loudness of reverberant sound
	Speed of music
Intimacy (audible)	Short initial–time-delay gap (eighteenth-century music room)
	Medium initial–time-delay gap (late nineteenth-century concert hall)
	Very long initial–time-delay gap (cathedral)
	Spatial responsiveness
Timbre and tone colour	Richness of bass
	Richness of treble
	Tonal distortion
	Texture
	Balance
	Blend
	Diffusion in hall
	Attack
Ensemble	Musicians' ability to hear each other
Dynamic range	Loudness of fortissimo; relation of background noise to loudness of pianissimo; this is about 50 dB for the strings and about 30 dB for the woodwind

Too much reverberation produces a loss of clarity, inordinate loudness and makes blending difficult; too little imparts a dry (i.e. lack of fullness of tone) and responsive acoustical character to the hall. For music it is more critical than for speech to have a rise in the reverberation time at the low frequencies; this increase in bass response contributes towards "warmth of tone". Reviewing a performance of a Mozart Piano Concerto (No. 21 in C major K467) at the Salzburg Festival in the Mozarteum Concert Hall, Kenneth Loveland wrote in *The Times* of 27 August, 1974: "... *the acoustic of the Mozarteum tends to overweigh orchestral sound and to obliterate the niceties of balance which one knew were within these performances.*"

The values given in Table 10.6 based on successful halls gives a rough guide. Figure 10.15a shows a more extensive range of values of reverberation times in several concert halls (with audience) and broadcasting studios throughout the world averaged over the frequency range 125–4000 Hz based on data from Somerville and Gilford (1957), Beranek (1962), Pancholy *et al.* (1965, 1966, 1967) and Sugden (1967), whereas Figs. 10.15b and c is a plot of the data presented by Doelle (1972) for concert halls and opera houses throughout the world.

Kuhl (1954) recorded three pieces of music in many concert halls and broadcasting studios with widely varying reverberation times. The recordings were replayed to audiences which included musicians, acousticians and recording engineers. The results are summarised in Table 10.7.

TABLE 10.6. APPROXIMATE RANGE OF REVERBERATION TIMES FOR VARIOUS TYPES OF
MUSIC

Purpose of hall	$\left(\dfrac{(T_{500\,Hz} + T_{1000\,Hz})}{2}\right)$ (s)	$\left(\dfrac{T_{125\,Hz} + T_{250\,Hz}}{T_{500\,Hz} + T_{1000\,Hz}}\right)$
Orchestral music	1·8–2·2	1·25–1·50
Opera, chamber music, drama	1.6–1.8	times the values in
Small music halls	1.5	the first column

This experiment was interesting but not realistic of concert-hall listening or
performing. In opera houses a compromise is necessary in deciding the design
reverberation time because some listeners want to hear and understand vocal diction
especially in recitative passages; a similar problem is posed by halls in which Lieder
recitals are to be given, or by multi-purpose halls in which speeches and music
performances are to take place. Beranek (1962) shows us that much can be learnt
from past experience which over time has won acclaim. For example, La Scala
Opera House in Milan has a reverberation time of 1·2 s, whereas that for the
Festspielhaus at Bayreuth is 1·55 s; both values apply to fully occupied opera houses
and to a frequency range of 500–1000 Hz. Winckel (1967) reports that he circulated a
questionnaire in 1948 to leading conductors all over the world asking which concert
halls were acoustically the best for performances by large symphony orchestras. The

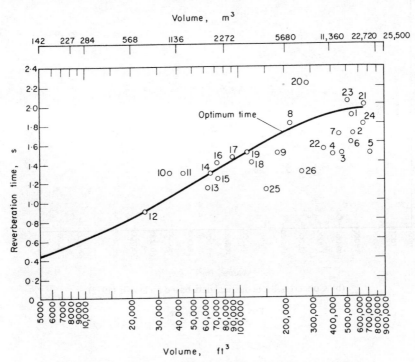

(a) Suggested optimum reverberation time curve for concert halls.

FIG. 10.15. Reverberation times sound absorption and volumes of various concert halls and opera houses
over the world.

	Name and year of completion	Floor shape*	Seating capacity	Volume per audience seat (m³ per seat)
1	Grande Salle de Concerts, Paris, 1963	F	937	12·8
2	Konserttisali, Turku, Finland, 1953	F	1002	9·6
3	Salle Musica, La Chaux-de-Fonds, Switzerland, 1955	R	1032	6·6
4	Radiohuset, Studio 1, Copenhagen, 1945	F	1093	10·9
5	Herkulessaal, Munich, 1953	R	1200	11·7
6	Konzertsaal, Musikhochschule, Berlin, 1954	R	1340	7·2
7	Konserthus, Göteberg, Sweden, 1935	F	1371	8·7
8	Beethovenhalle, Bonn, 1959	I	1407	11·2
9	Kulttuuritalo, Helsinki, 1957	F	1500	6·7
10	Grosser Tonhallesaal, Zurich, 1895	R	1546	7·4
11	Grosser Musikvereinssaal, Vienna, 1870	R	1680	8·9
12	Tivoli Koncertsal, Copenhagen, 1956	F	1789	7·1
13	Concert Hall, Chiba, Japan, 1965	I	1800	7·8
14	Severance Hall, Cleveland, Ohio, 1930	H	1890	8·2
15	Philharmonic Hall, Liverpool, England, 1939	F	1955	6·9
16	Liederhalle, Gosser Saal, Stuttgart, 1956	I	2000	8·1
17	Neues Festspielhaus, Salzburg, Austria, 1960	F	2160	6·0
18	Colston Hall, Bristol, England, 1951	R	2180	6·2
19	Concertgebouw, Amsterdam, 1887	R	2206	8·5
20	Philharmonie, Berlin, 1963	I	2218	11·3
21	Grote Zaal, De Doelen, Rotterdam, 1966	I	2232	12·5
22	St. Andrew's Hall, Glasgow, Scotland, 1874	R	2500	9·2
23	Free Trade Hall, Manchester, England, 1951	R	2569	6·0
24	Symphony Hall, Boston, Mass., 1900	R	2631	7·1
25	F.R. Mann Concert Hall, Tel Aviv, Israel, 1957	F	2715	7·8
26	Jubilee Auditoriums, Edmonton, Alberta, and Calgary, Alberta, 1957	F	2731	7·9
27	Carnegie Hall, New York, 1891	H	2760	8·8
28	Queen Elizabeth Theater, Vancouver, British Columbia, 1959	I	2800	5·6
29	Philharmonic Hall, Lincoln Center, New York, 1962	I	2836	8·7
30	Kleinhans Music Hall, Buffalo, N.Y., 1940	F	2839	6·4
31	Academy of Music, Philadelphia, 1857	H	2984	5·3
32	Royal Festival Hall, London, 1951	R	3000	7·3
33	Binyanei Ha'oomah, Jerusalem, Israel, 1960	F	3142	7·9
34	Tanglewood Music Shed, Lenox, Mass., 1938		6000	7·1
35	Royal Albert Hall, London, 1871	C	6080	14·2

*R = rectangular, F = fan-shaped, H = horseshoe-shaped, C = curvilinear, I = irregular.

(b) Concert halls: acoustical data based on Doelle (1972).

FIG. 10.15. (*Continued*).

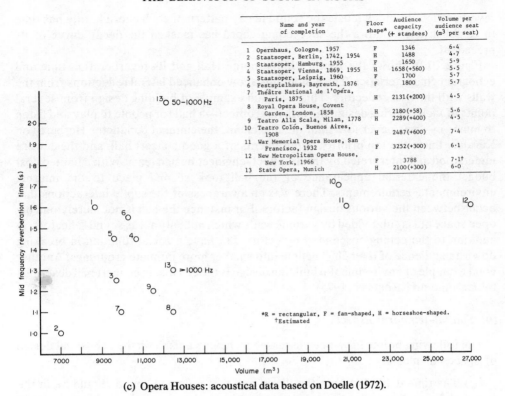

	Name and year of completion	Floor shape*	Audience capacity (+ standees)	Volume per audience seat (m³ per seat)
1	Opernhaus, Cologne, 1957	F	1346	6·4
2	Staatsoper, Berlin, 1742, 1954	H	1488	4·7
3	Staatsoper, Hamburg, 1955	F	1650	5·9
4	Staatsoper, Vienna, 1869, 1955	H	1658(+560)	5·5
5	Staatsoper, Leipzig, 1960	F	1700	5·7
6	Festspielhaus, Bayreuth, 1876	F	1800	5·7
7	Théâtre National de l'Opéra, Paris, 1875	H	2131(+200)	4·5
8	Royal Opera House, Covent Garden, London, 1858	H	2180(+58)	5·6
9	Teatro Alla Scala, Milan, 1778	H	2289(+400)	4·5
10	Teatro Colón, Buenos Aires, 1908	H	2487(+600)	7·4
11	War Memorial Opera House, San Francisco, 1932	F	3252(+300)	6·1
12	New Metropolitan Opera House, New York, 1966	F	3788	7·1†
13	State Opera, Munich	H	2100(+300)	6·0

*R = rectangular, F = fan-shaped, H = horseshoe-shaped.
†Estimated

(c) Opera Houses: acoustical data based on Doelle (1972).

FIG. 10.15. (*Continued*).

Musikvereinsaal in Vienna, the Concertgebouw in Amsterdam, the Symphony Hall in Boston and the Teatro Colón in Buenos Aires came out best. All except the Teatro Colón are rectangular in shape; all have plaster on wood, coffered or plaster on metal lath ceilings, and architectural features such as balconies and columns, which diffuse the sound; the reverberation times were in a closeband of 1·7–2·1s at 500–1000 Hz when the halls were occupied; similarly, the volume per seat fell in a narrow range of 7·1–8·9 m³ per seat; plaster walls and wood floors were used.

Research by Atal *et al.* (1965) and others shows that it is mainly the initial portion of a sound decay process which is responsible for the subjective impressions of

TABLE 10.7. OPTIMUM REVERBERATION TIMES IN CONCERT HALLS (KUHL, 1954)

Piece	Composer	Preferred reverberation time (s)	Range of opinion
First movement Jupiter Symphony K 551	Mozart	1·5	Good agreement
Fourth movement Symphony No 4 in *E minor*	Brahms	2·1	Large divergence
Danse Sacrale from *Le Sacré du Printemps*	Stravinsky	1·5	Good agreement

reverberation. In much music the sound decay pattern of each chord hardly has time to establish itself before the succeeding chord has masked the decay curve of its precedent.

Figure 10.16 shows the Berlin Philharmonie Hall and its reverberation time and echogram characteristics. The echograms show enhanced lateral reflections from the walls with the new reflectors. This is a good example of building design from several points of view. During design the principal object—a hall for people to play and listen to music in—was never lost sight of. The client, the eminent conductor Herbert von Karajan, knew the top priority requirements of a good concert hall, and these were understood and interpreted by the architect–engineer design team. With a limited cost budget the external appearance of the hall took second place to the internal environmental requirements. There was an awareness of the subtle interactions that occur between the various design factors. For instance the hall is not merely one big open space but is subdivided by various walls which not only act as sound reflectors, in addition to the ceiling suspended reflectors, but have a social function in breaking down an audience of over 2000 people into smaller more intimate groupings. Another good example of environmental building design is the de Doelan concert hall described by Croome and Roberts (1975).

(b) Spatial–temporal aspects

The following points have been assessed as being valuable in the acoustical design of spaces of music.

(i) The time delay between the direct and first reflected sound should be in the order of 20 ms; this is related to acoustical intimacy, clarity, brilliance, attack and texture. A good balance between the reverberant, reflected and direct sound fields is required for fullness of tone. The first reflections of sound coalesce with the direct sound if they are received by the listener within about 20–35 ms of each other. For clarity the amplitudes of the direct sound and the first reflections must be greater than that of the reverberant sound following. Since the direct sound decreases with the square of the distance and even more rapidly over the audience, reflected sound must reinforce the direct sound at long distances from the music source. Beranek (1962) established that the warmth of tone depends on the ratio:

$$\frac{\epsilon_D}{\epsilon_R} = \frac{\text{energy in the direct and first reflected sound } (\epsilon_D)}{\text{energy in the reverberant sound } (\epsilon_R)}. \tag{10.43}$$

Subjective tests showed that a difference of 7 dB in this ratio level (i.e. $10 \lg (\epsilon_D/\epsilon_R)$) distinguished a good hall from one of poor quality.

(ii) For clarity the direct sound must be loud enough at each seat, and also the reflected sound should not mask the direct sound.

(iii) Spatial responsiveness refers to the subjective impression of the way in which the sequence of sound reflections arrive at the ears of the listener after the direct sound; recent research suggests that five or more evenly spaced reflections within 60 ms is a good criterion to aim for. Further, the shape of the hall should allow lateral reflections (i.e. from walls) to precede ceiling reflections and reflections from various surfaces should not mask one another.

(a) Berlin Philharmonic Orchestra and audience await their conductor, Herbert von Karajan (15 October 1963).

FIG. 10.16.

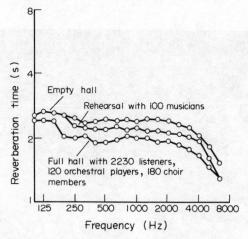

(b) Reverberation time characteristic of Berlin Philharmonie Hall (Cremer, 1965).

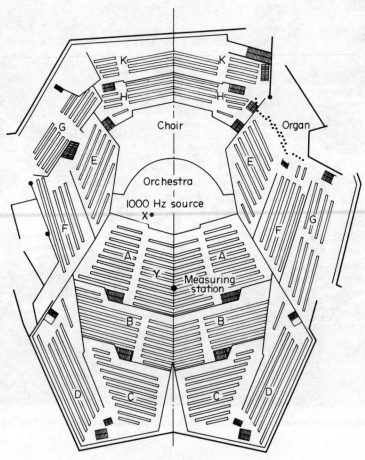

(c) Plan of Berlin Philharmonie Hall.

FIG. 10.16. (*Continued*).

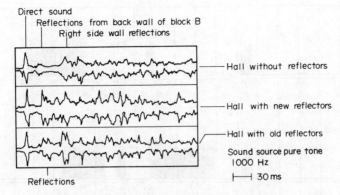

(d) Echogram for 1000 Hz tone at X received at Y (Cremer, 1965).

FIG. 10.16. (*Continued*).

(iv) Meyer and Kuttruff have emphasised that macrodiffusion. of the sound is essential to preserve a balance between clarity and fullness of tone. Reflecting elements are used to achieve this in the Beethovenhalle in Bonn, the Philharmonie in Berlin and the Stadthalle in Göttingen. Intelligibility of speech is even possible in spaces with a high reverberation time if the diffusion is good.

Marshall (1967b) has shown the importance of the shape of the concert hall. The property—*spatial responsiveness* or *spaciousness*—he interpreted as being related to loudness attributes, and generating a sense of envelopment for the listener and ensuring that the hall responds spatially to the music. He argued that the reverberant field provides an acoustical context in which significant acoustical events are perceived besides giving an acoustical feedback for performers. Echograms taken in two halls of different dimensions are shown in Fig. 10.17. In hall X the auditory view is limited to overhead reflections, and they are widely spaced in time so that integration of the acoustical events by the ear is less likely, whereas in hall Y the echo patterns are more evenly distributed and more ordered so that the listener is more likely to have a sense of envelopment in the sound.

To avoid ceiling reflections masking the direct sound and preceding the wall reflections, the minimum height H for a hall is given by

$$\frac{H}{2} > W + \frac{w}{2}, \qquad (10.44)$$

where W is effective hall width ($W/2$ for a music source in the centre of the hall), and w is the width of the performers' area (Marshall, 1967b, 1968). If insufficient reflections are perceived by the listeners and performers, partly due to masking and partly due to the weak lateral reflections, the acoustical feedback will be poor and the hall will "lack response".

Macrodiffusion can be achieved by using coffered ceilings, or by suspended reflecting elements as in the Philharmonie Hall, Berlin. Experience in Germany and Great Britain suggests that rectangular-shaped coffering (hard surfaces) using

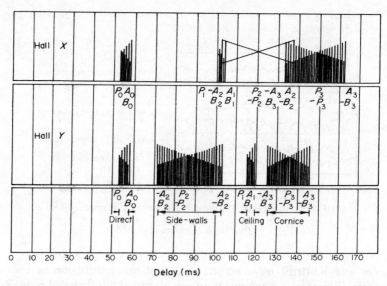

FIG. 10.17. Echograms for two concert halls; hall X has been judged to be unsuccessful and hall Y has been judged to be successful from the acoustic point of view (Marshall, 1967b).

minimum dimensions of about $1 \text{ m} \times 1 \text{ m} \times 0.5 \text{ m}$ deep and some degree of randomness is best, although pyramidal-shaped coffering has been used in the Stadthalle in Göttingen with equal effectiveness. The coffering assists in:

(i) decreasing the strength of the main ceiling reflection and thus lowering the masking level of the direct sound and also the lateral reflections;
(ii) possibly increasing the acoustical feedback by the macrodiffusion of sound.

Control of the reverberation time using electronic methods such as *assisted resonance* and *ambiophony* have been much under discussion in recent years (Parkin and Morgan 1971), but unless composers such as Stockhausen insist, it is unlikely that this will be a normal feature of concert-hall design but will be rather more a corrective remedy if the initial design does fail. These techniques do, however, permit a variable degree of reverberation to be achieved; this is necessary in a multi-purpose hall.

Attempts have been made to understand the concept of spatial responsiveness in more depth (e.g. Damaske, 1967–8; Burgtorf, 1967–8). These observations lead Kuttruff (1973) to make the conclusion that to establish a subjectively diffuse sound field, and hence an acoustical impression of space, it is sufficient to have sound incidence from not more than a few substantially different directions provided that the associated components are incoherent. Generally two signals will only be partially incoherent. The degree of coherence between two sound pressures $p_1(t)$ and $p_2(t)$ can be expressed by the correlation coefficient ψ, where

$$\psi = \frac{\overline{p_1 p_2}}{(\overline{p}_1^2 \overline{p}_2^2)^{1/2}}. \tag{10.45}$$

For coherency $\psi = \pm 1$, whereas $\psi = 0$ signifies total incoherency. But subjective impressions of diffusion depend on other factors—reverberation time, and the relative intensities of the direct and reverberant sound fields. The latter has been characterised by the ratio

$$\Delta L' = 10 \lg \left(\frac{\epsilon_d}{\epsilon_r} \right), \qquad (10.46)$$

where ϵ_d is the energy of the direct sound and ϵ_r is the total energy of all the reflections in a room. Note the similarity of this expression with the ratio of Beranek given as equation (10.43). Reichardt and Schmidt (1966) have investigated the effect of $\Delta L'$ on changes in perception levels of spaciousness (Fig. 10.18a). Listeners were asked to indicate the changes in the parameter $\Delta L'$ which caused a just-perceptible change in the subjective impression of spaciousness. This offers a quantifiable way of comparing the spatial–temporal patterns of sound fields produced in different spaces. Work by Schmidt (1968) as shown in Fig. 10.18b suggests that, within certain limits, the reverberation time and the logarithmic ratio $\Delta L'$ may be substituted for one another. When $\Delta L' = 0$ (i.e. $\epsilon_d = \epsilon_r$) the subjectively experienced reverberation time is equivalent to the actual reverberation time, but as $\Delta L'$ increases from zero (i.e. $\epsilon_d > \epsilon_r$) the reverberation time is felt to be less than the actual value; the converse happens when $\Delta L'$ falls below zero (i.e. $\epsilon_r > \epsilon_d$). This clearly is not only significant in natural room acoustics but also in the cases where electroacoustic installations have to be used.

Incidentally developments in hi-fi equipment such as quadraphonic sound are attempting to recreate the listeners' sense of spaciousness as perceived in a concert hall.

Thiele (1953) and Haas (1951) stated that reflections with delay times of less than 50 ms are useful but those delayed by a time interval greater than this are detrimental. Thiele (1953) speaks of the *definition* of a sound signal in a room quantified by integrating the energy contained in the useful reflections and dividing it

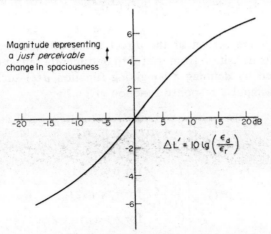

(a) Spaciousness of a sound field as a function of level difference between direct sound and reverberant sound; one ordinate unit corresponds to a just-perceivable change in spaciousness (Reichardt and Schmidt, 1966).

FIG. 10.18. Subjective experiments on spatial responsiveness.

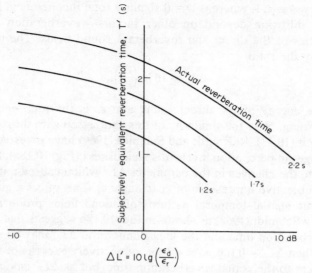

$$\Delta L' = 10 \lg\left(\frac{\epsilon_d}{\epsilon_r}\right)$$

(b) Subjectively equivalent reverberation time T' of a reference sound field, as a function of $\Delta L' = 10 \lg (\epsilon_d/\epsilon_r)$ and actual reverberation time of a synthetic sound field which is compared to the reference sound field directly (Schmidt, 1968).

FIG. 10.18. (*Continued*).

by the total energy in the direct and total reflected sound, thus

$$D = \frac{\displaystyle\int_0^{50\,\text{ms}} \{g(t)\}^2\, dt}{\displaystyle\int_0^{\infty} \{g(t)\}^2\, dt} \qquad (10.47)$$

where $t = 0$ denotes the arrival of the direct sound. It has become common to assume a continuous transition from useful to detrimental reflections, and this may be easily formulated by defining a weighting function $a(t)$ such that the useful reflections have an impulse response function given by

$$N = \int_0^{\infty} a(t)\{g(t)\}^2\, dt \qquad (10.48)$$

and for a linear transition

$$a(t) = \begin{cases} 1 & \text{for } 0 \leqslant t < t_1, \\ \dfrac{t_2 - t}{t_2 - t_1} & \text{for } t_1 \leqslant t \leqslant t_2, \\ 0 & \text{for } t > t_2. \end{cases} \qquad (10.49)$$

Niese (1956) suggests $t_1 = 17$ ms and $t_2 = 33$ ms for use in evaluating $a(t)$. He defines detrimental sound S as that which is delayed by more than 33 ms; S can be calculated by integrating the reverberant sound energy under the sound decay curve with a lower limit of $t = 30$ ms. The decay curve can be expressed in terms of the

"Doris, I think I've finally got the acoustics sorted out."

Stereo Types by Honeysett (with acknowledgements to *Punch* 7 November 1973).

reverberation time T by

$$h(t) = \begin{cases} 0 & \text{for } 0 < t < 30 \text{ ms,} \\ \left(\dfrac{2N}{30}\right)^{1/2} \exp\dfrac{-6\cdot9(t-30)}{T} & \text{for } t > 30 \text{ ms.} \end{cases} \tag{10.50}$$

Kuttruff (1973) then states the *echo degree* as

$$E = \frac{S}{S+N}. \tag{10.51}$$

According to Kuttruff (1973) there is a good correlation between the echo degree and syllable intelligibility. The logarithmic ratio

$$\Delta L = 10 \lg \left(\frac{N}{S}\right) \tag{10.52}$$

has been found by Beranek to be related to the acceptability of concert halls, whereas Lochner and Burger (1960) have found good agreement between ΔL and speech intelligibility.

Schroeder (in Mackenzie, 1975) describes some recent research into the most important objective and subjective factors of concert hall design. When the reverberation time fell below 2·2 s, or rose above 2·4 s, it became one of the most important parameters. Definition as defined by equation 10.47 was also found to be of high subjective importance and an inverse correlation was found to exist between it and reverberation time. A low value of definition is better for music but for speech this trend would probably be very different. A new parameter of importance was *inter-aural coherence* which Schroeder defines as the amplitude of the maximum of the inter-aural cross-correlation function; it is a binaural measure and should have a low value. If the cross-correlation function between the two ears is at zero delay then direct sound predominates over sound from other directions; as the peak value increases, sound from one direction predominates over that from another direction and this can probably be interpreted to mean that the sound diffusion is low. Schroeder continues to demonstrate how diffusion can be increased by using $\lambda/4$ stepped metal surfaces with designed inter-space distances. It can be shown (Jordan, 1975b) that definition D, and inter-aural coherence C, are both related to early energy;

$$D = \frac{\epsilon_{NL} + \epsilon_L}{\epsilon} \quad \text{and} \quad C = \frac{\epsilon_{NL}}{\epsilon_{NL} + \epsilon_L}$$

where ϵ denotes the total energy; ϵ_L the lateral, early energy; ϵ_{NL} the non-lateral, early energy. The time interval for early energy giving the best correlation with subjective tests has been found to be 100 m sec. The closer the sound growth and decay curves approximate to exponential functions the better the subjective rating becomes.

Finally Wilkens and Plenge (in Mackenzie, 1975) describe further subjective work using music by Mozart (Jupiter Symphony, first movement), Brahms (First Symphony, fourth movement) and Bartok (Concerto for Orchestra, first movement) listened to in four different parts of six German concert halls. All of the music was played by the Berlin Philharmonic Orchestra under their conductor Herbert von Karajan.

Evaluation of subjective opinions always has the difficulty of accounting for

Hall	Form	Seats	Volume	$T_{500-1000\,Hz}$
Berlin Philharmonie	Arena	2200	25,000 m³	2·0 s
Musikhalle, Hamburg	Rectangular	1980	11,600	2·2
Stadthalle, Hannover	Circular	3660	34,000	2·0
Stadthalle, Braunschweig	Hexagonal	2166	19,000	1·9
Rheinhalle, Düsseldorf	Circular	1842	33,000	2·5
Stadthalle, Wuppertal	Rectangular	1614	25,000	2·7

differences in taste, some people preferring a full sound—others a light, transparent sound. Wilkens and Plenge concluded that volume of sound, sensation of definition and clearness, and perception of tonal quality accounted for 89% of the judgement variance in their tests; perceived loudness was an essential influence on sound quality judgements at the place where it was heard. The best hall had the smallest deviations between reverberation time and early delay time.

This area of architectural acoustics has shown rapid progress in the last twenty-five years. Great activity continues to be evident. It is not yet possible to pin-point exactly *the* set of objective measures which cover the full range of subjective evaluation and which enable the shape, the form and the materials of a concert hall to be mapped and varied allowing the architect a full creative rein and yet guaranteeing a completely satisfying acoustic environment. Like other aspects of building environmental engineering, acousticians are not looking for the one hall with the optimal environmental conditions, but are seeking the boundaries which limit the optimal range of physical data relevant to subjective response. Many halls with different sizes, shapes and materials can accomplish these requirements.

Table 10.8 summarises the principal factors which have been reported as making significant contributions to human satisfaction as should be judged by people performing and those listening to music.

TABLE 10.8. ACOUSTIC PARAMETERS OF SIGNIFICANCE IN THE DESIGN OF CONCERT HALLS

Author	Parameters
Beranek (1962, 1965)	Initial time delay. Reverberation time rise at low frequencies. Ratio of early to reverberant sound.
West (1966)	$2(H/W)$ ratio important in determining early lateral and non-lateral reflections for a hall height H, width W.
Marshall (1967, 1968a,b)	Spatial responsiveness dependent on $H/2 \geq W + w$ (see equation 10.44). Early lateral reflections important. Order of reflections from different surfaces important. Distinguish between stage and audience conditions.
Keets (1968)	Apparent source width (ASW) related to fraction of incoherent lateral energy within first 50 m sec; ASW related to $(1 - C_{50})$ where C_{50} is cross-correlation coefficient for first 50 m sec. $(1 - C_{50})$ related linearly to subjective scale of spatial impression.

TABLE 10.8. (*cont.*)

Author	Parameters
Barron (1971)	Spatial impression or broadening of source and loudness are only positive subjective effects. Initial time delay of 10–80 m sec unimportant. Lateral reflections important whereas ceiling reflections produce coloration but order of arrival of these reflections unimportant to spatial impression. Degree of spatial impression measured by ratio of lateral to non-lateral energy arriving within first 80 m sec.
Kürer (1969)	Family of criteria all interrelated and may be deduced from same integrated pulse curve. Criteria include definition, steepness, ratio of early to late energy, early decay time, rise time, point of gravity time.
Lehmann (1972)	Initial reverberation (0–10, 0–15 or 0–20 dB). Point of gravity time and rise time, 50 m sec criteria.
Gottlob (1973)	Reverberation time $\simeq 2$ s. Definition. Interaural coherence. Rise time and point of gravity time. Directional distribution of early energy is important.
Schroeder (1975)	Reverberation time (when $2\cdot2$ s $> T > 2\cdot4$ s). Definition. Interaural coherence.
Wilkens and Plenge (1975)	Volume of sound. Definition. Perception of tonal quality.
Jordan (1959, 1969, 1970, 1973, 1975a,b).	Inversion index. Variation and level of early delay time. Definition or interaural coherence. Rise time or point of gravity time. Some other criterion yet undefined.

10.8. Case Study: Sydney Opera House

I am indebted to Dr. V. L. Jordan for providing the technical acoustical data for this section (Jordan, 1973, 1975).

This famous building came into existence over a period of sixteen years; the inauguration took place in 1973. The final cost of the project was over ten times the original cost figure. Construction of the interior spaces began in 1969; much of the time previous to this was spent developing the ideas of the first designs conceived by the architect, Jørn Utzon. The second design period covered the years 1966–73 and was under the direction of the principal architect Peter Hall; it was mainly concerned with the interior design, whereas thought and effort in the first design period were mainly expended on the exterior. The roof shells and the situation for the complex in Sydney Harbour are both brilliant concepts stemming from the Utzon design period; the external view of the complex is breath-taking (Fig. 10.19a). The structural problems associated with the roof design caused many headaches but successful solutions were found. Much criticism has been made about the functional values of the complex—no car parks, the stage and the orchestral pits in the Opera Theatre both too small—but we wish to trace here the acoustical design in order to feel the magnitude of the task and also to see how other factors interact with those used to design the aural environment in a building.

When the second design period began the principal question to be answered by the building design team was: Given the empty spaces below the roof shells, what does

Sydney really need to accommodate in its arts centre? This question was resolved and the complex was then designed to consist of the following spaces:

CONCERT HALL	(about 2800 seats)
OPERA THEATRE	(about 1600 seats)
ORCHESTRA REHEARSAL AND RECORDING STUDIO	
DRAMA THEATRE	(about 550 seats)
CHAMBER MUSIC/CINEMA/EXHIBITION AREA	

  (in practice this hall is exclusively used as
  a cinema but has been called the Music Room)
 RECITAL HALL

The title of this arts centre has remained the Sydney Opera House, but it is misleading to people who are not acquainted with the background to this project, and is perhaps disappointing to opera performers and enthusiasts who were expecting the Opera Theatre to have a larger seating capacity and a much more ambitious stage design.

There then remained the following major questions concerning the acoustical

(a) The Sydney Opera House photographed during the official opening-day celebrations on 20 October 1973; the complex was opened by Her Majesty the Queen. Inaugural performances commenced in the complex on 28 September 1973.

FIG. 10.19. The Sydney Opera House (with acknowledgements to the General Manager of the Sydney Opera House).

(b) The Concert Hall of the Sydney Opera House.

FIG. 10.19. (*Continued*).

design of these spaces:

(a) Would the boundaries of the interior spaces encompassed by and defined by the roof shells allow a satisfactory sound field to occur? For example, would the ceiling shape produce an even distribution of reflected sound from the stage over the audience? Would the nearly vertical and parallel side wall produce a diffuse field of lateral reflections?

(b) What acoustic design criteria should be adopted; after all there have been failures as well as successes in modern concert-hall designs?

(c) Would the outer shell provide sufficient sound isolation from exterior noise? Would the inner building wall and floor structures provide sufficient insulation between interior spaces that may be in operation simultaneously?

Each of these major questions give rise to a number of other questions. Some answers to questions (a) and (b) could be sought from the data and references in this Chapter, whereas some solutions to question (c) were suggested in Chapters 4 and 9. The approaches that were actually used for this case study are described by Jordan (1973, 1975).

10.8.1. GENERAL APPROACH

The theoretical and empirical evidence for this section was discussed in section 10.2.5. Scale models were used to evaluate the acoustics of the Concert Hall and Opera Theatre. A 1:10 scale was used, guided by the fact that air absorption increases at high frequencies, such that for model frequencies above 30–40 kHz a sophisticated air dehumidification arrangement is needed to limit this absorption. (It should be noted that in this country the BBC have done this to evaluate the acoustics of their Maida Vale Studios for example—see Harwood, 1970, also Burd and Day in Mackenzie, 1975). The models were constructed of thick plywood and varnished on the interior surfaces. Seats were simulated by continuous neoprene strips and people by neoprene blocks topped with small cardboard squares to represent the heads.

A high-tension spark source was used to generate pulses of very short duration. Their passage around the inside of the model were measured by 6 mm ($\frac{1}{4}$ in.) diameter condenser microphones and recorded on a tape-recorder with a 1:10 speed turndown ratio. Model sound decay rates and reverberation times were then compared with those calculated.

Early decay time (EDT) correlates with subjective impressions of spatial responsiveness (see section 10.2.5 and Atal *et al.*, 1965). A working hypothesis was adopted that *the values of early decay time should not be much lower, no more than 10–20% lower, than the statistical reverberation time*, i.e. $T_{15} > 0.8$ to $0.9\ T_{15-30}$. The *variation* in early decay time indicates the degree of sound diffusion, the least variation gives the most homogeneous hall. These criteria were applied to individual locations throughout the model.

The early decay time was measured in the audience area and compared with the average value measured on the stage by a coefficient termed the *inversion index* defined as:

$$I = \frac{\text{early decay time in audience area}}{\text{early decay time on stage}}.$$

The criterion used was that $I \geq 1$ so that the sound fields build up more rapidly on the stage to the benefit of the musicians. I is related to auditorium shape and the amount of lateral reflection.

The model work enabled the early decay times to be measured and the inversion index then to be calculated. Several modifications to the shape of the model were made until satisfactory values of early decay time, reverberation time and inversion index were obtained for the performing and the audience areas. For example, model tests of the first design of the Opera Theatre (Fig. 10.20) showed that the early decay time had a value above that of reverberation time, but the inversion index was higher in the pit than on the stage where it was below $1\cdot0$. The ceiling was moved upwards

(a) Plan

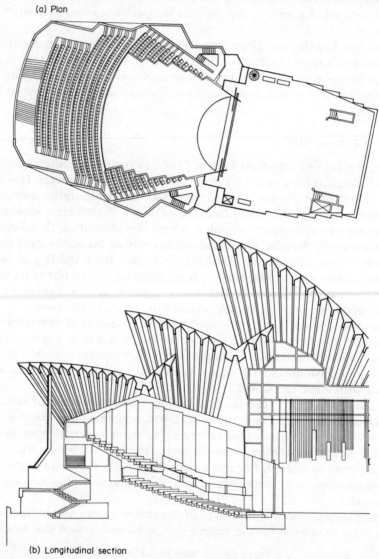

(b) Longitudinal section

FIG. 10.20. First design of the Sydney Opera House (Jordan, 1973, 1975).

as far as the shell structure permitted. This resulted in the inversion index being greater than 1·0 in both cases. Suspended reflectors were also placed above the orchestra seating area.

10.8.2. APPRAISAL OF ACOUSTICAL CONDITIONS IN THE COMPLETED CONCERT HALL AND OPERA THEATRE

Acoustical tests were carried out in the completed Concert Hall and Opera Theatre during 1972–3. A view of the inside of the Concert Hall is given in Fig. 10.19b.

Two testing methods were chosen and used in the hall when empty and also with near capacity audiences.

(a) To measure the statistical reverberation time, the opening bars of the Coriolanus Overture by Beethoven were played and recorded at a number of locations in the Concert Hall, the locations corresponding to those used in the model (Fig. 10.21). Frequency analysis (1/3 octave and 1/1 octave) gave the sound decay rate at various frequencies,

$$-\left(\frac{dL_p}{dt}\right) dB/s,$$

and since the reverberation time T is the time for the sound pressure level to decay by 60 dB

$$\frac{60}{T} = -\left(\frac{dL_p}{dt}\right),$$

T may be found at each frequency.

(b) Pistol shots were fired in various parts of the Concert Hall and the Opera Theatre and recorded at locations again corresponding to those used in the model tests. The tapes were analysed by filtering, reverse recording of the tape, and by integrating the signal. The integrated signal was fed to a level recorder and the values of early decay time measured from the curves recorded. The same method minus the integration was used to obtain the values of reverberation time.

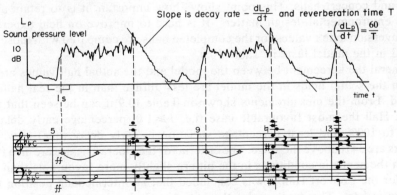

FIG. 10.21. The opening of the Coriolanus Overture by Beethoven and the associated reverberation process.

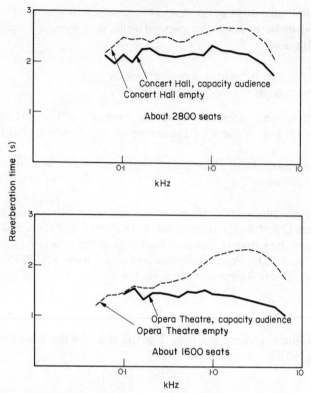

FIG. 10.22. Reverberation time frequency characteristic in one-third octave values.

The reverberation time characteristics for the hall are shown in Fig. 10.22a and b. Notice that there is no appreciable rise in the reverberation time in the lower frequencies as recommended earlier in this chapter; Jordan (1973) considers that the frequency dependence of the medium frequencies to the higher frequencies is a more important issue. He suggests that a "certain suppression" of reverberation time in the 400–1000 Hz with a corresponding emphasis above 1000 Hz has merit for studios and concert halls. This point shows how important it is to retain an open mind in environmental design matters. It is hard to improve on field experience.

The inversion index values for the completed halls are compared with those values obtained in the model tests in Table 10.9.

In general the agreement between the model and the actual hall values are good, although the sound fields in the model are less diffuse than in the real halls when occupied. From the measurements shown in Table 10.9 it can be seen that for the Concert Hall the most favourable case (i.e. $I \geqslant 1 \cdot 0$, percentage early delay time nearest to 100% and least percentage variation in early delay time) is when the reflectors are 34 ft above the stage. The criteria for the Opera Theatre are also fulfilled although the acoustic conditions in the pit are slightly more favourable than on the stage. For the Concert Hall it was concluded that an intermediate position for the reflectors is best. The reason why the inversion index increases with increasing elevation of the reflectors when the hall is empty but decreases when the hall is almost

TABLE 10.9. COMPARISON OF CONCERT HALL AND MODEL CRITERIA (JORDAN, 1975a)

Criterion and Conditions	Model (2–16 kHz)	Hall (250–2000 Hz)	
		empty	capacity audience
Inversion Index (Ideal: ≥1·0)			
Reflectors at soffit level	1·11	1·14	0·96
Reflectors just below crown	1·05	1·10	1·06
Reflectors 34 ft above stage	—	—	1·05*
Percentage 'good' E.D.T. values (Ideal: = 100%)			
Reflectors at soffit level	94	96	85
Reflectors just below crown	83	100	85
Reflectors 34 ft above stage	—	—	94*
Percentage overall variation of E.D.T. values (Ideal: = 0%)			
Reflectors at soffit level	20	13	17·5
Reflectors just below crown	21	13·5	17·5
Reflectors 34 ft above stage	—	—	10*

* Means frequency range only 500–2000 Hz.

COMPARISON OF OPERA THEATRE AND MODEL CRITERIA (JORDAN, 1975a)

Source at	Model		Hall			
			empty		capacity audience	
	Pit	Stage	Pit	Stage	Pit	Stage
Inversion Index (Ideal: ≥1·0)	1·22	1·13	1·52	1·09	1·22	1·10
Percentage 'good' E.D.T. values (Ideal: 100%)	100	100	100	100	100	96
Percentage overall variation (Ideal: 0%)	38	30	21	15	9	19

full, has not yet been explained; Jordan (1973) points out that the orchestral musicians were not simulated adequately in the model tests. Toroidal-shaped reflectors were found to give better sound diffusion than curved ones.

10.8.3. DRAMA THEATRE

The average reverberation time was about 0·9 s and considered ideal for this 550-seat theatre.

10.8.4. ORCHESTRA REHEARSAL HALL AND RECORDING STUDIO

This Hall has a volume of about 5200 m^3 and an average reverberation time of 2·0 s

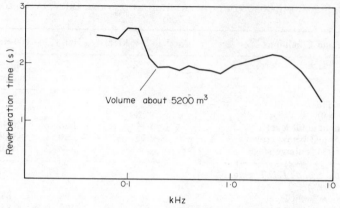

FIG. 10.23. Reverberation time frequency characteristic for Orchestra Rehearsal/Recording Studio in one-third octave values (Jordan, 1973)

(Fig. 10.23). Some people would argue that it is better to rehearse in the same hall as the one in which the performance takes place, not only because the acoustics can be made similar when the hall is full or empty by using seats which have the same absorption values whether occupied or unoccupied but because the musicians will be attuned to the total space.

Comparing the reverberation time characteristic in Fig. 10.23 and that for the Concert Hall with an audience (Fig. 10.22a) it can be seen that the characteristics are similar except below 300 Hz where the reverberation time in the Rehearsal Hall rises to as high as 2·6 s at 100 Hz. Thick, slot-perforated plywood with a mineral wool backing was used for the panelling. Curtains are used to achieve the lower reverberation times necessary when the Rehearsal Hall is used as a studio.

The optimum reverberation time for a music *sound* studio is about 1·6 s for a volume of 5200 m³, and for a *television* studio it is only about 0·6 s (Fig. 10.24). Compare this to the 2·0 s value desired when this space is used for concert rehearsals. Note that for adjacent listening and control rooms the maximum reverberation time should be 0·4 s up to 250 Hz falling to 0·3 s at 8000 Hz (Burd *et al.*, 1966).

10.8.5. SOUND INSULATION

The ceiling has an interior cocoon of sprayed concrete on a supported steel mesh under the roof shells. This structure forms an effective sound isolation barrier; on the inside of this is mounted the plaster–plywood panelling. Laminated glass totalling 18 mm thickness was included in the walls. The effectiveness of this was tested using a helicopter hovering at 200 ft (Fig. 10.25a).

The difficult problem of acoustically separating the Concert Hall from the Rehearsal Hall underneath it was solved by using double slabs of heavy concrete separated by neoprene pads. The sound reduction index of this construction is shown in Fig. 10.25b.

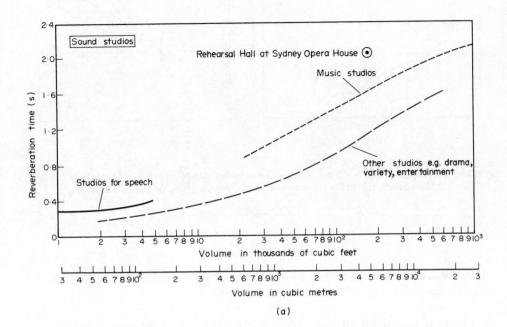

(a)

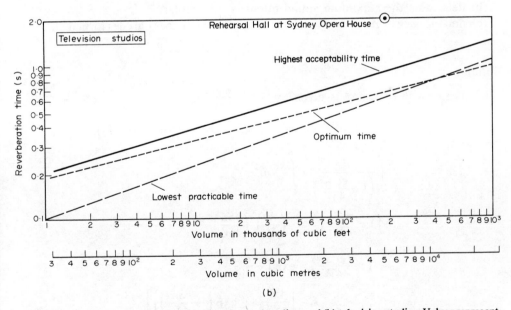

(b)

FIG. 10.24. Optimum reverberation time for (a) sound studios, and (b) television studios. Values represent maximum reverberation time in the frequency range 500–2000 Hz based on preferred BBC studios. Listening and control rooms should have a maximum reverberation time of 0·4 at 250 Hz falling to 0·3 s at 8000 Hz (Burd *et al.*, 1966).

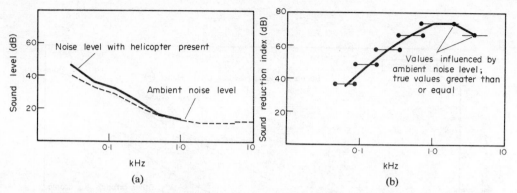

(a) Sound levels in the Concert Hall due to helicopter outside the building at 200 ft (61 m) above sea-level.
(b) Sound reduction index between Rehearsal/Recording Studios and Concert Hall.

FIG. 10.25. Sound insulation tests (Jordan, 1973, 1975).

10.9. Examples

1. A room of volume 86 m^3 has a total sound absorption of 10 metric Sabines. If a sound source having 10 μW sound output is turned on:

(a) what is the sound intensity level inside the room at the end of 0·2 s?
(b) determine the maximum sound intensity attainable;
(c) find the decay rate of the sound intensity level when the source is turned off.

Solution

(a) The growth of sound is given by

$$\frac{d(\epsilon V)}{dt} = W - \frac{\epsilon c}{4} S\bar{\alpha} \quad \text{(see equation (10.21b))},$$

$$\int_0^{\epsilon_t} \frac{d\epsilon}{\left(1 - \frac{\epsilon c S\bar{\alpha}}{4W}\right)} = \int_0^t \frac{W}{V}\, dt,$$

$$\ln\left(1 - \frac{\epsilon c S\bar{\alpha}}{4W}\right)_0^{\epsilon_t} = -\left(\frac{Wt}{V}\right)_0^t \left(\frac{cS\bar{\alpha}}{4W}\right),$$

$$\left(1 - \frac{\epsilon c S\bar{\alpha}}{4W}\right) = \exp\left(-\frac{cS\alpha t}{4V}\right),$$

$$\epsilon_t = \left[\frac{4W}{cS\bar{\alpha}}\left\{1 - \exp\left(-\frac{cS\bar{\alpha}t}{4V}\right)\right\}\right].$$

Now $I = \frac{\epsilon c}{4}$ and $\epsilon = \frac{p^2}{\rho c^2}$.

Therefore
$$I_t = \frac{W}{S\bar{\alpha}}\left\{1 - \exp\left(-\frac{cS\bar{\alpha}t}{4V}\right)\right\}$$

also
$$p_i^2 = \frac{4W\rho c}{S\bar{\alpha}}\left\{1 - \exp\left(-\frac{cS\bar{\alpha}t}{4V}\right)\right\}.$$

Now $W = 10 \times 10^{-6}$ W, $S\bar{\alpha} = 10$ m^2 Sabines, $c = 343$ m/s, $t = 0\cdot2$ s, $V = 86$ m^3

$$I_{0\cdot2} = \frac{10 \times 10^{-6}}{10}\left\{1 - \exp\left(-\frac{343 \times 10 \times 0\cdot2}{4 \times 86}\right)\right\}$$
$$= 86\cdot4 \times 10^{-8}\,\text{W/m}^2.$$

Therefore
$$L_1 = 10\lg\left(\frac{86\cdot4 \times 10^{-8}}{10^{-12}}\right) = 59\cdot36\,\text{dB re }10^{-12}\,\text{W/m}^2.$$

(b)
$$I_{max} = \frac{W}{S\bar{\alpha}} = \frac{10 \times 10^{-6}}{10} = 10^{-6}\,\text{W/m}^2\,(60\,\text{dB}).$$

Using equation (10.24) the decay rate (dB/s)
$$= \frac{10cS\bar{\alpha}}{4V2\cdot3}$$
$$= \frac{(10)(343)(10)}{(4)(86)(2\cdot3)}$$
$$= 43\cdot4\,\text{dB/s}.$$

A typical value of voice speech power is about 1 μW which in loud speech rises to about 1 mW; during the 1974 Scarborough Shouting Contest a woman obtained a sound pressure level reading of 112 dB (approaching 7 mW at a distance 1 m)! The sound power of a musical instrument lies in the range 10 μW to 100 mW, whereas a large orchestra may generate 10 W in very loud passages.

---

2. A reverberation chamber measures $2 \times 2 \times 3$ m and $\bar{\alpha}$ is 0·04.
(a) If a sound source of 1 μW is tested, find the maximum sound pressure level.
(b) Find the new sound pressure level if a person goes into the chamber to make measurements assuming the average absorption for a person is 0·88 metre Sabines.

Solution

(a) From the previous question
$$p_{max} = \sqrt{\frac{4W\rho c}{S\bar{\alpha}}}$$
$$= \sqrt{\frac{4(1 \times 10^{-6})(343 \times 1\cdot2)}{0\cdot04\{(2 \times 4) + (8 \times 3)\}}}$$
$$= 0\cdot036\,\text{Pa}.$$
$$L_{p_{max}} = 20\lg\left(\frac{0\cdot036}{2 \times 10^{-5}}\right)$$
$$= 65\cdot1\,\text{dB}.$$

(b) The total absorption is modified, hence
$$p_{max} = \sqrt{\frac{4 \times 10^{-6} \times 412}{\{(0\cdot04 \times 32) + 0\cdot88\}}}$$

$$= \sqrt{7\cdot64 \times 10^{-4}}$$
$$= 0\cdot0276 \text{ Pa.}$$

$$L_{P_{max}} = 20 \lg \left(\frac{0\cdot0276}{2 \times 10^{-5}} \right)$$

$$= 62\cdot8 \text{ dB.}$$

This forms the basis of a method for measuring the sound absorption of the room surfaces, or one unknown surface material, using a known source of acoustic power.

---

3. Ten persons are talking in a room which has a total sound absorption of 0·975 metre Sabines and $\bar{\alpha} = 0\cdot05$. If each person produces a sound power output of 10 μW compare the background sound pressure level of the reverberant sound with the direct sound pressure level at 0·3 m from the nearest speaker.

Solution

Equation 10.30 gives the reverberant sound pressure as

$$p_R^2 = \frac{4\rho c W}{R}; \quad \text{where} \quad R = \frac{S\bar{\alpha}}{1 - \bar{\alpha}},$$

$$p_R = \sqrt{\frac{(4)(412)(10 \times 10 \times 10^{-6})}{\left(\frac{0\cdot975}{1 - 0\cdot05} \right)}},$$

$$p_R = 0\cdot4 \text{ Pa.}$$

Hence the sound pressure level is

$$L_p = 20 \lg \left(\frac{0\cdot4}{2 \times 10^{-5}} \right)$$

$$= 86 \text{ db due to reverberant sound field.}$$

Equation 10.12 gives the direct sound pressure as

$$p^2 = \frac{W\rho c}{4\pi r^2},$$

$$p = \sqrt{\frac{(10 \times 10 \times 10^{-6})412}{4\pi (0\cdot3)^2}},$$

$$p = 0\cdot0604 \text{ Pa.}$$

This gives a sound pressure level

$$L_p = 20 \lg \left(\frac{0\cdot0604}{2 \times 10^{-5}} \right)$$

$$= 89\cdot6 \text{ dB due to direct sound field.}$$

Note: If the direct and reverberant field energies are equated, then, neglecting directivity,

$$\frac{W\rho c}{4\pi r^2} = \frac{4 W\rho c}{R}.$$

Therefore $\qquad r = 0.14\sqrt{R}$ metres,

direct field radius $\qquad r = 0.14\sqrt{\dfrac{S\bar{\alpha}}{1-\bar{\alpha}}},$

$$r = 0.14\sqrt{\dfrac{0.975}{0.95}},$$

$$r = 0.144 \text{ m}.$$

For sound absorption to effectively reduce the reverberant field the distance between the sound source and the absorbent must be greater than 0.144 m.

---

4. The sound pressure level in a reverberation chamber $3 \times 4 \times 5$ m is 70 dB re 2×10^{-5} Pa and the reverberation time is 4 s. Find the acoustic power output of the machine.

Solution

$$p_{max} = \sqrt{\dfrac{4W\rho c}{S\bar{\alpha}}} \quad \text{(see question 1)},$$

$$T = \dfrac{0.16V}{S\bar{\alpha}} \quad \text{(see equation (10.20))}.$$

Now $\qquad S\bar{\alpha} = \dfrac{0.16V}{T}.$

Therefore $\qquad p_{max} = \sqrt{\dfrac{4W\rho cT}{0.16V}}.$

Hence $\qquad W = p_{max}^2 \dfrac{0.16(3 \times 4 \times 5)}{4 \times 412 \times 4},$

$$W = 5.83 \ \mu\text{W}.$$

---

5. A room has dimensions $4 \times 5 \times 8$ m. Determine:

(a) the mean free path of a sound wave;
(b) the number of reflections per second made by sound waves with the walls of the room;
(c) the decay rate of the sound in the room assuming $\bar{\alpha} = 0.1$.

Solution

(a) From equation (10.22) *et seq.* the mean free path is $4V/S = 4(4 \times 5 \times 8)/184 = 3.48$ m and is the average distance a sound wave travels through the air between two successive encounters with any two surfaces.

(b) $\qquad n = \dfrac{cS}{4V}$ reflections/s

$$= \dfrac{343}{3.48}$$

$$= 98.5 \text{ per s}.$$

(c) The decay rate is given by equation (10.24):

$$\frac{L_{p_m} - L_p}{t} = \frac{10cS\bar{a}}{2 \cdot 3 \times 4V}$$

$$= \frac{10 \times 98 \cdot 5 \times 0 \cdot 1}{2 \cdot 3}$$

$$= 42 \cdot 6 \text{ dB/s.}$$

---

6. The reverberation time of a room changes from 1·3 s to 1·2 s as the audience changes from 200 to 400 people. Determine the lower and higher limits of seating capacity if the reverberation time limits are 1·0 and 1·40 s. (Noise Control Conference UWIST, 1973.)

Solution

The reverberation time can be written as the formula for a room volume V m³:

$$T = \frac{0 \cdot 16V}{\Sigma SA + Nb},$$

where ΣSA is the total sound absorption at the surfaces, N is the number of people in the audience and b is an absorption factor related to the difference in absorption between an occupied and empty seat. This formula can be rearranged in terms of general constants X and Y thus:

$$\frac{1}{T} = X + YN.$$

In this case we have data for $N = 200$, $T = 1 \cdot 3$ and $N = 400$, $T = 1 \cdot 2$.

Therefore

$$\frac{1}{1 \cdot 3} = X + 200Y,$$

$$\frac{1}{1 \cdot 2} = X + 400Y.$$

These equations give values of $X = 55/78$ and $Y = 1/3120$; hence if $T = 1 \cdot 0$ s, N will be 920 and for $T = 1 \cdot 4$ s N will be 29.

---

7. A hall has a volume of 30000 m³ and a mid-frequency reverberation time of 2·8 s. Determine the Sabine absorption in m² which reduces the noise level by 5 dB. (Noise Control Conference UWIST, 1973.)

Using equation (10.30) the sound pressure due to the reverberant field is

$$p_R^2 = \frac{4W\rho c}{R}$$

for a sound source of power W operating in a room having a room constant $R = (S\bar{a}/1 - \bar{a})$. In terms of sound pressure level,

$$L_{p,R} = L_W + 10 \lg \left(\frac{1}{R}\right) + \text{constant.}$$

For a constant power level any change in sound level due to a change in absorption will be given by

$$\Delta L_{p,R} = 10 \lg \left(\frac{R_2}{R_1}\right).$$

Assuming the ratio $(1 - \bar{\alpha}_1)/(1 - \bar{\alpha}_2) \simeq 1$, in this case

$$5 = 10 \lg \left\{\frac{(S\bar{\alpha})_2}{(S\bar{\alpha})_1}\right\}.$$

Using Sabine's formula,

$$(S\bar{\alpha}_1) = \frac{(0 \cdot 161)(30\,000)}{2 \cdot 8} = 1725 \text{ m}^2 \text{ Sabines}.$$

Hence $(S\bar{\alpha}_2) = 5455 \text{ m}^2$ Sabines.

Absorption added to reduce sound level by 5 dB is 3730 m^2 Sabines.

---

8. An office has the following characteristics:

	Area	Sound absorption coefficient
Floor	100 m^2	0·03
Ceiling	100 m^2	0·03
Walls	240 m^2	0·04
16 Desks	each	0·10
16 Occupants	each	0·35

If the ceiling is treated with acoustic tile having an average absorption coefficient of 0·75, determine the noise reduction produced under reverberant conditions. (Noise Control Conference UWIST, 1973.)

Floor absorption

$$= 100 \times 0 \cdot 03$$
$$= \quad 3 \text{ m}^2 \text{ Sabines}$$

Ceiling absorption†

$$= 100 \times 0 \cdot 03$$
$$= \quad 3 \text{ m}^2 \text{ Sabines}$$

Wall absorption

$$= 240 \times 0 \cdot 04$$
$$= \quad 9 \cdot 6 \text{ m}^2 \text{ Sabines}$$

Desk absorption

$$= \quad 16 \times 0 \cdot 10$$
$$= \quad 1 \cdot 6 \text{ m}^2 \text{ Sabines}$$

Occupant's absorption

$$= \quad 16 \times 0 \cdot 35$$
$$= \quad 5 \cdot 6 \text{ m}^2 \text{ Sabines}$$

Total
$$= 22 \cdot 8 \text{ m}^2 \text{ Sabines}$$

---

†The added ceiling absorption is the difference between the new value and that before treatment $= (100 \times 0 \cdot 75) - 3 = 72 \text{ m}^2$ Sabines.

Total absorption after treatment $= 94 \cdot 8 \text{ m}^2$ Sabines hence

noise reduction $= 10 \lg (94 \cdot 8/22 \cdot 8) = 4 \cdot 1$ dB.

CHAPTER 11

THE TRANSMISSION OF SOUND
THROUGH STRUCTURES

ACOUSTICAL energy can enter a space from a direct airborne source, such as an outlet of an airconditioning system, or it can originate from sound transmitted via a structure, radiating directly into the space causing items to vibrate and rattle.

Structural sound arises from:

(i) equipment directly vibrating the structure, or another common cause, is the impact sound due to footsteps. The fundamentals of mechanical vibrations and some methods of controlling structureborne sound transmission were described in Chapter 9 and further reference to control was made in Chapter 4;

(ii) the surface of structures may be set into vibration by incident airborne sound.

The various direct and indirect sound pathways are shown diagrammatically in Fig. 11.1.

11.1. The Direct Transmission of Airborne Sound Through a Massive Wall

Consider the passage of an airborne sound wave incident on a massive, non-distorting wall of mass per unit surface area m. Assume the air on each side of the wall has the same density ρ. The incident sound energy ϵ_i is reflected at and transmitted through the surface so that

$$\epsilon_i = \epsilon_r + \epsilon_t. \tag{11.1}$$

This neglects any flanking transmission paths. The *sound transmission coefficient* τ is defined as

$$\tau = \frac{\epsilon_t}{\epsilon_i}. \tag{11.2}$$

Expressing the energy in terms of the respective acoustic pressure amplitudes (i.e. $|p_i|$, $|p_r|$ and $|p_t|$) and the specific acoustic impedance ρc,

$$\tau = \frac{|p_t|^2/\rho c^2}{|p_i|^2/\rho c^2}.$$

Therefore

$$\tau = \frac{|p_t|^2}{|p_i|^2}. \tag{11.3}$$

If the wall oscillates with a uniform displacement ϕ, then the acoustic pressure

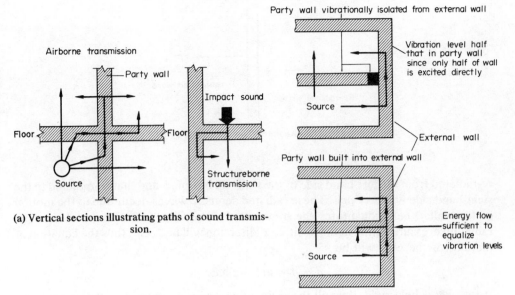

(a) Vertical sections illustrating paths of sound transmission.

(b) Flanking transmission with heavy single-leaf party and external walls.

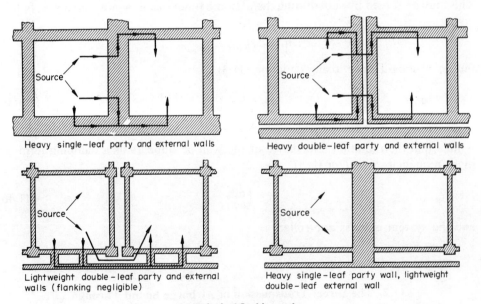

Heavy single-leaf party and external walls

Heavy double-leaf party and external walls

Lightweight double-leaf party and external walls (flanking negligible)

Heavy single-leaf party wall, lightweight double-leaf external wall

(c) Principal flanking paths.

FIG. 11.1. Sound transmission pathways through building structures.

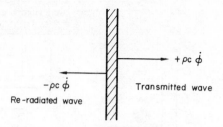

FIG. 11.2. Radiated sound waves at a wall.

re-radiated from the left-hand side of the wall will be $-\rho c\dot{\phi}$ and that re-radiated to the right-hand side will be $+\rho c\dot{\phi}$. These radiated acoustic waves occur due to the motion of the wall; $+\rho c\dot{\phi}$ constitutes the transmitted wave (Fig. 11.2). The total acoustic radiation pressure acting in the positive x-direction will be $2\rho c\dot{\phi}$; thus the equation of motion for the wall will be

$$p_i + p_r = m\ddot{\phi} + 2\rho c\dot{\phi}.$$

If the wall is held rigid, then all the incident pressure is reflected so that $p_i = p_r$ and pressure doubling has occurred, hence

$$2p_i = m\ddot{\phi} + 2\rho c\dot{\phi}.$$

This may be solved by substituting the periodic functions $\dot{\phi} = \dot{\phi}_0 e^{j\omega t}$ and $p_i = |p_i|e^{j\omega t}$, giving

$$2|p_i| = (j\omega m + 2\rho c)\dot{\phi}_0$$

but $|p_t| = \rho c\dot{\phi}_0$. Hence, using equation (11.3),

$$\tau = \frac{1}{\left(1 + \dfrac{j\omega m}{2\rho c}\right)^2}. \tag{11.4a}$$

Notice that the radiation impedance of the wall ρc appears as a damping term. At high frequencies $(\omega m/2\rho c) \gg 1$.

Hence $$\tau = \left(\frac{2\rho c}{\omega m}\right)^2 \tag{11.4b}$$

and the system is mass controlled.

11.2. The Direct Transmission of Airborne Sound Through a Sprung Wall

In practice, walls are made of materials which exert stiffness and viscous damping. Our model is more realistic if the wall is imagined to be restrained by a spring–damping system (Fig. 11.3). The resultant equation of motion is now

$$2p_i = m\ddot{\phi} + (2\rho c + r)\dot{\phi} + k\phi.$$

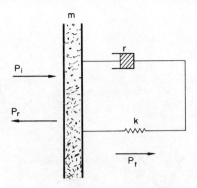

FIG. 11.3. Spring-damper model of wall.

Solving in a similar manner as previously,

$$\tau = \frac{(2\rho c)^2}{(r + 2\rho c)^2 + \{\omega m - (k/\omega)\}^2}.$$ (11.5)

At very low frequencies $k/\omega \gg \omega m$ and also $\gg (r + 2\rho c)$, hence the system is stiffness controlled $\{\tau = (2\rho c \omega/k)^2\}$; mass control is evident at high frequencies when $\tau = (2\rho c/\omega m)^2$, and resonance occurs when $\omega m = k/\omega$ and the damping term dominates the acoustic transmission.

11.3. Resonance

A panel will resonate at certain frequencies depending on its size, mass, stiffness and edge conditions. For a thin rectangular panel, size $a \times b$ and thickness h (each in metres), supported at its four edges, the resonant frequencies are given by

$$f_{m,n} = \frac{\pi}{2} \left\{ \frac{Yh^2}{\rho_p\, 12(1 - \sigma^2)} \right\}^{1/2} \left(\frac{m^2}{a^2} + \frac{n^2}{b^2} \right) \text{Hz.}$$ (11.6a)

For example, glass has a value for Young's modulus of elasticity Y of 7×10^{10} Pa, and a density ρ_p equal to $2 \cdot 5 \times 10^3$ kg/m^3, the Poisson's ratio σ being $0 \cdot 22$, hence

$$f_{m,n} = 2 \cdot 5 \times 10^3 h \left(\frac{m^2}{a^2} + \frac{n^2}{b^2} \right).$$ (11.6b)

Double-wall systems treat the intervening air space as a spring vibrated by the masses on either side of it. For a double wall in which one of the walls is not heavier than the other wall by a multiple, the resonant frequency is

$$f = \frac{600}{\sqrt{\dfrac{10\, m_1 m_2 d}{(m_1 + m_2)}}},$$ (11.6c)

m_1 and m_2 are in kg/m^2 units whereas the interspacing d is in millimetres.
If $m_1 = m_2$,

$$f = \frac{850}{\sqrt{10\, md}}.$$ (11.6d)

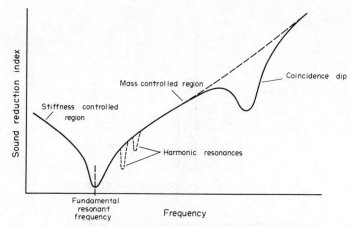

(a) Theoretical sound reduction index–frequency curve.

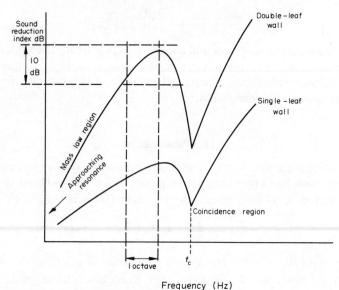

(b) Comparison of shapes of typical sound insulation curves for single- and double-leaf walls.

FIG. 11.4. Sound reduction characteristics of structures.

If a limit is set for the resonant frequency then combinations of m_1, m_2 and d can be estimated to achieve this. Note that for a light wall consisting of two panels, the area weight and the elasticity of the two panels should differ so that both panels obtain a variable critical frequency of the wave coincidence effect; coincidence is discussed in section 11.4.1.

11.4. Sound Reduction Index

The acoustic transmission characteristics of building materials is usually quoted in terms of the *sound reduction index* R (formerly called the *transmission loss* TL), where R is defined as the difference between the sound pressure level† measured on the incident side and that on the transmitted side of the wall

$$R = 10 \lg \left(\frac{|p_i|^2}{|p_0|^2} \right) - 10 \lg \left(\frac{|p_t|^2}{|p_0|^2} \right),$$

$$R = 10 \lg \left(\frac{|p_i|^2}{|p_t|^2} \right) \text{ dB.}$$

Hence
$$R = -\lg \tau. \tag{11.7}$$

Using equation (11.5),

$$R = 20 \lg \left[\frac{\{(r + 2\rho c)^2 + (\omega m - k/\omega)^2\}^{1/2}}{2\rho c} \right]. \tag{11.8}$$

If $2\rho c \gg r$,

$$R = 10 \lg \left\{ 1 + \frac{(\omega^2 m - k)^2}{(2\rho c \omega)^2} \right\}. \tag{11.9}$$

Equation (11.9) is shown diagrammatically in Fig. 11.4a.
At low frequencies R is stiffness controlled and

$$R = 20 \lg \left(\frac{k}{2\rho c \omega} \right). \tag{11.10}$$

At high frequencies the *mass law* applies when

$$R = 20 \lg \left(\frac{\omega m}{2\rho c} \right) \tag{11.11}$$

and a doubling in mass or frequency will increase the insulation by about 6 dB (see Fig. 11.4). In practice the relationship between R and m is given by

$$R = 10 + 14 \cdot 5 \lg m, \tag{11.12}$$

where m is the surface density in kg/m²; this gives an increase in R of 4·4 dB when m is doubled. Besides the effect of *resonance* and also *flanking transmissions*, deviations in the mass law occur due to *coincidence, sound leakage, room absorption* and the *degree of randomness in the incident sound field*; these will now be discussed.

11.4.1. COINCIDENCE EFFECT

Thin homogeneous materials that possess low damping can be set into flexural vibration. At certain angles of incidence the bending (or flexural) and incident waves are in phase and the transmission of sound is high. For coincidence the sound wave of wavelength λ, incident upon a surface at angle θ, must match the frequency of the bending wave travelling in the panel having wavelength λ_b (Fig. 11.5), thus

$$\frac{\lambda}{\sin \theta} = \lambda_b,$$

---

†Sound pressure levels were defined in section 2.2.1.

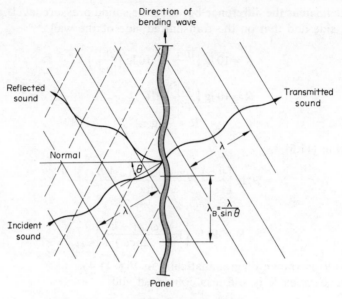

FIG. 11.5. The coincidence effect.

but comparing the velocities of sound for the incident and bending waves $c = f_c \lambda$ and $c_b = f \lambda_b$, (f_c denoting the coincidence frequency), for frequency matching

$$\frac{\lambda}{\lambda_b} = \sin \theta = \frac{c}{c_b}.$$

Now the velocity of sound for a bending wave can be expressed in terms of the elastic properties of the panel and

$$c_b = (2\pi f_c)^{1/2} \left\{ \frac{Yh^2}{12\rho_p (1 - \sigma^2)} \right\}^{1/4} \text{ m/s}, \tag{11.13}$$

where the symbols have their usual meaning (see section 11.3). But $c = c_b \sin \theta$, hence the coincidence frequency is

$$f_c = \frac{c^2}{2\pi \sin^2 \theta} \left\{ \frac{12\rho_p (1 - \sigma^2)}{Yh^2} \right\}^{1/2}. \tag{11.14}$$

Using the constants already given in Section 11.3 for glass, the coincidence frequency for this material is

$$f_c = \frac{12}{h \sin^2 \theta}. \tag{11.15a}$$

The lowest frequency at which coincidence occurs will be for grazing incident sound ($\theta = 90°$),

$$f_c = \frac{12}{h} \tag{11.15b}$$

or, if h is expressed in mm,

$$f_c = \frac{12000}{h} \text{ Hz}. \tag{11.15c}$$

Note that the coincidence frequency can also be expressed in the form

$$f_c = \frac{c^2}{1\cdot 8h \sin^2 \theta}\left(\frac{\rho_p}{Y}\right)^{1/2}, \tag{11.15d}$$

which is useful if values of Poisson's ratio are not to hand. For a homogeneous plate or bar the elastic modulus Y is related to the bending stiffness per unit width B by $B = Yh^3/12$ in N m units.

Cremer (1942) expresses the transmission of an infinite panel in the form

$$\frac{1}{\tau} = \left\{1 + \frac{\omega m_s}{2\rho c}r\left(\frac{f}{f_c}\right)^2 \cos\theta \sin^4\theta\right\}^2 + \frac{\omega^2 m_s^2}{4\rho^2 c^2}\cos^2\theta\left\{1 - \left(\frac{f}{f_c}\right)^2 \sin^4\theta\right\}^2, \tag{11.16}$$

where m_s is the surface mass kg/m^2 (note $\rho_p h = m_s$) and r is the damping factor. When $f \ll f_c$, then the sound reduction index is

$$R = 10 \lg\left\{1 + \left(\frac{\omega m_s \cos\theta}{2\rho c}\right)^2\right\}, \tag{11.17}$$

and hence the theoretical mass law appears as already derived which is not dependent on damping. When $f = f_c$,

$$\frac{1}{\tau} = \left(1 + \frac{\omega m_s r}{2\rho c}\cos\theta \sin^4\theta\right)^2 + \frac{\omega m_s^2 \cos^2\theta}{4\rho^2 c^2}(1 - \sin^4\theta)^2 \tag{11.18}$$

and the transmission depends on r and θ, $1/\tau$ being a maximum at normal incidence $(\theta = 0°)$ when

$$\frac{1}{\tau} = 1 + \frac{\omega m_s^2}{4\rho^2 c^2} \tag{11.19}$$

and a minimum (or maximum transmission) at grazing incidence $(\theta = 90°)$ when

$$\frac{1}{\tau} = 1.$$

In between $0° < \theta < 90°$ damping helps to limit the effect of coincidence. Table 11.1 shows coincidence frequencies for some materials.

TABLE 11.1. APPROXIMATE CRITICAL FREQUENCIES

Material	Thickness (mm)	Surface mass (kg/m^2)	Critical frequency (Hz)
Brick	215	400	100
Brick	102·5	200	200
Lead	18	200	15 000
Plasterboard	10	9	2900–4500
Plasterboard	20	18	1400–2300
Glassfibre reinforced gypsum	10	18	2000
Plywood	10	7·5	1300
Steel	3	25	4000
Glass (grazing incidence)	24		500
	12		1000
	6		2000

In practice coincidence frequencies which fall in speech frequency range 500–3000 Hz are important and are prevalent where structures have surface masses in the range 30–100 kg/m². They should be avoided in the case of light partitions (~ 30 kg/m²) by making the partition less stiff, or decreasing the partition thickness, and consequently increasing the coincidence frequency: for heavier walls (~ 100 kg/m²) reduce the coincidence frequency to under 100 Hz by stiffening. This decrease in the sound reduction index applies to a limited range of frequencies at a given angle of incidence.

11.4.2. LEAKAGE OF SOUND THROUGH HOLES AND CRACKS

The maximum sound that can be expected for a partition of surface area S, sound transmission τ, having a hole of crack area S_0 and sound transmission τ_0, may be found by considering the resultant transmission coefficient τ_R since

$$(S + S_0)\tau_R = S\tau + S_0\tau_0.$$

The resultant sound reduction index will be

$$R = 10 \lg \frac{1}{\tau_R}.$$

Hence
$$R = 10 \lg \left\{ \frac{(S + S_0)}{(S_0\tau_0 + S\tau)} \right\}. \tag{11.20}$$

For the hole or crack $\tau_0 = 1$, but for maximum sound reduction (i.e. minimum sound transmission), $S_0 = 0$, $\tau_0 = 0$ and $\tau \to 0$. In the case $S_0 \neq 0$, $\tau_0 = 1$, $\tau = 0$ and assuming $S/S_0 \gg 1$,

$$R_{max} = 10 \lg \left(\frac{S}{S_0} \right). \tag{11.21}$$

Westerberg (1971) has investigated the effect of holes on the sound insulation of walls. For small holes and slits in thick walls the sound transmitted is a maximum at resonance frequencies determined by

$$n\frac{\lambda}{2} = t \tag{11.22}$$

for a wall thickness t. For circular holes the first resonance occurs at

$$f = \frac{c}{2(t + 2\alpha)}, \tag{11.23}$$

where α is the acoustic length correction for one end of an open pipe and varies between 0·35 and 0·41 for the cases considered by Westerberg. Figure 11.6 shows how the sound reduction index of a 150 mm concrete wall deteriorates as the hole diameter increases. Gomperts (1968), Gösele (1969), Ingerslev and Nielsen (1944a, b), Fasold (1963), Kihlman (1963) and Sauter and Soraka (1970) report more measurements made on the effect of holes in walls on sound transmission.

Sealants are available which can easily be applied to the joints between room surfaces.

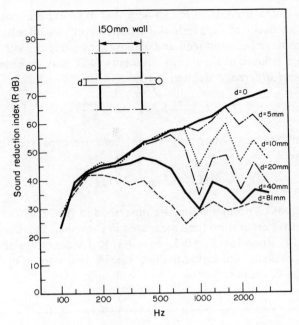

FIG. 11.6. Sound reduction index of the wall with and without one open circular hole of varying diameter. $R = 54$ (walls without leaks), 52, 49, 45, 39 and 34 dB with increasing diameter of the hole (Westerberg, 1971).

11.4.3. THE EFFECT OF ROOM ABSORPTION ON THE SOUND FIELD

The sound energy in the source room ϵ_s is distributed via the direct transmission path (transmission coefficient τ, area S), reflection and absorption at the room surfaces (reflection coefficient r, absorption coefficient α), and by indirect paths (flanking transmission coefficient β), thus

$$\alpha + r + \tau + \beta = 1.$$

The sound energy in the receiving room ϵ_R will be the resultant of that due to τ and β, the sound level will be modified by the total absorption of the receiver room surfaces A. Hence

$$A\epsilon_R = S\epsilon_s\tau + \beta\epsilon_s.$$

Neglecting β

$$\frac{\epsilon_R}{\epsilon_S} = \frac{S\tau}{A},$$

taking logs

$$10 \lg\left(\frac{\epsilon_R}{\epsilon_S}\right) = 10 \lg S + 10 \lg \tau - \lg A.$$

Now $R = 10 \lg(1/\tau)$ and expressing the energy level in terms of sound pressure levels, the sound insulation will be

$$L_{p,S} - L_{p,R} = R - 10 \lg \frac{S}{A}. \tag{11.24}$$

If $L_{p,s}$ and $L_{p,R}$ are measured, then R can be found in an experimental set-up where A is known. The accuracy of the method will depend on how effectively the flanking transmission paths have been reduced and on how accurately the absorption value of the receiving room is known. Field measurements of R may be compared by using the *normalised level difference* defined as

$$D_N = (L_{p,S} - L_{p,R}) + 10 \lg \left(\frac{A_0}{A}\right), \tag{11.25}$$

where A_0 is $10\,\text{m}^2$ Sabines. Alternatively, D_N may be expressed in terms of the reverberation time

$$D_N = (L_{p,s} - L_{p,R}) + 10 \lg \left(\frac{T}{T_0}\right). \tag{11.26}$$

These two methods of normalisation are described in BS 2750. T_0 is taken as $0\cdot5$ s based on values of reverberation time measured in furnished living rooms. Note that since $A_0 = 0\cdot16V/T_0$, then for $T_0 = 0\cdot5$, $A_0 = 0\cdot32V$. Evidence for these figures has been obtained by Jackson and Leventhall (1972). In their survey of fifty rooms the following data was collected:

	Range of volume (m³)	Average volume (m³)	Reverberation time (s)		
			125 Hz	1000 Hz	8000 Hz
Living rooms	25–76	44	0·69	0·51	0·40
Kitchens	8–78	23	0·76	0·68	0·61

11.4.4. RANDOM INCIDENT SOUND FIELD

For a sound field incident at angle θ, equation (11.9) needs modifying to

$$R_\theta = 10 \lg \left\{1 + \frac{(\omega^2 m - k)^2 \cos^2 \theta}{(2\rho c \omega)^2}\right\}.$$

Over the mass law range,

$$R_\theta = 10 \lg \left\{1 + \left(\frac{\omega m \cos \theta}{2\rho c}\right)^2\right\}. \tag{11.27}$$

For normal incident sound when $\theta = 0°$ the sound reduction index is a maximum and

$$R = 10 \lg \left\{1 + \left(\frac{\omega m}{2\rho c}\right)^2\right\}.$$

For random incident sound R is approximately

$$R_r = R - 10 \lg (0\cdot23R). \tag{11.28}$$

In practice sound is not incident at all angles the limiting angle being about 80°, thus

$$R_r = R - 5 \text{ (dB)}. \tag{11.29}$$

11.5. Multi-layer Structures

Sound insulation depends on mass, stiffness and damping. In recent years much research has been orientated towards finding efficient sound-insulating structures that have low mass, hence making the structure easier to handle but having values of stiffness and damping which compensate for any loss in insulation due to a decrease in mass. The use of laminated or sandwiched airgap structures has become common in building design today.

Multi-layer structures present problems. The resonance of the structure can be amplified if coupling occurs between the resonant frequency of each panel; air cavities can house standing waves; coupling of coincidence frequencies gives an increased coincidence effect; two-dimensional standing waves can occur in the plane of the panel. These effects can be avoided by careful design and attention to the use of absorbent materials in cavities, layering structures of different thicknesses, and the fixing of damping strips at the panel edges.

London (1950) showed the resonant frequency of a double panel can be expressed as

$$f = \frac{1}{2\pi \cos \theta} \left(\frac{2\rho c^2}{md}\right)^{1/2}.$$ (11.30)

where m is the total mass of the panels in kg/m^2 and d is the cavity width in m. If the panels have one layer of mass m_1 and the other layer of mass m_2,

$$f = \frac{c}{\pi \cos \theta} \left\{\frac{\rho}{(m_1 + m_2)d}\right\}^{1/2}.$$ (11.31)

For normal incident sound onto a double glass panel,

$$f = 120 \left\{\frac{1}{(m_1 + m_2)d}\right\}^{1/2}.$$ (11.32)

Cavity resonance will occur when standing waves are formed according to

$$f_n = \frac{nc}{2d \cos \theta}.$$ (11.33)

Two-dimensional standing waves will act in the plane of the panel if

$$f_{m,n} = \frac{c}{2}\left\{\left(\frac{m}{a}\right)^2 + \left(\frac{n}{b}\right)^2\right\}^{1/2}$$ (11.34)

is valid when the wavelength of the sound $> 2d$.

Work by Mulholland et al. (1968), Crocker and Price (1969) and Sharp and Beauchamp (1969) has attempted to predict the sound reduction index of multi-layer structures. Sharp and Beauchamp (1969) show that the transmission coefficient τ and the reflection coefficient r for the composite structure can be expressed in the matrix form

$$\binom{\tau}{0} = C\binom{1}{r},$$ (11.35)

where the matrix element C is given by

$$C = \prod_{i=1}^{N} \frac{1}{\tau_i} C^{(i)}$$

for an N-layered structure. For a two-layer structure the element $C^{(i)}$ is

$$C^{(i)} = \begin{pmatrix} C_{11} & C_{12} \\ C_{21} & C_{22} \end{pmatrix}, \tag{11.36}$$

whereas a three-layer structure will have

$$C^{(i)} = \begin{pmatrix} C_{11} & C_{12} & C_{13} \\ C_{21} & C_{22} & C_{23} \\ C_{31} & C_{32} & C_{33} \end{pmatrix} \tag{11.37}$$

The element C links τ and r with the elastic properties of the panel. Equation 11.35 has been solved by computer over a range of incidence angles and the sound reduction index calculated. Errors become significant when the airgap is large. The theory has been extended to treat laminated structures that have no slip between the individual layers. A comparison between theory and practice is shown in Fig. 11.7.

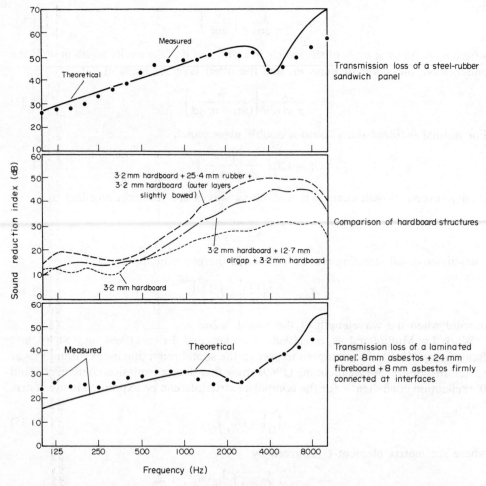

FIG. 11.7. Theoretical and empirical comparisons of sound transmission through structures (Sharp and Beauchamp, 1969).

11.6. Examples

1. Derive an expression for the sound reduction index R of the test panel in the diagram assuming a constant and uniform sound energy (W) distribution and no flanking transmission.

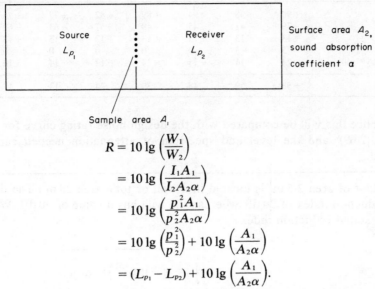

Sample area A_1

$$R = 10 \lg \left(\frac{W_1}{W_2} \right)$$

$$= 10 \lg \left(\frac{I_1 A_1}{I_2 A_2 \alpha} \right)$$

$$= 10 \lg \left(\frac{p_1^2 A_1}{p_2^2 A_2 \alpha} \right)$$

$$= 10 \lg \left(\frac{p_1^2}{p_2^2} \right) + 10 \lg \left(\frac{A_1}{A_2 \alpha} \right)$$

$$= (L_{p_1} - L_{p_2}) + 10 \lg \left(\frac{A_1}{A_2 \alpha} \right).$$

Note that in practice $T_2 = 0 \cdot 16 V_2 / A_2 \alpha$ and the receiver room is designed to have a given reverberation time and is calibrated. In practice flanking transmission paths deteriorate the design value of the sound reduction index for the partition, e.g. sound transmitted via suspended ceilings.

In general, speech will be clearly heard through a 30 dB partition, remains intelligible through a 40 dB one, is unintelligible through a 50 dB one and inaudible through a 60 dB one. People tend to speak about 5–10 dB higher when on the telephone or when there is a high background level in a space, e.g. several other people talking. Remember that there is a high personal variation not only in voice sound power level but also in the peakiness of the sound frequency spectrum.

Flanking paths must be avoided. In practice partitions may have an adequate sound reduction index, but the effectiveness of this can be defeated by an inadequate suspended ceiling for instance. Suspended ceilings should have a minimum mass of 30 kg/m².

---

2. A committee room $5 \times 7 \times 4$ m adjoins a plant-room having a 230 mm plastered brick wall of area 5×4 m. Find the resulting sound pressure level in the committee room L_{p_2} if the sound pressure level in the plant-room is L_{p_1}.
Given:

Frequency (Hz):	63	125	250	500	1000	2000	4000	8000
T (seconds) in committee room	2	2	1	1	1	1	1	1
R (brickwork) (dB)	39	41	45	49	57	63	62	65
L_{p_1} (plant-room) (dB)	90	98	94	89	82	77	75	70

Solution

Using $L_{p_2} = L_{p_1} - R + 10 \lg S + 10 \lg (T/0\cdot16 V)$, where $S = 20\,\text{m}^2$ and $V = 140\,\text{m}^3$ (see equations (11.24), (11.25) and (11.26)):

$+ L_{p_1}$	+ 90	+ 98	+ 94	+ 89	+ 82	+ 77	+ 75	+ 70
$- R$	− 39	− 41	− 45	− 49	− 57	− 63	− 62	− 65
$+ 10 \lg 20$	+ 13	+ 13	+ 13	+ 13	+ 13	+ 13	+ 13	+ 13
$+ 10 \lg T$	+ 3	+ 3	0	0	0	0	0	0
$- 10 \lg (0\cdot16 \times 140)$	− 14	− 14	− 14	− 14	− 14	− 14	− 14	− 14
L_{p_2}	+ 53	59	48	39	24	13	12	4

In practice this will be compared with the design noise rating curve for the space (see Fig. 2.69), and the level and spectrum of attenuation needed can then be estimated.

---

3. A door of area $2\cdot5\,\text{m}^2$ is located in a wall of total area $28\,\text{m}^2$. The door has a sound reduction index of 20 dB whereas the wall has a value of 30 dB. What is the resultant sound reduction index?

	R	$\tau \left\{ \text{NB: } R = 10 \lg \left(\frac{1}{\tau} \right) \right\}$	Area
Door	20	0·01	2·5
Wall	30	0·001	25·5

$$\text{Resultant } R = 10 \lg \left\{ \frac{28}{(2\cdot5 \times 0\cdot01) + (25\cdot5 \times 0\cdot001)} \right\},$$

$$R = 27\cdot47\,\text{dB}.$$

---

4. The space under a solid door, area A, is 1% of the total door area. If the sound level outside the room is 90 dBA, find the sound level inside the room with the door closed. Assume the door does not transmit any sound ($\tau = 0$).

If τ for the crack is 1, then

$$R = 10 \lg \left\{ \frac{A + 0\cdot01A}{(0 \times A) + (1 \times 0\cdot1A)} \right\},$$

$$R = 20\,\text{dB};$$

hence the sound level in the room is 70 dB. Note repeating the calculation for a crack of $0\cdot1\%A$ gives $R = 30\,\text{dB}$.

For a building with an external wall area S and a glazing area S_g, brick area S_b ($S = S_g + S_b$), the effect of glazing with sound transmission τ_g on the resultant sound reduction index R can be calculated from

$$R = 10 \lg \left\{ \frac{1}{\left(\frac{S_g}{S} \right) \tau_g + \left(\frac{S_b}{S} \right) \tau_b} \right\}.$$

LIST OF BRITISH STANDARDS AND CODES OF PRACTICE RELATING TO SOUND AND VIBRATION

THE author wishes to extend grateful acknowledgement to Dr. Martin V. Lowson, formerly Rolls-Royce Reader in Fluid Mechanics, Department of Transport Technology, Loughborough University of Technology and now Chief Research Scientist at Westland Aircraft Co., who prepared this list.

BS 415:1967	Safety requirements for mains-operated sound and vision equipment (substantially agrees with IEC 65)
BS 661:1969*	Glossary of acoustical terms
BS 848:1968*	Fan noise testing
BS 880:1950	Musical pitch (substantially agrees with ISO/R 16); amendment PD 6033 February 1967
BS 1568:1960	Magnetic tape sound recording and reproduction; amendment PD 5962 December 1966 (IEC 94)
BS 1927:1953	Dimensions of circular cone diaphragm loud-speakers
BS 1928:1965	Processed disc records and reproducing equipment
BS 1988:1953	Measurement of frequency variation in sound recording and reproduction
BS 2042:1953	An artificial ear for the calibration of earphones of the external type; amendment PD 1795 January, 1954
BS 2475:1964	Octave and one-third octave band-pass filters; amendment PD 5536 May 1965 (substantially agrees with IEC 225)
BS 2497:—	A reference zero for the calibration of pure-tone audiometers 2497 Pt. 1 1968 Data for earphone coupler combinations maintained at certain standardising laboratories 2497 Pt. II 1969 Data for certain earphones used in commercial practice
BS 2498:1954	Recommendations for ascertaining and expressing the performance of loudspeakers by objective measurements
BS 2750:1956*	Recommendations for field and laboratory measurement of airborne and impact sound transmission in buildings; amendment PD 5065 October 1963 (substantially agrees with ISO/R 140)
BS 2829:1957	35 mm magnetic sound recording azimuth alignment test films; amendment PD 2766 April 1957
BS 2831:1971*	Methods of test of air filters used in airconditioning and general ventilation (relevant to flow resistance measurements)
BS 2980:1958	Pure-tone audiometers
BS 2981:1958	The dimensional features of magnetic sound recording on perforated film; amendments PD 4788 January 1963, PD 4977 July 1963, PD 5104 December 1963; amendment 388 December 1969; (substantially agrees with ISO/R 162)
BS 3015:1958	Glossary of terms used in vibration and shock testing
BS 3045:1958	The relation between the sone scale of loudness and the phon scale of loudness level (agrees with ISO/R 131)
BS 3154:1959	Frequency characteristics for magnetic sound recording on film; amendment 389 December 1969

* Publications of particular interest to the building designer.

BS 3171:1968	Methods of test of air conduction hearing aids (differs from IEC 126)
BS 3383:1961	Normal equal-loudness contours for pure tones and normal threshold of hearing under free-field listening conditions (substantially agrees with ISO/R 226)
BS 3425:1966*	Method of the measurement of noise emitted by motor vehicles; amendment PD 6024 February 1967, (substantially agrees with ISO/R 362)
BS 3489:1962*	Sound level meters (industrial grade); amendment PD 4825 February 1963 (substantially agrees with IEC 123)
BS 3499:	School Music Equipment

3499: Part 1: 1962 Percussion Instruments (amendments PD 4646 September 1962; PD 5058 October 1963; PD 5981 December 1966 and 381 December 1969)

3499: Part 2: Woodwind instruments

3499: Part 2a: 1964 Recorders

3499: Part 2b: 1968 Clarinets

3499: Part 3a: 1965 Trumpet in B. Flat, amendment 192 January 1969

3499: Part 4: Keyboard instruments

    Part 4a: 1964 Pianofortes: amendment PD 5884 July 1966 PD 5273 How to look after a piano

3499: Part 5: 1964 Recommendations to purchasers of stringed instruments, amendment 318 September 1969

    PD 5286: 1964 Care and maintenance of stringed instruments

3499: Part 6: 1968 Methods of test for musical instruments

3499: Part 7a: 1963 Music stands: amendments PD 5885 July 1966; PD 6253 September 1967

3499: Part 7b: 1963 Percussion instrument trolleys; amendment PD 6254 September 1967

3499: Part 8a: 1966 Record players, VHF tuners, amplifiers and loudspeakers: amendments PD 6025 February 1967; PD 6377 April 1968

3499: Part 8b: 1969 Magnetic tape recording and reproducing equipment

3499: Part 9: Electronic musical instruments

    Part 9a: 1966 Recommendations to purchasers of electric guitars

    Part 9b: 1966 Amplifiers suitable for use with musical instruments; amendment 164 December 1968

    Part 9c: 1969 Electronic organs with pedal boards

BS 3539:1962	Sound level meters for the measurement of noise emitted by motor vehicles: amendment AMD 22 July 1968
BS 3593:1963*	Recommendation on preferred frequencies for acoustical measurements (agrees with ISO/R 266)
BS 3638:1963*	Method for the measurement of sound absorption coefficients (ISO) in a reverberation room: amendment PD 5127 January 1964 (agrees with ISO/R 354)
BS 3675:1963	Performance of 16 mm portable sound-and-picture cinematograph projectors
BS 3860:1965	Methods for measuring and expressing the performance of audio-frequency amplifiers for domestic, public address and similar applications
BS 3942:1965	Markings of control settings on hearing aids (substantially agrees with IEC 224)
BS 4009:1966	An artificial mastoid for the calibration of bone vibrators used in hearing aids and audiometers
BS 4054:1966	Methods for measuring and expressing the performance of radio receivers; receivers for a.m. and f.m. sound broadcast transmissions
BS 4078:1966	Cartridge operated fixing tools (recoil noise and expelled gases)
BS 4142:1967*	Method of rating industrial noise affecting mixed residential and industrial areas
BS 4196:1967*	Guide to the selection of methods of measuring noise emitted by machinery (ISO/R 495)
BS 4197:1967*	Precision sound level meters (IEC 179)
BS 4198:1967*	Method for calculating loudness (agrees with ISO/R 532)
BS 4297:1968	Characteristics and performance of a peak programme meter
BS 4298:1968	Dimensional features of six-track magnetic sound recording on 70 mm release prints
BS 4527:1969	Dimensional features of magnetic sound recording on one to four tracks on 35 mm release prints (supersedes part of BS 2981:1958)
BS 4528:1969	Dimensional features of single track magnetic sound recording on 35 mm motion picture film (supersedes part of BS 2981: 1958)
BS 5429:1969	Dimensional features of magnetic sound recording on 8 mm (Type S) picture film
BS 4530:1969	Frequency characteristics for magnetic sound recording on 8 mm (Type S) perforated film

BS 4599:1970   Dimensional features of optical sound recording on 8 mm (Type S) motion picture film with picture.

BS 4612:1970   An identification code for synchronised 6·25 mm audiotape used in conjunction with motion picture film.

BS 4668:1971   An acoustic coupler (IEC reference type) for calibration of earphones used in audiometry (metric units).

BS 4669:1971   An artificial ear of the wide-band type for the calibration of earphones used in audiometry (metric units).

BS 4675:1971*   Recommendations for a basis for comparative evaluation of vibration in machinery.

BS 4718:1971*   Methods of test for silencers for air distribution systems

BS 5228: 1975   Code of Practice for noise control on construction and demolition sites.

BS CP 3:1960*   Chapter III Sound insulation and noise reduction (see 1972 revision).

BS CP 327:1964*   Part 3: Sound distribution systems.

BS CP 352: 1968*   Mechanical ventilation and airconditioning in buildings—(contains a section on sound-proofing and anti-vibration devices).

BS CP 1972*   Reducing the exposure of employed persons to noise.

NOTE: ISO/R   Recommendations issued by the International Organization for Standardization.

IEC   Recommendation issued by the International Electrotechnical Commission.

HEALTH AND SAFETY AT WORK

THE following booklets on health and safety at work have been written by the Department of Employment and the HM Factory Inspectorate and are published by HMSO. Booklet 25 is specifically concerned with the problem of noise and its effect on the health of the factory worker but has been superseded by the Code of Practice for Reducing the Exposure of Employed Persons to Noise obtainable from the HMSO.

1. Lifting and carrying
2. Canteens, messrooms and refreshment services
3. Safety devices for hand and foot operated presses
4. Safety in the use of abrasive wheels
5. Cloakroom accommodation and washing facilities
6A Safety in construction work: general site safety practice
6B Safety in construction work: roofing
6C Safety in construction work: excavations
6D Safety in construction work: scaffolding
6E Safety in construction work: demolitions
6F Safety in construction work: system building
9. Safety in the use of machinery in bakeries
10. Fire fighting in factories
11. Guarding of hand-fed platen machines
12. Safety at drop forging hammers
13. Ionising radiations: precautions for industrial users
14. Safety in the use of mechanical power presses
15. Dry cleaning plant: precautions against solvent risks
16. The structural requirements of the Factories Act
17. Improving the foundry environment
18. Industrial dermatitis: precautionary measures
19. Safety in laundries
20. Drilling machines: guarding of spindles and attachments
21. Organisation of industrial health services
22. Dust explosions in factories
23. Hours of employment of women and young persons
24. Electrical limit switches and their applications
25. Noise and the worker
27. Precautions in the use of nitrate salt baths
28. Plant and machinery maintenance
29. Carbon monoxide poisoning: causes and prevention
30. The storage of liquefied petroleum gas at factories
31. Safety in electrical testing
32. Repair of drums and small tanks: explosion and fire risk
33. Safety in the use of guillotines and shears
34. Guide to the use of flame arresters and explosion reliefs
35. Basic rules for safety and health at work

36. First aid in factories
37. Liquid chlorine
38. Electric arc welding
39. Lighting in offices, shops and railway premises
40. Means of escape in case of fire in offices, shops and railway premises
41. Safety in the use of woodworking machinery
42. Guarding of cutters of horizontal milling machines
43. Safety in mechanical handling
44. Asbestos: health precautions in industry
45. Seats for workers in factories, offices and shops
46. Evaporating and other ovens
47. Safety in the stacking of materials
48. First aid in offices, shops and railway premises

APPENDIX III

SELECTED REFERENCE BIBLIOGRAPHY

Report of the Wilson Committee on the problem of noise, HMSO, London, 1963.

A. G. ALDERSEY-WILLIAMS, *Noise in factories*, Factory Building Studies, No. 6, HMSO, London, 1960.

P. H. PARKIN and H. R. HUMPHREYS, *Acoustics, noise and buildings* (1958), second edition in SI units, Faber & Faber, London, 1968.

C. M. HARRIS (editor), *Handbook of noise control*, McGraw-Hill, New York, Toronto and London, 1957.

C. M. HARRIS and C. E. CREDE (editors), *Shock and vibration handbook*, 3 volumes, McGraw-Hill, 1961.

NATIONAL PHYSICAL LABORATORY, *Symposium No. 12—The control of noise measurement*, McGraw-Hill, USA, 1957.

INDUSTRIAL WELFARE SOCIETY, *Industrial noise*, Industrial Welfare Society, London, 1961.

Guide for conservation of hearing in noise, American Academy of Ophthalmology and Otolaryngology, 1957.

Industrial noise manual, American Industrial Hygiene Association, Detroit, 1966.

R. S. F. SCHILLING (editor), Modern trends in occupational health, Chapter 17 in *Noise* by W. Burns and T. S. Littler, Butterworths, London, 1960.

INDUSTRIAL HYGIENE FOUNDATION OF AMERICA INC., *An annotated bibliography on noise, its measurement, effect and control*, Mellon Institute, Pittsburgh, 1955.

A. PETERSON and E. GROSS, *Handbook of noise measurement*, General Radio Company (UK) Ltd., Bourne End, Bucks, 1967.

W. BURNS, *Noise and man*, John Murray, London, 1968.

K. D. KRYTER, *The effects of noise on man*, Academic Press, 1970.

R. A. PIESSE, *Ear protectors*, Commonwealth Acoustic Laboratories, Sydney, Australia, Report CAL No. 21, 1962.

A. BELL, *Noise: An occupational hazard and public nuisance*, World Health Organization, Geneva, obtainable from HMSO, London, 1966.

E. N. BAZLEY, *The airborne sound insulation of partitions*, HMSO, London, 1966.

E. J. EVANS and E. N. BAZLEY, *Sound absorbing materials*, HMSO, London, 1960.

J. T. BROCH, *Acoustic noise measurements*, B. & K. Laboratories Ltd., Hounslow, 1971.

W. D. WARD and J. E. FRICKE, *Noise as a public health hazard*, American Speech and Hearing Association, Washington DC, 1969.

W. BURNS and D. W. ROBINSON, *Hearing and noise in industry*, HMSO, London, 1970.

L. L. DOELLE, *Environmental Acoustics*, McGraw-Hill, 1972.

BRITISH OCCUPATIONAL HYGIENE SOCIETY, *Hygiene standard for wide band noise*, Pergamon Press, 1971.

DE, HM FACTORY INSPECTORATE, Technical Data Note No. 12, *Notes for the guidance of designers on the reduction of machinery noise*, obtainable free of charge from any office of HM Factory Inspectorate.

V. O. KNUDSEN and C. M. HARRIS, *Acoustical Designing in Architecture*, chapter 9, John Wiley, 1950.

R. BRUCE LINDSAY, *Acoustics—Historical and Physical Development*, Dowden, Hutchinson & Ross Inc., Stroudsburg, Pennsylvania, 1973.

Institution of Heating and Ventilating Engineers Guide, 1970, Books A (section A1) and B (section B12).

American Society of Heating Refrigeration and Airconditioning Engineers Guide, 1972, Fundamentals Volume, chapter 6, and *1973, Systems Volume*, chapter 35.

L. L. BERANEK, *Noise and Vibration Control*, McGraw-Hill, 1971.

L. L. BERANEK, *Music acoustics and architecture*, John Wiley, 1962.

L. F. YERGES, *Sound, Noise and Vibration Control*, van Nostrand Reinhold Co., 1969.

W. C. SABINE, *Collected papers in acoustics*, Dover.

APPENDIX IV

Noise Advisory Council Publications published by HMSO up to May 1976.

Aircraft noise: Flight routeing near airports 1971
Neighbourhood noise, 1971
Traffic noise: The vehicle regulations and their enforcement, 1972
Aircraft noise: Should the noise and number be revised? 1972
Aircraft noise: Selection of runway sites for Maplin, 1972
A guide to noise units, 1974
Aircraft engine noise research, 1974
Noise in the next ten years, 1974
Aircraft noise: Review of aircraft departure routeing policy, 1974
Noise in Public Places, 1974
Bothered by Noise? 1975
Noise Units, 1975

Noise Advisory Council
  Department of the Environment
  Queen Anne's Chambers
  28 Broadway
  London SW1 H9JU

REFERENCES

NOTE that in 1974 The British Acoustical Society and the Acoustics Group of the Institute of Physics were amalgamated to form the Institute of Acoustics, the present address of which is 47 Belgrave Square, London SW1.

The Heating and Ventilating Research Association in Bracknell, Berkshire has been renamed the Building Services Research and Information Association.

The Chartered Institution of Building Services has now replaced the I.H.V.E (see pages 10 and 26).

ABEY-WICKRAMA, I., *et al.* (1969) *Lancet*, 13 December, 1275–7.
ADRIAN, E. D. (1928) *The Basis of Sensation*, Christophers, London.
ALDERSEY-WILLIAMS, A. G. (1973a) *Lancet*, 13 December, 1275–7.
ALDERSEY-WILLIAMS, A. G. (1973b) Acoustics around the world, British Acoustical Society Conference, London, 15 May.
ALDERSEY-WILLIAMS, A. G. (1973c) Personal communication (see Bains, 1973).
ALEXANDRE, A., BARDE, J. Ph., *et al.* (1975) *Road Traffic Noise*; Applied Science Pub. Ltd.
ALLEN, C. H. (1957) *Noise Control* 3, 1.
ALMIDA, M. R. de (1950) *Archs. Otoloar.* **51**, 215–22.
ANDO, Y., HATTORI, H. (1970) *J. Acoust. Soc. Am.*, **47**, 1128–1130.
ANDO, Y., HATTORI, H. (1973) *J. Sound Vib.*, **27**, 101–110.
ANDREWS, B., and FINCH, D. M. (1957) *Highway Res. Board Proc.* **31**, 456–65.
ANDRIUKIN, A. A. (1961) *Cor vasa* **3**, 285–93.
ANON. (1959) Report 22, 637 and Appendix, General direktion PTT, Switzerland.
ANTHROP, D. F. (1973) *Noise Pollution*, Lexington Books (D. C. Heath & Co.).
ANTONI, N., ROSENBERG, K., and WELIN, A. (1972) Swedish National Institute for Building Research Report R50/69, A. B. Svensk Byggtjänst, Box 1403, 111-84, Stockholm.
ARVIDSSON, O., *et al.* (1965) *Nord. hyg. Tidskr.* **xlvii**, 153–86.
ASHLEY, C. (1971) *Symposium on Environmental Engineering Aspects of Pollution Control, London,* 22–23 June, Society of Environmental Engineers.
ÅSTRAND, P. O., and RODAHL, K. (1970) *Textbook of Work Physiology*, McGraw-Hill, New York, London and Tokyo.
ATAL, B. S., *et al.* (1962) *Proceedings of the Fourth International Congress on Acoustics, Copenhagen,* H31.
ATAL, B. S., SCHROEDER, M. R., and SESSLER, G. M. (1965) *Proceedings of the Fifth International Congress on Acoustics, Liège,* Paper G 32.
ATHERLEY, G. R. C., *et al.* (1970) *Performance under Sub-optimal Conditions*, (ed. Davis), Taylor and Francis, pp. 6–15.
ATTENBOROUGH, K. (1973) Personal communication (Open University).
AYRES, L. P. (1911) *Am. Phys. Educ. Rev.* **16**, 321–4.
BAINS, A. W. (1973) Personal communication, Engineering Design Consultants, London.
BAKAN, J. (1959) *Br. J. Psychol.* **50**, 325–32.
BAKER, J. K. (1974) Chapter 9 in *Noise and Vibration Control for Industrialists* (ed. S. A. Petrusewicz and D. K. Longmore, Elek Science).
BALL, E. F., WEBSTER, C. J. D. (1976) *The Building Services Engineer*, **44**, (2), 33–40.
BARBENZA, C. M. de, *et al.* (1970) *Sound* **4**, 75–79.
BARRON, M. (1971) *J. Sound Vib.*, **15**, 475.
BARTLETT, F. C. (1934) *The Problem of Noise*, Cambridge University Press.
BAUMANN, K. (1974) *Klima-Kälte Ing.* **3** (4) 147–52.
BAZLEY, E. N. (1966) *The Airborne Sound Insulation of Partitions*, National Physical Laboratory, HMSO.
BÉKÉSY, G. (1960) *Experiments in Hearing*, McGraw-Hill.

BÉKÉSY, G. (1964) *Handbook of Experimental Psychology* (ed. S. S. Stevens), chapter 27, Wiley.
BERANEK, L. L. (1947a) *Trans. ASME* **69**, 96–100.
BERANEK, L. L. (1947b) *Proc. Instn. Radio Engrs* **35**, 880–90.
BERANEK, L. L. (1954) *Acoustics*, McGraw-Hill.
BERANEK, L. L. (1956) *J. Acoust. Soc. Am.* **28**, 833–52.
BERANEK, L. L. (1960) *Noise Reduction*, McGraw-Hill.
BERANEK, L. L. (1962) *Music Acoustics and Architecture*, Wiley.
BERANEK, L. L. (1971) *Noise and Vibration Control*, McGraw-Hill.
BERANEK, L. L., SCHULTZ, T. J. (1965) *Acustica*, **15**, 307.
BERLYNE, D. E. (1960) *Conflict, Arousal and Curiosity*, McGraw-Hill.
BERLYNE, D. E., *et al.* (1966) *J. Exp. Psychol.* **72**, 1–6.
BERRY, G. (1971) *First National Symposium on Structural Insulation*, Paper 12, Structural Insulation Association, London.
BINDRA, D. (1959) *Motivation: A Systematic Reinterpretation*, Ronald Press Co. New York.
BISHOP, D. E. (1966) *J. Acoust. Soc. Am.* **40**, 108–22.
BITTER, C., and HORCH, C. (1958) TND Report 25.
BITTER, C., and van WEEREN, P. (1955), Report 24, Research Institute for Public Health Engineering, TNO, The Hague, Holland.
BLAZIER, W. E. (1972) *Am. Soc. Heat. Refrig. Air-condit. Engrs. J.* **14** (5) 44–50.
BLITZ, J. (1973) *Br. Acoust. Soc. Proc.* **2** (3); Institute of Acoustics Meeting, *Urban Noise Measurement and Evaluation*, Winchester, November 13–14.
BOALT, C. (1965) *Statens Inst. för Byggnadsforskning, Stockholm*, Report 21.
BOGGS, D. H., and SIMON, J. R. (1968) *J. Appl. Psychol.* **52**, 148–53.
BOJE, A. (1971) *Open-plan Offices*, Business Books Ltd.
BOMMES, L. (1961) Geräuschenwicklung bei Ventilatoren kleiner und mittlerer Umfangsgechwindigkeit, *Lärmbekampfung* **5** (5) 69.
BOMMES, L. (1968) *Trans. Am. Soc. Heat. Refrig. Air-condit. Engrs*, No 2057, 74.
BORG, E., and MØLLER, A. R. (1973) Vara omedvetna reaktioner pa buller, (Our unconscious reactions to noise), *Forsk. framsteg* **7**, 5–9.
BORING, E. G. (1942) *Sensation and Perception in the History of Experimental Psychology*, Appleton-Century, New York.
BORING, E. G., and STEVENS, S. S. (1936) *Proc. Natn. Acad. Sci. USA* **22**, 514–21.
BORSKY, P. N. (1964) *J. Sound Vibr.* **2** (4) 527.
BOTTOM, C. G. (1970) *Unification of noise criteria*, British Acoustical Society Meeting, 23 July.
BOTTOM, C. G., and CROOME, D. J. (1969) *Appl. Acoust.* **2**, 279–96.
BOWMAN, N. (1973) Personal communication and Ph.D thesis (London University).
BOYCE, P. R. (1974) Electricity Council Research Centre Report No. ECRC/M718, Job No. 4104.
BRANDT, O. (1954) *Tek. tidsk.* 1129.
BREGMAN, H. L., *et al.* (1972) *Development of a Noise Annoyance Sensitivity Scale*, NASACR-1954, National Aeronautics and Space Administration, Washington DC, USA.
BRIFFA, F. E. J., and KENNEDY, D. R. (1966–7) *Proc. Instn. Mech. Engrs.*, **181** (3c) 62–72.
BRIFFA, F. E. J., *et al.* (1973) *J. Inst. Fuel*, **xlvi** (5) 207–16.
BROADBENT, D. E. (1954) *Q. Jl. Exp. Psychol.* **6**, 1.
BROADBENT, D. E. (1957) Chapter 10, Effects of noise on behaviour, in *Handbook of Noise Control* (ed. by C. M. Harris), McGraw-Hill.
BROADBENT, D. E. (1958a) *Perception and Communication*, Pergamon.
BROADBENT, D. E. (1958b) *J. Acoust. Soc. Am.* **30**, 824–7.
BROADBENT, D. E. (1960) *Occup. Psychol.* **34**, 133–40.
BROADBENT, D. E. (1965) *Human Factors* **1**, 155.
BROADBENT, D. E. (1966) *New Society* 12–14, 3 March.
BROADBENT, D. E. (1971) Chapter 9, Noise and other stresses, in *Decision and Stress*, Academic Press, London.
BROOKES, M. J. (1972) *Appl. Ergonomics* **3**, (4), 224–36.
BROSIO, E. (1968) *Transmission Loss in the Proximity of a Partition Wall*, Instituto Electtrotecnico Nazionale Galileo Ferravis, Turin.
BROWN, R. L., *et al.* (1965) *Percept. Motor Skills* **20**, 794–854.
BRUCKMEYER, F., and LANG, J. (1967) *Österreichische Ing.-Z.* **10** (8) (9) and (10) 302–6, 338–44, 376–85 respectively.
BRUCKMEYER, F., and LANG, J. (1968) *Österreichische Ing. Z.* **11** (3) 73–77.
BRUNDRETT, G. W. (1973) Electricity Council Research Centre Report No. ECRC/M584, Job No. 019.
BRYAN, M. E. (1973a) *Physics Bull.* 89–91, February.
BRYAN, M. E. (1973b) *Noise and Loudness Evaluation*, British Acoustical Society Meeting, University of Salford, 25–26 October; also see *Br. Acoust. Soc. Proc.* **2** (3).

BRYAN, M. E. (1974) *Ergonomics* **17** (4) 138.
BRYAN, M. E., and TEMPEST, W. (1973) *Appl. Acoust.* **6**, 219–31.
BUGARD, P. (1951) Action biologique de bruits complexes de niveau éléve, *CR Soc. Biol. Paris* **145**, 807.
BURD, A. N., *et al.* (1966) *BBC Engineering Monograph* 64.
BURD, A. N., *et al.* (1968) *BBC Engineering Monograph* 73.
BURGTORF, W. (1961) *Acustica* **11**, 97.
BURGTORF, W., and OEHLSCHLÄGEL, H. K. (1964) *Acustica* **14**, 254.
BURGTORF, W., and WAGENER, B. (1967/8) *Acustica* **19**, 72.
BURKE, R. F., *et al.* (1969) *Elect. Rev.* 269–73, 21 February.
BURNS, W. (1968) *Noise and Man*, Murray.
BUTCHER, N. F. H. (1938) A comparative study of the effects of continuous and intermittent auditory distraction on the mental output of school children, MA thesis, University of London.
CANAC, R., and BLADIER, B. (1954) Sur la nuisance du bruit décelée par les potentiels recueillis en dérivation bi-occipitale, *CR Acad. Sci. Paris* **239**, 1313.
CANNON, W. B. (1929) *Bodily Changes in Pain, Hunger, Fear and Rage; an account of recent researches into the function of emotional excitement*, 2nd edn., New York.
CANTER, D. (1968) *Trans. Bartlett Soc.* **6**, 616.
CANTER, D. (1969) *Archs Assoc. Q.* **1** (2).
CANTER, D. (1974) *Psychology for Architects*, published by Applied Science.
CARLSON, J. A., *et al.* (1967) *Psychol. Reps.*, **20** 1021–2.
CARTER, F. W. (1932) *Engineering*, 4 and 11 November.
CAVANAUGH, W. J., *et al.* (1962) *J. Acoust. Soc. Am.* **34** (4) 475–92.
CHADDOCK, J. B. (1957) Report TIR-45, Bolt–Beranek–Newman Inc.
CHAPMAN, D. (1948) *A Survey of Noise in British Homes*, National Building Studies, Technical Paper No. 2, HMSO.
CHURCHER, B. G., and KING, A. J. (1937) *J. Instn Elect. Engrs.*, **81**, 59–90.
CLARK, R. J., and PETRUSEWICZ, S. A. (1970–1) Report 175, School of Engineering, Bath University.
COBLENTZ, A., and JOSSE, R. (1968) *Cahiers du Centre Scientifique et Technique du Bâtiment*, Paris, Report 92.
COERMANN, R. R. (1970) *Ergonomics and Physical Environmental Factors*, No. 21 in Occupational Safety and Health Series (International Labour Office, Geneva), pp. 17–41.
COHEN, A. (1968a) *Noise and Psychological State*, Technical Report RR-9, July, Occupational Health Programme, Cincinnati, Ohio, published by United States Dept. Health, Education and Welfare.
COHEN, A. (1968b) *Trans. NY Acad. Sci*, Series 2, **30**, 910–18.
COHEN, A., *et al.* (1966) Technical Report RR-4, January (from same sources as Cohen (1968)).
COLQUHOUN, W. P., EDWARDS, R. S. (1975) *Ergonomics*, **18**, (1), 81–87.
CONRAD, D. W. (1973) *Ergonomics* **16** (6) 739–47.
CORBEILLE, C., and BALDES, E. J. (1929) Respiratory responses to acoustic stimulation in intact and decerebrate animals **88**, 481.
CORCORAN, D. W. J. (1962) *Q. Jl Exp. Psychol.* **14**, 178–82.
CORCORAN, D. W. J. (1963a) Individual differences in performance after loss of sleep, PhD thesis, University of Cambridge.
CORCORAN, D. W. J. (1963b) *J. Appl. Psychol.* **47**, 412–15.
COWELL, J. R., and MOTTERAM, J. M. (1973) *Speech Privacy, Partitions and Sound Conditioning*, Sound Research Laboratories Ltd., Report.
CREMER, L. (1942) *Akust. Z.* **7**, 7–104, May.
CREMER, L. (1948) *Geometrische Raumakustik* **1**, S. Hirzelverlag, Stuttgart.
CREMER, L. (1953) *Acustica* **3**, 249.
CREMER, L. (1965) DBT **10** (2) 850–62.
CREMER, L., and HECKL, M. (1967) *Körperschall*, Springer-Verlag (English version with E. E. Ungar, 1973).
CRESTI, C. (1970) *Le Corbusier*, Hamlyn.
CRICKMAY, C. L. (1972) Open University Technology Foundation Course T100, Units 32–34, Design I, Imagination and Method.
CROCKER, M. J., and PRICE, A. J. (1969) *J. Sound. Vibr.* **9**, 469.
CROCKER, M. J. (editor) (1972) *Noise and Vibration Control Engineering*, pub. by Ray W. Herrick Laboratories, Purdue University, Indiana, USA.
CROOME, D. J. (1969a) *Acoustical Conditions in Landscaped Offices*, Loughborough Consultants, Loughborough University of Technology.
CROOME, D. J. (1969b) *J. Instn. Heat. Vent. Engrs.*, **38**, 32–41.
CROOME, D. J. (1969c) *Airconditioning System Design for Buildings* (ed. Sherratt), Elsevier 131–49.

CROOME, D. J. (1970) *Ergonomics and Physical Environmental Factors*, No. 21 in Occupational Safety and Health Series (International Labour Office, Geneva), 116–31.

CROOME, D. J. (1971) *The rôle of feedback in the building design process*, Conference on Integrated Design in Buildings, Loughborough University of Technology, 5–6 July.

CROOME, D. J. (1976) *Applied Acoustics*, 9, (4), 303–315.

CROOME, D. J., and STEWART, L. J. (1970) *J. Instn. Heat. and Vent. Engrs.*, 38, 239–52, 12.

CROOME, D. J., and BROOK, A. D. (1974) Application of the articulation index to the acoustical design of landscaped offices, *International Conference on Noise Control, Washington DC, USA*.

CROOME, D. J., and ROBERTS, B. M. (1975) *Airconditioning and Ventilation of Buildings*, Pergamon.

CUMMINGS, A. (1973) *Building Services Engr.*, 41 (9) 123–32.

CURLE, N. (1955) *Proc. R. Soc. A*, 231, 505.

DAMASKE, P. (1967/8) *Acustica* 19, 199.

DARROW, C. W. (1947) *Psychol. Rev.* 54, 157–68.

DAVIES, A. D. M., DAVIES, D. R. (1975) *Ergonomics*, 18, (3), 321–336.

DAVIES, D. R., and KRKOVIC, A. (1965), *Am. J. Psychol.* 78, 304–6.

DAVIES, D. R., and HOCKEY, G. R. J. (1966) *Br. J. Psychol.* 57, 381–9.

DAVIS, H. (1954) CHABA Report 4, St. Louis (see Broadbent, 1958).

DAVIS, H. (1958) *US Armed Forces Med. J.* 9, 1027–48.

DAVIS, R. C. (1932) Electrical skin resistance before, during and after a period of noise stimulation, *J. Exp. Psychol.* 15, 108.

DAVIS, R. C., and BERRY, T. (1964) *Psychol. Rep.* 15, 95–113.

DAVIS, R. C., et al. (1955) *Psychol. Monographs* 69 (20), No. 405.

DAY, B. F. (1973) *Br. acoust. Soc. Proc.* 2 (1).

DAY, H. I. (1967) *Ontario J. Educ. Res.* 9, 185–91.

DEAN, R. D., and McGLOTHLEN, C. L. (1962) The effect of environmental stress interactions on performance, *Aerospace Med.* 33, 333 (abstract).

DELANY, M. E. (1971) *Proc. Br. Acoust. Soc.* 1 (1), Paper 52.

DE LANGE, P. A. (1968) *Appl. Acoust.* 1, 157–73.

DIECKMANN, D. (1958) *Ergonomics* 2, 347.

DOAK, P. E. (1959) *Acustica* 9.

DOELLE, L. (1972) *Environmental Acoustics*, McGraw-Hill.

DUFFY, F. (1964) *Architectural Rev.*, February.

DYER, I., and MILLER, L. N. (1959) *Noise Control*, 180–3, May.

ECK, B. (1973) *Fans*, Pergamon Press.

EDBERG, G. (1974) National Swedish Building Res. Summaries S20.

EIJK, J. van den, and BITTER, C. (1971) *Seventh International Congress on Acoustics*, Budapest, Paper 20A 24, 113–6.

EINBRODT, M. J., and BECKMANN, M. (1969) *Arbeitsmed. Sozialmed. Arbeitshyg.* 2, 49.

EMBLETON, T. F. W. (1963) *J. Acoust. Soc. Am.* 35 (8) 119–25.

EMBLETON, T. F. W., and DAGG, I. R. (1962) *Sound* 1, 32–36.

ENDEJANN, R. H. (1975), *ASHRAE Journal*, 17, (6), 21–22.

ETHOLM, B., and EGENBERG, K. E. (1964) The influence of noise on some circulatory functions, *Acta oto-laryngol* 58, 209–13.

EYRING, C. T. (1930) *J. Acoust. Soc.*, 217–41, January.

EYSENCK, H. J., et al. (1968), Eysenck Personality Form B, London University Press Ltd.

FARRELL, R. (1971) *J. Audio Eng. Soc.* 19 (3) 197–201.

FASOLD, W. (1963) *Hochfreq. Tech. Elecktroakust.* 72, H377.

FASOLD, W. (1965) *Acustica* 15 (5) 271–84.

FELTON, J. S., and SPENCER, C. (1957) *Morale of Workers Exposed to High Levels of Occupational Noise*, University of Oklahoma School of Medicine, Norman, Oklahoma.

FERGUSON, D. (1973) *Ergonomics* 16 (5) 649–64.

FINKELSTEIN, W. (1974) *Building Services Engr.* 41 (3) 268–75.

FLETCHER, H. (1934) *J. Acoust. Soc. Am.* 6, 59.

FLETCHER, H. (1953) *Speech and Hearing*, 2nd edn., van Nostrand.

FLETCHER, H. (1958) *Speech and Hearing Communication*, van Nostrand.

FLETCHER, H., and MUNSON, W. A. (1933) *J. Acoust. Soc. Am.* 5.

FLETCHER, H., and MUNSON, W. A. (1937) *J. Acoust. Soc. Am.* 9, 1–10.

FLETCHER, H., and STEINBERG, J. C. (1930) *J. Acoust. Soc. Am.* 1 (2) 1.

FOG, H., et al. (1968) *Build Int.*, September–October, 55–57.

FORD, R. D., and LORD, P. (1968) *J. Acoust. Soc. Am.* 43, 1062–8.

FORD, R. D., et al. (1970) *Appl. Acoustics* 3, 69–84.

FOUDRAINE, J. (1974) *Not Made of Wood*, Quartet Books.

FRANCIS, J. G. (1969) An appraisal of the acoustic environment, Undergraduate Thesis, Loughborough University of Technology (see Croome and Roberts, 1974, Chapter 12).

FRANÇOIS, P. (1970) *Appl. Acoust.* **3** (1) 23–45.

FREEMAN, H. L. (1975) *Int. J. Mental Health*, **4**, (3), 6–14.

FREEMAN, J., and NEIDT, C. O. (1959) *J. Educ. Res.* **53**, 91–96.

FRENCH, N. R., and STEINBERG, J. C. (1947) *J. Acoust. Soc. Am.* **19**, 90–119.

GALLOWAY, W. J., and BISHOP, D. E. (1970) FAA Report, 70–79, August.

GARDNER, W., and LICKLIDER, J. C. R. (1959) *J. Am. Dental Assoc.* **59**, 1144–9.

GARDNER, W., *et al.* (1960) *Science* 32–33.

GARNER, W. R. (1954) *J. Acoust. Soc. Am.* **26**, 75–88.

GATONNI, F., TARNOPOLSKY, A. (1973) *Psychol. Med.*, **3**, 516.

GELDARD, F. (1972) *The Human Senses*, 2nd edn., Wiley.

GHAMAH-ZADEH, N., and BRIFFA, F. E. J. (1969) *Combustion Oscillations in Small Boilers*, Associazione Termotecnica Italiana, Milan.

GIBBONS, S. G. (1970) Physiological effects of noise, PhD thesis, University of Salford.

GILBERT, P. (1973) *Br. Acoust. Soc. Proc.* **2** (3); *Urban noise measurement and evaluation*, Institute of Acoustics Meeting, Winchester, 13–14 November.

GILGEN, A. (1970) *Lärm, Forschungsausschuss für Planungsfragen*, Inst. für Orts. Regional und Landesplanung der ETH Zürich.

GIVONI, B., and GOLDMAN, R. F. (1971) *J. appl. Physiol.* **30** (3) 429–33.

GLOAG, J. (1950) *Men and Buildings*, Chantry.

GOMPERTS, M. C. (1968) IG-TNO, Report 33, Delft, Holland.

GOMPERTS, M. C. (1972) Report R40, Research Institute for Public Health Engineering, TNO, Schoesmakerstraat 97, PO Box 214, Delft, Holland.

GORDON, C. G. (1969) *Am. Soc. Mech. Engrs.* 16–20, Los Angeles, California, USA.

GÖSELE, K. (1959) *Gesundheitsingenieur* **80**, 106–12.

GÖSELE, K. (1967) *Schweiz. Bauzeitung* **22**, 480–6.

GÖSELE, K. (1968) *Gesundheitsingenieur* **89**, 129, and 168–72.

GÖSELE, K. (1969) *Berichte H63 aus der Bauforschung*, W. Ernst and Son (Berlin).

GÖSELE, K., and VOITSBERGER, C. A. (1970) *Gesundheitsingenieur* **91** (4) 108–17.

GOTTLOB, D. (1973) *Vergleich objektiver akustischer Parameter mit Ergebnissen subjektiver Untersuchungen an Konzertsälen*, Thesis, Göttingen.

GOTTSCHALK, O. (1968) *Flexible Verwaltungsbauten*, Quickborn, Schelle.

GRAHAM, J. B. (1966) *Trans. Am. Soc. Heat. Refrig. Air-condit. Engrs.* **72**, 11.

GRANDJEAN, E. (1962) Biological effects of noise, *Fourth International Congress on Acoustics*, Copenhagen.

GRANDJEAN, E. (1973a) *Ergonomics of the Home*, Taylor & Francis.

GRANDJEAN, E. (1973b) *Städtehygiene* (6) and (8).

GRANDJEAN, E. (1974) Personal communication.

GRANDJEAN, E., and KRYTER, K. D. (1961) *Les Effects du Bruit sur l'Homme*, Geigy (Basle), in the series L'Homme et son Milieu.

GRANDJEAN, E., *et al.* (1966) *Int. Z. aug. Physiol. einschl. Arbeitsphysiol.* **23**, 191–202.

GRAY, P. G., and CARTWRIGHT, A. (1958) *Noise in Three Groups of Flats with Different Floor Insulations*, National Building Studies, Research Paper No. 27, HMSO.

GREIN, H. (1974) *Sulzer Research*, 87–111.

GRETHER, W. F., *et al.* (1971) Effects of combined heat, noise and vibration stress on human performance and physiological functions, *Aerospace Med.* **42** (10) 1092–7.

GRETHER, W. F., *et al.* (1972) Further study of combined heat, noise and vibration stress, *Aerospace Med.* **43** (6) 641–5.

GRIFFITHS, I. D., and LANGDON, F. J. (1968) *J. Sound Vibr.* **8**, 16.

GROSJEAN, J. (1974) Chapter 8 in *Noise and Vibration Control for Industrialists* (ed. S. A. Petrusewicz and D. K. Longmore), Elek Science Press.

GROSSMAN, S. P. (1967) *A Textbook of Physiological Psychology*, Wiley, New York.

GUBLER, F. (1967) *Industrielle Org.* **36**, 359.

GUIGNARD, J. C., and GUIGNARD, E. (1970) Inst. Sound and Vibration Research Memorandum 373, University of Southampton.

GUINOT, G. (1964) *Third Congress of the Association Internationale Contre le Bruit*, Paris, May.

GUIRAO, M., and STEVENS, S. S. (1964) *J. Acoust. Soc. Am.* **36**, 1176–82.

GUPTA, A. K. (1973) *J. Inst. Fuel*, **xlvi** (3) 119–23.

HAAS, H. (1951) *Acustica* **1**, 49.

HALE, H. B. (1952) *Amer. J. Physiology* **171**, 732.

HALL, J. C. (1952) *J. Educ. Res.* **45**, 451.

HANFLING, O. (1973) Body and Mind, Units 1–2 of Open University Third Level Arts Course, Problems of Philosophy.

HARDY, H. C. (1963) *Trans. Am. Soc. Heat. Refrig. Air-condit. Engrs.* **69**, 85; also *Am. Soc. Heat. Refrig. Air-condit. Engrs. J.* **5** (1) 95–100.

HARMAN, D. M. (1971) *Building Acoustics* (ed. Smith *et al.*), pp. 5–18, Oriel Press.

HARMON, F. L. (1933) *Archs. Psychol.* No. 147 (New York).

HARRIS, C. M. (ed.) (1957a) *Handbook of Noise Control*, McGraw-Hill.

HARRIS, C. M. (1957b) *Handbook of Noise Control*, Chapter 18 by H. J. Sabine.

HARRIS, C. M., and CREDE, C. E. (1961) *Shock and Vibration Handbook*, volume 3, 18–21 to 18–25.

HARRIS, J. D. (1952) *J. Acoust. Soc. Am.* **24**, 750–5.

HARWOOD, H. D. (1970) *Acoustics of auditoria*, Acoustics Group of Inst. Physics and Physical Society Meeting, University of Aston in Birmingham, 1 July (BBC Research Dept., Tadworth, Surrey).

HATHAWAY, S. R., *et al.* (1951) *The Minnesota Multiphasic Personality Inventory*, the Psychological Corporation, New York.

HAWKES, R. J., and DOUGLAS, H. (1970) *Art* 1/2, November, 34–45.

HAY, B. (1972) *Building Services Engr.* **40**, 105–6.

HAY, B. (1973) *Phys. Bull.*, February, 87–88.

HAY, B., and KEMP, M. F. (1972) *J. of Sound Vibr.* **23** (3).

HEGVOLD, L. W. (1971) *Acoustical Design of Open Planned Offices*, Canadian Building Digest 139, Nat. Res. Council of Canada, Divn. Building Research.

HELIES, J., and CADIERGUES, R. (1965) Publication 8 (R), COSTIC, St. Rèmy, France.

HELIES, J., and CADIERGUES, R. (1967) *Inds. Therm. et Aerauliques*, April, 235–250.

HELLMAN, R., and ZWISLOCKI, J. J. (1961) *J. Acoust. Soc. Am.* **33**, 687–94.

HELLMAN, R., and ZWISLOCKI, J. J. (1963) *J. Acoust. Soc. Am.* **35**, 856–65.

HELLMAN, R., and ZWISLOCKI, J. J. (1964) *J. Acoust. Soc. Am.* **36**, 1618–27.

HELLMAN, R., and ZWISLOCKI, J. J. (1968) *J. Acoust. Soc. Am.* **43**, 60–63.

HELPER, M. M. (1957) Report 270, US Army Med. Res. Labs., Fort Knox, Kentucky.

HENDERSON, J. (1959) *VVS Jl (Sweden)*, October.

HENNENBERG, F. (1962) *The Leipzig Gewandhaus Orchestra*, VEB edition, Leipzig.

HERRIDGE, C. F. (1972) *Sound*, **6**, 32–36.

HERRIDGE, C. F. (1974) *J. Psychosomatic Research*, **18**, 239–243.

HERZBERG, F. (1959) *The Motivation to Work*, Wiley, New York.

HESS, R., Jr. (1954) Der Schlaf; Merkmale, Wesen und Bedeutung, *Umschau* **53**, 452.

HESSELGREN, S. (1971) *Experimental Studies on Arch. Perception*, Nat. Swedish Bldg. Research Document, D2.

HEUSSER, M. (1968) *Büro Verkauf* **37**, 452.

HIRD, J. B. (1967) *Royal Society of Health Journal*, **87**, (3), May-June, 171–172.

HOAGLAND, H. (1957) *Hormones, Brain Function and Behaviour*, Academic Press, New York.

HOCKEY, G. R. J. (1959) *New Scientist*, 1 May, 244–6.

HOCKEY, G. R. J. (1970a) *Q. Jl. Exp. Psychol.* **22**, 37–42.

HOCKEY, G. R. J. (1970b) *Br. J. Psychol.* **61**, 473–80.

HOFFMAN, J. E. (1966) The effect of noise on intellectual performance as related to personality and social factors in upper division high school students, unpublished doctoral dissertation, University of Southern California.

HOLMBERG, J., THYLEBRING, H. (1974) *National Swedish Building Research Summary*, R65: 1974.

HOLMES, M. J. (1973a) Laboratory Report No. 75, Heating and Ventilating Research Association, Bracknell, Berkshire.

HOLMES, M. J. (1973b), Laboratory Report No. 78, Heating and Ventilating Research Association, Bracknell, Berkshire.

HONIKMAN, B. (1971) *Proceedings of the Arch. Psychology Conference, Kingston Polytechnic, 1–4 September 1970*, published by Kingston Polytechnic and RIBA Publications Ltd.

HOOD, J. D. (1968) *J. Acoust. Soc. Am.* **44**, 959–64.

HOOGENDOORN, K. (1973) *Building Res. Practice*, 202–6.

HOUSTON, B. K. (1968) Inhibition and the facilitating effect of noise on interference tasks, *Percept. Motor Skills* **27**, 947–50.

HOUSTON, B. K., and JONES, T. M. (1967) Distraction and Stroop Color-word Performance, *J. Exp. Psychol.* **74**, 54–56.

HOWES, D. H. (1950) *Am. J. Psychol.* **63**, 1–30.

HUME, W. F. (1966) *J. Psychosom. Res.* **9**, 383–91.

HUMES, J. F. (1941) The effects of occupational music on scrappage in the manufacture of radio tubes, *J. Appl. Psychol.* **25**, 573–87.

HUMPHREYS, H. R., MELLUISH, D. J. (1971) *Sound Insulation in Building (Cost and Performance of Walls and Floors)*, HMSO.

HUNDERT, A. T., and GREENFIELD, N. (1969) Physical space and organisational behaviour, *Proceedings of 77th Annual Convention of American Psychological Association, Washington.*

HURLE, I. R., *et al.* (1968) *Proc. Royal Society* A, **303**, 409.

HURTY, W. C., and RUBENSTEIN, M. F. (1964) *Dynamics of Structures*, Prentice-Hall.

INGARD, U., and OPPENHEIM, A. (1965) Final Report ASHRAE project NR. RP. 37, August.

INGARD, U., *et al.* (1968) *Trans. Am. Soc. Heat. Refrig. Air-condit. Engrs.* **74** (2068).

INGERSLEV, R., and NIELSEN, A. K. (1944a) *Ingeniørvidensk. Skr.* **5**.

INGERSLEV, F., and NIELSEN, A. K. (1944b) Publication 1, Acoustical Laboratory, Academy of Technical Science, Copenhagen.

IREDALE, R. A. (1973) *Noise from prime movers*, British Acoustical Society Meeting, Southampton University, 9–10 July; also see *Br. Acoust. Soc. Proc.* **2** (3).

ISVR (1974) *Institute of Sound and Vibration Research Review*, Southampton University, Autumn (11) 5.

IZUMIYAMA, M. (1964) *Tohoku Psychol. Fol.* **23**.

JACKSON, G. M., and LEVENTHALL, H. G. (1972) *Appl. Acoust.* **5** (4) 265–77.

JAMES, W. (1892) *Textbook of Psychology*, Macmillan, London.

JANSEN, G. (1961) *Stahl Eisen* **81**, 217–20.

JANSEN, G. (1964) *Int. Z. Angew. Physiol.* **20**, 233–9.

JANSEN, G. (1967) *Zur Nervösen Belastung durch Lärm*, Steinkopff Verlag, Darmstadt.

JANSEN, G., and KLENSCH, H. (1964) *J. Appl. Physiol.* **20**, 258–70.

JANSEN, G., and SCHULZE, J. (1964) *Klin. Wschr.* **3**, 132–4.

JENSEN, P. (1973) *J. Air Pollut. Control Ass.* **23** (12) 1028–34.

JERISON, H. J. (1957) *J. Acoust. Soc. Am.* **29**, 1163.

JOHANNSON, C. R., *et al.* (1970) Unpublished report, Dept. Psychology, University, Lund, Sweden.

JOHNS, D. J. (ed.) (1968) *Wind effects on buildings and structures*, Proceedings of Conference, Loughborough University of Technology.

JOHNS, D. J. (1970) *Structural dynamics*, Proceedings of Conference, Loughborough University of Technology.

JORDAN, V. L. (1959), Proc. 3rd *Int. Cong. Acoustics*, **II**, 922.

JORDAN, V. L. (1969) *Appl. Acoust.* **2**, 77–99.

JORDAN, V. L. (1970) *J. Acoust. Soc. Am.*, **47**, 408.

JORDAN, V. L. (1973) *J. Proc. R. Soc. NSW* **106**, 33–53.

JORDAN, V. L. (1975a) in *Auditorium Acoustics*, edited by Mackenzie, Applied Science Press.

JORDAN, V. L. (1975b) *Applied Acoustics*, **8**, (3), 217–235.

JOUVET, M. (1963) *Electroenceph. Clin. Neurophysiol. Suppl.* **24**, 133–57.

JUNGK, R. (1968) *The Big Machine*, Charles Scribners & Sons, New York.

KALSBEEK, J. W. H., *et al.* (1965) *British Psychological Society Conference, London.*

KAMBER, F. (1968) *Gesundheitstechnik* **7**, 208–13; also 15–22.

KARRASCH, K. (1952) *Zeitblätter Arbeitswiss*, Nos. 517 and 416, **6**, 177 (Westdeutscher Verlag).

KEEFE, F. B., and JOHNSON, L. C. (1970) Cardiovascular responses to auditory stimuli, *Psychonom. Sci.* **19** (6) 335–7.

KEETS, W. de V. (1968) Proc. 6th Int. Cong. Acoustics, **III**, Section E, 49.

KEIGHLEY, E. C. (1965), Building Research Station, Note EN 54.

KEIGHLEY, E. C. (1966) *J. Sound Vibr.* **4** (1) 73.

KENNEDY, D. R. (1965) *J. Inst. Fuel* **38**, 30.

KENNEDY, F. (1936) *New York St. J. Med.* **36**, 1927.

KERRICK, J. S., *et al.* (1969) *J. Acoust. Soc. Am.* **45**, 1014–7.

KIHLMAN, T. (1963) *Byggforskningen*, Report 93, Stockholm.

KING, A. J. (1965) *The Measurement and Suppression of Noise with Special Reference to Electrical Machines*, Chapman & Hall.

KINGSBURY, H. F., and TAYLOR, D. W. (1970) *J. Sound Vibr.*, 19–23 May.

KIRK, R. E., and HECHT, E. (1963) Maintenance of vigilance by programmed noise, *Percept. Motor Skills* **16**, 553–60.

KLUGE, E. and FRIEDEL, B. (1953) Über die Einwirkung einförmiger akustischer Reize auf den Funktionszustand des Gehirns, *Z. Psychother. Med. Psychol.* **3**, 212.

KOCH-EMMERY, W. (1964) *Heiz.–Lüft.–Haustech.* **15**, 11.

KÖHLER, W. (1910) *Z. Psychol.* **58**, 59–140.

KONZ, S. A. (1964) The effect of background music on productivity of four tasks, Unpublished doctoral dissertation, University of Illinois.

KORN, J. (1971) *Heat. Pip. Air Condit.*, April, 84–86.

KOSTEN, C. W., and van OS, G. J. (1962) Community reaction criteria for external noises, *NPL Symposium*, No. 12, HMSO.

KOVRLGIN, S. D., and MIKEYER, H. P. (1965) Report N. 65-28297, Joint Publications Research Service, Washington DC.

KRISTENSEN, J. (1964) *Ingeniøren* **23**, 710–7.

KRYTER, K. D. (1950) The effects of noise on man: (1) I, Effects of noise on behaviour, *J. Speech Dis. Mon.* Suppl. 1.

KRYTER, K. D. (1958) The effect of noise on man, Monograph Supplement 1, *J. Speech Hear. Disorders*, American Speech and Hearing Association, Washington DC.

KRYTER, K. D. (1959) *J. acoust. Soc. Am.* **31**, 1415.

KRYTER, K. D. (1962) *J. acoust. Soc. Am.* **34** (11) 1689–97.

KRYTER, K. D. (1966) Report CR442, Nat. Aeronautics and Space Admin., Washington DC.

KRYTER, K. D. (1969) *Methods for the Calculation of the Articulation Index*, Standard 53.5-1969, National Standards Institute, New York.

KRYTER, K. D. (1970) *The Effects of Noise on Man*, Academic Press.

KRYTER, K. D., and GRANDJEAN, E. (1961) *Les Effets du Bruit sur l'Homme*, in the series L'Homme et son Milieu, Geigy, Basle.

KUHL, W. (1954) *Acustica* **4**, 618.

KURTZE, G. (1964) *Physik und Technik der Lärmbekämpfung*, G. Braun, Karlsruhe.

KÜPER, R. (1969) *Acustica*, **21**, 370.

KUTTRUFF, H. (1973) *Room Acoustics*, Applied Science Publishers Ltd.

KYBURZ, W. (1968) *Büro Verkauf* **37**, 433.

LAIRD, D. A. (1929) Experiments on the physiological cost of noise, *J. Nat. Inst. Industr. Psychol.* **4**, 251.

LAIRD, D. A. (1930) The effects of noise: a summary of experimental literature, *J. Acoust. Soc. Am.* **1**, 256.

LAIRD, D. A. (1932) Experiments on the influence of noise upon digestion, and the counteracting effects of various food agencies, *Med. J. Rec.* **135**, 461.

LAIRD, D. A. (1933) *J. Appl. Psychol.* **17**, 320–30.

LAIRD, D. A. and COYE, K. (1929) *J. Acoust. Soc. Am.* **1**, 158–63.

LAIRD, D. A. et al. (1930–1) *J. Acoust. Soc. Am.* **2**, 94.

LAMURE, C., and ANZOU, S. (1964) *Cah. Cent. Scient. Tech. Bâtim.* **71**, 599.

LAMURE, C., and ANZOU, S. (1966) *Cah. Cent. Scient. Tech. Bâtim.* **78**, 1–18.

LAMURE, C., and BACELON, M. (1967) *Cah. Cent. Scient. Tech. Bâtim.* **88**, 761–68.

LANE, H. L. (1971) *Proceedings of the 7th International Congress of Acoustics*, Paper 24C14, Budapest, 221–4.

LANGDON, F. J. (1966) *Research on Noise Nuisance*, CIB Symposium, Commission W45.

LANGDON, F. J. and SCHOLES, W. E. (1968) Building Research Station Current Paper 38/68.

LAPPAT, A. (1969) *Bauen Wohn. Ravensh.* **24**, 1.

LARGE, J. B. (1971) *Phys. Bull.* **22**.

LARGE, J. B. (1972) *Phys. Bull.* **23**.

LARGE, J. B. (1973) *Phys. Bull.* **24**, 729.

LAWRENCE, A. (1970) *Architectural Acoustics*, Elsevier.

LEE, V., et al. (1972) Open University Course E281, *Personality Growth and Learning*, Units 1–4 (Creativity, Learning Styles).

LEHMANN, D. W., et al. (1965) *J. School Hlth.* **35**, 212.

LEHMANN, G. (undated) *Sick People and Noise*, Max-Planck Institut für Arbeitsphysiologie, Dortmund.

LEHMANN, G., and TAMM, J. (1956) *Int. Z. Angew. Physiol.* **16**, 217.

LEHMANN, G., and MEYER-DELIUS, J. (1958) *Forschungsberichte des Wirtschafts und Verkehrsministeriums Nordrhein-Westfalen*.

LEHMANN, P. (1972) *Auswertungen von Impulsmessungen in verschiedenen Konzertsälen*, DAGA, Stuttgart.

LEIDEL, W. (1969) *Einfluss von Zungenabstand und Zungenradius auf Kennlinie und Geräusch eines Radialventilators*, DLR Forschungsbericht, 69-16.

LEONARD, J. A. (1959) MRC App. Psych. Unit Report 326, Cambridge, England.

LESKOV, E. A. (1970) *Experimental investigation on absorption attenuators*, personal communication USSR.

LESKOV, E. A. et al. (1970) *App. Acoustics*, **3**, 47.

LEVI, L. (1966) *Forsvarsmed.* **2**, 3.

LEVI, L. (1967) In *An Introduction to Clinic Neuroendocrinology* (ed. E. Bajusz), S. Karger, Basel, Switzerland.

LEVI, L. (1972) Stress and distress in response to psychosocial stimuli, *Acta Med. Scand.* Suppl. No. 528 (also published by Pergamon, Oxford).

LEWIS, P. T. (1970) *Real windows*, Conference on Building Acoustics, British Acoustical Society, April 1970.

LEWIS, P. T. (1971) *Building Acoustics* (ed. T. SMITH *et al.*), Chapter 7, pp. 116–54, Paper 72/73, Oriel Press.

LEWIS, P. T. (1973a) *Br. Acoust. Soc. Proc.* 2 (1).

LEWIS, P. T. (1973b) *Noise and Vibration Control Conference*, University of Cardiff, April, 16–18.

LIBBEY–OWENS–FORD GLASS CO., Manufacturer's data.

LIGHTHILL, M. J. (1952) *Proc. R. Soc.* A, 211, 564–87.

LJUNGGREN, S. (1973) Nat. Swedish Building Research Summary D10.

LOCHER, K., and NASSENSTEIN, G. (1971) *Schweiz. Bl. Heiz. Lüft.* 1, 22–31.

LOCHNER, J. P. A., and BURGER, J. F. (1960) *Acustica* 10, 394.

LOCHNER, J. P. A., and BURGER, J. F. (1961) *Acustica* 11, 195.

LONDON, A. (1950) *J. Acoust. Soc. Am.* 22, 270.

LORD, P. (1969a) *Sound insulation in office and domestic buildings*, Conference on Systems Noise in Buildings, Edinburgh, September.

LORD, P. (1969b) *Reducing noise nuisance to the public*, Noise Control in Industry Conference, May 1970, Production Engineering Research Association, Melton Mowbray.

LOSCH, W. (1970) *Sanitär Heizungstechnik* 35 (11) 712–16.

LOTTERMOSER, W. (1968) *Instrumßau-Z.* 23 (4, 5).

LOVELOCK, E. C. (1974) *Building Services Engr.* 42 (11) 183–94.

LUKAS, J. S., and KRYTER, K. D. (1969) NASA Contract NASI-7592, Stanford Research Institute, Menlo Park, California.

LUNDBERG, B. (1963) *Second International Acoustics Congress, Austria, 1962.*

McBAIN, W. N. (1961) Noise, the 'arousal hypothesis' and monotonous work, *J. Appl. Psychol.* 45, 309–17.

McCALLUM, C. (1969) *Attention in Neurophysiology*, edited by Evans and Mulholland, published by Butterworths.

MacEWEN, M. (1974) *Crisis in Architecture*, RIBA Publications Ltd.

McGEHEE, W., and GARDNER, J. E. (1949) Music in a complex industrial job, *Personnel Psychol.* 2, 405–17.

McGRATH, J. E. (1970) *Social and Psychological Factors in Stress*, 10–21, Holt, Rinehart & Winston, New York.

MACKENZIE, R. (1975) *Auditorium Acoustics*, published by Applied Science.

McKENNELL, A. C. (1963) UK Government Social Survey Report SS, 337, Cmnd 12056, HMSO.

McKENNELL, A. C., and HUNT, A. E. (1966) UK Government Social Survey Report SS, 332.

McKENNELL, A. C., and HUNT, A. E. (1970) *Transportation Noises* (ed. Chalupnik) 228, University of Washington Press.

MacKINNON, D. W. (1961) The personality correlates of creativity: study of American architects, in *Proceedings of the XIXth International Congress of Applied Psychology, Copenhagen* (ed. Nielson) 2, 11–39.

MacKINNON, D. W. (1967) *Am. Psychol.* 17, 484–94.

MacKINNON, D. W. (1970) as 1961, Chapter 22 in *Creativity* (ed. by Vernon), Penguin.

McK.NICHOLL, A. G. (1974) Personal communication, Ergolab, Stockholm, Sweden.

McNAIR, H. P. (1972) Gas Council Report, No. OO/T, R D/72/5.

MACPHERSON, R. K. (1962) *Br. J. Ind. Med.* 19, 151–64.

MacPHERSON, R. K. (1973) *Ergonomics* 16 (5) 611–22.

MACKWORTH, N. H. (1950) *Researches on the Measurement of Human Performance*, Special Report Series No. 268, Medical Research Council, published by HMSO.

MAEKAWA, Z. (1968) *Appl. Acoust.* 1, 157–73.

McROBERT, H. (1973) *Noise and loudness evaluation*, The British Acoustical Society Proc. 2 (3), meeting held at Salford University, 25 and 26 October.

MAGNUS, K. (1965) *Vibrations*, Blackie.

MALMO, R. B. (1959) Activation: a neuro-psychological dimension, *Psychological Review* 66, 367–86.

MANNING, P. (1967) *The Primary School; an Environment for Education*, Pilkington Research Unit.

MARINER, T., and HEHMAN, H. W. W. (1967) *J. Acoust. Soc. Am.* 41 (1) 206–14.

MARKUS, T., *et al.* (1972) *Building Performance by Building Performance Research Unit*, Strathclyde University, Applied Science.

MARSH, J. (1971a) *Appl. Acoust.* 4 (1) 55–70.

MARSH, J. (1971b) *Appl. Acoust.* 4 (2) 131–54.

MARSH, J. (1971c) *Appl. Acoust.* 4 (3) 174–92.

MARSH, J. A. (1971) *Building Acoustics* (ed. T. SMITH *et al.*), Chapter 5, pp. 71–86, Paper 72/73, Oriel Press.

MARSHALL, A. H. (1967a) PhD thesis, Institute of Sound and Vibration Research, University of Southampton.

MARSHALL, A. H. (1967b) *J. Sound Vibr.* **5** (1) 100–12.

MARSHALL, A. H. (1968) *Archs. Sci. Rev.* **11** (3) 81–87.

MARSHALL, A. H. (1968) *J. Sound Vibr.*, **7**, 116.

MARUYAMA, K. (1964) *Tohoku Psychol. Fol.* **23**.

MARVET, B. H. (1959) *Trans. ASHRAE* **65**, 613–26.

MEHRABIAN, A. (1971) *Silent Messages*, Wadsworth Publishing Co., Belmont, California.

MEHRABIAN, A., and RUSSELL, J. A. (1974) *An Approach to Environmental Psychology*, Massachusetts Institute of Technology Press.

MEISTER, F. J., and RUHRBERG, W. (1959a) *VDI-Z (Germany)* **101** (13) 527–35.

MEISTER, F. J., and RUHRBERG, W. (1959b) *Lärmbekämpfung* **3** (1) 5–11.

MELZACK, R. (1961) *Scient. Am.* **204**, 41–49.

MEYER, E., and SCHODDER, G. R. (1952) *Nachr. Akad. Wiss. Göttingen, Math.-Phys. K1.* **6**, 31.

MEYER, E., *et al.* (1965) *Acustica* **15**, 175–8.

MEYER-DELIUS, J. (1957) *Auto-tech. Z.* **10**.

MILLER, G. A. (1947) *Psychol. Bull.* **44**, 105–29.

MILLER, G. A., and NICELY, P. E. (1955) *J. Acoust. Soc. Am.* **27**, 338–52.

MILLS, C. H. G., and ROBINSON, D. W. (1961) *Engineer* **211**, 1070–4.

MOREIRA, N. M., and BRYAN, M. E. (1972) *J. Sound Vibr.* **12** (4) 449–62.

MORGAN, C. T. (1965) *Physiological Psychology*, 3rd edn., McGraw-Hill, Kogakusha.

MORGAN, J. J. B. (1916) The overcoming of distraction and other resistances, *Archs Psychol. N.Y.* No. 35.

MORGENSTEIN, F. S., *et al.* (1974) *Ergonomics* **17** (2) 211–20.

MORSE, P. M. (1939) *J. Acoust. Soc. Am.* **11**, 205.

MORSE, P. M. (1948) *Vibration and Sound*, McGraw-Hill.

MULHOLLAND, K. A., PRICE, A. J., and PARBROOK, H. D. (1968) *J. Acoust. Soc. Am.* **43**, 1432.

MÜLLER, G. (1966) *Sanitär Heizungstechnik* **31** (12) 855–60.

MUNCEY, R. W., NICKSON, A. F. B., and DUBOUT, P. (1953) *Acustica* **3**, 168.

MUNSON, W. A. (1957) In *Handbook of Noise Control* (ed. Harris), Chapter 5, 5-17 to 5-19.

NAGATSUKA, Y. (1964) *Tohoku Psychol. Fol.* **23**.

NEMECEK, J., and GRANDJEAN, E. (1971) *Industrielle Org.* **40** (5) 233.

NEMECEK, J., and GRANDJEAN, E. (1973a) *Human Factors* **15** (2) 111–24.

NEMECEK, J., and GRANDJEAN, E. (1973b) *Appl. Ergonomics*, March, 19–22.

NEWLAND, D. E., and LIQUORISH, A. D. (1974) Personal communication.

NIESE, H. (1956) *Hochfreq. Tech. Elektroakust.* **65**, 4.

NIESE, H. (1961) *Acustica* **11**, 199.

NIESE, H. (1967) *Hochfreq. Tech. Elektroakust.* **66**, 70.

NORRIS, C. H., *et al.* (1959) *Structural Design for Dynamic Loads*, McGraw-Hill.

OLLERHEAD, J. B. (1973) *Appl. Ergonomics*, September, 130–8.

OLSEN, J., and NELSON, E. N. (1961) *Minn. Med.* **44**, 527–9.

OLTMAN, P. K. (1964) Field dependence and arousal, *Percept. Motor Skills* **19**, 441.

OOSTING, W. A. (1967) Report 706.007, Department Building Acoustics, TNO, Delft, Holland.

OPPLIGER, G. C., and GRANDJEAN, E. (1959) Vasomotorische Reaktionen der Hand auf Lärmreize, *Helv. Physiol. Pharmac. Acta* **17**, 275.

ORNSTEIN, R. (1974) BBC Radio 4 broadcast, Sunday June 16th 1974, New Maps of the Mind, produced by Michael Totton; also see book *The Psychology of Consciousness*, W. H. Freeman.

PANCHOLY, M., *et al.* (1965) *Indian Concrete J.* **39** (12) 474.

PANCHOLY, M., *et al.* (1966) *J. Music Acad.* Madras **37**, 144–62.

PANCHOLY, M., *et al.* (1967) *J. Instn Engrs. (India)* **67** (9), Pt. GE3, 427–45.

PARK, J. F., Jr., and PAYNE, M. C., Jr. (1963) Effects of noise level and difficulty of task in performing division, *J. Appl. Psychol.* **47**, 367–8.

PARKIN, P. H., and HUMPHREYS, H. R. (1969) *Acoustics, Noise and Buildings*, Faber & Faber.

PARKIN, P. H., and MORGAN, K. (1971) Building Research Station Current Paper CP17/71.

PARKIN, P. H., *et al.* (1968) *London Noise Survey*, HMSO.

PEPLER, R. D. (1959) *J. Comp. Physiol. Psychol.* **52**, 446–50.

PERL, E. R., GALAMBOS, R., and GLORIG, A. (1953) The estimation of hearing threshold by electroencephalography, *Electroenceph. Clin. Neurophysiol.* **5**, 501.

PETRUSEWICZ, S. A., and LONGMORE, D. K. (1974) *Noise and Vibration Control for Industrialists*, Elek Science, London.

PEUTZ, V. M. A. (1973) *Br. Acoust. Soc. Proc.* **2** (1), Paper 72/53.

PHILLIPS, D. (1971) *Knowledge from What?*, Rand-McNally, New York.

PILKINGTON BROS. LTD. (1961) AIRO Report No. 8/294.

PILKINGTON BROS. LTD. (1964) *Glass and Windows Bulletin* No. 3, February.
PILKINGTON BROS. LTD. (1970) *The Airborne Sound Insulation of Glass* (4), June.
PIRN, R. (1971) *J. Acoust. Soc. Am.* **49**, 1339–45.
PLENGE, G. (1965/6) *Acustica* **16**, 269.
PLUTCHIK, R. (1959) The effects of high intensity intermittent sound on performance, feeling, and physiology, *Psycholog. Bull.* **56**, 133–51.
POMFRET, B. (1973) *Noise Control and Vibration Reduction*, Nov.–Dec., 260–6.
POOLE, R., GOETZINGER, C. P., and ROUSEY, C. L. (1966) A Study of the effects of auditory stimuli on respiration, *Acta Oto-lar.* **61**, 143–52.
PORT, E. (1963) *Acustica* **13**.
POULTON, E. C. (1965) *Ergonomics* **8**, 69–76.
POULTON, E. C. (1970) *Environment and Human Efficiency*, Thomas, Illinois.
POULTON, E. C. (1971) *Psychology at Work* (ed. P. B. Warr), Chapter 3, Penguin.
POULTON, E. C., *et al.* (1974) *Ergonomics* **17** (1) 59–73.
POWELL, J. A. (1971) *Building Acoustics* (ed. T. Smith *et al.*), Chapter 10, pp. 197–211, Oriel Press Ltd.
RAAB, W. (1968) *Psychosom. Med.* **30** (6) 809–18.
RATHÉ, E. (1966) *Acustica* **17**, 268–77.
RATHÉ, E. J. (1969) *J. Sound Vibr.* **10** (3) 472–9.
RATHBONE, R. A. (1967) *Proc. Instn. Mech. Engrs.* **182** (3E) 20.
RAYLEIGH, J. W. S. (1945) *Theory of Sound*, volumes I and II, Dover.
REASON, J. T. (1968) *Br. J. Psychol.* **59**, 385–93.
RECHTSCHAFFEN, A., *et al.* (1966) *Percept. Motor Skills* **22**, 927–42.
REESE, J. A., and GUPTA, A. K. (1974) *Am. Soc. Heat. Refrig. Air-condit. Engrs. Jl.* **16** (9) 59–63.
REICHARDT, W., and SCHMIDT, W. (1966) *Acustica* **17**, 75.
REICHOW, H. B. (1963) *Architects' J.* 13 Feb., 357–60.
REILLY, N. (1959) *Med. J. Australia* **1**, 2000.
RETTINGER, M. (1959) *Noise Control*, July, 212–14.
RIBNER, H. S. (1964) *Adv. Appl. Mech.* **8**, 103.
RICE, C. G. (1973) *Urban noise measurement and evaluation*, Institute of Acoustics Symposium, Winchester, November 13–14; also see *Br. Acoust. Soc. Proc.* **2** (3).
RICHARDS, E. J. (1966) *Conference on World Airports—The Way Ahead*, Institution of Civil Engineers, London.
RICHTER, H. R. (1966a) Disturbed EEG sleep patterns by traffic noise, *National Noise Abatement Society, International Conference*, London.
RICHTER, H. R. (1966b) *Rev. Neurol. Paris* **115**, 592–95 (Séances May 27, 28 held by Societé D. Électroencéphalographie de Langue Françoise).
RICHTER, H. R. (1971) *Universitäs* **26** (4) 403–10.
RICHTER, H. R. (undated) *Sleep Disturbances Which We Are Not Aware of Caused by Traffic Noise*, EEG Station of the Neurological University Clinic, Basel, Switzerland.
RIDLEY, G. K. (1963) Supplement to *AEI Eng.* October.
RIESCZ, R. R. (1928) *Phys. Rev.* **31**, 867–75.
RILAND, L. H. (1970) *A Resurvey of Employee Reactions to the Landscape Environment One Year After Initial Occupation*, Eastman–Kodak, New York, March.
ROBERTS, J. P., LEVENTHALL, H. G. (1975) *J. Inst. of Fuel*, March, 45–48.
ROBINSON, D. W. (1957) *Acustica* **7**, 217–33.
ROBINSON, D. W. (1969) *The Concept of Noise Pollution Level*, NPL Aero Report AC 38.
ROBINSON, D. W. (1971) *J. Sound Vibr.* **14**.
ROBINSON, D. W., and DADSON, R. S. (1956) *Br. J. Appl. Phys.* **7**, 166–181.
ROBINSON, D. W. *et al.* (1961) *Engineer* **211**, 493–7.
ROBINSON, D. W., *et al.* (1963a) *Acustica* **13**, 324–36.
ROBINSON, D. W., *et al.* (1963b) *Noise, Final Report*, Cmd 2056, HMSO, London.
ROSEN, S. (1962) *Annals of Otology, Rhiniology and Laryngology*, **71**, 727.
ROSSI, L., OPPLIGER, G. C., and GRANDJEAN, E. (1959) Gli effeti neurovegetativi sull'uomo di rumori sovrapposti ad un rumore di fondo, *Med. Lavoro* **50**, 332.
ROTHLIN, E., GERLETTI, A., and EMMENEGGER, H. (1956) Experimental psycho-neurogenic hypertension and its treatment with hydrogenated ergot alkaloids, *Acta Med. Scand.* **154**, and Suppl. **312**, 27.
ROWLAND, V. (1957) *Electroenceph. Clin. Neurophysiol.* **9**, 585–94.
RÜCKWARD, W. (1970) *Gesundheitsingenieur* **91** (5) 141–8.
SABINE, W. C. (1923) *Collected Papers on Acoustics*, Harvard University Press.
SAKAMOTO, H. (1957) Studies on the mechanism of adaptation of organisms exposed to noise, with special reference to the reaction of pituitary-adrenocortical system, *J. Sci. Labour*, Tokyo **33**, 93, 175.

SAKAMOTO, H. (1959) *Mie Med. J.* ix, 39.

SAUNDERS, A. F. (1961) *Ergonomics* 4, 253–8.

SAUTER, A., and SORAKA, W. (1970) *J. Acoust. Soc. Am.* 47, 5.

SCHₐEBER, J. P. (1967) *Étude analytique en laboratoire de l'influence du bruit sur le sommeil*, Report, Faculté de Medicine, Strasbourg University.

SCHMIDT, R. (1967) *Industrielle Org.* 36, 348.

SCHMIDT, W. (1967) *Licht Beleucht.* 20, 5.

SCHMIDT, W. (1968) *Hochfreq. Tech. Elektroakust.* 77, 37.

SCHOLES, W. E., et al. (1971) *Building Acoustics* (ed. T. Smith, et al.), Chapter 6, pp. 87–115, Paper 72/73, Oriel Press.

SCHOPENHAUER, A. (1890) *Studies in Pessimism*, Allen & Unwin.

SCHREIBER, L. (1971)*Heiz.–Lüft.–Haustech.* 22 (1) 9–13.

SCHROEDER, M. R. (1965) *J. Acoust. Soc. Am.* 37, 409.

SCHROEDER, M. R., et al. (1962) *J. Acoust. Soc. Am.* 34, 76.

SCHROEDER, M. R. (1966) *J. Acoust. Soc. Am.* 40, 549–51.

SCHUBERT, P. (1969) *Hochfreq. Tech. Elektroakust.* 78, 230.

SCHULTZ, T. J. (1970) Bolt, Beranek and Newman Report, 2005.

SCHULTZ, T. J. (1971a) *Building Acoustics* (ed. T. Smith, et al.), 168–72, Paper 72/3, Oriel Press.

SCHULTZ, T. J. (1971b) *Building Acoustics* (ed. T. Smith, et al.), 209–10, Paper 72/3, Oriel Press.

SCHULTZ, T. J. (1972) *Community Noise Ratings*, Applied Science.

SCHULTZ, T. J., and WATTERS, B. G. (1964) *J. Acoust. Soc. Am.* 36, 885.

SCHUMACHER, E. F. (1973) *Small is Beautiful*, Blond and Briggs.

SCHWARTZ, A. K. (1967) Effects of white noise on bar-press rate in rats, *Psychonomic Sci.* 8, 93–94.

SCHWETZ, F., von, et al. (1970) *Mschr. Ohrenheilk. LarRhinol.* 4, 162–7.

SCHWIRTZ, W. A. M. (1969) *Tech. Phys. Dienst*, TNO Report 806, 102 Marrt.

SELYE, H. J. (1950) Stress, *Acta Montreal*.

SEMCZUK, B. (1968) Studies on the influence of acoustic stimuli on resp. movements, *Polish Med. J.*, 7 (5) 1090–6.

SERAPHIM, H. P. (1958) *Acustica* 8, 280.

SERAPHIM, H. P. (1961) *Acustica* 11, 80.

SERAPHIM, H. P. (1963) *Acustica* 13, 75.

SESSLER, G. M., and WEST, J. E. (1964) *J. Acoust. Soc. Am.* 36, 1725.

SESSLER, S. M. (1973) *Am. Soc. Heat. Refrig. Air-condit. Engrs. Jl.* 15 (10) 39–46.

SETO, W. W. (1964) *Theory and Problems of Mechanical Vibrations*, McGraw-Hill (in the Schaums Outline Series); also 1971 *Acoustics*.

SHARLAND, I. (1969) *Fan noise*, British Acoustical Society Meeting, Edinburgh University, September.

SHARP, B. H. S., and BEAUCHAMP, J. W. (1969) Building Research Establishment Current Paper CP 31/69.

SHATALOV, N. N., et al. (1962) Report T-411-R, N65-15577, Defence Research Board, Toronto, Canada.

SHELTON, S. J. (1968) *J. Instn. Heat. Vent. Engrs.* 36, 131.

SHEPHERD, M. (1975) *Lancet*, 1, 322.

SILINCK, K., et al. (1941) *Čas. Lék. česk.* 1701–2 (Der psychothermische Reflex).

SIMON, C. W., and EMMONS, W. H. (1954) *A Critical Review of Learn While you Sleep Sources*, Report P-534, The Rand Corporation, Santa Monica, California.

SIMPSON, G. C. et al. (1974) *Ergonomics*, 17, (4), 481–487.

SKIPP, B. O. (ed.) (1966) *Vibration in Civil Engineering*, Butterworths.

SLATER, B. R. (1968) *J. Educat. Psychol.* 59, 239–43.

SMITH, H. C. (1947) *Music in Relation to Employee Attitudes, Piece-work Production, and Industrial Accidents*, Applied Psychology Monographs No. 14.

SMITH, K. R. (1951) Intermittent loud noise and mental performance, *Science* 114, 132–3.

SMITH, T. J. B., and KILHAM, J. K. (1963) *J. Acoust. Soc. Am.* 35, 715.

SMITH, W. A. S. (1961) Effects of industrial music in a work situation requiring complex mental activity, *Psych. Reports* 8, 159–62.

SMITHSON, R. N., and FOSTER, P. J. (1965) *Combust. Flame* 9, 426.

SMOLLEN, L. E. (1966) *J. Acoust. Soc. Am.* 40, 195–204.

SNOW, W. B. (1936) *J. Acoust. Soc. Am.* 8, 14.

SOMERVILLE, T., and GILFORD, C. L. S. (1957) *Proc. Instn. Elect. Engrs.* Pt. B, No. 14, March, 85–97.

SOROKA, W. W. (1970) *Appl. Acoust.* 3, 309–21.

STARKIE, D. N. M., JOHNSON, D. M. (1975) *The Economic Value of Peace and Quiet*, Saxon House, (D. C. Heath Limited).

STEINMANN, B., and JAGGI, U. (1955) Über den Einfluss von Geräuschen und Lärm auf den Blutdruck des Menschen, *Cardiologia*, Basel 27, 223.

STEINICKE, G. (1957) *Forschungsberichte des Wirtschafts und Verkehrsministeriums Nordrhein-Westfalen.*

STENNET, R. G. (1957) The relationship of performance level to level of arousal, *Journal of Experimental Psychology* **54**, 54–61.

STEPHENS, R. W. B., and BATE, A. E. (1966) *Acoustics and Vibrational Physics*, Arnold.

STEPHENS, R. W. B. (editor), (1975) *Acoustics 1974*, Chapman and Hall.

STEPHENS, S. D. G. (1970) *Q. Jl Exp. Psychol.* **22**, 9–13.

STERN, R. M. (1964) *Psychol. Rep.* **14**, 799–802.

STEVENS, J. C., and MARKS, L. E. (1965) *Proc. Natn. Acad. Sci. USA* **54**, 407–11.

STEVENS, S. S. (1934) *Proc. Natn. Acad. Sci. USA* **20**, 457–9.

STEVENS, S. S. (1935) *J. Acoust. Soc. Am.* **6**, 150.

STEVENS, S. S. (1937) *J. Acoust. Soc. Am.* **8**, 191–5.

STEVENS, S. S. (1941) *The Effects of Noise and Vibration on Psycho-Motor Efficiency*, Harvard University, OSRD Report 32, 274.

STEVENS, S. S. (1953) *Science* **118**, 576.

STEVENS, S. S. (1957) *Psychol. Rev.* **64**, 153–81.

STEVENS, S. S. (1959) *J. Acoust. Soc. Am.* **31**, 995–1003.

STEVENS, S. S. (1966) *Perception Psychophys.* **1**, 5–8.

STEVENS, S. S. (1969) *Perception Psychophys.* **6**, 251–6.

STEVENS, S. S. (1972a) *J. Sound Vibr.* **23** (3) 297–306.

STEVENS, S. S. (1972b) *J. Acoust. Soc. Am.* **51**, 575–601.

STEWART, L. J. (1969) *Noise in ventilating and airconditioning systems*, British Acoustical Society Meeting, Edinburgh University, September.

STEWART, W. F. R. (1970) *Children in flats: a family study*, National Soc. for the Prevention of Cruelty to Children.

STIKAR, J., and HLAVAC, S. (1963) The impact of noise on continued activity, *Čslká Psychol.* **7**, 246–51.

STRAHLE, W. C. (1972) *J. Sound Vibration* **23** (1) 113–25.

STRAKHOV, A. B. (1966) Report N67-11646, Joint Publications Res. Service, Washington DC.

STRUMPF, F. M. (1967) *Water installation units and their noise controls*, Conference on Acoustic Noise and its Control, Institution of Electrical Engineers, London.

STRUMPF, F. M., and OVENSEN, K. (1968) *Hot Water Systems and their noise production*, Paper F-5-9, Sixth International Congress on Acoustics, Tokyo.

SUGDEN, D. (1967) *Arup J.* **1** (4) 2–27.

SUGGS, C. W., and SPLINTER, W. E. (1961) Some physiological responses of man to workload and environment, *J. Appl. Physiol.* **16** (3) 413–20.

SUTHERLAND, A. (1969) Gas Council Res. Communication GC 161.

SUTHERLAND, L. C. (ed.) (1968) *Sonic and Vibration Environments for Ground Facilities*, Wyle Laboratories Research Report WR 68-2, USA.

SYLWAN, O. B. (1973) British Acoustical Society Meeting on *Viscoelasticity in Solids*, London, July 6th; also see *Br. Acoust. Soc. Proc.* **2** (3).

TARRIÈRE, C., and WISNER, A. (1962) *Travail hum.* **25**, 1–28.

TARZI, N. (1976) Personal Communication.

TAYLOR, C. (1973) Heating and Ventilating Research Association Information Circular, No. 25.

TAYLOR, J. R. (1973) *Refrig. Air Condit.*, December, 40–47.

TEICHNER, W. H., AREES, E., and REILLY, R. (1963) Noise and human performance, a psychophysiological approach, *Ergonomics* **6** (1) 83–97.

TERRACE, H. S., and STEVENS, S. S. (1962) *Am. J. Psychol.* **75**, 596–604.

THIELE, R. (1953) *Acoustica* **3**, 391.

THOLÉN, O. (1975) National Swedish Building Research Summary S45: 1975.

THOMAS, A., and WILLIAMS, G. T. (1966) *Proc. R. Soc. A*, **294**, 449.

THOMAS, R. J. (1969) *Traffic noise—the performance and economics of noise reducing materials*, British Acoustical Society Conference, 27–28 March, Newcastle-upon-Tyne.

TICO (1967) *Consult. Engr.* June.

TOPE, H. G. (1970) *Die Maschine* (4).

TOPE, H. G., and ROBINSON, J. J. (1971) *Running noise of high performance flat belts*, Second International Power Transmission Conference, June, London, Paper 6.

TWIGWELL, K. M. (1970) *Int. Conf. on the Working Environment*, Reading Junior Chamber of Commerce, University of Reading.

UGLOW, W., *et al.* (1937) Über die Wirkung von Lärm und Erschütterung auf den Gasaustausch, *Arbeitsphysiologie* **9**, 387.

UHRBROCK, R. S. (1961) Music on the job, its influence on worker morale and production, *Personnel Psychol.* **14**, 9–38.

UNGAR, E. E. (1971) Chapter 14 in *Noise and Vibration Control* (ed. L. L. Beranek), McGraw-Hill.

UTLEY, W. A. (1971) Personal communication with Building Research Station.

UTLEY, W. A. (1973) *Environmental design of buildings subject to traffic noise*, Symposium held at Building Research Establishment, July 4, 1973, Notes ref. 405/73.

UTLEY, W. A., and FLETCHER, B. L. (1969) *Appl. Acoust.* **2**, 131.

UTLEY, W. A., *et al.* (1969) *J. Sound Vibr.* 9 (1) 90–96.

VÉR, I. L., and HOLMER, C. I. (1971) Chapter 11 in *Noise and Vibration Control* (ed. L. L. Beranek), McGraw-Hill.

VERNON, P. E. (1970) *Creativity*, Penguin Education.

VITELES, M. S., and SMITH, K. R. (1946) An experimental investigation of the effect of change in atmospheric conditions and noise on human performance, *ASHVE Trans.* **52**, 167–82.

VOIGT, P., *et al.* (1974) *Traffic noise and annoyance in a laboratory condition*, Swiss Federal Institute of Technology, Zürich.

VULKAN, G. H., and STEPHENSON, R. J. (1970) *Unification of noise criteria*, British Acoustical Society Meeting, 23 July.

WALKER, C. (1971) *Building Acoustics* (ed. T. Smith, *et al.*), Chapter 9, 176–96, Oriel Press.

WALKER, J. H., and KERRUISH, N. (1960) *Proc. IEE* **107A**, 36.

WALLER, R. A. (1968) *J. Sound Vibr.* 8 (2) 177–85.

WALLER, R. A. (1969) *Appl. Acoust.* **2**, 121–30.

WALLER, R. A. (1970) *Regional Studies* **4**, 177–91.

WALLER, R. A. (1971) *7th International Congress on Acoustics*, Budapest, Paper 245, 10.

WARD, J. S. (1974) *Ergonomics* **17** (2) 233–40.

WARNAKA, G. E., *et al.* (1972) *Am. Ind. Hyg. Assoc. J.* (1) 1–11.

WATSON, J. H. (1968) *Refrig. Air Condit. Heat.* August, 30–33.

WEBSTER, J. C. (1969) *Sound and Vibr.* **3** (8) 22–26.

WEBSTER, J. C., and KLUMPP, R. G. (1963) *J. Acoust. Soc. Am.* **35**, 339–1344.

WEINSTEIN, A., and MACKENZIE, R. S. (1966) *Percept. Motor Skills* **22**, 498.

WELFORD, A. T. (1973) *Ergonomics* **16** (5) 567–80.

WELLS, R. J. (1958) *Noise Control* **4** (4) 9.

WENZEL, H. G. (1970) Indoor climatic conditions; physiological aspects, evaluation and optimum levels, *ILO handbook on Occupational Safety and Health*, 21, International Labour Office, Geneva, 287–322.

WEST, J. E. (1966) *J. Acoust. Soc. Am.*, **40**, 1245.

WEST, L. J. (1967) Psychopathology produced by sleep deprivation in *Proceedings of the Association for Research in Nervous and Mental Disease, Sleep and altered States of Consciousness* (ed. S. S. Ketu), Williams and Wilkins, Baltimore, Maryland.

WEST, M. (1973a) *Br. Acoust. Soc. Proc.* **2** (1), Paper 72/53.

WEST, M. (1973b) *Building Res. Practice*, July–August, 207–9.

WESTERBERG, G. (1971) *Appl. Acoust.* **4**, 115–29.

WESTON, H. C., and ADAMS, S. (1935) Industrial Health Research Board, Report 70, 1, London.

WHALE, W. E. (1966a) *Heat. Pip. Air Condit.* **38** (1) 122.

WHALE, W. E. (1966b) *Heat. Pip. Air Condit.* **38** (2) 97.

WHALE, W. E. (1966c) *Heat. Pip. Air Condit.* **38** (3) 118.

WIENER, F. M., and KEAST, D. M. (1959) *J. Acoust. Soc. Am.* **31** (6) 724–33.

WILKSTRÖM, B. (1964) Beitrag zur zweckmässigen Bestimmung und Darstellung des Ventilator-geräusches als akustische Berechnung von Lüftungsanlagen, dissertation to Technical University, Berlin.

WILHEMSEN, A., and LARSSON, B. (1973) National Swedish Building Research Summary D3, 1973.

WILKINSON, R. T. (1958) MRC, App. Psychol. Res. Unit, Report No. 323/58.

WILKINSON, R. T. (1961) *J. Exp. Psychol.* **62**, 263–71.

WILKINSON, R. T. (1963) *J. Exp. Psychol.* **66**, 332–7.

WILKINSON, R. T. (1964) *Ergonomics* **7**, 175–86.

WILKINSON, R. T., *et al.* (1966) *Psychol. Sci.* **5**, 471–2.

WILKINSON, R. T. (1969) Some factors influencing the effect of environmental stressors upon performance, *Psych. Bull.* **72** (4) 260–72.

WILKINSON, R. T. (1972) *New Society* **21**, 13 July, 73–75.

WILKINSON, R. T. (1974) *Ergonomics*, **17**, (6), 745–756.

WILKINSON, R. T., and HAINES, E. (1970) *Acta Psychol.* **33**, in *Attention and Performance* III (ed. A. F. Saunders), 402–13, North-Holland, Amsterdam.

WILLIAMS, H. L., *et al.* (1964) *Electroenceph. Clin. Neurophysiol.* **16**, 269–79.

WILSON, A. H. (1963) *Noise*, HMSO.

WINCKEL, F. (1967) *Music, Sound and Sensation*, Dover Publications.

WINDER, A. F. (1976) Personal Communication.

WISE, A. F. E. (1973) Building Research Establishment Current Paper CP 6/73.

WOLGERS, B., and WIEDLING, K. (1971) Study of Office Environment, Nat. Swedish Bldg. Res., Summary T 15:1971 and Report 0033/70.

WOODHEAD, M. M. (1966) An effect of noise on the distribution of attention, *J. Appl. Psychol.* **50**, 296–9.

WOODS, R. I. (1972) *Noise Control in Mechanical Services* (ed. R. I. Woods), Sound Attenuators Ltd. and Sound Research Laboratories Ltd.

WOODS, R. I., and SMITH, T. J. B. (1969) *J. Instn. Heat. Vent. Engrs.* **37**, A34.

WOOLLEY, R. M. (1967) *Glass Age* **10**, 44.

WOOLS, R., and CANTER, D. (1970) *Appl. Ergonomics* **1**, (3), 144.

WYATT, S., and LANGDON, J. H. (1935) Industrial Health Research Board Report, 77, HMSO.

WYON, D. (1970) *Performance under Sub-optimal Conditions* (ed. Davis), 68–82, Taylor and Francis.

WYON, D., et al. (1968) *J. Hygiene* **66**, 229–48.

YANG, S. J. (1973) *Noise from prime movers*, British Acoustical Society Meeting, Southampton University, 9–10 July; also see *BAS Proceedings* **2** (3).

YERGES, J. F., and BOLLINGER, J. G. (1972) *Sound Vibr.* **6** (6) 19–24.

YOUNG, R. W. (1964) *J. Acoust. Soc. Am.* **36**, 289.

YOUNG, R. W. (1965) *J. Acoust. Soc. Am.* **38**, 524–30.

ZEEMAN, E. C. (1965) *J. Math. Computer Sci. Biol. Med.*, Medical Research Council, 277.

ZEITLIN, L. R. (1969) *Report to Organisation and Procedures Department*, Port of New York Authority, April.

ZELLER, W. (1964) *Lärmbekämpfung* **8**, 113–9.

ZELLER, W. (1974) See Finkelstein (1974).

ZWICKER, E. (1960) *Acustica* **10** (1).

AUTHOR INDEX

Abey-Wickrama, I. 34, 582
Acking 24
Adams, S. 161, 166, 595
Adrian, E.D. 38, 582
Alberti 15
Aldersey-Williams, A.G. 191, 198, 202, 203, 580, 582
Alexandre, A. 134, 582
Alfredon 464, 466
Allen, C.H. 223, 226, 582
Almida, M.R. de 161, 582
Ando, Y. 34, 582
Andrews, B. 129, 582
Andriukin, A.A. 90, 582
Anthrop, D.E. 32, 582
Antoni, N. 206, 582
Anzou, S. 140, 589
Arees, E. 594
Arup, Sir Ove 17, 27
Arvidsson, O. 160, 582
Ashley, C. 113, 582
Åstrand, P.O. 88, 582
Atal, B.S. 511, 519, 533, 547, 582
Atherley, G.R.C. 87, 88, 89, 93, 582
Attenborough, K. 122, 582
Awaya, K. 452
Ayres, I.P. 161, 582

Bacelon, M. 140, 589
Bach, J.S. 15, 90
Baines, A.W. 199, 582
Bakan, J. 164, 582
Baker, J.K. 421, 485, 582
Baldes, E.J. 88, 91, 584
Ball, E.F. 211, 212, 213, 582
Barbenza, C.M. de 66, 582
Barde, J. Ph. 582
Barron, M. 544, 582
Bartlett, F.C. 33, 116, 582
Bartok 542
Bate, A.E. 170, 173, 594
Baudelaire 528
Baumann, K. 226, 582
Bazley, E.N. 291, 580, 582
Beament, J. 73
Beauchamp, J.W. 571, 572, 593
Beckmann, M. 52, 585

Beethoven 86, 549
Békésy, G. 74, 84, 582
Bell, A. 580
Benjamen, Sir John 3
Benroy, E. 25
Beranek, L.L. 19, 47, 48, 138, 142, 146, 167, 168, 169, 172, 179, 180, 183, 204, 223, 226, 270, 342, 343, 382, 384, 385, 387, 388, 389, 390, 505, 529, 530, 531, 534, 539, 542, 543, 580, 583
Berg, A. 101
Berlyne, D.E. 36, 162, 583
Bernstein 115
Berry, G. 91, 583
Berry, T. 585
Bilsen 519
Bindra, D. 162, 583
Bishop, D.E. 130, 142, 583, 586
Bitter, C. 109, 119, 120, 121, 309, 310, 320, 322, 583, 585
Bladier, B. 92, 584
Blazier, W.E. 237, 238, 239, 241, 304, 305, 583
Blitz, J. 371, 583
Boalt, C. 122, 583
Boggs, D.H. 163, 583
Boje, A. 583
Bollinger, J.G. 174, 596
Boltzmann 39
Bommes, L. 218, 219, 220, 221, 223, 583
Borg, E. 88, 583
Boring, E.G. 583
Borsky, P.N. 152, 583
Bottom, C.G. 116, 145, 583
Boulez 5
Bowman, N. 179, 583
Boyce, P.R. 57, 59, 60, 61, 62, 186, 187, 188, 583
Brahms 542
Brandt, O. 291, 583
Bregman, H.L. 583
Briffa, F.E.J. 246, 247, 248, 250, 251, 254, 255, 583, 586
Brion, M. 520
Britten, B. 350
Broadbent, D.E. 40, 41, 88, 154, 155, 161, 162, 164, 583
Broch, J.T. 580
Brook, A.D. 164, 191, 197, 198, 199, 585
Brookes, M.J. 24, 55, 56, 57, 59, 62, 583

Brosio, E. 320, 583
Brown, R.L. 163, 583
Bruckmayer, F. 139, 141, 163, 583
Bruckner 523
Brüel 68
Brundrett, G.W. 583
Bryan, M.E. 116, 117, 118, 119, 129, 152, 156, 157, 158, 164, 583, 584, 591
Bugard, P. 88, 91, 584
Burd, A.N. 354, 355, 356, 547, 552, 553, 584
Burger, J.F. 542, 590
Burgtorf, W. 525, 538, 584
Burke, R.F. 233, 584
Burns, W. 85, 154, 273, 580, 584
Butcher, N.F.H. 162, 584

Cadiergues, R. 266, 270, 587
Canac, R. 92, 584
Cannon, W.B. 91, 584
Canter, D. 24, 44, 584, 596
Carlson, J.A. 163, 584
Carpenter, I.K. 23
Carter, F.W. 235, 584
Cartwright, A. 120, 586
Cavanaugh, W.J. 169, 170, 173, 487, 584
Chaddock, J.B. 266, 584
Chapman, D. 119, 122, 584
Chombart de Lauwe, P. 113, 300
Churcher, B.G. 65, 584
Clark, R.J. 181, 183, 184, 230, 231, 232, 233, 320, 584
Coblentz, A. 584
Coermann, R.R. 109, 113, 584
Cohen, A. 164, 584
Colquhoun, W.P. 155, 584
Conrad, D.W. 87, 88, 89, 92, 164, 584
Corbeille, C. 88, 91, 584
Corcoran, D.W.J. 36, 155, 156, 162, 584
Cowell, J.R. 178, 584
Coye, K. 116, 589
Crede, C.E. 475, 476, 484, 486, 487, 491, 492, 493, 580, 587
Cremer, L. 342, 490, 500, 504, 521, 525, 529, 536, 537, 567, 584
Cresti, C. 15, 584
Crickmay, C.L. 22, 584
Crocker, M.J. 212, 350, 464, 466, 571, 584
Croome, D.J. xvii, 24, 44, 52, 116, 131, 162, 163, 164, 180, 191, 197, 198, 199, 217, 309, 395, 396, 397, 400, 401, 402, 404, 534, 583, 584, 585
Cummings, A. 245, 249, 250, 585
Curle, N. 265, 585

Dadson, R.S. 65, 592
Dagg, I.R. 518, 519, 585
Damaske, P. 538, 585
Darrow, C.W. 88, 92, 585
Dart, S.L. 487
Davies, A.D.M. 165, 585
Davies, D.R. 80, 164, 585

Davis, H. 90, 585
Davis, R.C. 88, 91, 585
Day, B.F. 191, 192, 196, 197, 585
Day, H.I. 164, 585
de Doelan 524
De Lange, P.A. 294, 296, 585
de Odyngton, W. 422
Dean, R.D. 88, 585
Delany, M.E. 134, 455, 585
Dieckmann, D. 109, 585
Doak, P.E. 515, 585
Doelle, L.L. 274, 276, 277, 318, 319, 337, 338, 392, 530, 532, 533, 580, 585
Douglas, H. 587
Dubout, P. 591
Duffy, F. 585
Dunkerley 481
Dyer, I. 244, 585

Eck, B. 218, 222, 257, 316, 334, 340, 341, 405, 585
Edberg, G. 14, 15, 16, 17, 585
Edwards, R.S. 155, 584
Egenberg, K.E. 88, 585
Eijk, J. van den 320, 321, 322, 585
Einbrodt, M.J. 52, 585
Embleton, T.F.W. 285, 518, 519, 585
Emmenegger, H. 592
Emmons, W.H. 102, 593
Endejann, R.H. 211, 585
Etholm, B. 88, 585
Evans, E.J. 580
Eyring, C.T. 507, 585
Eysenck, H.J. 66, 88, 158, 585

Farrell, R. 174, 178, 585
Fasold, W. 314, 568, 585
Fechner 39
Felton, J.S. 161, 585
Ferguson, D. 34, 35, 585
Ffowcs-Williams 210
Finch, D.M. 129, 582
Finkelstein, W. 215, 216, 217, 218, 219, 221, 222, 223, 224, 225, 585
Fischer, B. 153
Fletcher, B.L. 595
Fletcher, H. 65, 67, 72, 167, 172, 585
Fog, H. 125, 126, 127, 129, 585
Ford, R.D. 124, 316, 585
Foster, P.J. 248, 593
Foudraine, J. xiii, 586
Francis, J.G. 131, 586
François, P. 228, 229, 230, 586
Freeman, H.L. 34, 586
Freeman, J. 161, 586
French, N.R. 167, 168, 586
Fricke, J.E. 580
Friedel, B. 88, 92, 588
Frye 487

Galambos, R.   591
Galloway, W.J.   142, 586
Gandhi   79
Gardner, J.E.   161, 590
Gardner, W.   107, 108, 586
Garner, W.R.   65, 586
Gatonni, F.   34, 586
Geldard, F.   39, 78, 586
Gerlach   508, 509
Gerletti, A.   592
Ghamah-Zadeh, N.   254, 255, 586
Gibbons, S.G.   87, 88, 91, 93, 586
Gierke   150
Gilbert, P.   142, 148, 586
Gilford, C.I.S.   530, 593
Gilgen, A.   144, 586
Gillot, J.   83
Givoni, B.   88, 586
Gloag, J.   5, 586
Glorig, A.   591
Goethe   49
Goetzinger, C.P.   592
Goldman, R.F.   88, 586
Golombek, H.   154
Gomperts, M.C.   294, 296, 297, 298, 568, 586
Goodman, P.   7, 487
Gordan, A.   25
Gordon, C.G.   209, 586
Göselle, K.   212, 568, 586
Gottlob, D.   544, 586
Gottschalk, O.   186, 189, 586
Graham, J.B.   221, 586
Grandjean, E.   52, 53, 54, 55, 59, 62, 83, 88, 92,
    99, 102, 108, 126, 127, 133, 134, 137, 140, 142,
    143, 144, 145, 147, 148, 153, 182, 185, 186, 187,
    189, 213, 281, 286, 287, 309, 310, 370, 371, 373,
    586, 589, 591, 592
Gray, P.G.   119, 120, 121, 320, 321, 586
Greenfield, N.   52, 588
Gregory   36
Greim, H.   210, 586
Grether, W.F.   155, 586
Griffiths, I.D.   133, 586
Grosjean, J.   483, 484, 586
Gross, E.   580
Grossman, S.P.   586
Gubler, F.   52, 586
Guignard, E.   113, 586
Guignard, J.C.   113, 586
Guinot, G.   163, 586
Guirao, M.   77, 78, 586
Gupta, A.K.   250, 406, 586, 592
Guth, E.   487

Haas, H.   525, 527, 528, 539, 586
Haines, E.   80, 81, 88, 92, 595
Hale, H.B.   92, 586
Hall, J.C.   162, 587
Hall, P.   544
Hamme   487
Hanfling, O.   85, 86, 587

Hardy, H.C.   266, 487, 587
Harman, D.M.   587
Harmon, F.L.   88, 91, 587
Harris, C.M.   333, 341, 475, 476, 484, 486, 487,
    491, 492, 493, 517, 580, 587
Harris, J.D.   75, 587
Harwood, H.D.   547, 587
Hathaway, S.R.   66, 88, 587
Hattori, H.   34, 582
Hawkes, R.J.   529, 587
Hay, B.   53, 133, 185, 204, 205, 393, 587
Hebbel, F.   101
Hecht, E.   162, 588
Heckl, M.   490, 584
Hegvold, L.W.   174, 587
Hehman, H.W.W.   314, 590
Helies, J.   266, 270, 587
Hellman, R.   65, 587
Helmholtz   74
Helper, M.M.   91, 587
Henderson, J.   212, 587
Hennenberg, F.   523, 587
Hepworth   5
Herridge, C.F.   35, 587
Hertz   149
Herzberg, F.   51, 587
Hess, R., Jr.   92, 587
Hesse, H.   101, 102
Hesselgren, S.   24, 44, 587
Heusser, M.   52, 587
Hird, J.B.   123, 587
Hirschorn   256
Hlavac, S.   161, 594
Hoagland, H.   88, 89, 92, 587
Hockey, G.R.J.   154, 155, 156, 164, 585, 587
Hockney   5
Hoffman, J.E.   163, 587
Holmberg, J.   323, 587
Holmer, C.I.   490, 595
Holmes, M.J.   261–70, 587
Honikman, B.   12, 24, 587
Hood, J.D.   66, 587
Hoogendoorn, K.   185, 187, 189, 190, 191, 201,
    587
Hoover   238
Horch, C.   320, 322, 583
Horder, Lord   33
Houston, B.K.   161, 162, 587
Howes, D.H.   71, 72, 587
Hume, W.F.   89, 587
Humes, J.F.   161, 587
Humphreys, H.R.   274, 307, 308, 580, 588, 591
Hundent, A.T.   52, 588
Hunt, A.E.   34, 44, 47, 122, 590
Hurle, I.R.   247, 248, 588
Hurty, W.C.   588
Huxley, J.   272

Ingard, U.   228, 229, 256–9, 262, 588
Ingerslev   151
Ingerslev, F.   280, 588

Ingerslev, R.   568, 588
Iredale, R.A.   255, 588
Izumiyama, M.   163, 588

Jackson, G.M.   570, 588
Jaggi, U.   88, 91, 593
James, W.   36, 37, 588
Jansen, G.   88–92, 95, 97, 98, 100, 102, 105, 106, 588
Jensen, P.   108, 588
Jerison, H.J.   154, 588
Johannson, C.R.   160, 588
Johns, D.J.   149, 588
Johnson, D.M.   350, 593
Johnson, L.C.   88, 588
Jones, T.M.   161, 587
Jordan, V.L.   510, 511, 513, 542, 544, 547, 548, 550, 551, 552, 554, 588
Josse, R.   584
Jouvet, M.   92, 105, 588
Jungk, R.   3, 19, 25, 588

Kalsbeek, J.W.H.   88, 588
Kamber, F.   211, 214, 588
Karrasch, K.   88, 91, 588
Kathé, E.J.   592
Keast, D.M.   285, 595
Keefe, F.B.   88, 588
Keets, W. de V.   543, 588
Keighley, E.C.   130, 588
Kemp, M.F.   204, 205, 587
Kennedy, D.R.   251, 255, 583, 588
Kennedy, F.   88, 91, 588
Kerrick, J.S.   115, 588
Kerruish, N.   235, 595
Kihlman, T.   568, 588
Kilham, J.K.   246, 248, 249, 593
King, A.J.   65, 234, 584, 588
Kingsbury, H.F.   172, 174, 177, 588
Kirk, R.E.   162, 588
Kjaer   68
Klensch, H.   588
Kluge, E.   88, 92, 588
Klumpp, R.G.   595
Knudsen, V.O.   580
Koch-Emmery, W.   221, 588
Köhler, W.   77, 588
Konz, S.A.   163, 588
Korn, J.   255, 588
Kosten, C.W.   180, 589
Kovrlgin, S.D.   161, 589
Kristensen, J.   211, 213, 589
Krkovic, A.   80, 585
Kryter, K.D.   65, 85, 88, 100, 102, 105, 107, 130, 134, 138, 142, 152, 161, 163, 169, 170, 174, 175, 580, 586, 589, 590
Kuhl, W.   530, 533, 589
Kuller   24
Küper, R.   589
Kürer   544

Kurtze, G.   213, 589
Kuttruff, H.   496–9, 503, 511, 519, 520, 522, 526, 529, 537, 538, 542, 589
Kyburz, W.   52, 54, 589

Laird, D.A.   88, 90, 91, 116, 161, 589
Lamure, C.   140, 589
Lane, H.L.   178, 589
Lang, J.   139, 141, 163, 583
Langdon, F.J.   130, 132, 133, 281, 586, 589
Langdon, J.H.   161, 596
Lappat, A.   52, 589
Large, J.B.   139, 589
Larsson, B.   356, 357, 595
Lawrence, A.   318, 319, 589
Lazan   487
Le Corbusier   5, 8, 12, 15, 16
Lee, V.   12, 589
Lehmann, D.W.   88, 163, 589
Lehmann, G.   90, 94, 102, 589
Lehmann, P.   544, 589
Leidel, W.   215, 216, 589
Leonard, J.A.   589
Leonardo da Vinci   15
Leskov, E.A.   339, 340, 589
Leventhall, H.G.   250, 570, 588, 592
Levi, L.   88, 89, 93, 589
Lewerentz   15
Lewis, P.T.   174, 175, 291, 294, 370, 590
Licklider, J.C.R.   107, 586
Lighthill, M.J.   210, 418, 590
Lindsay, R.B.   580
Liquorish, A.D.   487, 591
Ljunggren, S.   370, 590
Llewelyn-Davies, Lord   24
Locher, K.   245, 590
Lochner, J.P.A.   542, 590
London, A.   571, 590
Longmore, D.K.   464, 480, 591
Lord, P.   316, 317, 318, 585, 590
Losch, W.   245, 590
Lottermoser, W.   590
Loveland, K.   530
Lovelock, E.C.   24, 590
Lukas, J.S.   105, 107, 590
Luke, D.   49
Lundberg, B.   140, 590

McBain, W.N.   162, 590
McCallum, C.   80, 590
MacEwen, M.   5, 11, 590
McGehee, W.   161, 590
McGlothlen, C.L.   88, 585
McGrath, J.E.   79, 154, 590
McKennell, A.C.   34, 44, 47, 122, 590
Mackenzie, R.   508, 509, 521, 542, 547, 590
Mackenzie, R.S.   162, 595
Mackinnon, D.W.   14, 590
McK. Nicholl, A.G.   155, 590
Mackworth, N.H.   88, 590

McNair, H.P.   154, 590
MacPherson, R.K.   81, 108, 109, 590
McRobert, H.   66, 590
Maekawa, Z.   282, 283, 590
Magnus, K.   590
Mahler   523
Malmo, R.B.   590
Manning, P.   163, 590
Mariner, T.   314, 590
Marks, L.E.   40, 594
Markus, T.   10, 20, 23, 24, 590
Marsh, J.   294, 590
Marsh, J.A.   590
Marshall, A.H.   42, 391, 529, 537, 538, 543, 590, 591
Maruyama, K.   163, 591
Marvet, B.H.   266, 270, 591
Mehrabian, A.   24, 37, 49, 50, 51, 165, 591
Meister, F.J.   285, 591
Melluish, D.J.   274, 588
Melzack, R.   108, 591
Menuhin Yehudi   7
Meyer, E.   520, 525, 527, 537, 591
Meyer-Delius, J.   88, 90, 94, 589, 591
Mikeyer, H.P.   161, 589
Miller, G.A.   171, 179, 591
Miller, I.N.   244, 585
Mills, C.H.G.   130, 591
Mitchell   162
Møller, A.R.   88, 583
Moore   5
Moreira, N.M.   116–19, 129, 152, 591
Morgan, C.T.   88, 591
Morgan, J.J.B.   88, 91, 591
Morgan, K.   529, 538, 591
Morgenstein, F.S.   43, 591
Morse, P.M.   270, 341, 442, 591
Motteram, J.M.   178, 584
Mulholland, K.A.   571, 591
Müller, G.   349, 591
Muncey, R.W.   527, 591
Munson, W.A.   65, 67, 72, 73, 585, 591
Murray   142

Nagatsuka, Y.   163, 591
Nassenstein, G.   245, 590
Neidt, C.O.   161, 586
Nelson, E.N.   107, 591
Nemecek, J.   52–55, 59, 62, 185, 186, 187, 189, 591
Nervi   11
Newland, D.E.   487, 591
Nicely, P.E.   171, 179, 591
Nickson, A.F.B.   591
Nielsen, A.K.   280, 568, 588
Niese, H.   512, 540, 591
Norris, C.H.   149, 591

Oehlschlägel, H.K.   526, 584
Oldham, G.   6

Ollerhead, J.B.   136–9, 285, 591
Olsen, J.   107, 591
Oltman, P.K.   162, 591
Oosting, W.A.   293, 591
Oppenheim, A.   256–9, 262, 588
Oppliger, G.C.   88, 92, 99, 591, 592
Ornstein, R.   12, 13, 17, 591
Ovensen, K.   594

Pancholy, M.   530, 591
Parbrook, H.D.   591
Park, J.F., Jr.   163, 591
Parkin, P.H.   148, 307, 308, 529, 538, 580, 591
Paul, J.   520
Payne, M.C., Jr.   163, 591
Pepler, R.D.   155, 156, 591
Perl, E.R.   92, 591
Peterson, A.   580
Petrusewicz, S.A.   181, 183, 184, 230–3, 320, 464, 480, 584, 591
Peutz, V.M.A.   191–4, 197, 198, 202, 203, 591
Phillips, D.   47, 591
Piesse, R.A.   142, 580
Pilkington Bros. Ltd.   293, 294, 591, 592
Pirn, R.   174, 592
Plato   36
Plenge, G.   519, 542, 543, 544, 592
Plutchik, R.   162, 592
Pomfret, B.   234, 235, 236, 592
Poole, R.   88, 592
Port, E.   70, 592
Poulton, E.C.   40, 41, 155, 156, 592
Powell, J.A.   184, 592
Price, A.J.   571, 584, 591

Raab, W.   88, 93, 592
Rashid, Shaikh   8
Rathbone, R.A.   46, 592
Rathé, E.   280, 281, 283, 592
Rayleigh   481
Rayleigh, J.W.S.   592
Rayleigh, Lord   39
Reason, J.T.   66, 592
Rechtschaffen, A.   592
Resse, J.A.   406, 592
Reichardt, W.   539, 592
Reichow, H.B.   281, 592
Reilly, N.   90, 592
Reilly, R.   594
Rettinger, M.   283, 284, 592
Ribner, H.S.   248, 592
Rice, C.G.   152, 153, 592
Richards, E.J.   136, 285, 350, 592
Richter, H.R.   88, 92, 102, 592
Ridley, G.K.   235, 592
Ridpath   380
Riescz, R.R.   39, 592
Riland, L.H.   52, 592
Roberts, B.M.   402, 534, 585
Roberts, J.P.   250, 592

Robinson, D.W.   65, 125, 126, 130, 144, 146, 580, 591, 592
Robinson, J.J.   594
Rodahl, K.   88, 582
Rogers, C.R.   12
Rosen, S.   85, 592
Rosenberg, K.   582
Rossi, L.   92, 592
Rothlin, E.   90, 592
Rousey, C.L.   592
Rowland, V.   92, 105, 592
Rubenstein, M.F.   149, 588
Rückward, W.   212, 592
Ruhrberg, W.   285, 591
Russell, J.A.   24, 37, 49, 50, 51, 165, 591

Sabine, W.C.   117, 207, 333, 517, 522, 523, 580, 592
Sakamoto, H.   88, 92, 592, 593
Sapcote, B.   25
Saunders, A.F.   162, 593
Sauter, A.   568, 593
Schieber, J.P.   92, 105, 593
Schilling, R.S.F.   580
Schmidt   52
Schmidt, R.   593
Schmidt, W.   539, 540, 592, 593
Schodder, G.R.   525, 527, 591
Scholes, W.E.   281, 283, 589, 593
Schopenhauer, A.   118, 593
Schreiber, L.   163, 593
Schroeder, M.R.   498, 513, 542, 544, 582, 593
Schubert, P.   526, 527, 593
Schuder   212
Schultz, T.J.   47, 124, 138, 152, 191, 201, 202, 314, 316, 520, 583, 593
Schulze, J.   102, 105, 588
Schumacher, E.F.   79, 593
Schwartz, A.K.   162, 593
Schweiz   310
Schwetz, F. von   85, 593
Schwirtz, W.A.M.   314, 593
Selye, H.J.   88, 89, 93, 593
Semczuk, B.   88, 593
Seraphim, H.P.   519, 525, 593
Sessler, G.M.   520, 582, 593
Sessler, S.M.   236, 238, 239, 242, 593
Seto, W.W.   483, 593
Severn, K.   25
Sharland, I.   216, 593
Sharp, B.H.S.   571, 572, 593
Shatalov, N.N.   90, 593
Shaw, George Bernard   25
Shawe-Taylor, D.   523, 524
Shelton, S.J.   593
Shepherd, M.   34, 593
Silinck, K.   88, 593
Simon, C.W.   102, 593
Simon, J.R.   163, 583
Simpson, G.C.   164, 593
Skipp, B.O.   149, 593

Slater, B.R.   163, 593
Smith, H.C.   162, 593
Smith, K.R.   88, 155, 593, 595
Smith, T.   590
Smith, T.J.B.   246, 248, 249, 266, 593, 596
Smith, W.A.S.   163, 593
Smithson, R.N.   248, 593
Smollen, L.E.   482, 593
Snow, W.B.   72, 593
Somerville, T.   530, 593
Soraka, W.   568, 593
Soroka, W.W.   257, 260, 593
Spencer, C.   161, 585
Splinter, W.E.   88, 594
Starkie, D.N.M.   350, 593
Steinberg, J.C.   167, 168, 585, 586
Steinicke, G.   92, 108, 594
Steinmann, B.   88, 91, 593
Stennet, R.G.   594
Stephens, R.W.B.   150, 151, 170, 173, 210, 283, 594
Stephens, S.D.G.   66, 594
Stephenson, R.J.   163, 165, 595
Stern, R.M.   88, 91, 594
Stevens, J.C.   594
Stevens, S.S.   39, 40, 41, 65–69, 71–74, 76, 77, 78, 88, 91, 135, 583, 586, 594
Stewart, L.J.   256, 585, 594
Stewart, W.F.R.   123, 594
Stikar, J.   161, 594
Stockhausen   5
Strahle, W.C.   248, 594
Strakhov, A.B.   90, 594
Strauss, Richard   101, 523
Stringer   12
Strumpf, F.M.   214, 594
Sugden, D.   530, 594
Suggs, C.W.   88, 594
Sutherland, A.   250, 594
Sutherland, L.C.   149, 594
Sylwan, O.B.   493, 594

Tamm, J.   90, 94, 589
Tancho   422
Tarnopolsky, A.   34, 586
Tarrièr, C.   162, 594
Tarzi, N.   35, 594
Taylor   243, 244, 479, 485
Taylor, C.   594
Taylor, D.W.   172, 174, 177, 588
Taylor, J.R.   594
Teichner, W.H.   88, 163, 594
Tempest, W.   152, 584
Terrace, H.S.   76, 77, 594
Thiele, R.   539, 594
Tholén, O.   370, 594
Thomas, A.   247, 594
Thomas, R.J.   594
Thylebring, H.   323, 587
Tico   488, 594
Toepler   39

Tope, H.G. 227, 594
Twigwell, K.M. 594
Tyzzer, F.G. 487

Uglow, W. 88, 91, 594
Uhrbrock, R.S. 594
Ungar, E.E. 489, 490, 493, 595
Utley, W.A. 294, 378, 595
Utzon, J. 544

van Os, G.J. 19, 180, 389
van Weeren, P. 109, 119, 120, 121, 309, 310, 320, 583
Vér I.L. 490, 595
Vernon, P.E. 12, 17, 595
Viteles, M.S. 88, 155, 595
Vitruvius 522
Vivaldi 15, 115
Voigt, P. 153, 595
Voitsberger, C.A. 212, 586
Volkman 73
von Karajan, Herbert 534, 535, 542
Vulkan, G.H. 163, 165, 595

Wagener, B. 584
Wagner 523
Wahl 487
Walker, C. 327, 329, 595
Walker, J.H. 235, 595
Waller, R.A. 179, 204, 350, 378, 595
Walters, H.B. 421
Ward, J.S. 323, 324, 595
Ward, W.D. 580
Warnaka, G.E. 422, 595
Watson, J.H. 595
Watters, B.G. 520, 593
Weber 39
Webster, C.I.D. 211, 212, 213, 582
Webster, J.C. 47, 179, 595
Weinstein, A. 162, 595
Welford, A.T. 78, 79, 80, 595

Welin, A. 582
Wells, R.J. 339, 595
Wenzel, H.G. 88, 595
West, J.E. 520, 543, 593, 595
West, L.J. 102, 595
West, M. 191, 193, 194, 195, 595
Westerberg, G. 568, 569, 595
Weston, H.C. 161, 166, 595
Whale, W.E. 595
Wiedling, K. 54, 596
Wiener, F.M. 285, 595
Wilhemsen, A. 356, 357, 595
Wilkens 542, 543, 544
Wilkinson, R.T. 80, 81, 88, 92, 154, 155, 156, 160, 595
Wilkström, B. 223, 224, 226, 595
Williams, G.T. 247, 595
Williams, H.L. 92, 102, 104, 105, 595
Wilson, A.H. 122, 124, 127, 142, 183, 595
Winckel, F. 64, 74, 524, 531, 595
Winder, A.F. 35, 595
Wise, A.F.E. 378, 380, 381, 382, 596
Wisner, A. 162, 594
Wolgers, B. 54, 596
Woodhead, M.M. 163, 596
Woods, R.I. 266, 486, 494, 495, 596
Woods, S. 7
Woolley, R.M. 291, 596
Wools, R. 44, 596
Wyatt, S. 161, 596
Wyon, D. 81, 163, 596

Yang, S.J. 233, 234, 596
Yerges, J.F. 174, 596
Yerges, L.F. 580
Young, R.W. 182, 387, 596

Zeeman, E.C. 43, 44, 596
Zeitlin, L.R. 52, 596
Zeller, W. 186, 189, 221, 257, 596
Zwicker, E. 68, 596
Zwislocki, J.J. 65, 587

SUBJECT INDEX

Absorbent linings 291
Absorbent materials 206
Absorption 391, 508
  effect on sound field 569
Absorption attenuators 339
Absorption coefficient 196, 200, 201, 333, 518, 569
Absorption materials 199, 200, 316
Acclimatisation 38
Acoustic enclosures 307
Acoustic energy 452
Acoustic filter 464
Acoustic impedance 63, 429, 452
Acoustic impedance function 429
Acoustic pressure 427, 428, 445, 447, 448
  Pascal as unit of xiv
Acoustic pressure wave 465
Acoustic reactance 429
Acoustic resistance 427, 428, 452
Acoustic splitters 306
Acoustical criteria 363
Acoustical design briefs 363
Acoustical engineering undergraduate course xiv
Acoustical mass 428
Acoustics 29
  physical and subjective qualities and quantities in 61
Acoustics laboratory 302
Activity system 21
Adaptation 33, 35, 38, 100, 121, 152
Adrenaline 87, 100
Adrenaline/noradrenaline release 89, 93
Adrenocorticotrophic hormone (ACTH) 89
Aesthetics and sound absorption 337
Air distribution 352
Air-distribution networks, sound generation 256–65
Air-handling diffuser 204
Air jet, sound power spectrum 266
Air supply rate 379
Air terminal devices 265–71
  acoustics 265
  high-pressure 271
Airborne sound 149, 417
  direct transmission through massive wall 560–2
  direct transmission through sprung wall 562–3

Airconditioning 182, 205, 206, 384, 393–413, 457, 516
  background sound pressure level 405–13
  sound generation 214
Aircraft noise xiv, xvi, 126, 127, 129, 134, 137–40, 142–4, 147, 284, 285, 364
  see also Sonic boom
Airflow rates 257
Airflow systems 208
  attenuators 339–46
  design 401
Airflow velocity effects 256
Airports. See Aircraft noise
Airstream, solid bodies in 228
Alcohol and noise 155–6
Ambiophony 538
Amplitude function 437
Analogue computer 434
Analogues 430–5
Analysis and imagination 19
Anechoic rooms 457
Angular frequency 438
Annoyance 87, 114–16, 119, 122, 124, 130, 136, 139, 150–2
Annoyance assessment 119
Annoyance index 153
Antinodes 440
Anxiety-depression syndrome 89
Appraisal of buildings 23
Architect
  and engineer 11, 12
  excessive demands on 6
  image of 14
Architectural heritage 3
Architecture
  perception of 3
  present problems of 3
Arousal level 36, 37, 44, 46, 79, 87, 100, 113, 154, 156, 158, 159, 185
Arousal theory 82
Articulation 320
Articulation index 168, 169, 170, 173–8, 180, 197, 198
Articulation testing 167
Asphalted felts 491
Atmosphere, sound propagation through 455–7
Attention sphere 44
Attention switching 197

Attenuation 202, 374, 399, 402, 436
Attenuation coefficient 451
Attenuation curves 193, 195
Attenuation equipment 333
Attenuation function 197
Attenuation rate 197–9, 203
Attenuation rate function 199
Attenuators
    absorption 339
    airflow systems 339–46
    dynamic insertion loss ratings 345
    general comments 344
    interference 341
    noise reduction ratings 345
    reflection, relaxation and resonance 341
Audience absorption 520
Audioanalgesia 107
Audiometric rooms 362
Auditoria 391–3
Auditory system 83
Autonomic arousal 37

Background noise 130, 202, 204, 280, 390, 394
Barriers 281–3
Bearings 214, 233
Bedrooms
    internal partitions 100
    sound levels 108
Behavioural arousal 37
Behavioural response 154
Bell casting 421–2
Belt-drive system for fans 226
Bending frequency 349
Bends, attenuation 407
Bessel equation 442
Bezold–Brücke effect 73
Binomial theorem 447
Blood circulation 90, 96, 97, 105
Boiler fans 358
Boiler noise 245, 322
Brain 38, 40–42, 79, 83, 106
Brain research 12
Branches, attenuation 407
British Standards and Codes 575–7
Brownian motion 443
Brush noise 233
Building design
    integrated 11
    present and future 7
    shortcomings of 10–11
    traditional process 8
Building environmental engineer 25–27, 61, 208
Building environmental engineering 3–29
    undergraduate courses xiv
Building fabrics 287
Building façade 374
Building orientation 373
Building Regulations 310, 311–14, 354
Building services 26
Building services engineer 25, 26
Building structure 9, 286–96

external 377–82
selection of 287, 377–82
Building systems 21
Buildings
    appraisal of 23
    lightweight and heavyweight types 379
    noise control within 300
    orientation 285
    space distribution within 307–8
    tall 285
    vibration design criteria 494–5
Bulk modulus 420, 445, 447
Burner noise 246, 247, 249, 250, 255
Burner transfer function 249
Bürolandschaft concept 51, 57, 174
Butterfly dampers 257

Capital cost budget 10
Cardiovascular response 90, 94
Casing-radiated noise 271
Cavitation 210, 211, 230
Cavities 274, 291, 316, 571
Ceiling diffusers 409
Ceiling materials 201
Ceiling reflection 200, 202
Ceiling reflectors 391, 393
Ceiling treatments 190
Ceilings 314
    absorbent 189, 194, 201, 202, 206
    acoustimetal suspended 194
    coffered 537
    composite 201
    designing 201
    egg-crate 201
    resilient 327, 328
    sculpted 201
    suspended 317
    Sydney Opera House 552
Central stimulus 44
Channel resistance factor 82
Characteristic time constant 468
Chess-playing 153
Children 119, 123
Chillers 236–9
Chimney 245, 255
Civil Aviation Act 1971 xiv
Classical impedance analogy 430, 431
Classrooms 174
Cochlea 84
Cocktail party effect 197
Coherence between two sound pressures 538–9
Coincidence effect 565–8
Coincidence frequency 566, 567, 568
Colour warmth 50
Combustion chamber geometry 250
Combustion chamber transfer function 249
Combustion equipment 245–55
Combustion noise, turbulent 245–9
Comfort 24, 25, 33, 35, 44–51, 154
    dynamic model of 44
    factors relevant to 33

Comfort diagram 46, 79, 124
Comfort model 49, 113, 115
Complementary function 469
Complex absorption coefficients 197
Complex amplitude spectrum 438
Complex pressure reflection coefficients 197
Complex signals 438–9
Composite noise rating (CNR) 142
Composite structures 571–2
Compression 443, 444
Compressors 236–9
Conceptual diagram 20
Conceptual model 20
Concert halls 498, 499, 512, 521–4, 528–38, 542, 543, 549, 550, 552
Condensation 294
Consonance 75
Continental practice 373
Contingent negative variation 80
Continuous Performance Test (CPT) 165
Control of Pollution Act xv, xiv
Convergence-divergence 13, 17
Cooling towers 244–5, 359
Corrected noise level (CNL) 144, 150
Correlation coefficient 538
Cost-benefit analysis 9, 10
Coulomb damping 426–7, 493
Coulomb friction 466
Cracks 379
  sound leakage through 568
Creativity 13
Critical bandwidth 71, 75
Critical damping coefficient 425
Critical frequency 273, 294
Cross-correlation function 542
Crossroads 373
Cross-talk 178, 179
Cross-talk attenuator 345
Cumulative noise level 153

Dampers 257, 268, 270, 475
Damping 291, 296, 421–7, 450, 489, 571
  Coulomb 426–7, 493
  critical 467
  degrees of 467, 471
  heavy 467
  in polymers 423
  in structure 423
  internal 422–3
  light 467
  material 316, 426
Damping capacity 421, 422
Damping coefficient 428
Damping constant 423, 424, 498
Damping energy 423
Damping factor 294, 474
Damping forces 417, 423, 466
Damping ratio 425, 426
Damping tapes 492
Darcy's formula 402
Daylight factors 23

dBA measure 124, 128
Deafness 84–86
Decay curve 512, 513
Decay rate 425
Decay time 547, 548
Decision making 22
Deductive logic 19
Delay time 510, 525, 539
Design, internal space problems 382–4
Design criteria 124, 413
Design failure 23
Design procedure 386–7
Design team 8, 9, 17, 23
Dieckmann values 113
Diesel generator housing 361
Difference tones 75
Diffuse reflections 499
Diffusers 266, 409
Dipole 209
Dirac delta function 438
Directivity factor 192, 198, 411, 502
Directivity index 200, 503
Directivity patterns 502
Displacement amplitude 475
Displacement-frequency function 471
Dissonance 75
Distraction 46, 113, 115, 127, 152
Distractive effects 114
Distribution network 208
Disturbance 114, 122, 124, 127–9, 133, 139, 160
Dom-Ino project 12
Door sealing 319
Doors 303, 318, 319
Double-skin construction 316
Double-wall systems 563
Dreaming 103
Drehfehler phenomenon 226, 396
Dual-duct unit 346
Duct attenuator 339, 358, 359
Duct lengths 261
Duct resonances 401
Duct wall
  insulation 345
  noise-break-out through 345
Ducts 255–65, 394, 398
  air distribution 396
  attenuation 406
  circular 260
  influence of flexible connections on casing sound radiation from dual duct controller 271
  rectangular 261
  sound generation 261
  straight 256
  surface vibration 270
  wave propagation 441–2
Dynamic hysteresis 490
Dynamic matrix 483

Ear 64–78
  differential frequency discrimination 75

Ear (cont'd)
  mechanism 84
Echoes 515, 527, 542
Education 24, 27
Effective perceived noise levels 135, 138
Elastic modulus 421
Electric motors 232–5, 394
  design aspects 234–5
  hum 234
  no-load noise ratings 235
  vibration 233, 234
Electrical analogues 430–5
Electrocortical arousal 37
Electroencephalographic recordings (EEG) 37, 102, 104
Emission exposure 129
Emotional effects 114
Emotional sensation level 46
Emotional sensitivity bipolar scale 114
Emotional sensitivity level 113
End reflection, attenuation 408
Energy
  in sound waves 451
  in standing waves 452–3
  saving 9
Energy crisis 9
Energy expenditure 24
Energy mean noise level 145
Energy transfer 427
Engineering 28
Engineers 28
  and architects 11, 12
Ensemble average 513
Environment
  balance of 33
  human interaction with 49
  man's response to 85
  total 51
Environment susceptibility 48
Environmental control 24
Environmental criteria 377
Environmental design 45
Environmental noise 150, 151
Environmental perception 85
Environmental performance 26
Environmental system 21
Equation of motion 450, 463, 468
Equivalence constant 128
Equivalent daytime disturbance number (EDD) 140
Equivalent energy level 133
Equivalent level of sustained noise 128, 141
Equivalent noise level 153
Equivalent sound level 133
Examination rooms 165
Excitation pattern 71
Exciting force function 468
Eye-pupil size 91, 98
Eyring formula 507, 508

Factor G 389

Fan plenum 360
Fan speed 400, 402
Fans 212–26, 306, 394, 486
  belt-drive system 226
  boiler 358
  broadband and vortex noise 216
  efficiency and noise output 220
  frequency curve 223
  mechanical noise 216
  predicted sound levels with change in operating point 219
  rotational noise 215, 222, 226
  selection 218
  sound intensity level 222
  sound power level 223, 226, 406
  sound spectra 214, 217, 222
  vector diagrams 218
Fatigue 38
Feedback 38, 82
Finger-pulse activity 106
Finger-pulse amplitude measurements 95
Flame transfer function 249
Flames, thermo-acoustic efficiency of 247
Flanking paths 274, 312
Flanking transmission coefficient 569
Flanking transmissions 316, 565
Flats 320
Floors 314, 315, 320, 323
  floating 304, 325, 326, 328
  lath and plaster 327
  wood-joist 327
Fluctuations 209
Flutter echo 515
Force frequency 484
Fourier theorem 438
Fourier transforms 438, 439, 519
Free field 457
Free-field radius 516
Frequency 438
Frequency analysis 370
Frequency ranges 443
Frequency ratio 486
Frequency transmission characteristics 430
Fresnel zone number 282
Friction, Coulomb 466
Fuel savings 10
Furniture density 193

Gas turbines 255
Gases
  sound propagation in 445–7
  wave equation for 448
Gaussian distribution 146
General wave equation 448–50, 454, 466
Glass panel, insulation value 294
Glazing 287, 288, 294, 379–83
  double 288, 294, 383
  single 288, 296, 298, 308
Goodness factor 153
Grade I insulation 310
Grade II insulation 311

Grilles   266, 408
Ground absorption   372
Ground surfaces   372, 457

Haas effect   527
Harmonics   61–62, 70, 75
Health (booklets)   578–9
Health and Safety at Work Act of 1974   xv
Hearing of neonates   84–85
Hearing damage   85
Hearing loss   85, 151
Heat transfer coils   228
Heating and ventilating engineer   26
Heating equipment, sound generation   214
Helmholtz resonator   336, 342, 464
High rise buildings   123
Holes, sound leakage through   568
Housing   378
Human figure   15, 16
Human mind, patterns of   11
Hysteresis phenomena   490

Ideas   15
Imagination and analysis   19
Impedance model   427
Impulse response analysis of rooms   519
Impulse response function   540
Impulsive sounds, loudness of   70
Induction units   270, 409
Inductive logic   19
Industrial noise   xv, 84, 144, 147
Industrial premises   284
Inertance   428
Inertia matrix   483
Influence coefficients   482
Information rate   50
Innovation, obstacles to   22
Installation work   398
Integrated building design   11
Intelligence   13, 160
Intelligence tests   15, 156
Inter-aural coherence   542
Interdisciplinary working   25
Interference   151
Interference attenuators   341
Internal friction   422
Inverse square law   281, 282, 457
Inversion index   550
IQ scores   156, 158

Job satisfaction   51

K-complex waves   102, 105
K-values   113
17-Ketosteroid   87, 89
Kirchhoff's laws   430, 431
Kitchens   323
Knowledge   36

L_{10} levels   364, 370, 371, 378, 402
L_{50} levels   371
L_{90} levels   366, 370
Lagrange's equation   482
Laminated structures   317, 318
Landscaped offices   52–61, 133, 165, 174, 175, 185, 187, 191–6, 200–5
Landscaping   283
Laplacian (or Del) operator   449
Lecture rooms   165, 308, 393, 398
Lighting systems   353
Lightweight construction   316
Logarithmic decrement   426, 467
Lombard voice reflex   172, 178, 187
Loss angle   424
Loss factor   424–6, 490, 492, 493
Loudness   64
　and pitch   74
　calculation   66–67
　estimation   68
　evaluation   64
　measuring   65
　of impulsive sounds   70
　of multicomponent tones   72
　perceived   73
Loudness curve   65, 172
Loudness judgements   66
Low-frequency noise   396, 400
Low-frequency vibrations   109–13

Magnitude estimation   64
Magnitude production   65
Magnitude scaling   434
Mark VII procedure   68
Masking   72
Mass   419–20, 571
Mass law   273, 274, 277, 288, 565, 567
Mastics   490
Matrix notation   482–3
Maxwell's reciprocal theorem   482
Mean energy level   128, 129, 133, 153
Mean sound level   133
Meatus   84
Mechanical engineering services   207
Mel   71
Membrane absorbers   343
Mental health   34–35, 100, 116
Mental tasks   159
Mind   85
Mobility analogy   430
Model tests   548
Modulor concept   15
Monopole   209
Motor Vehicles (Construction and Use) Regulations of 1973   xiv
Multi-layer structures   571–2
Music   87, 90, 107, 108, 115, 161
　designing spaces for   520–43
　see also Concert halls
Musical acoustical quality   529, 530
Musical instruments   61, 62

N-complex somatic responses 89
Natural frequency of motion 463–4
Neighbourhood noise xvi
Neighbourhood satisfaction factor 144
Neonates, hearing of 84–85
Neuroses 34–35
Newton's Second Law of Motion 419, 420, 448, 449, 482
Newton's Third Law of Motion 454
Nodes 440
Noise
  complaints of 33
  definition 418
  disturbing effects 33
  effects of high levels 34
  in public places xvi
  increase in 147
  physiological response 83–100
  see also Sound
Noise Abatement Act of 1960 xiv
Noise Advisory Council xv, xvi
  publications 581
Noise and number index (NNI) 138–40, 142, 150
Noise animosity factor 144
Noise assessment 146
  external environment 364
  room design aspects 187, 189
Noise control 7, 10
  basic principles 272–3
  procedure 208
  within buildings 300
Noise control equipment
  requirements of 387
  selection of 387
Noise criteria 34, 115, 146, 356, 382
  external 378
  future 150
  internal 378
  speech communication 179–85
Noise criteria (NC) curves 180, 181, 382, 384
Noise disturbance 53
Noise dose 129
Noise Exposure Forecast (NEF) 142
Noise indices 150
Noise insulation. *See* Sound insulation
Noise levels 130, 147, 148
  background. *See* Background noise
  designation 69–70
  in factories 84
  noise rating as function 117
  variations in 368, 369
  see also Aircraft noise; Industrial noise; Railway noise; Traffic noise
Noise measurement 364
Noise-measuring truck 32
Noise nuisances xiv, 116, 119
Noise pollution level 144–6, 153
Noise rating 382
  as function of noise level 117
Noise rating (NR) criteria 182
Noise rating curves 180, 181, 183, 390, 394

Noise recording sample 366
Noise reduction 149
Noise scales 129, 134, 150
Noise sensitivity 44, 47, 118
Noise sources 47, 117–23, 148, 207–71, 394
  distance of buildings from 280
  distracting 53
  external 324
  general aspects 207
  multiple 516
Noise susceptibility 118, 119, 152
Noise tolerance 47, 132
Noise units, guide to 144
Normalised level difference 570
Noy values 134–5
Nuisance. *See* Noise nuisance

Objective system 21
Occupational density 187
Occupational noise xvi
Offices 308, 387–90
  design 187
  environment 53
  landscaped 52–61, 133, 165, 174, 175, 185–205, 389
  soundproof 352–4
Ohm's acoustical law 71
Open-planning 174
Opera houses 521, 530, 531, 533, 544–52
Optimum start control systems 10
Orthogonality principle 483
Out-of-balance forces 214, 226
Overcrowding 57, 187
Overheating 287

Particular integral 469
Partitions 277, 316–18, 388, 568
Party-wall grade 310, 311
Pascal as unit of acoustic pressure xiv
Perceived noise decibels (PNdB) 134–5, 138
Perceived noise level (PNL) 69, 134, 135, 138
Perceived noisiness 152
Performance 153–66
  and noise interactions 161–5
  meaning of 154
Period T 438
Periodic excitation in linear systems 468–72
Periodic function 437, 438
Periodic oscillation noise 245, 249
Peripheral stimuli 44
Personality 13, 44, 66, 85, 118, 124, 152, 160, 164
Personality tests 14
Phase angle 470
Phase function 437, 471
Physical sensation level 46
Physical sensitivity level 113
Physiological measures 100
Physiological response to noise 83–100
Pinna 84

Pipes, wave propagation in 441–2
Pitch 72–76
  and loudness 74
  and sound pressure level 73
Pitch number 75
Planning 386, 398
Plant-room 207–8, 300–7, 394
  basement 300
  construction techniques 304
  design schemes 304
  doors 302
  equipment 306
  intermediate floor 300
  roof-level 301
Plasterboard 318
Plethysmographic studies 92, 99
Poisson distribution 79
Pollution Control: Progress and Problems xv
Porosity 334
Powerhouse 301
Preferred octave speech interference level
    (PSIL) 179
Presbycusis 85
Privacy 55, 57, 168–79, 200, 202, 363, 387–90
Privacy number 388, 389
Probability function 42
Problems, identification matrix 382
Productivity 161
Propagation constant 438, 450
Proportionality constant 265
Pseudosound sources 418
Psyche function 38, 81, 86
Psyche states 100
Psychiatric breakdown 35
Psychological factors 24, 51
Psychological response 113–49
Psychophysiological factors 185
Pump noise 184, 229–32, 320, 322
Pumps, vibration isolation 230

$\bar{Q}$ function 139, 141
Quadrupole 210
Quality design 10

R index 140, 142
Radiators 184, 320–2, 323
Railway noise 142
Railways animosity factor 144
Rapid eye movements (REM) 103–5
Rarefaction 443, 444
Ray tracing techniques 391
Rayleigh's method 481
Reflection attenuators 341
Reflection coefficient 569, 571
Reflective surfaces 392
Refrigeration 236–9
Refuse chutes 313–14
Regression analysis 152
Reiher–Meister sensitivity scale 113
Relative humidity 155

Relativity, theory of 420
Relaxation attenuators 343
Resilient layers 325
Resilient materials 487
Resonance 429, 493, 563–4, 571
  assisted 538
Resonance attenuators 342
Resonance frequency 273, 294
Resources system 22
Response, scale 129
Response function 101, 105
Reverberant field 501, 504, 515
  diffuse 502
Reverberant field energy 514–15
Reverberant sound field 456
Reverberant sound pressure 516–18
Reverberation 171–3, 189, 191, 199–200, 333,
    339, 392, 509
Reverberation time 426, 498, 506, 511, 515,
    519–20, 527–42, 550–3
Reverberation time equation 177, 191, 196
Reynolds number 250
Rise time 510, 511
Road traffic. *See* Traffic noise
Road Traffic Act of 1974 xiv
Room acoustic characteristics 411
Room constant 411, 514–15
Room design aspects of noise assessment 187,
    189
Room effect 198
Rooms
  impulse response analysis 519
  low-height 516
  sound in 496–559
  sound pressure level 515–16
Royal Commission on Environmental Pol-
    lution xv

Sabine formula 507, 508
Safety (booklets) 578–9
Scalar wave function 450
Schools 308
  open-plan 174, 206
Screens and screening 200–2, 374–7
Sealed buildings 288
Sealed window units 288
Sealing 291
Sensory mechanisms 42
Sensory pathways 37
Sensory perception 36
Sensory system 40
Shock waves 354
Signal-to-noise ratio 41
Site investigation 364, 366, 370
Skin resistance 87, 89, 91
Sleep 38, 87, 101–9, 150
  noise-induced 107
Sleep-dream pattern 104
Sleep loss 155
Snape Concert Hall, Aldeburgh 350
Snell's law 456

Social factors   51
Social problems   123
Solar gain factor   380
Solar radiation   379, 380
Solids, wave equation for   449
Somatic responses   100
Sonic boom   354–7
Sonoluminescence   210
Soul   85
Sound
  definition   443
  direct   496, 512, 525–7, 534, 539, 540
  reverberant   496, 512, 526–7, 539, 540
  spatial-temporal patterns   61
  subjective qualities   76
  velocity   443, 448, 449, 455, 466
  see also Noise
Sound absorbent layer thickness   340–1
Sound absorption   200, 333–6
  applications   337–9
Sound absorption coefficients   336, 340, 345
Sound absorption material   394
Sound coloration   515, 519
Sound control. See Noise control
Sound decay   193, 196, 506, 509, 511
Sound decay curve   540
Sound density   77, 78
Sound diffusion   391
Sound distribution   191, 516
Sound duration   75
Sound energy   569
Sound energy curves   512
Sound energy density   501
Sound fields   496
  growth and decay   505–14
  random incident   570
Sound generation   400
  airconditioning   214
  air-distribution networks   256–65
  airflow systems   209
  heating equipment   214
  mechanisms   209–10
  waterflow systems   210–14
Sound growth   507, 509, 511
Sound growth curves   510, 513
Sound image formation   196
Sound image sources   499
Sound insulation   119, 123, 273–80, 379, 552,
    569, 571
  regulations   xv
  specifications   310
  terms used   309
Sound intensity   63, 451, 455, 456, 500, 501, 505,
    506
Sound leakage through holes and cracks   568
Sound level meter   124
Sound levels   52, 62, 100, 118, 145, 150, 165
  L   183
  measuring   65
  variation   368
Sound patterns   512–13
Sound perception arriving at ears via direct and

reflected pathways   525
Sound plenum attenuation unit   339
Sound power function   192
Sound power level   62, 265
Sound pressure   451, 452, 501, 502
  coherence between two   538–9
Sound pressure levels   63, 66, 70, 72, 132, 135,
    191, 192, 198, 398, 402, 503
  and pitch   73
  in rooms   515–16
Sound propagation   443–62, 508
  in gases   445–7
  through atmosphere   455–7
Sound protection   148, 149
Sound quality   527
Sound reduction index   277–80, 298, 299, 307,
    310, 316, 320, 354, 379, 388, 564–71
Sound sources
  and pathways   417, 427–30
  fundamental nature of simple pulsating spheri-
    cal   453–5
Sound stimulation   51
Sound transmission
  through structures   560–74
  see also Airborne sound; Structureborne
    sound
Sound transmission coefficient   560
Sound volume   76
Sound waves   436, 443, 496, 499
  energy in   451
  longitudinal   450
Soundproof booths   352
Space distribution within buildings   307–8
Spatial ability   12
Spatial function   192
Spatial orientation   12
Spatial responsiveness   525, 534, 537, 538
Speaker-listener orientation   198
Speaker-receiver orientation   177
Specific acoustical impedance   455, 560
Specific damping capacity   424
Specific damping energy   424
Specifications   363, 365
Spectral function   438
Speech   166, 187, 512, 524, 527, 568
Speech communication   166–206, 189
  noise level criteria   179–85
Speech frequencies   168, 170
Speech intelligibility   168, 169, 170
Speech interference level   165, 180
Speech sound pressure level   180
Spring systems   463–6, 475, 480
Standard deviation   145, 146
Standards   144, 147, 153, 494
Static hysteresis   490
Steam turbines   255
Stiffness   420, 429, 571
Stiffness factor   421
Stiffness function   475
Stiffness matrix   483
Stiffness ratio   484
Stochastic dominance   42

Stochastic-dominance model 43
Stochastic-dominance probability function 42, 82
Stress and stress effects 34, 35, 78–83, 121, 130, 150, 152, 154, 159–60
  indicators 87
  interactions 155
  quantitative picture of 80
  reactions to 80
Stress-arousal model 79
Stress combinations 159, 160
Stress-strain patterns 81
Stress-strain relationship 79, 87
Strouhal number 228
Structureborne sound 149, 417, 463–95, 560–74
Summation tones 75
Sydney Opera House 544–52
  acoustical appraisal 549
  arts centre complex 545
  drama theatre 551
  general design approach 547
  model tests 548
  orchestra rehearsal hall and recording studio 551
  sound insulation 552
Sympatho-adrenomedullary interaction 89
Systems design thinking 19

Tapping machine 314
Tartini's tones 75
Technologists 28
Technology, history of 28
Temperature, indoor variation 379
Temperature gradients 456
Terotechnology 23, 24
Theatres 521
Thermal stimulation 50
Thermal stress 154
Thermo-acoustic efficiency of flames 247
Thermoneutrality 154
Thermoregulatory system 90
Thinking continuum, divergent-convergent 17
Thought
  creative and analytical 17
  orientation of 16
Timbre 61
Time delay 526–8, 534
Time scaling 434
Town planning 284
Traffic density 281
Traffic flow 371
Traffic lights 373
Traffic management 284
Traffic noise xv, xvi, 123, 127–9, 133, 139–41, 144, 147, 153, 308, 374
Traffic noise criteria 133–4
Traffic noise index (TNI) 133
Transient peak index (TPI) 130, 132
Transient sounds 515
Transition pieces 257
Transmissibility curves

  absolute 477
  relative 478
Transmissibility function 473
Transmission coefficient 202, 568, 569, 571
Transmission function 439, 519
Transmission loss 564
Tube banks 228
Turbulent combustion noise 245–9

Universal Gas Constant 449
Urban planning 280

Vasoconstriction 89, 90, 105
Vasodilation 90
Vector wave function 450
Velocity amplitude 370
Velocity potential 453
Venetian blind 383
Ventilation 402
  variable-speed system 398
Ventilation rate 379, 380
Ventilator units 288
Vibration 291, 346–9, 396
  coupled modes 481
  duct surfaces 270
  due to cavitation 211
  electric motors 233, 234
  flexural 565
  ground 370
  horizontal 113
  low-frequency 109–13, 348
  mechanical 417–27
  normal modes 481
  pumps 230
  rotating machinery 214
  sources of 417
  tolerance limits 113
  undamped, free 463–6
  vertical 113
  viscous damped, free 466
Vibration control 306
  by reducing exciting forces 494
Vibration design criteria for buildings 494–5
Vibration force 417
Vibration isolation
  active and passive systems 472
  damped active system 473
  efficiency 472, 474
  forced, undamped system 473
  limitations of theory 475
  multi-degrees of freedom systems 480–4
  non-linear systems 475
  practice 485–94
  recommended efficiencies 494
  standards 494
  system design 486
  theory 472–5
    limitations of 475–84
  transmissibility 472
Vibration isolators 475, 485–8

Viscoelastic materials   489
Visualisation   12
Vocality   77
Voice levels   178, 180
  see also Speech
Voice spectra   172
Volume velocity   427, 429
Vortex noise   216

Walls   314
  double   316
Water flow systems, sound generation   210–14
Water flow velocity   211
Water flushing noise   213
Water turbulence   230
Waterhammer   210
Wave amplitude   437
Wave behaviour in bounded medium   439–41
Wave equation   208, 445–51
  for gases   448
  for solids   449
Wave function   436, 443, 447–50
Wave modes   441–2
Wave motion   209, 436–42
Wave patterns   496
Wave propagation   497
  in pipes and airconditioning ducts   441–2

Wavefront, law of first   525
Wavelength   438, 565
Waves
  axial   441
  linear   439
  mathematical representation   436
  oblique   441
  periodic   437
  spherical   454
  standing (or stationary)   439–41, 501, 571
    energy in   452–3
  tangential   441
Weber–Fechner law   39, 40
Weighting function   540
White noise   62
Wilcoxon matched-pair signed rank test   183
Wilson Committee   138, 183, 378
Wind gradients   457
Window factor   294
Windows   288, 380
  acoustical properties of   294
Working efficiency   154
Working spaces   35

Yerkes–Dodson law   159
Young's modulus   420, 449